Ken Binmore and Joan Davies

Calculus
Concepts and Methods

CAMBRIDGE

Calculus

KEN BINMORE held the Chair of Mathematics for many years at the London School of Economics, where this book was born. In recent years, he has used his mathematical skills in developing the theory of games in a number of posts as a professor of economics, both in Britain and the USA. His most recent exploit was to head the team that used mathematical ideas to design the UK auction of telecom frequencies that made 35 billion dollars for the British taxpayer.

JOAN DAVIES has a doctorate in mathematics from the University of Oxford. She is committed to bringing an understanding of mathematics to a wider audience. She currently lectures at the London School of Economics.

Dedications

To Jim Clunie

Ken Binmore

To my family
Roy, Elizabeth, Sarah and Marion

Joan Davies

Calculus

Ken Binmore
University College, London

Joan Davies
London School of Economics

CAMBRIDGE
UNIVERSITY PRESS

PUBLISHED BY THE PRESS SYNDICATE OF THE UNIVERSITY OF CAMBRIDGE
The Pitt Building, Trumpington Street, Cambridge, United Kingdom

CAMBRIDGE UNIVERSITY PRESS
The Edinburgh Building, Cambridge CB2 2RU, UK
40 West 20th Street, New York, NY 10011-4211, USA
10 Stamford Road, Oakleigh, VIC 3166, Australia
Ruiz de Alarcón 13, 28014, Madrid, Spain
Dock House, The Waterfront, Cape Town 8001, South Africa

http//:www.cambridge.org

First published 2001

Printed in the United Kingdom at the University Press, Cambridge

Typeface Times 10/12pt.   *System* LATEX 2_ε [DBD]

A catalogue record of this book is available from the British Library

Library of Congress Cataloguing in Publication data
ISBN 0 521 77541 8   hardback

Contents

Contents

Preface

Ancient accountants laid pebbles in columns on a sand tray to help them do their sums. It is thought that the impression left in the sand when a pebble is moved to another location is the origin of our symbol for zero.

The word calculus has the same source, since it means a pebble in Latin. Nowadays it means any systematic way of working out something mathematical. We still speak of a calculator when referring to the modern electronic equivalent of an ancient sandtray and pebbles. However, since Isaac Newton invented the differential and integral calculus, the word is seldom applied to anything else. Although there are pebbles on its cover, this book is therefore about differentiating and integrating.

Students who don't already know what derivatives and integrals are would be wise to start with another book. Our aim is to go beyond the first steps to discuss how calculus works when it is necessary to cope with several variables all at once.

We appreciate that some readers will be rusty on the basics, and others will be doubtful that they ever really understood what they can remember. We therefore go over the material on the calculus of one variable in a manner that we hope will offer some new insights even to those rare souls who feel confident of their mathematical prowess. However, we strongly recommend against using this material as a substitute for a first course in calculus. It goes too fast and offers too much detail to be useful for this purpose.

It should be emphasised that this is not a cook book containing a menu of formulas that students are expected to learn by rote in order to establish their erudition at examination time. We see no point in turning out students who can write down the formal derivation of the Slutsky equations, but have no idea what the mathematical manipulations they have learned to reproduce actually mean. When one teaches *how* things are done without explaining *why*, one does worse than fill the heads of the weaker students with mumbo-jumbo, one teaches the stronger students something very wrong – that mathematics is a list of theorems and proofs that have no practical relevance to anything real.

The attitude that mathematics is a menu of formulas that ordinary mortals can only admire from afar is very common among those who know no mathematics at all. Research mathematicians are often greeted with incredulity when they say what they do for a living. People think that inventing a new piece of mathematics would be like inventing a new commandment to be added to the ten that Moses brought down from the mountain. Such awe of mathematics creates a form of hysterical paralysis that must be overcome before a student can join the community of those of us who see mathematics as an ever changing box of tools that educated people can use to make sense of the world around them.

Within this community, a model is not expressed in mathematical form to invite the applause of those who are easily impressed, or to obfuscate the issues in order to immunise the model from criticisms by uninitiated outsiders. Instead the community we represent is always anxious to find the *simplest* possible model that captures how a particular aspect of some physical or social process works.

For us, mathematical sophistication is pointless unless it serves to demystify things that we would not otherwise be able to understand. We do not see mathematical modelling as some grandiose activity that can only be carried out by professors at the blackboard. Mathematical modelling is what *everybody* should do when seeking to make sense of a problem. Of course, beginners will only be able to construct very simple models – but a good teacher will only ask them to solve very simple problems.

Such an attitude to solving problems is not possible with students whose intellectual processes freeze over at the mention of an equation. The remedy for this species of mathematical paralysis lies in teaching that mathematics is something one *does* – not something that one just appreciates. Rather than offering them a cook book, one needs to teach students to put together simple recipes of their own. We need to build their confidence in their own ability to think coherent mathematical thoughts *all by themselves.*

Such confidence comes from involving students in the mathematics as it is developed, using the traditional method of demanding weekly answers to carefully chosen sets of problems. The problems must not be too hard – but nor must they be too easy. Nobody gets their confidence boosted by being asked to jump through hoops held too low. On the contrary, if we only ask students to solve problems that they can see are trivial, we merely confirm to them that their own low opinion of their mathematical ability is shared by their teacher. Some hoops have to be held high enough that students get to feel they have achieved something by jumping through them.

We feel particularly strongly on this latter point, having watched the confidence of our student intake gradually diminish over the years as cook book teaching has taken over our schools under the pretence that old fashioned rote learning is being replaced by progressive methods that emphasise the underlying concepts.

As this successor to the first version of *Calculus* shows, mathematicians don't mind adjusting the content of their courses as school syllabuses develop over time. We are even willing to welcome less mathematics being taught in school if this means that more children become numerate. But the kind of cook book teaching that leaves students helpless if a problem does not fit one of a small number of narrow categories seems to us inexcusable.

Our hostility to cook book teaching should not be taken to imply that this book is a rigorous work of mathematical analysis. It is a how to do it book, which contains no formal theorems at all. But we always explain why the methods work, because there is no way anybody can know *how* to use a method to tackle a new kind of problem unless they know *why* the method works.

Our approach to explaining why a method works is largely geometrical. To this end, the book contains an unusually large number of diagrams – even more than in the first version. The availability of colour and computer programs like *Mathematica* means that the diagrams are also better.

In addition, the already large number of examples and exercises has been augmented, with a view to increasing understanding by illustrating some of the things that can go wrong in cases that are usually passed over without comment.

Finally, we include examples of how the mathematics we are teaching gets used in practice. The mathematics is the same wherever it is applied, but the applications to economics on which we concentrate are often particularly instructive because of the need to be especially careful about what is being kept constant during a differentiation. Students whose prime interest is in the hard sciences may find that these applications are a lot more fun than examples from physics that they will have seen in some form before.

With our attitudes to the teaching of mathematics, it will come as no surprise that we view the exercises as an integral part of the text. There is no point in trying to read this or any other mathematics book without making a serious commitment to tackle as many of the exercises as time allows. Indeed, being able to solve a fair number of the exercises without assistance is the basic test of whether you understand the material. You may think you understand the concepts, but if you can't do any of the exercises, then you don't. You may think you *don't* understand the concepts, but if you can do most of the exercises, then you do. Either way, you need to attempt the exercises to find out where you stand.

A lot of work went into tailoring this successor to *Calculus* to the needs of today's students. Our readers will have to work equally hard to enjoy its benefits. We hope that they will also share our feeling of having done something genuinely worthwhile.

Ken Binmore
Joan Davies

Acknowledgements

Joan Davies is grateful to the following:

firstly to Michele Harvey for her support and formative suggestions during the early stages of the project,

to Gillian Colkin and especially to David Cartwright for their invaluable comments on drafts of the manuscript and for their help in testing exercises,

to Mark Baltovic for a student's view of the book and to Dean Ives and James Ward for useful comments,

to Roy Davies for help in coping with the idiosyncrasies of various software packages.

Ken Binmore gratefully acknowledges the support of the Leverhulme Foundation and the Economic and Social Research Council.

One

Matrices and vectors

This book takes for granted that readers have some previous knowledge of the calculus of real functions of one real variable. It would be helpful to also have some knowledge of linear algebra. However, for those whose knowledge may be rusty from long disuse or raw with recent acquisition, sections on the necessary material from these subjects have been included where appropriate. Although these revision sections (marked with the symbol $\diamond$) are as self-contained as possible, they are *not* suitable for those who have no acquaintance with the topics covered. The material in the revision sections is surveyed rather than explained. It is suggested that readers who feel fairly confident of their mastery of this surveyed material scan through the revision sections quickly to check that the notation and techniques are all familiar before going on. Probably, however, there will be few readers who do not find something here and there in the revision sections which merits their close attention.

The current chapter is concerned with the fundamental techniques from linear algebra which we shall be using. This will be particularly useful for those who may be studying linear algebra concurrently with the present text.

Algebraists are sometimes neglectful of the geometric implications of their results. Since we shall be making much use of geometrical arguments, particular attention should therefore be paid to §1.3 onwards, in which the geometric relevance of various vector notions is explained. This material will be required in Chapter 3. Those who are not very confident of their linear algebra may prefer leaving §1.10 until Chapter 5.

1.1 Matrices$^\diamond$

A matrix is a rectangular array of numbers – a notation which enables calculations to be carried out in a systematic manner. We enclose the array in brackets as in the examples below:

$$A = \begin{pmatrix} 4 & 1 \\ 0 & -1 \\ 3 & 2 \end{pmatrix} \qquad B = \begin{pmatrix} 1 & 0 & -1 \\ 2 & 1 & 0 \end{pmatrix}$$

A matrix with m rows and n columns is called an $m \times n$ matrix. Thus A is a 3×2 matrix and B is a 2×3 matrix.

A general $m \times n$ matrix may be expressed as

$$C = \begin{pmatrix} c_{11} & c_{12} & \cdots & c_{1n} \\ c_{21} & c_{22} & \cdots & c_{2n} \\ \vdots & \vdots & \ddots & \vdots \\ c_{m1} & c_{m2} & \cdots & c_{mn} \end{pmatrix}$$

where the first number in each subscript is the row and the second number in the subscript is the column.

For example, c_{21} is the entry in the second row and the first column of the matrix C. Similarly, the entry in the third row and the first column of the preceding matrix A can be denoted by a_{31}:

$$\text{column 1} \\ \downarrow$$

$$\text{row 3} \longrightarrow \begin{pmatrix} 4 & 1 \\ 0 & -1 \\ 3 & 2 \end{pmatrix} \qquad a_{31} = 3$$

† See Chapter 13.

We call the entries of a matrix **scalars**. Sometimes it is useful to allow the scalars to be complex numbers[†] but our scalars will always be *real numbers*. We denote the set of real numbers by $\mathbb{R}$.

Scalar multiplication

One can do a certain amount of algebra with matrices and under this and the next few headings we shall describe the mechanics of some of the operations which are possible.

The first operation we shall consider is called **scalar multiplication**. If A is an $m \times n$ matrix and c is a scalar, then cA is the $m \times n$ matrix obtained by multiplying each entry of A by c. For example,

$$2A = 2 \begin{pmatrix} 4 & 1 \\ 0 & -1 \\ 3 & 2 \end{pmatrix} = \begin{pmatrix} 2 \times 4 & 2 \times 1 \\ 2 \times 0 & 2 \times -1 \\ 2 \times 3 & 2 \times 2 \end{pmatrix} = \begin{pmatrix} 8 & 2 \\ 0 & -2 \\ 6 & 4 \end{pmatrix}$$

Similarly,

$$5B = 5 \begin{pmatrix} 1 & 0 & -1 \\ 2 & 1 & 0 \end{pmatrix} = \begin{pmatrix} 5 & 0 & -5 \\ 10 & 5 & 0 \end{pmatrix}$$

Matrix addition and subtraction

If C and D are two $m \times n$ matrices, then $C + D$ is the $m \times n$ matrix obtained by adding corresponding entries of C and D. Similarly, $C - D$ is the $m \times n$ matrix obtained by subtracting corresponding entries. For example, if

$$C = \begin{pmatrix} 1 & -1 & 0 \\ -2 & 3 & 1 \\ 4 & 1 & 0 \end{pmatrix} \quad \text{and} \quad D = \begin{pmatrix} 2 & 1 & 5 \\ -1 & -3 & 4 \\ -3 & 2 & 1 \end{pmatrix}$$

then

$$C + D = \begin{pmatrix} 1+2 & -1+1 & 0+5 \\ -2-1 & 3-3 & 1+4 \\ 4-3 & 1+2 & 0+1 \end{pmatrix} = \begin{pmatrix} 3 & 0 & 5 \\ -3 & 0 & 5 \\ 1 & 3 & 1 \end{pmatrix}$$

and

$$C - D = \begin{pmatrix} 1-2 & -1-1 & 0-5 \\ -2+1 & 3+3 & 1-4 \\ 4+3 & 1-2 & 0-1 \end{pmatrix} = \begin{pmatrix} -1 & -2 & -5 \\ -1 & 6 & -3 \\ 7 & -1 & -1 \end{pmatrix}$$

Note that

$$C + C = \begin{pmatrix} 1+1 & -1-1 & 0+0 \\ -2-2 & 3+3 & 1+1 \\ 4+4 & 1+1 & 0+0 \end{pmatrix} = \begin{pmatrix} 2 & -2 & 0 \\ -4 & 6 & 2 \\ 8 & 2 & 0 \end{pmatrix} = 2C$$

and that

$$C - C = \begin{pmatrix} 1-1 & -1+1 & 0-0 \\ -2+2 & 3-3 & 1-1 \\ 4-4 & 1-1 & 0-0 \end{pmatrix} = \begin{pmatrix} 0 & 0 & 0 \\ 0 & 0 & 0 \\ 0 & 0 & 0 \end{pmatrix}$$

The final matrix is called the 3×3 **zero matrix**. We usually denote any zero matrix by 0. This is a little naughty because of the possibility of confusion with other zero matrices or with the scalar 0. However, it has the advantage that we can then write

$$C - C = 0$$

for any matrix C.

Note that it makes no sense to try to add or subtract two matrices which are not of the same shape. Thus, for example,

$$A + B = \begin{pmatrix} 4 & 1 \\ 0 & -1 \\ 3 & 2 \end{pmatrix} + \begin{pmatrix} 1 & 0 & -1 \\ 2 & 1 & 0 \end{pmatrix}$$

is an entirely meaningless expression.

Matrix multiplication

If A is an $m \times n$ matrix and B is an $n \times p$ matrix,

$$A = \overset{\scriptstyle n}{\underset{m}{\begin{pmatrix} a_{11} & a_{12} & \cdots & a_{1n} \\ a_{21} & a_{22} & \cdots & a_{2n} \\ \vdots & \vdots & \ddots & \vdots \\ a_{m1} & a_{m2} & \cdots & a_{mn} \end{pmatrix}}} \qquad B = \overset{\scriptstyle p}{\underset{n}{\begin{pmatrix} b_{11} & b_{12} & \cdots & b_{1p} \\ b_{21} & b_{22} & \cdots & b_{2p} \\ \vdots & \vdots & \ddots & \vdots \\ b_{n1} & b_{n2} & \cdots & b_{np} \end{pmatrix}}}$$

then A and B can be multiplied to give an $m \times p$ matrix AB:

$$A \times B = AB$$

To work out the entry c_{jk} of AB which appears in its jth row and kth column, we require the jth row of A and the kth column of B as illustrated below.

$$\xymatrix{}$$

The entry c_{jk} is then given by

$$c_{jk} = a_{j1}b_{1k} + a_{j2}b_{2k} + a_{j3}b_{3k} + \cdots + a_{jn}b_{nk}$$

Example 1

We calculate the product AB of the matrices

$$A = \begin{pmatrix} 0 & 1 & 2 \\ 2 & 0 & 1 \end{pmatrix} \qquad B = \begin{pmatrix} 1 & 0 \\ 2 & 1 \\ 0 & 2 \end{pmatrix}$$

Since A is a 2×3 matrix and B is a 3×2 matrix, their product AB is a 2×2 matrix:

$$AB = \begin{pmatrix} 0 & 1 & 2 \\ 2 & 0 & 1 \end{pmatrix} \begin{pmatrix} 1 & 0 \\ 2 & 1 \\ 0 & 2 \end{pmatrix} = \begin{pmatrix} a & b \\ c & d \end{pmatrix}$$

To calculate c, we require the second row of A and the first column of B. These are indicated in the matrices below:

$$\begin{pmatrix} 0 & 1 & 2 \\ 2 & 0 & 1 \end{pmatrix} \begin{pmatrix} 1 & 0 \\ 2 & 1 \\ 0 & 2 \end{pmatrix} = \begin{pmatrix} a & b \\ c & d \end{pmatrix}$$

We obtain

$$c = 2 \times 1 + 0 \times 2 + 1 \times 0 = 2 + 0 + 0 = 2$$

Similarly,

$$\begin{aligned} a &= 0 \times 1 + 1 \times 2 + 2 \times 0 = 0 + 2 + 0 = 2 \\ b &= 0 \times 0 + 1 \times 1 + 2 \times 2 = 0 + 1 + 4 = 5 \\ d &= 2 \times 0 + 0 \times 1 + 1 \times 2 = 0 + 0 + 2 = 2 \end{aligned}$$

Thus

$$AB = \begin{pmatrix} 0 & 1 & 2 \\ 2 & 0 & 1 \end{pmatrix} \begin{pmatrix} 1 & 0 \\ 2 & 1 \\ 0 & 2 \end{pmatrix} = \begin{pmatrix} 2 & 5 \\ 2 & 2 \end{pmatrix}$$

Note again that it makes no sense to try to calculate AB unless the number of columns in A is the same as the number of rows in B. Thus, for example, it makes no sense to write

$$\begin{pmatrix} 1 & 0 \\ 2 & 1 \\ 0 & 2 \end{pmatrix} \begin{pmatrix} 0 & 1 & 2 \\ 2 & 0 & 1 \\ 2 & 1 & 3 \end{pmatrix}$$

Identity matrices

An $n \times n$ matrix is called a **square matrix** for obvious reasons. Thus, for example,

$$A = \begin{pmatrix} 1 & 2 & 3 \\ 3 & 1 & 2 \\ 2 & 3 & 1 \end{pmatrix}$$

is a square matrix. The main **diagonal** of a square matrix is indicated by the shaded entries in the matrix below:

$$A = \begin{pmatrix} a_{11} & a_{12} & \cdots & a_{1n} \\ a_{21} & a_{22} & \cdots & a_{2n} \\ \vdots & \vdots & \ddots & \vdots \\ a_{n1} & a_{n2} & \cdots & a_{nn} \end{pmatrix}$$

† Note that an identity matrix *must* be square. Just as a zero matrix behaves like the number 0, so an identity matrix behaves like the number 1.

The $n \times n$ **identity matrix**[†] is the $n \times n$ matrix whose main diagonal entries are all 1 and whose other entries are all 0. For convenience, we usually denote an identity matrix of any order by I. The 3×3 identity matrix is

$$I = \begin{pmatrix} 1 & 0 & 0 \\ 0 & 1 & 0 \\ 0 & 0 & 1 \end{pmatrix}$$

Generally, if I is the $n \times n$ identity matrix and A is an $m \times n$ matrix, B is an $n \times p$ matrix, then

$$AI = A \quad \text{and} \quad IB = B$$

Example 2

(i) $\begin{pmatrix} 0 & 1 & 2 \\ 2 & 0 & 1 \end{pmatrix} \begin{pmatrix} 1 & 0 & 0 \\ 0 & 1 & 0 \\ 0 & 0 & 1 \end{pmatrix} = \begin{pmatrix} 0 & 1 & 2 \\ 2 & 0 & 1 \end{pmatrix}$

(ii) $\begin{pmatrix} 1 & 0 & 0 \\ 0 & 1 & 0 \\ 0 & 0 & 1 \end{pmatrix} \begin{pmatrix} 1 & 0 \\ 2 & 1 \\ 0 & 2 \end{pmatrix} = \begin{pmatrix} 1 & 0 \\ 2 & 1 \\ 0 & 2 \end{pmatrix}$

Determinants

‡ There is some risk of confusing this notation with the modulus or absolute value of a real number. In fact, the determinant of a square matrix may be positive or negative.

With each *square* matrix there is associated a scalar called the **determinant** of the matrix. We shall denote the determinant of the square matrix A by $\det(A)$ or by $|A|$.[‡]

A 1×1 matrix $A = (a)$ is just a scalar, and $\det(A) = a$.

The determinant of the 2×2 matrix

$$A = \begin{pmatrix} a & b \\ c & d \end{pmatrix}$$

is given by

$$\det(A) = \begin{vmatrix} a & b \\ c & d \end{vmatrix} = ad - bc$$

To calculate the determinant for larger matrices we need the concepts of a minor and a cofactor.

The **minor** M corresponding to an entry a in a square matrix A is the determinant of the matrix obtained from A by deleting the row and the column containing a.

In the case of a 3×3 matrix

$$A = \begin{pmatrix} a_{11} & a_{12} & a_{13} \\ a_{21} & a_{22} & a_{23} \\ a_{31} & c_{32} & a_{33} \end{pmatrix}$$

we calculate the minor M_{23} corresponding to the entry a_{23} by deleting the second row and the third column of A as below:

$$A = \begin{pmatrix} a_{11} & a_{12} & a_{13} \\ \cancel{a_{21}} & \cancel{a_{22}} & \cancel{a_{23}} \\ a_{31} & c_{32} & a_{33} \end{pmatrix}$$

The minor M_{23} is then the determinant of what remains – i.e.

$$M_{23} = \begin{vmatrix} a_{11} & a_{12} \\ a_{31} & a_{32} \end{vmatrix} = a_{11}a_{32} - a_{12}a_{31}$$

If we alter the sign of the minor M of a general $n \times n$ matrix according to its associated position in the checkerboard pattern illustrated below,

$$\begin{pmatrix} + & - & + & - & + & \cdots \\ - & + & - & + & - & \cdots \\ + & - & + & - & + & \cdots \\ \vdots & \vdots & \vdots & \vdots & \vdots & \end{pmatrix}$$

the result is called the **cofactor** corresponding to the entry a.

In the case of a 3×3 matrix A, the cofactor corresponding to the entry a_{23} is equal to $-M_{23}$, since there is a minus sign, '$-$', in the second row and the third column of the checkerboard pattern:

$$\text{row 2} \longrightarrow \begin{array}{c} \quad\quad\quad\quad\text{column 3} \\ \quad\quad\quad\quad\downarrow \\ \begin{pmatrix} + & - & + \\ - & + & - \\ + & - & + \end{pmatrix} \end{array}$$

The **determinant** of an $n \times n$ matrix A is calculated by multiplying each entry of one row (or one column) by its corresponding cofactor and adding the results. The value of $\det(A)$ is the same whichever row or column is used.

Example 3

The determinant of the 1×1 matrix $A = (3)$ is simply $\det(A) = 3$.

The determinant of the 2×2 matrix

$$\begin{pmatrix} 1 & 2 \\ 3 & 4 \end{pmatrix}$$

is

$$\det(A) = \begin{vmatrix} 1 & 2 \\ 3 & 4 \end{vmatrix} = 1 \times 4 - 2 \times 3 = 4 - 6 = -2$$

We find the determinant of the 3×3 matrix

$$A = \begin{pmatrix} 1 & 2 & 3 \\ 3 & 1 & 2 \\ 2 & 3 & 1 \end{pmatrix}$$

in two ways: first using row 1 and then using column 2.

The entries of row 1 are

$$a_{11} = 1 \qquad a_{12} = 2 \qquad a_{13} = 3$$

We first find their corresponding minors,

$$M_{11} = \begin{vmatrix} 1 & 2 \\ 3 & 1 \end{vmatrix} = 1 - 6 = -5$$

$$M_{12} = \begin{vmatrix} 3 & 2 \\ 2 & 1 \end{vmatrix} = -1 \qquad M_{13} = \begin{vmatrix} 3 & 1 \\ 2 & 3 \end{vmatrix} = 7$$

and then alter the signs according to the checkerboard pattern

$$\begin{pmatrix} + & - & + \\ - & + & - \\ + & - & + \end{pmatrix}$$

to obtain the corresponding cofactors,

$$+M_{11} = -5 \qquad -M_{12} = -(-1) = 1 \qquad +M_{13} = 7$$

Then

$$\det(A) = a_{11}(M_{11}) + a_{12}(-M_{12}) + a_{13}(M_{13})$$
$$= 1(-5) + 2(1) + 3(7)$$
$$= -5 + 2 + 21 = 18$$

We calculate the determinant again, this time using column 2:

$$a_{12} = 2 \qquad a_{22} = 1 \qquad a_{32} = 3$$

The corresponding cofactors are

$$-M_{12} = -\begin{vmatrix} 3 & 2 \\ 2 & 1 \end{vmatrix} = 1 \qquad +M_{22} = \begin{vmatrix} 1 & 3 \\ 2 & 1 \end{vmatrix} = -5$$

$$-M_{32} = -\begin{vmatrix} 1 & 3 \\ 3 & 2 \end{vmatrix} = -(2-9) = 7$$

Thus

$$\begin{aligned}
\det(A) &= a_{12}(-M_{12}) + a_{22}(M_{22}) + a_{32}(-M_{32}) \\
&= 2(1) + 1(-5) + 3(7) \\
&= 2 - 5 + 21 = 18
\end{aligned}$$

Inverse matrices

We have dealt with matrix addition, subtraction and multiplication and found that these operations only make sense in certain restricted circumstances. The circumstances under which it is possible to *divide* by a matrix are even more restricted.

[†] An invertible matrix is also called nonsingular.

We say that a *square* matrix A is **invertible**[†] if there is another matrix B such that

$$AB = BA = I$$

In fact, if A is invertible there is precisely *one* such matrix B which we call the **inverse matrix** to A and write $B = A^{-1}$.

If A is an $n \times n$ matrix, then A^{-1} is an $n \times n$ matrix as well (*otherwise the equation would make no sense*).

Thus an invertible matrix A has an inverse matrix A^{-1} which satisfies

$$AA^{-1} = A^{-1}A = I$$

It can be shown that

A is invertible if and only if $\det(A) \neq 0$

If A is *not* square or if A is square but its determinant is *zero* (i.e. A is not invertible), then A does *not* have an inverse in the above sense.

Transpose matrices

In describing how to calculate the inverse of an invertible matrix, we shall need the idea of a transpose matrix. This is also useful in other connections.

If A is an $m \times n$ matrix, then its **transpose** A^{T} is the $n \times m$ matrix whose first row is the first column of A, whose second row is the second column of A, whose third row is the third column of A and so on.[‡]

[‡] Alternative notations for the transpose are A' and A^t.

$$A = \begin{array}{c} \\ m \end{array} \overset{\xleftarrow{\hspace{1em}} n \xrightarrow{\hspace{1em}}}{\left(\begin{array}{cccc} a_{11} & a_{12} & \cdots & a_{1n} \\ a_{21} & a_{22} & \cdots & a_{2n} \\ \vdots & \vdots & \ddots & \vdots \\ a_{m1} & a_{m2} & \cdots & a_{mn} \end{array} \right)} \qquad A^{\mathrm{T}} = \begin{array}{c} \\ n \end{array} \overset{\xleftarrow{\hspace{1em}} m \xrightarrow{\hspace{1em}}}{\left(\begin{array}{cccc} a_{11} & a_{21} & \cdots & a_{m1} \\ a_{12} & a_{22} & \cdots & a_{m2} \\ \vdots & \vdots & \ddots & \vdots \\ a_{1n} & a_{2n} & \cdots & a_{mn} \end{array} \right)}$$

An important special case occurs when A is a square matrix for which $A = A^T$. Such a matrix is called **symmetric** (about the main diagonal).

Example 4

(i) If $A = \begin{pmatrix} 4 & 1 \\ 0 & -1 \\ 3 & 2 \end{pmatrix}$, then $A^T = \begin{pmatrix} 4 & 0 & 3 \\ 1 & -1 & 2 \end{pmatrix}$.

Note that

$$(A^T)^T = \begin{pmatrix} 4 & 0 & 3 \\ 1 & -1 & 2 \end{pmatrix}^T = \begin{pmatrix} 4 & 1 \\ 0 & -1 \\ 3 & 2 \end{pmatrix} = A$$

(ii) If $A = \begin{pmatrix} 1 & 3 & 5 \\ 3 & 2 & 0 \\ 5 & 0 & 4 \end{pmatrix}$, then $A^T = \begin{pmatrix} 1 & 3 & 5 \\ 3 & 2 & 0 \\ 5 & 0 & 4 \end{pmatrix}$.

Thus $A = A^T$ and so A is symmetric.

The cofactor method for A^{-1}

The inverse of an invertible matrix A can be calculated from its determinant and its cofactors. In the case of 1×1 and 2×2 matrices, one might as well learn the resulting inverse matrix by heart.

A 1×1 invertible matrix $A = (a)$ is just a nonzero scalar and

$$A^{-1} = \left(\frac{1}{a}\right) \text{ provided } a \neq 0$$

A 2×2 invertible matrix

$$A = \begin{pmatrix} a & b \\ c & d \end{pmatrix}$$

is one for which $\det(A) = ad - bc \neq 0$. Its inverse is given by

$$A^{-1} = \frac{1}{ad - bc} \begin{pmatrix} d & -b \\ -c & a \end{pmatrix}$$

The formulas can easily be confirmed by checking that AA^{-1} and $A^{-1}A$ actually are equal to I. For example in the 2×2 case

$$A^{-1}A = \frac{1}{ad - bc} \begin{pmatrix} d & -b \\ -c & a \end{pmatrix} \begin{pmatrix} a & b \\ c & d \end{pmatrix}$$

$$= \frac{1}{ad - bc} \begin{pmatrix} da - bc & 0 \\ 0 & -bc + ad \end{pmatrix} = \begin{pmatrix} 1 & 0 \\ 0 & 1 \end{pmatrix} = I$$

The cofactor method gives an expression for the inverse of an invertible matrix A in terms of its cofactors.

In the case of an invertible 3×3 matrix for which $\det(A) \neq 0$

$$A = \begin{pmatrix} a_{11} & a_{12} & a_{13} \\ a_{21} & a_{22} & a_{23} \\ a_{31} & a_{32} & a_{33} \end{pmatrix}$$

the inverse is given by

$$A^{-1} = \frac{1}{\det(A)} \begin{pmatrix} +M_{11} & -M_{12} & +M_{13} \\ -M_{21} & +M_{22} & -M_{23} \\ +M_{31} & -M_{32} & +M_{33} \end{pmatrix}^{\mathrm{T}}$$

For a general invertible $n \times n$ matrix

$$A = \begin{pmatrix} a_{11} & a_{12} & \cdots & a_{1n} \\ a_{21} & a_{22} & \cdots & a_{2n} \\ \vdots & \vdots & \ddots & \vdots \\ a_{n1} & a_{n2} & \cdots & a_{nn} \end{pmatrix}$$

we have

$$A^{-1} = \frac{1}{\det(A)} \begin{pmatrix} +M_{11} & -M_{12} & \cdots & (-1)^{n-1}M_{1n} \\ -M_{21} & +M_{22} & \cdots & (-1)^{n}M_{2n} \\ \vdots & \vdots & \ddots & \vdots \\ (-1)^{n-1}M_{n1} & (-1)^{n}M_{n2} & \cdots & +M_{nn} \end{pmatrix}^{\mathrm{T}}$$

Thus, each entry of A is replaced by the corresponding minor and the sign is then altered according to the checkerboard pattern to obtain the corresponding cofactor. The inverse matrix A^{-1} is then obtained by multiplying the *transpose* of the result by the *scalar* $(\det(A))^{-1}$.

Example 5

We calculate the inverses of the invertible matrices of Example 3.

(i) $A = (3), \quad \det(A) = 3 \neq 0.$

$A^{-1} = \left(\frac{1}{3} \right)$

(ii) $A = \begin{pmatrix} 1 & 2 \\ 3 & 4 \end{pmatrix}, \qquad \det(A) = -2 \neq 0.$

$$A^{-1} = \frac{1}{-2} \begin{pmatrix} 4 & -2 \\ -3 & 1 \end{pmatrix} = \begin{pmatrix} -2 & 1 \\ \frac{3}{2} & -\frac{1}{2} \end{pmatrix}$$

(iii) $A = \begin{pmatrix} 1 & 2 & 3 \\ 3 & 1 & 2 \\ 2 & 3 & 1 \end{pmatrix}, \qquad \det(A) = 18 \neq 0.$

We begin by replacing each entry of A by the corresponding minor and obtain

$$\begin{pmatrix} -5 & -1 & 7 \\ -7 & -5 & -1 \\ 1 & -7 & -5 \end{pmatrix}$$

Next, the signs are altered according to the checkerboard pattern

$$\begin{pmatrix} + & - & + \\ - & + & - \\ + & - & + \end{pmatrix}$$

to yield the result

$$\begin{pmatrix} -5 & +1 & +7 \\ +7 & -5 & +1 \\ +1 & +7 & -5 \end{pmatrix}$$

We now take the transpose of this matrix – i.e.

$$\begin{pmatrix} -5 & 1 & 7 \\ 7 & -5 & 1 \\ 1 & 7 & -5 \end{pmatrix}^{\mathrm{T}} = \begin{pmatrix} -5 & 7 & 1 \\ 1 & -5 & 7 \\ 7 & 1 & -5 \end{pmatrix}$$

Finally, the inverse is obtained by multiplying by the scalar $(\det(A))^{-1} = \dfrac{1}{18}$. Thus

$$A^{-1} = \frac{1}{18}\begin{pmatrix} -5 & 7 & 1 \\ 1 & -5 & 7 \\ 7 & 1 & -5 \end{pmatrix}$$

1.2 Exercises

1. The matrices A, B, C and D are given by

$$A = \begin{pmatrix} 6 & 1 & 0 \\ 1 & 0 & 3 \end{pmatrix}, \qquad B = \begin{pmatrix} 3 & 1 \\ 1 & 0 \\ 0 & 2 \end{pmatrix}$$

$$C = \begin{pmatrix} 4 & 1 & 1 \\ 0 & 1 & 0 \\ -2 & 0 & 1 \end{pmatrix}, \qquad D = \begin{pmatrix} 1 \\ 2 \\ 3 \end{pmatrix}$$

Decide which of the following expressions make sense, giving reasons. Evaluate those expressions which make sense.

(i)* $2D$	(ii) $A + B$	(iii) $B - C$
(iv)* $C + D$	(v) $2C - 3D$	(vi) $2A - 3B^{\mathrm{T}}$
(vii)* $B + 3I$	(viii) $2C - 5I$	(ix) A^{-1}
(x)* AB	(xi) BA	(xii) AC
(xiii)* CA	(xiv) BC	(xv) CB
(xvi)* CD	(xvii) DC	(xviii)* $\det(A)$
(xix)* $\det(B)$	(xx) $\det(C)$	(xxi)* $\det(AB)$
(xxii)* $\det(CD)$	(xxiii) $A^{\mathrm{T}}A$	(xxiv)* AA^{T}
(xxv)* $D^{\mathrm{T}}A$	(xxvi) $D^{\mathrm{T}}C$	(xxvii)* $B^{\mathrm{T}}C$
(xxviii)* AC^{-1}	(xxix) $C^{-1}D$	(xxx)* $(C - 3I)^{-1}$.

2. The matrices A, B, C and D are given by

$$A = \begin{pmatrix} 2 & 1 \\ 4 & 3 \end{pmatrix}, \qquad B = \begin{pmatrix} 1 & 2 \\ 2 & 4 \end{pmatrix}$$

$$C = \begin{pmatrix} 1 & 2 & 3 \\ 4 & 1 & 2 \\ 3 & 5 & 1 \end{pmatrix}, \qquad D = \begin{pmatrix} 1 & 2 & 1 \\ 2 & 1 & 2 \\ 3 & 3 & 3 \end{pmatrix}$$

Which of these square matrices are invertible? Find the inverse matrices of those which are invertible and check your answers by multiplying the inverses and the matrices from which they were obtained.

3. Evaluate the following determinants using the cofactor expansion along an appropriate row or column.

(i) $\begin{vmatrix} 6 & 1 & 0 \\ 1 & 0 & 3 \\ -1 & 2 & 5 \end{vmatrix}$

(ii)* $\begin{vmatrix} 3 & 1 & -1 & 2 \\ 5 & 0 & 0 & 0 \\ 7 & 2 & 0 & 0 \\ 19 & 45 & 13 & -1 \end{vmatrix}$

(iii)* $\begin{vmatrix} 1 & 0 & 7 \\ t & 2 & -1 \\ 3 & 4 & -1 \end{vmatrix}$

(iv) $\begin{vmatrix} 2 & 1 & -1 & 0 \\ 5 & 3 & 1 & 0 \\ 0 & 2 & -2 & 1 \\ 0 & 1 & 1 & -1 \end{vmatrix}$

(v) $\begin{vmatrix} 0 & 1 & 0 & 0 \\ -3 & 0 & 0 & 0 \\ 0 & 0 & -5 & 0 \\ 0 & 0 & 0 & 15 \end{vmatrix}$

(vi)* $\begin{vmatrix} 0 & 0 & 0 & 0 & -2 \\ 0 & 0 & 0 & 1 & 7 \\ 0 & 0 & -5 & 3 & 1 \\ 0 & -3 & 0 & 7 & 4 \\ 3 & 2 & 0 & -3 & 8 \end{vmatrix}$

 * For what value(s) of t is the determinant (iii) equal to zero?

4.* For what value(s) of t is the matrix $\begin{pmatrix} 3-t & 3 \\ 5 & 1-t \end{pmatrix}$ not invertible?

5. Even if the expressions AB and BA both make sense, it is *not* necessarily true that $AB = BA$. Check this fact using the matrices A and B of Exercise 2.

6. It is *always* true that $A(BC) = (AB)C$, provided that each side makes sense. Check this result in the case when

$$A = (2 \quad 4) \qquad B = \begin{pmatrix} 1 & 2 \\ 3 & 4 \end{pmatrix} \qquad C = \begin{pmatrix} 3 \\ 5 \end{pmatrix}$$

7. It is *always* true that
$$(AB)^{-1} = B^{-1}A^{-1}$$

provided that A and B are $n \times n$ invertible matrices. Provided that AB makes sense, it is also *always* true that

$$(AB)^{\mathrm{T}} = B^{\mathrm{T}}A^{\mathrm{T}} \quad \text{and} \quad \det(AB) = \det(A)\det(B)$$

Check all these results in the case when

$$A = \begin{pmatrix} 2 & 1 \\ 4 & 3 \end{pmatrix} \qquad B = \begin{pmatrix} 1 & 2 \\ 3 & 4 \end{pmatrix}$$

The matrices need not be square for the result $(AB)^{\mathrm{T}} = B^{\mathrm{T}} A^{\mathrm{T}}$ to hold. Verify this result for the matrices

$$A = \begin{pmatrix} -2 & 5 \\ 0 & 1 \end{pmatrix} \qquad B = \begin{pmatrix} 3 \\ 4 \end{pmatrix}$$

8.* If A and B are both $n \times 1$ column matrices, verify that the matrix products $A^{\mathrm{T}} B$ and $B^{\mathrm{T}} A$ are both defined and find their orders.

What is the relationship between $A^{\mathrm{T}} B$ and $B^{\mathrm{T}} A$?

Verify that the matrix products AB^{T} and BA^{T} are both defined and find their orders.

What is the relationship between AB^{T} and BA^{T}?

9.* If A is a 3×3 matrix with $\det(A) = 7$, evaluate

(i) $\det(2A)$ (ii) $\det(A^2)$ (iii) $\det(3A^{-1})$ (iv) $\det((3A)^{-1})$

10. Let A, B and C be $n \times n$ matrices with the property that $AB = I$ and $CA = I$. Prove that $B = C$.

11.♣ Show that the matrix $A = \begin{pmatrix} -2 & 1 \\ -4 & 2 \end{pmatrix}$ has the property that $A^2 = 0$.

* Find the form of all 2×2 matrices with this property.

12. Let

$$A = \begin{pmatrix} 4 & 2 \\ 1 & -3 \end{pmatrix} \qquad B = \begin{pmatrix} 3 & 0 \\ 1 & -2 \end{pmatrix}$$

Show that there is no nonzero matrix C such that $AC = BC$.

13. Let

$$A = \begin{pmatrix} -3 & 1 \\ 6 & -2 \end{pmatrix} \qquad B = \begin{pmatrix} 1 & -1 \\ 2 & -5 \end{pmatrix} \qquad C = \begin{pmatrix} 0 & 2 \\ -1 & 4 \end{pmatrix}$$

Verify that $AB = AC$ although $B \neq C$. Verify also that the matrices $B - C$ and A are both not invertible.

14.* If A, B, C are nonzero square matrices with

$$AB = AC \qquad \text{and} \qquad B \neq C$$

prove that the matrices $B - C$ and A are both not invertible.

15.* If A is an $n \times n$ matrix and X is an $n \times 1$ nonzero column matrix with

$$AX = 0$$

show, by assuming the contrary, that $\det(A) = 0$.

This is an important result that is used extensively.

16.* Let A be an $n \times n$ matrix, X an $n \times 1$ nonzero column matrix and λ a scalar. Show that the *scalar* equation which gives the values of λ that satisfy the matrix equation $AX = \lambda X$ is

$$\det(A - \lambda I) = 0$$

Write $\lambda X = \lambda I X$ and use Exercise 15.

1.3 Vectors in $\mathbb{R}^2$

The set $\mathbb{R}$ of real numbers can be represented as the points along a horizontal line. Alternatively, we can think of the real numbers as representing displacements along such a line. For example, the point 2 can be displaced to the point 5 by the operation of adding 3. Thus, 3 can represent a displacement of three units to the right. Similarly, -2 can represent a displacement of two units to the left.

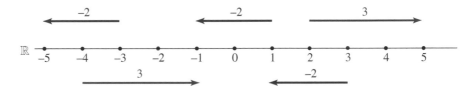

The set $\mathbb{R}^2$ consists of all 2×1 column matrices or **vectors**. If a vector $\mathbf{v} \in \mathbb{R}^2$, it is therefore an object of the form

$$\mathbf{v} = \begin{pmatrix} v_1 \\ v_2 \end{pmatrix}$$

where the two **components** v_1 and v_2 are real numbers, which we call **scalars** when vectors are also being discussed. To save space it is often convenient to use the transpose notation of §1.1, and write $\mathbf{v} = (v_1, v_2)^\mathrm{T}$.

Just as a real number can be used to represent either the position of a point or a displacement along a line, so a two dimensional vector can be used to represent either the position of a point,[†] or a displacement in a plane. For this purpose, as well as a horizontal x axis pointing to the right, we need a vertical y axis pointing upwards. The **cartesian plane** is simply an ordinary plane equipped with such axes.

[†] Hence it makes sense to write the point as $(v_1, v_2)^\mathrm{T}$ rather than the more usual (v_1, v_2).

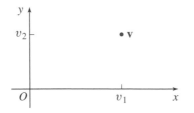
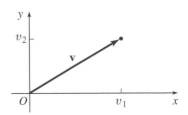

For example, $(3, 2)^{\mathrm{T}}$ can represent either the point obtained by displacing the origin three units horizontally to the right and two units vertically upwards, or it can represent the displacement itself.

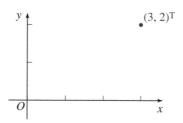

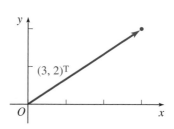

When thinking of a vector as a displacement, the arrow we use to represent the vector need not have its initial point at the origin. All arrows with the same length and direction, wherever they are located, represent the *same* displacement vector.

To make sure that we do not confuse a vector with a scalar, it is common to print vectors in boldface type, and to underline them when writing by hand.

The definitions of matrix addition and scalar multiplication given in §1.1 apply to vectors. Thus, as illustrated in the diagram, we define the **vector addition** of

$$\mathbf{v} = \begin{pmatrix} v_1 \\ v_2 \end{pmatrix} \quad \text{and} \quad \mathbf{w} = \begin{pmatrix} w_1 \\ w_2 \end{pmatrix} \quad \text{by} \quad \mathbf{v} + \mathbf{w} = \begin{pmatrix} v_1 + w_1 \\ v_2 + w_2 \end{pmatrix}$$

Observe that $\mathbf{w}$ is the displacement required to shift the point $\mathbf{v}$ to $\mathbf{v} + \mathbf{w}$.

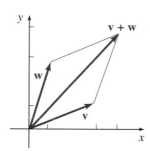

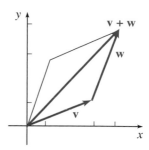

For obvious reasons, the rule for adding two vectors is called the **parallelogram law**, but it is often more useful to draw triangles than parallelograms, as when several vectors are added together.

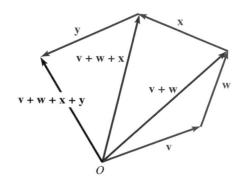

Two $n \times 1$ column vectors cannot be multiplied as matrices (unless $n = 1$). But we can always multiply a matrix by a scalar. In particular, **scalar multiplication** of a vector by a scalar $\alpha \in \mathbb{R}$ is defined by

$$\alpha \mathbf{v} = \begin{pmatrix} \alpha v_1 \\ \alpha v_2 \end{pmatrix}$$

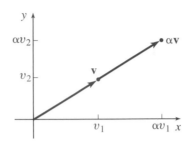

By Pythagoras' theorem, the **length** (or **norm**) of $\mathbf{v}$ is the scalar quantity

$$\|\mathbf{v}\| = \sqrt{(v_1^2 + v_2^2)}$$

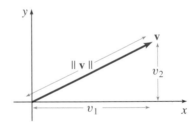

Vectors $\mathbf{u}$ with $\|\mathbf{u}\| = 1$ are called **unit vectors**.

With the exception of the zero vector $\mathbf{0} = (0, 0)^T$, all vectors determine a direction. When using a nonzero vector $\mathbf{v}$ to specify only a direction, it is convenient to replace $\mathbf{v}$ by the unit vector

$$\mathbf{u} = \frac{\mathbf{v}}{\|\mathbf{v}\|}$$

which points in the direction of $\mathbf{v}$.

For example, the vector $\mathbf{v} = (3, 4)^T$ is *not* a unit vector because

$$\|\mathbf{v}\| = (3^2 + 4^2)^{1/2} = 5.$$

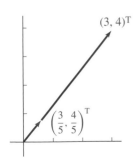

Its direction can be given by the vector

$$\mathbf{u} = \frac{\mathbf{v}}{\|\mathbf{v}\|} = \frac{1}{5}\mathbf{v} = \left(\frac{3}{5}, \frac{4}{5}\right)^T$$

a unit vector which points in the same direction as $\mathbf{v}$.

Vectors $\mathbf{w}$ and $\mathbf{v}$ are **parallel** if $\mathbf{w} = \alpha\mathbf{v}$ for some scalar $\alpha \in \mathbb{R}$ with $\alpha \neq 0$. This implies that they have the same or opposite directions, but not necessarily the same length.

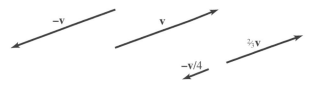

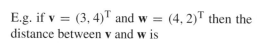

The vector $\mathbf{w} - \mathbf{v}$ is the vector from the point $\mathbf{v}$ to the point $\mathbf{w}$. The **distance** between two points $\mathbf{v}$ and $\mathbf{w}$ is defined by

$$\|\mathbf{w} - \mathbf{v}\| = \sqrt{\{(w_1 - v_1)^2 + (w_2 - v_2)\}^2} = \|\mathbf{v} - \mathbf{w}\|$$

E.g. if $\mathbf{v} = (3, 4)^{\mathrm{T}}$ and $\mathbf{w} = (4, 2)^{\mathrm{T}}$ then the distance between $\mathbf{v}$ and $\mathbf{w}$ is

$$\|(4, 2)^{\mathrm{T}} - (3, 4)^{\mathrm{T}}\|$$

$$= \sqrt{\{(4 - 3)^2 + (2 - 4)^2\}}$$

$$= \sqrt{5}$$

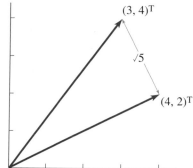

Although the matrix product of two vectors $\mathbf{v}$ and $\mathbf{w}$ makes no sense, we have

$$\mathbf{v}^{\mathrm{T}}\mathbf{w} = (v_1 \ v_2)\begin{pmatrix} w_1 \\ w_2 \end{pmatrix} = v_1 w_1 + v_2 w_2.$$

[†] The scalar product is also known as the inner product or the dot product, which is written as $\mathbf{v} \cdot \mathbf{w}$.

This scalar quantity is called the **scalar product**[†] of $\mathbf{v}$ and $\mathbf{w}$ and written as

$$\langle \mathbf{v}, \mathbf{w} \rangle = v_1 w_1 + v_2 w_2$$

e.g.

$$\left\langle \begin{pmatrix} 3 \\ 4 \end{pmatrix}, \begin{pmatrix} 4 \\ 2 \end{pmatrix} \right\rangle = 3 \times 4 + 4 \times 2 = 20$$

[‡] See Exercise 1.4.4.

It is easy to check that scalar products have the following properties:[‡]

(i) $\langle \mathbf{v}, \mathbf{v} \rangle = \|\mathbf{v}\|^2$ $(\mathbf{v} \in \mathbb{R}^2)$

(ii) $\langle \mathbf{v}, \mathbf{w} \rangle = \langle \mathbf{w}, \mathbf{v} \rangle$ $(\mathbf{v}, \mathbf{w} \in \mathbb{R}^2)$

(iii) $\langle \alpha\mathbf{u} + \beta\mathbf{v}, \mathbf{w} \rangle = \alpha\langle \mathbf{u}, \mathbf{w} \rangle + \beta\langle \mathbf{v}, \mathbf{w} \rangle$ $(\alpha, \beta \in \mathbb{R}, \ \mathbf{u}, \mathbf{v}, \mathbf{w} \in \mathbb{R}^2)$

The angle between vectors

The geometric interpretation of the scalar product gives rise to a formula for the angle between vectors. The cosine rule from trigonometry which states that

$$c^2 = a^2 + b^2 - 2ab\cos C$$

is a generalisation of Pythagoras' theorem, to which it reduces when $C = \pi/2$.

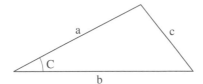 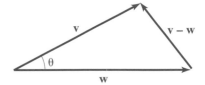

Rewriting the cosine rule in terms of vectors, we obtain

$$\|\mathbf{v} - \mathbf{w}\|^2 = \|\mathbf{v}\|^2 + \|\mathbf{w}\|^2 - 2\|\mathbf{v}\|\,\|\mathbf{w}\|\cos\theta \tag{1}$$

where θ is the angle between $\mathbf{v}$ and $\mathbf{w}$, chosen so that $0 \le \theta \le \pi$ to avoid ambiguity. Now

$$
\begin{aligned}
\|\mathbf{v} - \mathbf{w}\|^2 &= \langle \mathbf{v} - \mathbf{w}, \mathbf{v} - \mathbf{w}\rangle && \text{by (i)} \\
&= \langle \mathbf{v}, \mathbf{v}\rangle - \langle \mathbf{w}, \mathbf{v}\rangle - \langle \mathbf{v}, \mathbf{w}\rangle + \langle \mathbf{w}, \mathbf{w}\rangle && \text{by (ii) and (iii)} \\
&= \langle \mathbf{v}, \mathbf{v}\rangle - 2\langle \mathbf{v}, \mathbf{w}\rangle + \langle \mathbf{w}, \mathbf{w}\rangle && \text{by (ii)} \\
&= \|\mathbf{v}\|^2 + \|\mathbf{w}\|^2 - 2\langle \mathbf{v}, \mathbf{w}\rangle && \text{by (i)}
\end{aligned}
$$

Therefore

$$\|\mathbf{v} - \mathbf{w}\|^2 = \|\mathbf{v}\|^2 + \|\mathbf{w}\|^2 - 2\langle \mathbf{v}, \mathbf{w}\rangle \tag{2}$$

Comparing (1) and (2)

$$\langle \mathbf{v}, \mathbf{w}\rangle = \|\mathbf{v}\|\,\|\mathbf{w}\|\cos\theta$$

So θ can be calculated from the equation

$$\cos\theta = \frac{\langle \mathbf{v}, \mathbf{w}\rangle}{\|\mathbf{v}\|\|\mathbf{w}\|} \qquad 0 \le \theta \le \pi$$

E.g. if θ is the angle between $\begin{pmatrix} 2 \\ 0 \end{pmatrix}$ and $\begin{pmatrix} -1 \\ 1 \end{pmatrix}$,

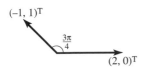

$$\cos\theta = \frac{\left\langle \begin{pmatrix} 2 \\ 0 \end{pmatrix}, \begin{pmatrix} -1 \\ 1 \end{pmatrix} \right\rangle}{\|(2, 0)^{\mathrm{T}}\|\,\|(-1, 1)^{\mathrm{T}}\|} = \frac{-2}{2\sqrt{2}} = -\frac{1}{\sqrt{2}}$$

Therefore $\theta = \dfrac{3\pi}{4}$.

$(-1, 1)^\text{T}$ $(1, 1)^\text{T}$

Nonzero vectors **v** and **w** are defined to be **orthogonal** (**perpendicular** or **normal**) if the angle between them is a right angle.

Recall that $\cos \theta = 0$ for $0 \leq \theta \leq \pi$ if and only if $\theta = \pi/2$, so that the angle between **v** and **w** is a right angle. It follows that **v** and **w** are orthogonal if and only if $\langle \mathbf{v}, \mathbf{w} \rangle = 0$.

E.g. the vectors $(1, 1)^\text{T}$ and $(-1, 1)^\text{T}$ are orthogonal. *To prove this, calculate the scalar product of the two vectors:*

$$\left\langle \begin{pmatrix} 1 \\ 1 \end{pmatrix}, \begin{pmatrix} -1 \\ 1 \end{pmatrix} \right\rangle = 1 \times (-1) + 1 \times 1$$

It is often useful to split a vector **v** into two vectors, one of which is in the direction of a vector **w** and the other in a direction orthogonal to **w**. The vector **v** is then the sum of a 'vector component along **w**' and a 'vector component orthogonal to **w**'. If θ is the angle between **v** and **w**, the *(scalar) component* of **v** in the direction of **w** is

$$\|\mathbf{v}\| \cos \theta = \frac{\langle \mathbf{v}, \mathbf{w} \rangle}{\|\mathbf{w}\|}$$

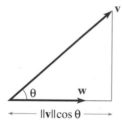
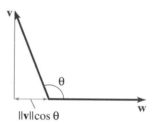

Observe that the component is negative when $\pi/2 < \theta \leq \pi$, because $\cos \theta$ is negative in this range.

Hence the **vector component of v along w** is

$$\left(\frac{\langle \mathbf{v}, \mathbf{w} \rangle}{\|\mathbf{w}\|} \right) \frac{\mathbf{w}}{\|\mathbf{w}\|} = \frac{\langle \mathbf{v}, \mathbf{w} \rangle}{\|\mathbf{w}\|^2} \mathbf{w}$$

Then the **vector component of v orthogonal to w** is $\mathbf{v} - \dfrac{\langle \mathbf{v}, \mathbf{w} \rangle}{\|\mathbf{w}\|^2} \mathbf{w}$.

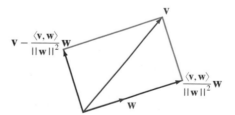

*To check that this vector is orthogonal to **w**, calculate their scalar product:*

$$\left\langle \mathbf{v} - \frac{\langle \mathbf{v}, \mathbf{w} \rangle}{\|\mathbf{w}\|^2} \mathbf{w}, \mathbf{w} \right\rangle = \langle \mathbf{v}, \mathbf{w} \rangle - \frac{\langle \mathbf{v}, \mathbf{w} \rangle}{\|\mathbf{w}\|^2} \|\mathbf{w}\|^2 = 0$$

E.g. the component of $(1, 4)^{\mathrm{T}}$ along $(1, 1)^{\mathrm{T}}$ is $\dfrac{\left\langle \binom{1}{4}, \binom{1}{1} \right\rangle}{\|(1, 1)^{\mathrm{T}}\|} = \dfrac{5}{\sqrt{2}}$.

The vector component of $(1, 4)^{\mathrm{T}}$ along $(1, 1)^{\mathrm{T}}$ is $\dfrac{5}{\sqrt{2}} \binom{1/\sqrt{2}}{1/\sqrt{2}} = \binom{5/2}{5/2}$.

The vector component of $(1, 4)^{\mathrm{T}}$ orthogonal to $(1, 1)^{\mathrm{T}}$ is

$$\binom{1}{4} - \binom{5/2}{5/2} = \binom{-3/2}{3/2}.$$

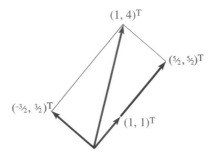

1.4 Exercises

1. Express the vectors in the diagram in the margin in component form.

    (i)   Find their sum and verify your answer by drawing them on graph paper along the sides of a polygon.

    *   Is the vector you obtain related to any of the given vectors?

    (ii)   Find the end point of $(7, 4)^{\mathrm{T}}$ if its initial point is $(-2, 1)^{\mathrm{T}}$.

    (iii)* Find the initial point of $(7, 4)^{\mathrm{T}}$ if its end point is $(3, -2)^{\mathrm{T}}$.

    (iv)   Find a vector of length one in the direction of the vector $(-5, 12)^{\mathrm{T}}$.

    (v)   Find a vector of length 7 in the direction opposite to $(-3, 4)^{\mathrm{T}}$.

2. Let $\mathbf{u} = (2, 1)^{\mathrm{T}}$, $\mathbf{v} = (8, -6)^{\mathrm{T}}$ and $\mathbf{w} = (-3, -4)^{\mathrm{T}}$

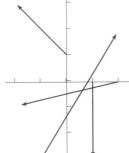

    (i)*   Draw the vectors $\mathbf{u}$, $\mathbf{v}$ and $\mathbf{w}$ on graph paper. Find the angle between $\mathbf{u}$ and $\mathbf{v}$ and between $\mathbf{v}$ and $\mathbf{w}$ and use your drawing to verify your answers.

    (ii)   Draw the vectors $\mathbf{u} + \mathbf{v}$, $\mathbf{v} - \mathbf{u}$, $\mathbf{u} + \mathbf{v} + \mathbf{w}$ and $2\mathbf{w} - \mathbf{u} - \mathbf{v}$.

    (iii)* Find the vectors $4(\mathbf{u} + 5\mathbf{v})$ and $\frac{1}{2}\mathbf{w} - 3(\mathbf{v} - \mathbf{u})$ in component form.

    (iv)*   Find $\|\mathbf{u} + \mathbf{v}\|$, $\|\mathbf{u}\| + \|\mathbf{v}\|$, $5\|\mathbf{u}\| + \|-2\mathbf{v}\|$ and $\|5\mathbf{u} - 2\mathbf{v}\|$.

    (v)   Find the vector $\mathbf{x}$ that satisfies the equation $\mathbf{u} - 3\mathbf{x} = \mathbf{v} + 2\mathbf{w} + \mathbf{x}$.

    (vi)*   Find the scalar component of $\mathbf{u}$ along $\mathbf{v}$. Express $\mathbf{u}$ as the sum of a vector along $\mathbf{v}$ and a vector perpendicular to $\mathbf{v}$.

(vii) Find the distance between the points **v** and **w**. Find also, the scalar product of **v** and **w** and the angle between them.

3.* Show that the distance between two points **v** and **w** increases as the angle θ between the vectors **v** and **w** increases. *Remember that* $0 \leq \theta \leq \pi$.

4.* Prove the three properties of the scalar product.

5.* If a and b are given real numbers which satisfy the equation

$$3a + 5b = 0$$

interpret the equation as the scalar product of two vectors, and write down at least two different pairs of vectors which are orthogonal.

6. Use vectors to prove the following elementary geometrical results:

(i)* The line segment joining the midpoints of two sides of a triangle is parallel to the third side and equal to half of it.

(ii) If one pair of opposite sides of a quadrilateral is parallel and equal, then so also is the other pair.

(iii) The midpoints of the sides of any quadrilateral are the vertices of a parallelogram.

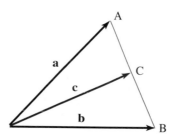

7. (i) If C is the midpoint of AB, find **c** in terms of **a** and **b** in the above figure.

(ii)* If C divides AB in the ratio of $m : n$ find **c**.

1.5 Vectors in $\mathbb{R}^3$

The set $\mathbb{R}^3$ consists of all 3×1 column matrices or vectors of the form

$$\mathbf{v} = \begin{pmatrix} v_1 \\ v_2 \\ v_3 \end{pmatrix}$$

where the components v_1, v_2 and v_3 are scalars or real numbers. Three dimensional vectors can be used to represent either positions or displacements in space.

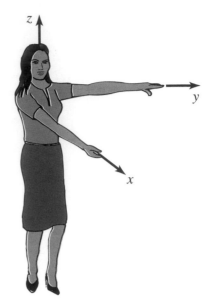

For this purpose we need three mutually orthogonal axes. The way that these are usually
drawn depends on the arms and body of a
person when these point in orthogonal directions, as shown in the figure. The right arm
then points in the direction of the first, or x
axis. The left arm points in the direction of
the second, or y axis. The head points in the
direction of the third, or z axis.

It is also conventional to use the **right hand rule**.

If the index finger of the right hand points in the
direction of the x axis and the middle finger in the
direction of the y axis, the thumb then points in the
direction of the z axis.

As with vectors in $\mathbb{R}^2$, a three dimensional vector can be used to represent either
a point or a displacement in $\mathbb{R}^3$.

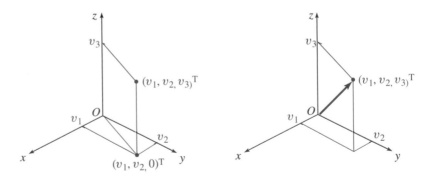

Vector addition and scalar multiplication are defined, just as in $\mathbb{R}^2$, by

$$\mathbf{v} + \mathbf{w} = \begin{pmatrix} v_1 + w_1 \\ v_2 + w_2 \\ v_3 + w_3 \end{pmatrix} \quad \text{and} \quad \alpha\mathbf{v} = \begin{pmatrix} \alpha v_1 \\ \alpha v_2 \\ \alpha v_3 \end{pmatrix} \quad \text{where } \alpha \in \mathbb{R}$$

E.g. let $\mathbf{v} = (1, 2, 3)^T$ and $\mathbf{w} = (2, 0, 5)^T$. Then

(i) $\quad \mathbf{v} + \mathbf{w} = \begin{pmatrix} 1 \\ 2 \\ 3 \end{pmatrix} + \begin{pmatrix} 2 \\ 0 \\ 5 \end{pmatrix} = \begin{pmatrix} 3 \\ 2 \\ 8 \end{pmatrix}$

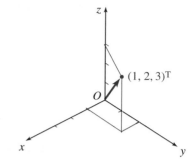

(ii) $\quad 2\mathbf{v} = 2 \begin{pmatrix} 1 \\ 2 \\ 3 \end{pmatrix} = \begin{pmatrix} 2 \\ 4 \\ 6 \end{pmatrix}$

Applying Pythagoras' theorem twice, the length of $\mathbf{v}$ is

$$\|\mathbf{v}\| = \sqrt{{v_1}^2 + {v_2}^2 + {v_3}^2}$$

and the distance between $\mathbf{v}$ and $\mathbf{w}$ is

$$\|\mathbf{w} - \mathbf{v}\| = \sqrt{(w_1 - v_1)^2 + (w_2 - v_2)^2 + (w_3 - v_3)^2}$$

E.g. let $\mathbf{v} = (1, 2, 3)^{\mathrm{T}}$ and $\mathbf{w} = (2, 0, 5)^{\mathrm{T}}$. Then the length of $\mathbf{v}$ is

$$\|\mathbf{v}\| = \{1^2 + 2^2 + 3^2\}^{1/2} = \sqrt{14}$$

and the distance between $\mathbf{v}$ and $\mathbf{w}$ is

$$\begin{aligned} \|\mathbf{w} - \mathbf{v}\| &= \{(2-1)^2 + (0-2)^2 + (5-3)^2\}^{1/2} \\ &= \{1 + 4 + 4\}^{1/2} = 3 \end{aligned}$$

The scalar product of $\mathbf{v}$ and $\mathbf{w}$ is defined by

$$\langle \mathbf{v}, \mathbf{w} \rangle = \left\langle \begin{pmatrix} v_1 \\ v_2 \\ v_3 \end{pmatrix}, \begin{pmatrix} w_1 \\ w_2 \\ w_3 \end{pmatrix} \right\rangle = v_1 w_1 + v_2 w_2 + v_3 w_3$$

This has the same properties as the scalar product for two dimensional vectors. The angle between vectors is given by the same formula and the condition for vectors $\mathbf{v}$ and $\mathbf{w}$ to be orthogonal is similarly $\langle \mathbf{v}, \mathbf{w} \rangle = 0$.

E.g. find the cosine of the angle θ between the vectors $\mathbf{v} = (1, 2, 3)^{\mathrm{T}}$ and $\mathbf{w} = (2, 0, 5)^{\mathrm{T}}$. We have

$$\|\mathbf{v}\| = \{1^2 + 2^2 + 3^2\}^{1/2} = \sqrt{14}$$
$$\|\mathbf{w}\| = \{2^2 + 0^2 + 5^2\}^{1/2} = \sqrt{29}$$
$$\langle \mathbf{v}, \mathbf{w} \rangle = 1 \times 2 + 2 \times 0 + 3 \times 5 = 17$$

Hence

$$\cos \theta = \frac{\langle \mathbf{v}, \mathbf{w} \rangle}{\|\mathbf{v}\|\,\|\mathbf{w}\|} = \frac{17}{\sqrt{14}\sqrt{29}}$$

The vector $\mathbf{v} = (1, 1, 2)^{\mathrm{T}}$ is *not* a unit vector because

$$\|\mathbf{v}\| = \{1^2 + 1^2 + 2^2\}^{1/2} = \sqrt{6}$$

But the vector

$$\mathbf{u} = \|\mathbf{v}\|^{-1}\mathbf{v} = \frac{1}{\sqrt{6}}\mathbf{v} = \left(\frac{1}{\sqrt{6}}, \frac{1}{\sqrt{6}}, \frac{2}{\sqrt{6}}\right)^{\mathrm{T}}$$

is a unit vector which points in the same direction as $\mathbf{v}$.

The vectors $\mathbf{v} = (1, 2, 3)^{\mathrm{T}}$ and $\mathbf{w} = (-3, 3, -1)^{\mathrm{T}}$ are orthogonal because

$$\langle \mathbf{v}, \mathbf{w} \rangle = 1 \times (-3) + 2 \times 3 + 3 \times (-1) = -3 + 6 - 3 = 0$$

1.6 Lines

Let $\mathbf{v} \in \mathbb{R}^3$, $\mathbf{v} \neq 0$. If $t \in \mathbb{R}$, $t\mathbf{v}$ is a vector in the direction of $\mathbf{v}$ if $t > 0$ and in the direction of $-\mathbf{v}$ if $t < 0$. As t varies over all real values, $t\mathbf{v}$ moves along the line drawn through the origin in the direction of $\mathbf{v}$. So the equation

$$\mathbf{x} = t\mathbf{v} \qquad (t \in \mathbb{R})$$

represents all points on the line through the origin in the direction of $\mathbf{v}$.

Note that $\boldsymbol{\xi}$ and $\mathbf{v}$ are constant, while $\mathbf{x}$ and t are variable.

If $\boldsymbol{\xi}$ and $\mathbf{v} \in \mathbb{R}^3$ are constant, $\mathbf{v} \neq 0$, the equation

$$\mathbf{x} = \boldsymbol{\xi} + t\mathbf{v} \qquad (t \in \mathbb{R})$$

represents all points on the line through $\boldsymbol{\xi}$ in the direction of $\mathbf{v}$.

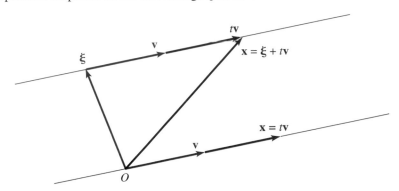

This is called the **parametric** equation of the line. Each value of the parameter t yields a unique point on the line and vice versa. If $\mathbf{v}$ is a *unit vector*, then

$$\|\mathbf{x} - \boldsymbol{\xi}\| = \|t\mathbf{v}\| = |t| \, \|\mathbf{v}\| = |t|$$

and so the distance between $\mathbf{x}$ and $\boldsymbol{\xi}$ is $|t|$.

Writing the equation in terms of components, we obtain

$$\begin{pmatrix} x \\ y \\ z \end{pmatrix} = \begin{pmatrix} \xi_1 \\ \xi_2 \\ \xi_3 \end{pmatrix} + t \begin{pmatrix} v_1 \\ v_2 \\ v_3 \end{pmatrix} \qquad (t \in \mathbb{R})$$

The corresponding scalar equations are

$$x = \xi_1 + tv_1$$
$$y = \xi_2 + tv_2$$
$$z = \xi_3 + tv_3$$

If none of the v_i are zero, solving for t in each equation yields equations involving the vector components, known as **cartesian equations**:

$$\frac{x - \xi_1}{v_1} = \frac{y - \xi_2}{v_2} = \frac{z - \xi_3}{v_3}$$

E.g. the equations of a line through $(6, -2, 0)^T$ in the direction $(4, 3, -1)^T$ are

$$\frac{x - 6}{4} = \frac{y + 2}{3} = \frac{z}{-1}$$

We can rewrite the equations as

$$\left.\begin{array}{rl} 3x \quad -4y & = 26 \\ y \quad +3z & = -2 \end{array}\right\} \quad \text{or as} \quad \left.\begin{array}{rl} x \quad\quad +4z & = 6 \\ y \quad +3z & = -2 \end{array}\right\}$$

or in many other equivalent ways.

A vector equation of the line is

$$\mathbf{x} = (6, -2, 0)^T + t(4, 3, -1)^T$$

To find if the point $(2, -5, 1)^T$ lies on the line, see if there exists a $t \in \mathbb{R}$ such that

$$\begin{pmatrix} 2 \\ -5 \\ 1 \end{pmatrix} = \begin{pmatrix} 6 \\ -2 \\ 0 \end{pmatrix} + t \begin{pmatrix} 4 \\ 3 \\ -1 \end{pmatrix}$$

$$\left.\begin{array}{l} 2 = 6 + 4t \\ -5 = -2 + 3t \\ 1 = 0 - t \end{array}\right\}$$

These three equations have a solution $t = -1$, so the point lies on the line. If the equations were inconsistent so that no solution could be found, we would conclude that $(2, -5, 1)^T$ does not lie on the line.

Two lines that lie in the same plane are **coplanar**. Such lines either intersect or are parallel. Two lines that are not coplanar are **skew**.

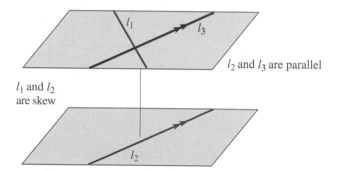

l_1 and l_2 are skew

l_2 and l_3 are parallel

The angle between two intersecting lines $\mathbf{x} = \boldsymbol{\xi} + t\mathbf{v}$ and $\mathbf{x} = \boldsymbol{\eta} + t\mathbf{u}$, is the angle between the vectors $\mathbf{u}$ and $\mathbf{v}$.

The lines intersect if and only if they have a point in common. This occurs when there exist scalars t_1, t_2 such that

$$\boldsymbol{\xi} + t_1 \mathbf{v} = \boldsymbol{\eta} + t_2 \mathbf{u}$$

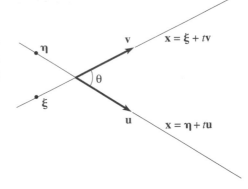

As t is a variable, the values it takes at the point of intersection will, in general, differ for the two lines.

E.g. to find whether the lines $\mathbf{x} = (1, 0, 1)^{\mathrm{T}} + t(-1, -1, -1)^{\mathrm{T}}$ and $\mathbf{x} = t(2, 0, 1)^{\mathrm{T}}$ intersect, see if there exist scalars t_1, t_2 such that

$$\begin{pmatrix} 1 \\ 0 \\ 1 \end{pmatrix} + t_1 \begin{pmatrix} -1 \\ -1 \\ -1 \end{pmatrix} = t_2 \begin{pmatrix} 2 \\ 0 \\ 1 \end{pmatrix}$$

$$1 - t_1 = 2t_2$$
$$-t_1 = 0$$
$$1 - t_1 = t_2$$

The equations are inconsistent, so there are no scalars t_1 and t_2 satisfying the above equations. Hence the lines do not intersect.

A line through points $\boldsymbol{\xi}$ and $\boldsymbol{\eta}$ has direction $\boldsymbol{\eta} - \boldsymbol{\xi}$ and so has equation

$$\mathbf{x} = \boldsymbol{\xi} + t(\boldsymbol{\eta} - \boldsymbol{\xi}) \qquad (t \in \mathbb{R})$$

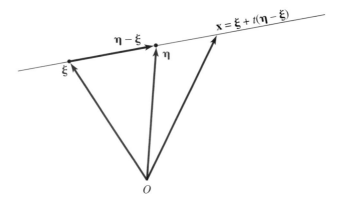

E.g. the line passing through $(1, 2, 3)^{\mathrm{T}}$ and $(3, 2, 1)^{\mathrm{T}}$ has the form

$$\begin{pmatrix} x \\ y \\ z \end{pmatrix} = \begin{pmatrix} 1 \\ 2 \\ 3 \end{pmatrix} + t \begin{pmatrix} 3 - 1 \\ 2 - 2 \\ 1 - 3 \end{pmatrix}$$

which can be expressed in the alternative form

$$\frac{x-1}{2} = \frac{z-3}{-2} \quad y = 2$$

Do not try to write the equation $y - 2 = t \times 0$ as $(y-2)/0 = t$. It is not valid to divide by 0.

It should be obvious that all of the above theory holds in $\mathbb{R}^2$, though of course, any two lines in $\mathbb{R}^2$ are always coplanar and there can be no skew lines in $\mathbb{R}^2$.

1.7 Planes

Consider the equation

$$\langle \mathbf{v}, \mathbf{x} \rangle = 0$$

where $\mathbf{v}$ is a fixed nonzero vector. In $\mathbb{R}^2$ the equation represents a line through the origin. In $\mathbb{R}^3$ it represents a plane through the origin. In both cases, the line or plane is orthogonal to $\mathbf{v}$. We say that $\mathbf{v}$ is a **normal** to the line or plane.

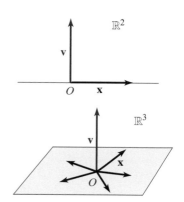

If $\boldsymbol{\xi}$ is a fixed vector, consider the equation

$$\langle \mathbf{v}, \mathbf{x} - \boldsymbol{\xi} \rangle = 0$$

In $\mathbb{R}^2$ it is the equation of a line through $\boldsymbol{\xi}$ orthogonal to $\mathbf{v}$. In $\mathbb{R}^3$ it is the equation of a plane through $\boldsymbol{\xi}$ with normal $\mathbf{v}$.

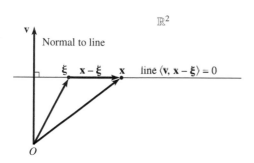
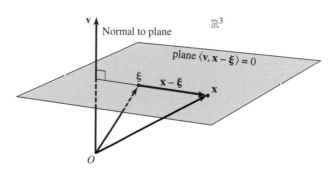

The equation may be written as

$$\langle \mathbf{v}, \mathbf{x} \rangle = \langle \mathbf{v}, \boldsymbol{\xi} \rangle$$

If θ is the angle between $\boldsymbol{\xi}$ and $\mathbf{v}$, and $\mathbf{v}$ is a unit vector so that $\|\mathbf{v}\| = 1$, we have

$$\langle \mathbf{v}, \boldsymbol{\xi} \rangle = \|\mathbf{v}\| \, \|\boldsymbol{\xi}\| \cos\theta = \|\boldsymbol{\xi}\| \cos\theta = c$$

Hence the equation becomes

$$\langle \mathbf{v}, \mathbf{x} \rangle = \|\boldsymbol{\xi}\| \cos\theta = c$$

When **v** points towards the plane, the distance from the origin to the plane is $c = \|\boldsymbol{\xi}\| \cos\theta$. When **v** points away from the plane, the distance from the origin to the plane is $-c = -\|\boldsymbol{\xi}\| \cos\theta$. Hence the distance from the origin to the plane is $|c|$.

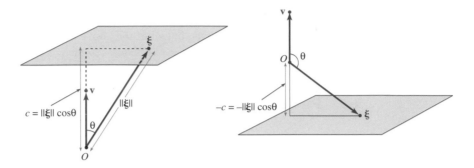

The equation $\langle \mathbf{v}, \mathbf{x} \rangle = c$ written in terms of the components is known as the **cartesian equation**:

$$v_1 x + v_2 y + v_3 z = c$$

Summarising, a single linear equation of the above form defines a plane in $\mathbb{R}^3$, with normal vector $\mathbf{v} = (v_1, v_2, v_3)^\mathrm{T}$. If **v** is a unit vector, $|c|$ is the distance of the plane from the origin.

Example 6 _____

In $\mathbb{R}^2$ the equation

$$3x + 4y = 5$$

is the equation of a line in two dimensions. The vector $(3, 4)^\mathrm{T}$ is *normal* to this line. A *unit* normal to the line is the vector $(\frac{3}{5}, \frac{4}{5})^\mathrm{T}$. If the equation is rewritten in the form

$$\frac{3}{5}x + \frac{4}{5}y = 1$$

we therefore conclude that the distance of the line from $(0, 0)^\mathrm{T}$ is 1.

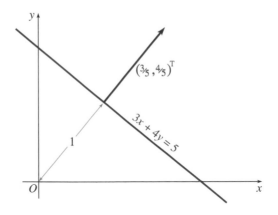

Example 7

The equation

$$x + 2y + 2z = 5$$

is the equation of a plane in $\mathbb{R}^3$. A normal to this plane is the vector $(1, 2, 2)^T$. A *unit* normal is the vector $(\frac{1}{3}, \frac{2}{3}, \frac{2}{3})^T$. If the equation is rewritten in the form

$$\frac{1}{3}x + \frac{2}{3}y + \frac{2}{3}z = \frac{5}{3}$$

we conclude that the distance from $(0, 0, 0)^T$ to the plane is $\frac{5}{3}$.

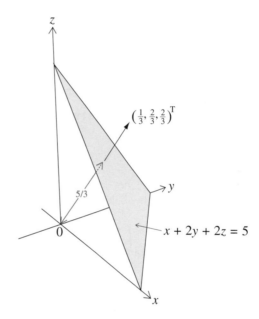

Example 8

A plane passes through the points $(1, 2, 3)^T$, $(3, 1, 2)^T$ and $(2, 3, 1)^T$. Suppose that we wish to find its distance from the origin.

Let the plane have equation $ux + vy + wz = c$. Then if $(u, v, w)^T$ is a unit vector, the distance of the plane from the origin is $|c|$. Substituting the components of the points in the equation

$$u1 + v2 + w3 = c$$
$$u3 + v1 + w2 = c$$
$$u2 + v3 + w1 = c$$

i.e.

$$\begin{pmatrix} 1 & 2 & 3 \\ 3 & 1 & 2 \\ 2 & 3 & 1 \end{pmatrix} \begin{pmatrix} u \\ v \\ w \end{pmatrix} = c \begin{pmatrix} 1 \\ 1 \\ 1 \end{pmatrix}$$

Hence

$$\begin{pmatrix} u \\ v \\ w \end{pmatrix} = \begin{pmatrix} 1 & 2 & 3 \\ 3 & 1 & 2 \\ 2 & 3 & 1 \end{pmatrix}^{-1} c \begin{pmatrix} 1 \\ 1 \\ 1 \end{pmatrix} = \frac{c}{18} \begin{pmatrix} -5 & 7 & 1 \\ 1 & -5 & 7 \\ 7 & 1 & -5 \end{pmatrix} \begin{pmatrix} 1 \\ 1 \\ 1 \end{pmatrix}$$

$$\begin{pmatrix} u \\ v \\ w \end{pmatrix} = \frac{c}{18} \begin{pmatrix} 3 \\ 3 \\ 3 \end{pmatrix} = \frac{c}{6} \begin{pmatrix} 1 \\ 1 \\ 1 \end{pmatrix}$$

We next impose the further condition that $(u, v, w)^{\mathrm{T}}$ be a *unit* vector – i.e. $u^2 + v^2 + w^2 = 1$. Then

$$1 = u^2 + v^2 + w^2 = \frac{c^2}{36} + \frac{c^2}{36} + \frac{c^2}{36} = \frac{c^2}{12}$$

Hence the required distance is

$$c = \sqrt{12}$$

The angle between two planes is the angle between their normals. So the angle between the planes

$$2x - y + z = 6 \quad \text{and} \quad x + 5y + 3z = 2$$

is the angle θ between the vectors $(2, -1, 1)^{\mathrm{T}}$ and $(1, 5, 3)^{\mathrm{T}}$. We can find θ by calculating the scalar product of these vectors:

$$\left\| \begin{pmatrix} 2 \\ -1 \\ 1 \end{pmatrix} \right\| \left\| \begin{pmatrix} 1 \\ 5 \\ 3 \end{pmatrix} \right\| \cos \theta = \left\langle \begin{pmatrix} 2 \\ -1 \\ 1 \end{pmatrix}, \begin{pmatrix} 1 \\ 5 \\ 3 \end{pmatrix} \right\rangle = 0$$

Hence $\theta = \dfrac{\pi}{2}$ and the planes are orthogonal.

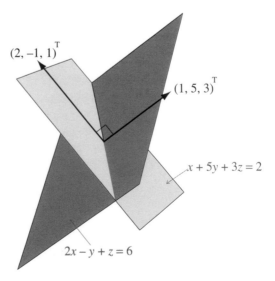

If a vector is orthogonal to the normal to a plane, we say that it is parallel to the plane. If two planes have the same normal, they are parallel.

Parametric equation of a plane

This is a natural extension of the parametric equation of a line. Let $\mathbf{u}$ and $\mathbf{v}$ be nonzero, nonparallel vectors. As s and t vary over all real values, the figure on the left below indicates that the vector $s\mathbf{u} + t\mathbf{v}$ varies over all points of the plane containing the directions of both $\mathbf{u}$ and $\mathbf{v}$.

If $\boldsymbol{\xi}$ is a fixed point of the plane and if $\mathbf{x}$ represents a general point of the plane, then the figure on the right below shows that the vector $\mathbf{x} - \boldsymbol{\xi}$ has the form $s\mathbf{u} + t\mathbf{v}$. Hence the equation

$$\mathbf{x} - \boldsymbol{\xi} = s\mathbf{u} + t\mathbf{v} \qquad (s, t \in \mathbb{R})$$

represents all points on the plane through the point $\boldsymbol{\xi}$ containing both of the vectors $\mathbf{u}$ and $\mathbf{v}$.

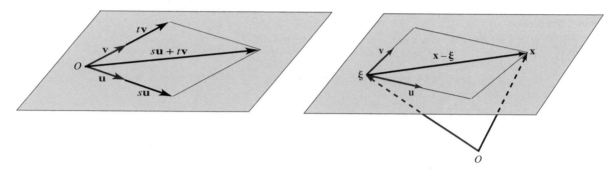

The vector product in $\mathbb{R}^3$

Moving all terms of the above equation to the left hand side and writing in terms of the components, we obtain

$$\begin{pmatrix} x - \xi_1 \\ y - \xi_2 \\ z - \xi_3 \end{pmatrix} - s \begin{pmatrix} u_1 \\ u_2 \\ u_3 \end{pmatrix} - t \begin{pmatrix} v_1 \\ v_2 \\ v_3 \end{pmatrix} = 0$$

This may be written as the matrix equation

$$\begin{pmatrix} x - \xi_1 & u_1 & v_1 \\ y - \xi_2 & u_2 & v_2 \\ z - \xi_3 & u_3 & v_3 \end{pmatrix} \begin{pmatrix} 1 \\ -s \\ -t \end{pmatrix} = 0$$

As the column matrix is nonzero, by Exercise 1.2.15, we have

$$\begin{vmatrix} x - \xi_1 & u_1 & v_1 \\ y - \xi_2 & u_2 & v_2 \\ z - \xi_3 & u_3 & v_3 \end{vmatrix} = 0$$

This gives the equation of the plane as

$$(u_2 v_3 - u_3 v_2)(x - \xi_1) + (u_3 v_1 - u_1 v_3)(y - \xi_2) + (u_1 v_2 - u_2 v_1)(z - \xi_3) = 0 \quad (1)$$

It follows that the vector normal to the plane, which is orthogonal to both **u** and **v** is

$$\begin{pmatrix} u_2 v_3 - u_3 v_2 \\ u_3 v_1 - u_1 v_3 \\ u_1 v_2 - u_2 v_1 \end{pmatrix}$$

This vector, called **u** × **v** is defined to be the **vector product** of the vectors **u** and **v** $\in \mathbb{R}^3$. Equation (1) of the plane then takes the form

$$\langle \mathbf{u} \times \mathbf{v}, \mathbf{x} - \boldsymbol{\xi} \rangle = 0$$

A simpler formula for the vector product can be given in terms of the vectors

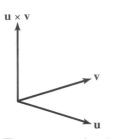

u × **v**

The vector product is also known as the **cross product**.

$$\mathbf{e}_1 = \begin{pmatrix} 1 \\ 0 \\ 0 \end{pmatrix} \quad \mathbf{e}_2 = \begin{pmatrix} 0 \\ 1 \\ 0 \end{pmatrix} \quad \text{and} \quad \mathbf{e}_3 = \begin{pmatrix} 0 \\ 0 \\ 1 \end{pmatrix}$$

Then in the notation of determinants

$$\mathbf{u} \times \mathbf{v} = \begin{vmatrix} \mathbf{e}_1 & \mathbf{e}_2 & \mathbf{e}_3 \\ u_1 & u_2 & u_3 \\ v_1 & v_2 & v_3 \end{vmatrix}$$

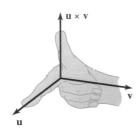

u × **v**

$$= \begin{vmatrix} u_2 & u_3 \\ v_2 & v_3 \end{vmatrix} \mathbf{e}_1 + \begin{vmatrix} u_3 & u_1 \\ v_3 & v_1 \end{vmatrix} \mathbf{e}_2 + \begin{vmatrix} u_1 & u_2 \\ v_1 & v_2 \end{vmatrix} \mathbf{e}_3$$

Since the entries of a determinant should be scalars, not vectors, the expression is merely a mnemonic to simplify the calculation of the vector product.

It is easily checked that

$$\mathbf{u} \times \mathbf{v} = -\mathbf{v} \times \mathbf{u}$$

Both **u** × **v** and **v** × **u** are orthogonal to **u** and **v**, but it can be checked that the direction of **u** × **v** is in accordance with the right hand rule, while that of **v** × **u** is the opposite direction.

We next prove that

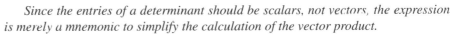
$$\|\mathbf{u} \times \mathbf{v}\| = \|\mathbf{u}\| \, \|\mathbf{v}\| \sin \theta \quad \text{where } \theta \text{ is the angle between } \mathbf{u} \text{ and } \mathbf{v}.$$

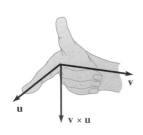

v × **u**

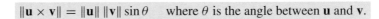
$$\|\mathbf{u}\| \, \|\mathbf{v}\| \sin \theta = \|\mathbf{u}\| \, \|\mathbf{v}\| \sqrt{1 - \cos^2 \theta}$$

$$= \|\mathbf{u}\| \, \|\mathbf{v}\| \sqrt{\left(1 - \frac{\langle \mathbf{u}, \mathbf{v} \rangle^2}{\|\mathbf{u}\|^2 \|\mathbf{v}\|^2} \right)}$$

$$= \sqrt{\left(\|\mathbf{u}\|^2 \|\mathbf{v}\|^2 - \langle \mathbf{u}, \mathbf{v} \rangle^2 \right)}$$

$$= \sqrt{((u_1^2 + u_2^2 + u_3^2)(v_1^2 + v_2^2 + v_3^2) - (u_1 v_1 + u_2 v_2 + u_3 v_3)^2)}$$

$$= \sqrt{((u_2 v_3 - u_3 v_2)^2 + (u_3 v_1 - u_1 v_3)^2 + (u_1 v_2 - u_2 v_1)^2)}$$

$$= \|\mathbf{u} \times \mathbf{v}\|$$

This establishes the result.

The figure on the right shows that the area of the parallelogram for which the vectors **u** and **v** are adjacent sides is $\|\mathbf{u}\|\,\|\mathbf{v}\|\sin\theta$ where θ is the angle between **u** and **v**.

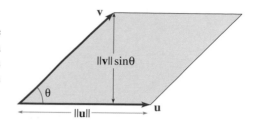

Hence

The area of the parallelogram whose adjacent sides are the vectors **u** and **v** is $\|\mathbf{u}\times\mathbf{v}\|$.

If we write $\mathbf{u}=(u_1,u_2,0)^{\mathrm{T}}$ and $\mathbf{v}=(v_1,v_2,0)^{\mathrm{T}}$ then

$$\mathbf{u}\times\mathbf{v} = \begin{vmatrix} \mathbf{e}_1 & \mathbf{e}_2 & \mathbf{e}_3 \\ u_1 & u_2 & 0 \\ v_1 & v_2 & 0 \end{vmatrix} = \begin{vmatrix} u_1 & u_2 \\ v_1 & v_2 \end{vmatrix}\mathbf{e}_3$$

$$= \begin{vmatrix} u_1 & v_1 \\ u_2 & v_2 \end{vmatrix}\mathbf{e}_3 = (u_1v_2 - u_2v_1)\mathbf{e}_3$$

Hence the length of $\mathbf{u}\times\mathbf{v}$ is the modulus of the determinant:

$$\|\mathbf{u}\times\mathbf{v}\| = \left|\det\begin{pmatrix} u_1 & v_1 \\ u_2 & v_2 \end{pmatrix}\right|$$

This gives us a useful geometric interpretation of a determinant:

The area of the parallelogram whose adjacent sides are the column vectors of a 2×2 matrix P is equal to $|\det P|$.

E.g. the area of the parallelogram on the right is

$$\left|\det\begin{pmatrix} 1 & 3 \\ 4 & 1 \end{pmatrix}\right| = |-11| = 11$$

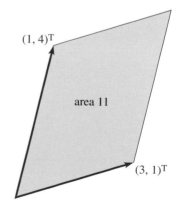

$(1,4)^{\mathrm{T}}$

area 11

$(3,1)^{\mathrm{T}}$

As a consequence of the above results, we can observe the relation between the areas A and B in the diagram below:

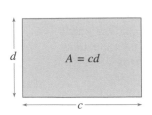 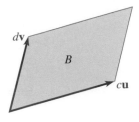

$A = cd$ and $B = \|c\mathbf{u} \times d\mathbf{v}\| = \|\mathbf{u} \times \mathbf{v}\|cd = |\det P|cd$. Hence

$$B = |\det P|A$$

The result also extends to the 3×3 case – the volume of a parallelepiped whose adjacent sides are the column vectors of a 3×3 matrix P is $|\det P|$.

Example 9

The vector product can be used to derive the cartesian or vector equations of a plane from the parametric form. In fact, when the equation of a plane is known in any form, the other forms of the equation can easily be derived from it, as illustrated by this example.

Suppose we wish to find a cartesian equation of the plane with parametric equation

$$\begin{pmatrix} x \\ y \\ z \end{pmatrix} = \begin{pmatrix} 1 \\ 1 \\ 0 \end{pmatrix} + s \begin{pmatrix} 2 \\ 1 \\ -2 \end{pmatrix} + t \begin{pmatrix} 1 \\ -2 \\ 2 \end{pmatrix} \qquad (s, t \in \mathbb{R})$$

The normal to the plane is $(2, 1, -2)^{\mathrm{T}} \times (1, -2, 2)^{\mathrm{T}}$, which can be calculated to be $(2, 6, 5)^{\mathrm{T}}$. As the point $(1, 1, 0)^{\mathrm{T}}$ lies on the plane, a vector equation for it is

$$\langle (2, 6, 5)^{\mathrm{T}}, \mathbf{x} - (1, 1, 0)^{\mathrm{T}} \rangle = 0$$

As $\langle (2, 6, 5)^{\mathrm{T}}, (1, 1, 0)^{\mathrm{T}} \rangle = 8$, the cartesian equation is

$$2x + 6y + 5z - 8 = 0$$

On the other hand, to find a parametric equation of the plane with cartesian equation

$$2x + 3y - z = 4$$

let $x = s$, $y = t$ so that $z = 2s + 3t - 4$. We can write the parametric form

$$\begin{pmatrix} x \\ y \\ z \end{pmatrix} = \begin{pmatrix} s \\ t \\ 2s + 3t - 4 \end{pmatrix} = \begin{pmatrix} 0 \\ 0 \\ -4 \end{pmatrix} + s \begin{pmatrix} 1 \\ 0 \\ 2 \end{pmatrix} + t \begin{pmatrix} 0 \\ 1 \\ 3 \end{pmatrix} \qquad (s, t \in \mathbb{R})$$

The vector $(2, 3, -1)^T$ is normal to the plane and $(0, 0, -4)^T$ lies on it and so a vector equation is

$$\langle (2, 3, -1)^T, \mathbf{x} - (0, 0, -4)^T \rangle = 0$$

1.8 Exercises

1.* Draw the line in $\mathbb{R}^2$

$$\mathbf{x} = (-3, -4)^T + t(2, 1)^T \qquad (t \in \mathbb{R})$$

and write down any two other vector equations of the line.

Note that there is an unlimited number of vector equations of the line.

2. Find a vector equation of the line through the points $(2, 2)^T$ and $(-1, 5)^T$. Prove that the point $(6, -2)^T$ lies on the line by finding a value of t that gives the point. Draw the line and use your drawing to verify that your value of t gives the point $(6, -2)^T$.

3.* Find the unit vector $\mathbf{u}$ in the direction of the vector $(-3, 4)^T$. Verify that the points $(4, -2)^T$ and $(-\frac{1}{5}, \frac{18}{5})^T$ lie on the line

$$\mathbf{x} = (1, 2)^T + t\mathbf{u}$$

For each of these points, find its distance from the point $(1, 2)^T$ and verify that this is equal to the modulus of the value of the parameter t which gives this point in the above equation.

4.* Find a vector in $\mathbb{R}^2$ which is orthogonal to the vector $(1, 2)^T$. Hence write down the equation of a general line in $\mathbb{R}^2$ which is orthogonal to the vector $(1, 2)^T$.

5. Find a vector in $\mathbb{R}^2$ in the direction of the line $2x + 7y + 3 = 0$. (*Consider its slope.*)

* Find two unit vectors parallel to the line and also, two unit vectors perpendicular to it.

6. (i) Find the equations of a line through the points $(-2, 4)^T$ and $(0, 2)^T$, and the line through $(0, 2)^T$ and the point $(5, -3)^T$. What can you conclude from these equations?

   (ii) Find the equation of the line through the point $(3, 2)^T$ in the direction of the vector $(2, -2)^T$. How is it related to the line through $(-2, 4)^T$ and $(0, 2)^T$?

   (iii) Find the equation of the line through $(5, -3)^T$ in the direction of the vector $(1, 1)^T$. How is it related to the line through $(-2, 4)^T$ and $(0, 2)^T$?

7. Which of the following vectors in $\mathbb{R}^3$ are unit vectors? Are any pair of these vectors orthogonal?

   (i) $(\frac{1}{6}, \frac{1}{3}, \frac{1}{2})^T$

(ii) $(\frac{1}{3}, \frac{1}{3}, -\frac{1}{3})^T$

(iii) $(\frac{1}{3}, \frac{2}{3}, -\frac{2}{3})^T$

8.* Using the components of the vectors $\mathbf{u} = (1, 0, -1)^T$ and $\mathbf{v} = (2, -1, 3)^T$ write the scalar product $\langle \mathbf{u}, \mathbf{v} \rangle$ as the matrix product $\mathbf{u}^T\mathbf{v}$. State the order of this matrix product and deduce the relationship between $\mathbf{u}^T\mathbf{v}$ and $\mathbf{v}^T\mathbf{u}$. Find the matrix product $\mathbf{u}\mathbf{v}^T$.

9. Find nonzero vectors $\mathbf{u}, \mathbf{v}, \mathbf{w}$ in $\mathbb{R}^3$ such that

$$\langle \mathbf{u}, \mathbf{v} \rangle = \langle \mathbf{u}, \mathbf{w} \rangle \qquad \text{but } \mathbf{v} \neq \mathbf{w}$$

10.* Find nonzero vectors $\mathbf{u}, \mathbf{v}, \mathbf{w}$ in $\mathbb{R}^3$ such that

$$\mathbf{u} \times \mathbf{v} = \mathbf{u} \times \mathbf{w} \qquad \text{but } \mathbf{v} \neq \mathbf{w}$$

11. Write down equations for the line through the points $(1, 2, 1)^T$ and $(2, 1, 2)^T$.

12. A line in three dimensional space is defined by the equations

$$\frac{x - 3}{1} = \frac{y - 1}{2} = \frac{z - 2}{1}.$$

Find two unit vectors parallel to this line.

13. Which of the following sets of points in $\mathbb{R}^3$ are collinear, that is, lie on a line?

(i)* $(2, 1, 4)^T, (4, 4, -1)^T, (6, 7, -6)^T$

(ii) $(1, 2, 3)^T, (-4, 2, 1)^T, (1, 1, 2)^T$

14.* Let l_1 be the line with equation $(x, y, z)^T = (1, -1, 2)^T + t(1, 2, 3)^T$, l_2 the line through $(5, 7, 4)^T$ and $(8, 13, 3)^T$ and l_3 the line through $(1, 17, 6)^T$ parallel to the vector $(7, -4, -3)^T$. Show that

(i) l_1 and l_3 are skew,

(ii) l_1 and l_2 intersect,

(iii) l_2 and l_3 intersect.

Find the points of intersection.

Determine whether each pair of intersecting lines is orthogonal. If not, find the angle between them.

15. For each of the following lines in $\mathbb{R}^3$ write down its direction and a point through which it passes:

(i) $\dfrac{x - 3}{2} = \dfrac{y + 1}{-5} = z$

(ii) $\dfrac{x - 1}{3} = \dfrac{z + 2}{2} \qquad y = 2$

16. Write down a vector which is normal to the plane $x + 2y + 3z = 6$. Write down any point which lies on this plane.

17.* Which of the planes $x + 4y - 3z = 5$ and $3x - y + 4z = 6$ lies further from the origin?

18. Find the equation of the plane which

(i) passes through $(3, 1, 2)^T$ and is parallel to the vectors $(1, 1, 1)^T$ and $(1, -1, 1)^T$,

(ii) passes through $(3, 1, 2)^T$ and $(1, 2, 3)^T$ and is parallel to the vector $(1, 1, 1)^T$.

19.* Find if the following planes are parallel, perpendicular or neither. If neither, find the angle between them.

(i) $4x + 3y - z = 0$ $2z = 8x + 6y + 1$

(ii) $x + y = 1$ $x + z = 1$

(iii) $5x - y + 7z = 0$ $-3x + 13y + 4z = 0$

20. Show that the line $\mathbf{x} = (0, 1, -1)^T + t(1, -1, 6)^T$ is

(i) parallel to the plane $-5x + 7y + 2z = 3$ and

(ii) perpendicular to the plane $3x + 18z = 3y - 2$.

Find its point of intersection with this second plane.

21. Find the point of intersection of the line

$$\begin{pmatrix} x \\ y \\ z \end{pmatrix} = \begin{pmatrix} 4 \\ -4 \\ 1 \end{pmatrix} + t \begin{pmatrix} 1 \\ 1 \\ 3 \end{pmatrix}$$

and the plane

$$\left\langle \begin{pmatrix} x \\ y \\ z \end{pmatrix}, \begin{pmatrix} 4 \\ 0 \\ -3 \end{pmatrix} \right\rangle = 7$$

22.* Find the cartesian equation of the plane through $(1, 0, 6)^T$ and $(-2, 3, -1)^T$ which is orthogonal to the plane

$$\left\langle \begin{pmatrix} x \\ y \\ z \end{pmatrix}, \begin{pmatrix} 5 \\ -1 \\ 1 \end{pmatrix} \right\rangle = 2$$

23. Write down the equation of a plane which passes through $(1, 2, 1)^T$ and which is normal to $(2, 1, 2)^T$. What is the distance of this plane from (i) the origin, and (ii) the point $(1, 2, 3)^T$?

24.* Write down the cartesian equations for the three coordinate planes: the xy plane, the yz plane and the zx plane.

Write down the general form for (i) a horizontal vector and (ii) a vertical vector in $\mathbb{R}^3$.

Write down the general form for the cartesian equation of (iii) a horizontal plane and (iv) a vertical plane.

25. Find vector and parametric equations for the plane with cartesian equation

$$2x - 5y + z = 3.$$

26. Find cartesian and parametric equations for the plane with vector equation

$$\langle \mathbf{x} - (1, 0, 1)^T, (2, -1, 1)^T \rangle = 0.$$

27. A plane passes through $(1, 1, 2)^T$, $(1, 2, 1)^T$, $(2, 1, 1)^T$. What is its distance from $(0, 0, 0)^T$?

28.* (i) The scalar product of vectors $\mathbf{u}$ and $\mathbf{v} \times \mathbf{w}$ is known as the **scalar triple product** of the vectors $\mathbf{u}$, $\mathbf{v}$ and $\mathbf{w}$ in $\mathbb{R}^3$. Show that it may be written in terms of their components as

$$\langle \mathbf{u}, \mathbf{v} \times \mathbf{w} \rangle = \begin{vmatrix} u_1 & u_2 & u_3 \\ v_1 & v_2 & v_3 \\ w_1 & w_2 & w_3 \end{vmatrix}$$

Hence show that the vectors $\mathbf{u}$, $\mathbf{v}$ and $\mathbf{w}$ are coplanar, that is, they lie in the same plane, if and only if this determinant is zero.

(ii) Find the constant if the vectors $(3, -1, 2)^T$, $(t, 5, 1)^T$ and $(-2, 3, 1)^T$ are coplanar.

29.* (i) Use the vector product to find the cartesian equation of the plane which contains the three points $(0, 4, -1)^T$, $(5, -1, 3)^T$ and $(2, -2, 1)^T$. *You will first need to find two vectors which lie along the plane.*

(ii) Alternatively, let the equation of the plane be

$$ax + by + cz + d = 0$$

Substitute the values of x, y and z for each of the three points lying on the plane. Now write the system of equations that you have obtained in the matrix form $A\mathbf{b} = \mathbf{0}$ where A is a 4×4 matrix and $\mathbf{b}$ and $\mathbf{0}$ are four dimensional vectors. Use Exercise 1.2.15 to show that the equation of the plane can be given in determinant form by

$$\begin{vmatrix} x & y & z & 1 \\ 0 & 4 & -1 & 1 \\ 5 & -1 & 3 & 1 \\ 2 & -2 & 1 & 1 \end{vmatrix} = 0$$

Do not evaluate the determinant.

1.9 Vectors in $\mathbb{R}^n$

It is not possible to visualise vectors of dimension n when $n > 3$, but most of the theory of vectors can be extended to n dimensions. An n dimensional vector $\mathbf{v}$ with components $v_1, v_2, \ldots, v_n \in \mathbb{R}^n$ is an $n \times 1$ column matrix

$$\mathbf{v} = \begin{pmatrix} v_1 \\ v_2 \\ \vdots \\ v_n \end{pmatrix} \quad \text{or} \quad (v_1, v_2, \ldots, v_n)^T$$

As usual, the definitions of matrix addition and scalar multiplication given in §1.1 apply. Thus

$$\mathbf{v} + \mathbf{w} = \begin{pmatrix} v_1 + w_1 \\ v_2 + w_2 \\ \vdots \\ v_n + w_n \end{pmatrix} \qquad \alpha\mathbf{x} = \begin{pmatrix} \alpha v_1 \\ \alpha v_2 \\ \vdots \\ \alpha v_n \end{pmatrix}$$

The **length** (or **norm**) of an n dimensional vector $\mathbf{v}$ is defined by

$$\|\mathbf{v}\| = \{v_1^2 + v_2^2 + \cdots + v_n^2\}^{1/2}$$

and the **distance** between two points $\mathbf{v}$ and $\mathbf{w}$ is defined by

$$\|\mathbf{w} - \mathbf{v}\| = \{(w_1 - v_1)^2 + (w_2 - v_2)^2 + \cdots + (w_n - v_n)^2\}^{1/2}$$

As before, we define the scalar $\mathbf{v}^T\mathbf{w}$ to be the **scalar product** of $\mathbf{v}$ and $\mathbf{w}$, written

$$\langle \mathbf{v}, \mathbf{w} \rangle = v_1 w_1 + v_2 w_2 + \cdots + v_n w_n$$

The usual properties of the scalar product hold.

(i) $\langle \mathbf{v}, \mathbf{v} \rangle = \|\mathbf{v}\|^2$

(ii) $\langle \mathbf{v}, \mathbf{w} \rangle = \langle \mathbf{w}, \mathbf{v} \rangle$

(iii) $\langle \alpha\mathbf{u} + \beta\mathbf{v}, \mathbf{w} \rangle = \alpha\langle \mathbf{u}, \mathbf{w} \rangle + \beta\langle \mathbf{v}, \mathbf{w} \rangle$.

Two vectors $\mathbf{v}$ and $\mathbf{w}$ are **orthogonal** (or **normal**) if and only if

$$\langle \mathbf{v}, \mathbf{w} \rangle = 0$$

Lines

Similarly, the equation

$$\mathbf{x} = \boldsymbol{\xi} + t\mathbf{v}$$

in which $\boldsymbol{\xi}$ and $\mathbf{v}$ are n dimensional vectors with $\mathbf{v} \neq \mathbf{0}$ and t is a scalar represents a line in $\mathbb{R}^n$, with $\mathbf{x}$ and $\boldsymbol{\xi}$ determining points and $\mathbf{v}$ determining a direction.

Writing the equation in full

$$\begin{pmatrix} x_1 \\ x_2 \\ \vdots \\ x_n \end{pmatrix} = \begin{pmatrix} \xi_1 \\ \xi_2 \\ \vdots \\ \xi_n \end{pmatrix} + t \begin{pmatrix} v_1 \\ v_2 \\ \vdots \\ v_n \end{pmatrix}$$

or

$$\frac{x_1 - \xi_1}{v_1} = \frac{x_2 - \xi_2}{v_2} = \cdots = \frac{x_n - \xi_n}{v_n}$$

which is an alternative form for the equations of a line through $(\xi_1, \xi_2, \ldots, \xi_n)^T$ in the direction $(v_1, v_2, \ldots, v_n)^T$ when none of the components of $\mathbf{v}$ are zero.

A line passing through two distinct points $\boldsymbol{\xi}$ and $\boldsymbol{\eta}$ has the parametric form

$$\mathbf{x} = \boldsymbol{\xi} + t(\boldsymbol{\eta} - \boldsymbol{\xi}) \qquad (t \in \mathbb{R})$$

E.g. the line passing through points $(7, 0, -2, 1)^T$ and $(3, 1, 0, -1)^T \in \mathbb{R}^4$ has the form

$$\begin{pmatrix} x_1 \\ x_2 \\ x_3 \\ x_4 \end{pmatrix} = \begin{pmatrix} 7 \\ 0 \\ -2 \\ 1 \end{pmatrix} + t \begin{pmatrix} 3-7 \\ 1-0 \\ 0-(-2) \\ -1-1 \end{pmatrix}$$

or

$$\frac{x_1 - 7}{-4} = \frac{x_2}{1} = \frac{x_3 + 2}{2} = \frac{x_4 - 1}{-2}$$

Hyperplanes

If $\boldsymbol{\xi}$ and $\mathbf{v}$ are n dimensional vectors and $\mathbf{v} \neq \mathbf{0}$, the equation

$$\langle \mathbf{v}, \mathbf{x} - \boldsymbol{\xi} \rangle = 0$$

is the equation of a **hyperplane** through the point $\boldsymbol{\xi}$ with **normal** $\mathbf{v}$. If we write $c = \langle \mathbf{v}, \boldsymbol{\xi} \rangle$, the equation takes the form

$$\langle \mathbf{v}, \mathbf{x} \rangle = c$$

Writing the equation in terms of components, it becomes

$$v_1 x_1 + v_2 x_2 + v_3 x_3 + \cdots + v_n x_n = c$$

Thus a hyperplane is defined by one 'linear' equation, in which $\mathbf{v} = (v_1, v_2, \ldots, v_n)^T$ is the normal to the hyperplane and, if $\mathbf{v}$ is a unit vector, $|c|$ is the distance from $\mathbf{0}$ to the hyperplane.

1.10   Flats

Let A denote an $m \times n$ matrix and let $\mathbf{c}$ be an m dimensional vector. Then the set of n dimensional vectors $\mathbf{x}$ which satisfy the equation

$$A\mathbf{x} = \mathbf{c}$$

is called a **flat**.

If the equation $A\mathbf{x} = \mathbf{c}$ is written out in full, it becomes

$$\begin{pmatrix} a_{11} & a_{12} & \cdots & a_{1n} \\ a_{21} & a_{22} & \cdots & a_{2n} \\ \vdots & \vdots & & \vdots \\ a_{m1} & a_{m2} & \cdots & a_{mn} \end{pmatrix} \begin{pmatrix} x_1 \\ x_2 \\ \vdots \\ x_n \end{pmatrix} = \begin{pmatrix} c_1 \\ c_2 \\ \vdots \\ c_m \end{pmatrix}$$

which is the same as the system of equations

$$\left. \begin{aligned} a_{11}x_1 + a_{12}x_2 + \cdots + a_{1n}x_n &= c_1 \\ a_{21}x_1 + a_{22}x_2 + \cdots + a_{2n}x_n &= c_2 \\ \cdots \\ a_{m1}x_1 + a_{m2}x_2 + \cdots + a_{mn}x_n &= c_m \end{aligned} \right\}$$

A flat is therefore the set of points satisfying m 'linear' equations.

In vector notation, if $\mathbf{a}_k = (a_{k1}, a_{k2}, \ldots, a_{kn})^T \neq \mathbf{0}$, the kth linear equation above is then the equation of the hyperplane $\langle \mathbf{a}_k, \mathbf{x} \rangle = c_k$. Our system of linear equations becomes

$$\left.\begin{aligned} \langle \mathbf{a}_1, \mathbf{x} \rangle &= c_1 \\ \langle \mathbf{a}_2, \mathbf{x} \rangle &= c_2 \\ &\cdots \\ \langle \mathbf{a}_m, \mathbf{x} \rangle &= c_m \end{aligned}\right\}$$

This shows that a flat consists of the points $\mathbf{x}$ which are common to m hyperplanes.

We illustrate some possibilities below:

(i) $m = 1, n = 3$. In this case A is a 1×3 matrix and the flat reduces to the single equation

$$a_1 x_1 + a_2 x_2 + a_3 x_3 = c_1 \quad \text{or} \quad \langle \mathbf{a}, \mathbf{x} \rangle = c_1$$

The flat is therefore, in general, a *plane*.

(ii) $m = 2, n = 3$. In this case A is a 2×3 matrix and the flat reduces to the pair of equations

$$\left.\begin{aligned} a_{11} x_1 + a_{12} x_2 + a_{13} x_3 &= c_1 \\ a_{21} x_1 + a_{22} x_2 + a_{23} x_3 &= c_2 \end{aligned}\right\} \quad \text{or} \quad \left.\begin{aligned} \langle \mathbf{a}_1, \mathbf{x} \rangle &= c_1 \\ \langle \mathbf{a}_2, \mathbf{x} \rangle &= c_2 \end{aligned}\right\}$$

The flat is therefore, in general, the intersection of two planes and hence is a *line*.

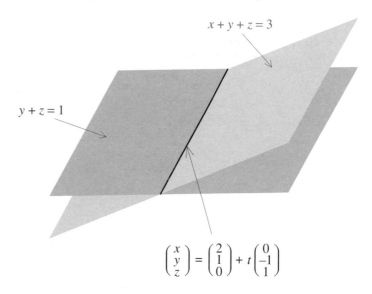

E.g. the flat

$$\left.\begin{aligned} x + y + z &= 3 \\ y + z &= 1 \end{aligned}\right\} \quad \text{or} \quad \left.\begin{aligned} \langle (1, 1, 1)^T, \mathbf{x} \rangle &= 3 \\ \langle (0, 1, 1)^T, \mathbf{x} \rangle &= 1 \end{aligned}\right\}$$

which is pictured above, is a line. To find the equation of the line, let $z = t$. Solve the equations

$$x + y = 3 - t$$
$$y = 1 - t$$

and obtain $y = 1 - t, x = 2$. Hence the equation of the line may be written as

$$\begin{pmatrix} x \\ y \\ z \end{pmatrix} = \begin{pmatrix} 2 \\ 1 \\ 0 \end{pmatrix} + t \begin{pmatrix} 0 \\ -1 \\ 1 \end{pmatrix} \qquad (t \in \mathbb{R})$$

showing that it passes through the point $(2, 1, 0)^{\mathrm{T}}$ in the direction $(0, -1, 1)^{\mathrm{T}}$.

(iii) $m = 3, n = 3$. In this case, A is a 3×3 matrix and the flat $Ax = c$ reduces to the system of equations

$$\left. \begin{array}{l} a_{11}x_1 + a_{12}x_2 + a_{13}x_3 = c_1 \\ a_{21}x_1 + a_{22}x_2 + a_{23}x_3 = c_2 \\ a_{31}x_1 + a_{32}x_2 + a_{33}x_3 = c_3 \end{array} \right\} \quad \text{or} \quad \left. \begin{array}{l} \langle \mathbf{a}_1, \mathbf{x} \rangle = c_1 \\ \langle \mathbf{a}_2, \mathbf{x} \rangle = c_2 \\ \langle \mathbf{a}_3, \mathbf{x} \rangle = c_3 \end{array} \right\}$$

The flat is therefore, in general, the intersection of three planes and hence is a *point*.

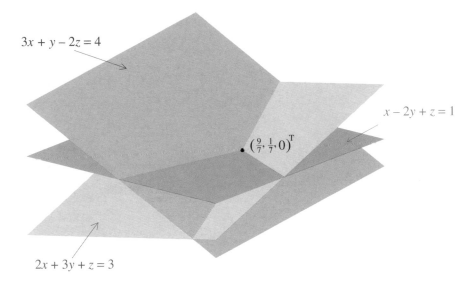

$3x + y - 2z = 4$

$x - 2y + z = 1$

$\left(\tfrac{9}{7}, \tfrac{1}{7}, 0\right)^{\mathrm{T}}$

$2x + 3y + z = 3$

E.g. the flat

$$\left. \begin{array}{r} 3x + y - 2z = 4 \\ x - 2y + z = 1 \\ 2x + 3y + z = 3 \end{array} \right\} \quad \text{or} \quad \left. \begin{array}{r} \langle (3, 1, -2)^{\mathrm{T}}, \mathbf{x} \rangle = 4 \\ \langle (1, -2, 1)^{\mathrm{T}}, \mathbf{x} \rangle = 1 \\ \langle (2, 3, 1)^{\mathrm{T}}, \mathbf{x} \rangle = 3 \end{array} \right\}$$

which is pictured above can be written in the matrix form

$$\begin{pmatrix} 3 & 1 & -2 \\ 1 & -2 & 1 \\ 2 & 3 & 1 \end{pmatrix} \begin{pmatrix} x \\ y \\ z \end{pmatrix} = \begin{pmatrix} 4 \\ 1 \\ 3 \end{pmatrix}$$

As the matrix is invertible the system has the unique solution

$$\begin{pmatrix} x \\ y \\ z \end{pmatrix} = \begin{pmatrix} 3 & 1 & -2 \\ 1 & -2 & 1 \\ 2 & 3 & 1 \end{pmatrix}^{-1} \begin{pmatrix} 4 \\ 1 \\ 3 \end{pmatrix} = \begin{pmatrix} \tfrac{9}{7} \\ \tfrac{1}{7} \\ 0 \end{pmatrix}$$

That is, the intersection of the three planes is the point $\left(\dfrac{9}{7}, \dfrac{1}{7}, 0\right)^{\mathrm{T}}$.

Observe that we have been careful to insert the words 'in general' in (i), (ii) and (iii) in order to signal that the result may fail in degenerate cases. In (ii), for example, things degenerate if the two planes are parallel and hence the flat will contain no points at all unless the two planes happen to be identical, in which case the flat will be this plane.

It is worth noting that when $m = n$, the matrix A is invertible if and only if the flat $A\mathbf{x} = \mathbf{c}$ consists of the single point $A^{-1}\mathbf{c}$.

1.11 Exercises

1. Prove that the point $(-1, -1, 5, 0)^{\mathrm{T}}$ in $\mathbb{R}^4$ lies on the line

$$\mathbf{x} = (-3, 0, 5, 2)^{\mathrm{T}} + t\mathbf{u}$$

where $\mathbf{u}$ is a unit vector in the direction of the vector $(2, -1, 0, -2)^{\mathrm{T}}$. Find the distance of this point from the point $(-3, 0, 5, 2)^{\mathrm{T}}$ and check that it is equal to the modulus of the parameter t which gives the point in the equation of the line.

2. Find the equation of the line in $\mathbb{R}^5$ through the points $(3, 0, -1, 2, 1)^{\mathrm{T}}$ and $(-1, 0, 0, 2, 0)^{\mathrm{T}}$.

3.* In $\mathbb{R}^4$ let l_1 be the line through the points $(0, 1, 1, 1)^{\mathrm{T}}$ and $(2, 0, -1, 0)^{\mathrm{T}}$, l_2 the line through the point $(3, 1, 0, 1)^{\mathrm{T}}$ in the direction of the vector $(-4, -1, 0, -1)^{\mathrm{T}}$ and l_3 the line through the origin parallel to the vector $(0, 1, 0, -1)^{\mathrm{T}}$. For each pair of lines, find if the following are true:

The lines (i) intersect, (ii) are parallel and (iii) have orthogonal directions.

4. Show that the line in $\mathbb{R}^4$ with equation

$$\mathbf{x} = (0, 1, 0, 3)^{\mathrm{T}} + t(1, 2, -1, 2)^{\mathrm{T}}$$

(i) lies in the hyperplane with equation

$$x_1 + x_2 + x_3 - x_4 + 2 = 0$$

(ii) is orthogonal to the hyperplane with equation

$$x_1 + 2x_2 - x_3 + 2x_4 = 5$$

Find its point of intersection with the second hyperplane.

5. Which of the hyperplanes in $\mathbb{R}^4$, $2x_1 - x_2 - 2x_4 = 3$ and $3x_3 = 5$, is further from the origin?

6. Find vectors $\mathbf{u}, \mathbf{v}, \mathbf{w}$ in $\mathbb{R}^4$ for which

$$\langle \mathbf{u}, \mathbf{v} \rangle = \langle \mathbf{u}, \mathbf{w} \rangle$$

but $\mathbf{v} \neq \mathbf{w}$.

7. Write the scalar product

$$\langle \mathbf{u}, \mathbf{v} \rangle = \langle (u_1, u_2, \ldots, u_n)^{\mathrm{T}}, (v_1, v_2, \ldots, v_n)^{\mathrm{T}} \rangle$$

in $\mathbb{R}^n$ as the matrix product $\mathbf{u}^{\mathrm{T}}\mathbf{v}$. How are $\mathbf{u}^{\mathrm{T}}\mathbf{v}$ and $\mathbf{v}^{\mathrm{T}}\mathbf{u}$ related? Find the matrix product $\mathbf{u}\mathbf{v}^{\mathrm{T}}$.

8. The flat defined by

$$\begin{pmatrix} 1 & 2 & 3 \\ 2 & 1 & 0 \\ 1 & -1 & -3 \end{pmatrix} \begin{pmatrix} x \\ y \\ z \end{pmatrix} = \begin{pmatrix} 1 \\ 1 \\ 0 \end{pmatrix}$$

is a line. Find the parametric equation of this line.

9. Show that the flat defined by

$$\begin{pmatrix} 2 & 4 & -1 \\ 1 & 5 & 0 \\ 0 & -2 & 0 \end{pmatrix} \begin{pmatrix} x \\ y \\ z \end{pmatrix} = \begin{pmatrix} 1 \\ 2 \\ 3 \end{pmatrix}$$

is a point. Find the components of this point.

10.* Show that the flats

(i) $$\begin{pmatrix} 1 & 1 & 1 \\ -1 & -1 & -1 \end{pmatrix} \begin{pmatrix} x \\ y \\ z \end{pmatrix} = \begin{pmatrix} 2 \\ -2 \end{pmatrix}$$ and

(ii) $$\begin{pmatrix} 1 & 1 & 1 \\ -1 & -1 & -1 \end{pmatrix} \begin{pmatrix} x \\ y \\ z \end{pmatrix} = \begin{pmatrix} 2 \\ 2 \end{pmatrix}$$

are, respectively, a plane and the empty set.

11. A flat is defined by

$$\begin{pmatrix} 1 & 2 & 3 \\ 2 & -1 & 1 \\ 4 & -3 & 1 \end{pmatrix} \begin{pmatrix} x \\ y \\ z \end{pmatrix} = \begin{pmatrix} 0 \\ 1 \\ -2 \end{pmatrix}$$

Prove that the matrix is not invertible and show that the flat is a line. Find the equation of this line.

12. Write the following systems of equations in the matrix form $A\mathbf{x} = \mathbf{c}$, where A is a 3×3 matrix and $\mathbf{x}$ and $\mathbf{c}$ are 3×1 column vectors.

Find if A is invertible, in which case find the unique point which is the flat. If A is not invertible, solve the equations by eliminating one of the variables. If a consistent solution can be found, give a formula for the flat stating its form. If the equations are inconsistent, the flat is the empty set.

$$(i)^* \begin{cases} x & -4y & -z & = 3 \\ 2x & +y & & = 2 \\ -3x & +2y & -z & = 5 \end{cases} \qquad (ii) \begin{cases} x & -4y & -z & = 3 \\ 2x & +y & & = 2 \\ 3x & -3y & -z & = 5 \end{cases}$$

$$(iii) \begin{cases} x & -4y & -z & = 3 \\ 2x & +y & & = 2 \\ -3x & +2y & -z & = 0 \end{cases} \qquad (iv) \begin{cases} x & -4y & -z & = 0 \\ 2x & +y & & = 0 \\ -3x & +2y & -z & = 0 \end{cases}$$

$$(v)^* \begin{cases} x & -4y & -z & = 3 \\ 2x & +y & & = 2 \\ 3x & -3y & -z & = 1 \end{cases} \qquad (vi)^* \begin{cases} x & -4y & -z & = 0 \\ 2x & +y & & = 0 \\ 3x & -3y & -z & = 0 \end{cases}$$

13. Find if each of the following flats consists of a single point, in which case, find the point.

$$(i)^* \begin{cases} x_1 & +2x_2 & & +x_4 & = 0 \\ 2x_1 & +x_2 & & & = 2 \\ x_1 & & & +x_4 & = 0 \\ -x_1 & & +x_3 & +x_4 & = 1 \end{cases}$$

$$(ii) \begin{cases} x_1 & & & +x_4 & = -1 \\ & 2x_2 & +x_3 & -x_4 & = 0 \\ x_1 & +2x_2 & +x_3 & & = 1 \\ & & -x_3 & +x_4 & = 1 \end{cases}$$

$$(iii)^* \begin{cases} & x_2 & & -x_4 & & = 0 \\ x_1 & & -x_3 & +3x_4 & & = 0 \\ x_1 & +x_2 & & & & = 1 \\ x_1 & & +2x_3 & & -x_5 & = 0 \\ & & -x_3 & & +x_5 & = 3 \end{cases}$$

$$(iv) \begin{cases} x_1 & & +2x_3 & & -x_5 & = 0 \\ x_1 & & & +x_4 & & = -1 \\ & x_2 & +x_3 & & & = 1 \\ x_1 & & +3x_3 & & & = 3 \\ & & x_3 & & +x_5 & = 3 \end{cases}$$

1.12 Applications (optional)

1.12.1 Commodity bundles

An important use for vectors is to represent *commodity bundles*. The owner of a small bakery who visits a supermarket and purchases the items on the shopping list

$$\left. \begin{array}{ll} \text{flour} & \text{40 kilograms} \\ \text{sugar} & \text{10 kilogram} \\ \text{milk} & \text{30 litres} \\ \text{yeast} & \text{1 kilogram} \end{array} \right\}$$

can be described by saying that he has acquired the commodity bundle

$$\mathbf{y} = \begin{pmatrix} y_1 \\ y_2 \\ y_3 \\ y_4 \end{pmatrix} = \begin{pmatrix} 40 \\ 10 \\ 30 \\ 1 \end{pmatrix}$$

Of course, one needs to know in considering $\mathbf{y}$ what kind of good each component represents and in what units this good is measured.

1.12.2 Linear production models

Suppose that $\mathbf{x}$ is an $n \times 1$ column vector, $\mathbf{y}$ is an $m \times 1$ column vector and A is an $m \times n$ matrix. We can then use the equation

$$\mathbf{y} = A\mathbf{x}$$

as a model for a simple production process. In this process, $\mathbf{y}$ represents the commodity bundle of raw materials (the **input**) required to produce the commodity bundle $\mathbf{x}$ of finished goods (the **output**). Such a production process is said to be **linear**.

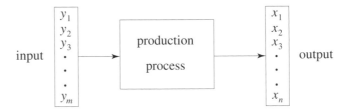

For example, suppose that the baker of §1.12.1 is interested in producing x_1 kg of bread and x_2 kg of cake. His output vector is therefore to be $\mathbf{x} = (x_1, x_2)^T$. In deciding how much of each type of ingredient will be required, he will need to consider his production process. If this is linear, his input vector $\mathbf{y} = (y_1, y_2, y_3, y_4)^T$ of required ingredients might, for example, be given by

$$\begin{pmatrix} y_1 \\ y_2 \\ y_3 \\ y_4 \end{pmatrix} = \begin{pmatrix} 10 & 10 \\ 0 & 2 \\ 0 & 3 \\ 1 & 0 \end{pmatrix} \begin{pmatrix} x_1 \\ x_2 \end{pmatrix}$$

1.12.3 Price vectors

The components of a *price vector*

$$\mathbf{p} = (p_1 \ p_2 \ \cdots \ p_n)^T$$

list the prices at which the corresponding commodities can be bought or sold. Given the price vector p, the *value* of a commodity bundle $\mathbf{x}$ is

$$\langle \mathbf{p}, \mathbf{x} \rangle = p_1 x_1 + p_2 x_2 + \cdots + p_n x_n$$

This is the amount for which the commodity bundle may be bought or sold.

If $\mathbf{p}$ is a fixed price vector and c is a given quantity of money, the commodity vectors $\mathbf{x}$ which lie on the hyperplane

$$\langle \mathbf{p}, \mathbf{x} \rangle = c$$

are those whose purchase costs precisely c.

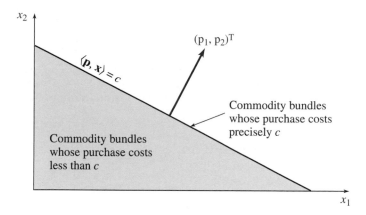

Alternatively, it may be that $\mathbf{x}$ is a fixed commodity bundle and that c is a given amount of money. In this case, the points $\mathbf{p}$ which lie on the hyperplane

$$\langle \mathbf{p}, \mathbf{x} \rangle = c$$

are the price vectors which ensure that the sale of $\mathbf{x}$ will realise an amount c.

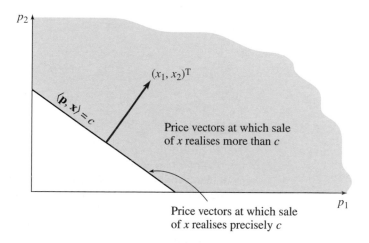

These two ways of looking at things are said to be **dual**. A buyer will be more interested in the former and a seller in the latter.

1.12.4 Linear programming

Suppose that, having acquired the commodity bundle $\mathbf{b}$ of ingredients, the baker of §1.12.1 decides to sell the result $\mathbf{x}$ of his baking. If the relevant price vector is $\mathbf{p}$, then the revenue he will acquire from the sale of $\mathbf{x}$ is

$$\langle \mathbf{p}, \mathbf{x} \rangle = \mathbf{p}^{\mathrm{T}}\mathbf{x}$$

His problem is to choose $\mathbf{x}$ so as to *maximise* this revenue. But he cannot choose $\mathbf{x}$ freely. In particular, he cannot bake more than his stock $\mathbf{b}$ of ingredients allows. To bake $\mathbf{x}$, he requires $\mathbf{y} = A\mathbf{x}$. His choice of $\mathbf{x}$ is therefore restricted by the *constraint*

$$A\mathbf{x} \leq \mathbf{b}$$

Note that $\mathbf{c} \leq \mathbf{d}$ means that $c_1 \leq d_1, c_2 \leq d_2, c_3 \leq d_3$, etc. Also, he cannot choose to bake negative quantities of bread or cakes. We therefore have the additional constraint

$$\mathbf{x} \geq \mathbf{0}$$

The baker's problem is therefore to find

$$\max\{\mathbf{p}^T\mathbf{x}\}$$

where $\mathbf{x}$ is subject to the *constraints*

$$A\mathbf{x} \leq \mathbf{b}$$
$$\mathbf{x} \geq \mathbf{0}.$$

Such a problem is called a **linear programming** problem. Observe how the use of matrix notation allows us to state the problem neatly and concisely. Written out in full, the problem takes the form: find

$$\max\{p_1x_1 + p_2x_2 + \cdots + p_nx_n\}$$

where $x_1, x_2, x_3, \ldots, x_n$ are subject to the constraints

$$a_{11}x_1 + a_{12}x_2 + \cdots + a_{1n}x_n \leq b_1$$
$$a_{21}x_1 + a_{22}x_2 + \cdots + a_{2n}x_n \leq b_2$$
$$\vdots$$
$$a_{m1}x_1 + a_{m2}x_2 + \cdots + a_{mn}x_n \leq b_n$$

and $x_1 \geq 0, x_2 \geq 0, \ldots, x_n \geq 0$.

1.12.5 Dual problem

A factory producing baked goods cannot compete with small bakers in respect of the quality of its products. Instead, it proposes to buy up our baker's stock of ingredients. What price vector $\mathbf{n}$ should it offer him?

If the baker bakes and sells $\mathbf{x}$, he will receive $\mathbf{p}^T\mathbf{x}$. If instead he sells the ingredients $\mathbf{y} = A\mathbf{x}$ at $\mathbf{q}$ he will receive $\mathbf{q}^T\mathbf{y} = \mathbf{q}^T A\mathbf{x} = (A^T\mathbf{q})^T\mathbf{x}$ (see Exercise 1.2.7). Selling the ingredients at $\mathbf{q}$ is therefore the same as selling the results of his baking at prices $A^T\mathbf{q}$. For these prices to be more attractive than the market prices, we require that

$$A^T\mathbf{q} \geq \mathbf{p}$$

We also, of course, require that

$$\mathbf{q} \geq \mathbf{0}$$

Assuming that the factory wishes to acquire the baker's stock $\mathbf{b}$ at minimum cost, it therefore has to find

$$\min\{\mathbf{q}^T\mathbf{b}\}$$

where $\mathbf{q}$ is subject to the *constraints*

$$A^T\mathbf{q} \geq \mathbf{p}$$
$$\mathbf{q} \geq \mathbf{0}$$

This linear programming problem is said to be **dual** to that of §1.12.4. It is more or less obvious that the minimum cost at which the factory can buy up the baker's ingredients is equal to the maximum revenue he can acquire by baking the ingredients and selling the results. This fact is the important **duality theorem** of linear programming. The prices **q** which solve the dual problem are called the **shadow prices** of the ingredients. These are the prices at which it is sensible to value the stock **b** given that an amount $\mathbf{x} = A\mathbf{y}$ of finished goods can be obtained costlessly from an amount **y** of stock and sold at prices **p**.

1.12.6 Game theory

In modern times, the economic theory of imperfect competition has become a branch of game theory. The competing firms are modelled as players who choose strategies in a game. The simplest kind of game is a two person, zero sum game. It is called zero sum because whatever one player wins, the other loses. Although zero sum games are seldom realistic models of real economic situations, it is instructive to look at how von Neumann proposed that they should be solved.

An $m \times n$ matrix A can be used as the **payoff matrix** in a **zero sum game**. We interpret the m rows of A as possible strategies for the first player and the n columns of A as possible strategies for the second player. The choice of a row and column by the two players determines an entry of the matrix. The first player then *wins* a payoff equal to this entry and the second player *loses* an equal amount.

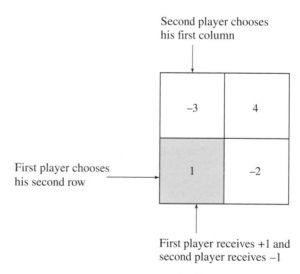

Second player chooses his first column

First player chooses his second row

First player receives +1 and second player receives −1

The game described above is called zero sum because the sum of the payoffs to the two players is always zero. For such games we are interested in what happens when the players simultaneously act in such a way that each player's action is a 'best reply' to the action of the other. Neither player will then have cause to regret his choice of action.

It turns out that the players should consider using 'mixed strategies' – i.e. instead of simply choosing a strategy, they should assign probabilities to each of their strategies and leave chance to make the final decision. This has the advantage that the opponent cannot then possibly predict what the final choice will be.

We use an $m \times 1$ vector $\mathbf{p} = (p_1, p_2, \ldots, p_m)^\mathrm{T}$ to represent a mixed strategy for the first player. The kth component p_k represents the probability with which the kth row is to be chosen. Similarly, an $n \times 1$ vector $\mathbf{q} = (q_1, q_2, \ldots, q_n)^\mathrm{T}$ represents a mixed strategy for the second player. If the two players independently choose the mixed strategies $\mathbf{p}$ and $\mathbf{q}$, then the expected gain to the first player (and hence the expected loss to the second player) is

$$\mathbf{p}^\mathrm{T} A \mathbf{q}$$

We illustrate this with the 2×2 payoff matrix considered above.

The expected gain to the first player is calculated by multiplying each payoff by the probability with which it occurs and then adding the results. The expected gain to the first player is therefore

$$(-3)p_1q_1 + 4p_1q_2 + 1p_2q_1 + (-2)p_2q_2$$

$$= (p_1, p_2) \begin{pmatrix} -3 & 4 \\ 1 & -2 \end{pmatrix} \begin{pmatrix} q_1 \\ q_2 \end{pmatrix} = \mathbf{p}^\mathrm{T} A \mathbf{q}$$

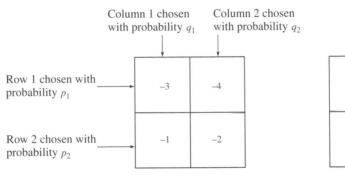

Possible outcomes   Corresponding probabilities

We assume that the first player chooses $\mathbf{p}$ in an attempt to maximise this quantity and the second player chooses $\mathbf{q}$ in an attempt to minimise it.

If each of the mixed strategies $\tilde{\mathbf{p}}$ and $\tilde{\mathbf{q}}$ is a best reply to the other, then

$$\tilde{\mathbf{p}}^\mathrm{T} A \tilde{\mathbf{q}} \geq \mathbf{p}^\mathrm{T} A \tilde{\mathbf{q}} \quad \text{(for all } \mathbf{p})$$

and

$$\tilde{\mathbf{p}}^\mathrm{T} A \tilde{\mathbf{q}} \leq \tilde{\mathbf{p}}^\mathrm{T} A \mathbf{q} \quad \text{(for all } \mathbf{q})$$

A pair of mixed strategies $\tilde{\mathbf{p}}$ and $\tilde{\mathbf{q}}$ with these properties is called a **Nash equilibrium** of the game. Von Neumann proved that every matrix A has such a Nash equilibrium and that it is sensible to regard this as the *solution* of the game.

In the case of the 2×2 matrix given above, the first player should choose the mixed strategy $(0.3, 0.7)^\mathrm{T}$ and the second player should choose the mixed strategy $(0.6, 0.4)^\mathrm{T}$. It is instructive to check that each of these is a best reply to the other. The resulting expected gain to the first player is -0.2 – i.e. this is a good game for the second player!

Two

Functions of one variable

This chapter introduces the notion of a function and the techniques of differentiation required for the analysis of functions of one variable. Standard functions are built up from elementary functions and their properties explored. This material is vital for nearly everything which follows in this book. Although some of it is revision, there is much that may be new, like the account of inverse functions, the use of Taylor's theorem and conic sections.

Throughout the chapter, foundations are laid in the one variable case for the investigation of functions of two or more variables that follows in subsequent chapters.

2.1 Intervals$^\diamond$

The set of real numbers is denoted by $\mathbb{R}$. A subset $I \subseteq \mathbb{R}$ is called an **interval** if, whenever it contains two real numbers, it also contains all the real numbers between them. An interval can be represented geometrically by a line segment. The interval consisting of all nonnegative real numbers is denoted by $\mathbb{R}_+$.

The geometrical representation and the notation used to describe typical intervals, are illustrated below:

$$[a, b] \qquad \{x \mid a \leq x \leq b\}$$

$$(a, b) \qquad \{x \mid a < x < b\}$$

$$(a, \infty) \qquad \{x \mid x > a\}$$

$$(-\infty, b] \qquad \{x \mid x \leq b\}$$

In these examples, a and b are end points, whether or not they belong to the interval they help to determine. Points of an interval that are not end points are called **interior points**. Remember that ∞ and $-\infty$ can be neither end points nor interior points, because they are not real numbers. They are simply a notational convenience.

2.2 Real valued functions of one real variable$^\diamond$

A **function** $f: X \to Y$ is a rule which assigns to *each* x in the set X a *unique* y in the set Y. We denote the element in Y assigned to the element x in X by

$$y = f(x)$$

For obvious reasons x is known as the **independent variable** and y as the **dependent variable**. The set X is called the **domain** of the function f. The set Y is called its **codomain**. When the domain and codomain of a function are both sets of real numbers, the function is said to be a **real valued** (or **scalar valued**) **function of one** (real) **variable**.

It is vital to appreciate that the independent variable x in the formula $f(x)$ is a 'dummy variable' in the sense that any variable symbol can be used in place of it.

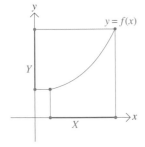

$$E.g. \text{ if } f(x) = x^2 \text{ then } f(p) = p^2.$$

A function can be illustrated by a **graph** in $\mathbb{R}^2$ as shown in the margin. In the figures below, (a) shows the case when $X = Y = \mathbb{R}$. The graph in (b) shows a function $f: (-\infty, b] \to \mathbb{R}$, whose domain is the set of all real numbers x satisfying $x \le b$. Its domain is not $\mathbb{R}$ because $f(x)$ is undefined when $x > b$. The graph in (c) does not represent a function since it does not give a unique y for each x.

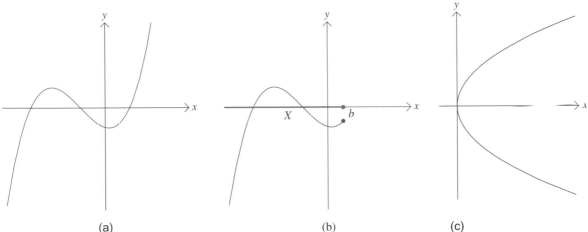

(a)   (b)   (c)

2.3 Some elementary functions$^\diamond$

We start by examining three elementary types of function which, between them, give rise to the functions commonly studied in calculus.

2.3.1 Power functions$^\diamond$

Let $\mathbb{N}$ be the set of positive integers, namely, $1, 2, 3, 4, \ldots$. The function $f: \mathbb{R} \to \mathbb{R}$ defined by

$$y = f(x) = x^n \qquad \text{where} \quad n \in \mathbb{N}$$

has a continuous graph that looks like the following figures for n odd and even:

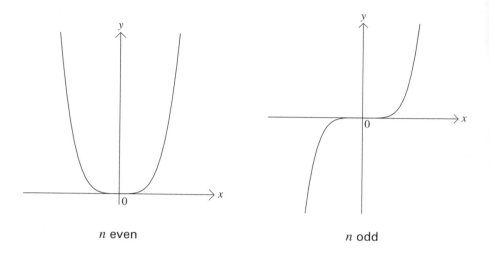

n even   *n* odd

2.3.2 Exponential functions[◇]

Let $\mathbb{Q}$ be the set of rational numbers or fractions, e.g. $1/2$, 3, $-7/5$. Given a real number $a > 0$, we can plot a graph of $y = a^r$ for each $r \in \mathbb{Q}$.

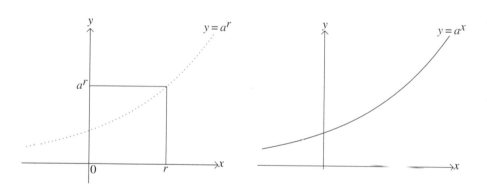

It is worth recognizing that a^x has a constant base and a variable exponent whereas x^n has a variable base and a constant exponent.

For irrational numbers (e.g. π, e), we assign values to a^x in such a way as to 'fill up the holes' in the graph of $y = a^r$. The function $f \colon \mathbb{R} \to (0, \infty)$ then defined by

$$y = f(x) = a^x$$

has a graph which is continuous for all x and is always increasing.

2.3.3 Trigonometric functions[◇]

In what follows, the angle x is always to be understood as measured in radians, where one **radian** is indicated by the figure on the right. Because the total circumference of a circle of radius 1 is 2π,

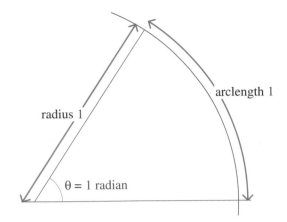

$$360° = 2\pi \text{ radians}$$
$$180° = \pi \text{ radians}$$
$$90° = \pi/2 \text{ radians}$$

The **sine** and **cosine** functions are shown for x satisfying $0 \le x \le 2\pi$, in the following circles of radius one.

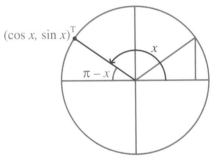

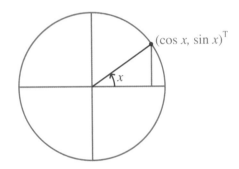

$$(\cos x, \sin x)^{\mathrm{T}} = (-\cos(\pi - x), \sin(\pi - x))^{\mathrm{T}}$$

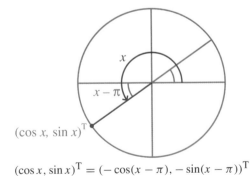

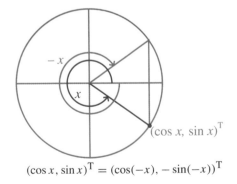

$$(\cos x, \sin x)^{\mathrm{T}} = (-\cos(x - \pi), -\sin(x - \pi))^{\mathrm{T}} \qquad (\cos x, \sin x)^{\mathrm{T}} = (\cos(-x), -\sin(-x))^{\mathrm{T}}$$

The top right diagram shows $\cos x$ and $\sin x$ for $0 \le x \le \frac{\pi}{2}$. As usual,

$$\cos x = \frac{\text{adjacent}}{\text{hypotenuse}} \quad \text{and} \quad \sin x = \frac{\text{opposite}}{\text{hypotenuse}}$$

The other diagrams show how one can always refer back to this diagram when faced with values of $\cos x$ and $\sin x$ with x outside the interval $[0, \frac{\pi}{2}]$.

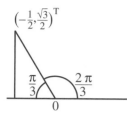

For example,

$$\left(\cos\frac{2\pi}{3}, \sin\frac{2\pi}{3}\right)^{\mathrm{T}} = \left(-\cos\frac{\pi}{3}, \sin\frac{\pi}{3}\right)^{\mathrm{T}} = \left(-\frac{1}{2}, \frac{\sqrt{3}}{2}\right)^{\mathrm{T}}$$

Since there are 2π radians in a full circle, the sine and cosine functions are periodic with period 2π (i.e. $\sin(x+2\pi) = \sin x$ and $\cos(x+2\pi) = \cos x$) with the following graphs:

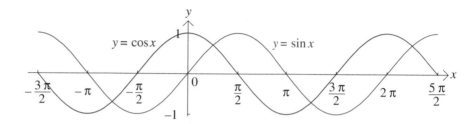

Observe from the graphs that $\sin x = \cos(x - \frac{\pi}{2})$, and from Pythagoras' theorem that $\cos^2 x + \sin^2 x = 1$.

2.4 Combinations of functions$^\diamond$

Functions may be combined in a variety of ways to generate other functions. The sum of functions is a function. So are the product of a function and a scalar and the product of two functions. Again, the quotient of functions is a function which is defined where the denominator is nonzero. Functions can also be composed as 'functions of functions' to produce new functions – e.g. $f(g(x)) = \sin(x^2)$.

The simplest type of combined or nonelementary function is a **polynomial** which is a combination of power functions of the form

$$P_n(x) = a_n x^n + a_{n-1}x^{n-1} + \cdots + a_1 x + a_0$$

where the a_i are real constants. If $a_n \neq 0$, the polynomial is said to be of **degree** n and has n **roots** $\xi_1, \xi_2, \ldots, \xi_n$, which means that it may be written as

$$P_n(x) = a_n(x - \xi_1)(x - \xi_2)\ldots(x - \xi_n)$$

Some of the roots may be repeated, that is, they may not all be different. Also, some of the roots may be complex numbers, in which case they occur in 'conjugate' pairs.[†] Alternatively, $P_n(x)$ may be expressed as the product of real linear and quadratic factors in x. E.g.

$$P_4(x) = x^4 - 2x^3 + 2x^2 - 2x + 1 = (x - 1)^2(x - \mathrm{i})(x + \mathrm{i}) = (x - 1)^2(x^2 + 1)$$

has degree 4 and roots 1 (repeated), i and $-\mathrm{i}$.

Linear and affine functions are polynomial functions of degree one, which are of special importance.

[†] See Chapter 13.

If $m \in \mathbb{R}$, the function $L: \mathbb{R} \to \mathbb{R}$ defined by

$$L(x) = mx$$

is said to be **linear**. Its graph $y = mx$ is a nonvertical line passing through the origin.

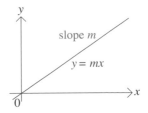

If $m \in \mathbb{R}$ and $c \in \mathbb{R}$, the function $A: \mathbb{R} \to \mathbb{R}$ defined by

$$A(x) = mx + c$$

is said to be **affine**. Its graph $y = mx + c$ is a nonvertical line.

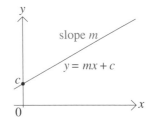

A common mistake is to confuse an affine function with a linear function because both have straight line graphs. But the graph of a linear function must pass through the origin.

A **rational function** R is the quotient of two polynomials – i.e.

$$R(x) = \frac{P(x)}{Q(x)}$$

where P and Q are polynomials. If P and Q have no common factor, the function R is not defined at the roots of Q, which must be excluded from the domain of R.

Note that in such a case $R(x)$ tends to ∞ or $-\infty$ as x tends to a root ξ of Q. We then say that the line $x = \xi$ is a **vertical asymptote** of the graph of $R(x)$.

For instance, the function

$$R(x) = \frac{x^2 - 7x + 8}{x^2 - x}$$

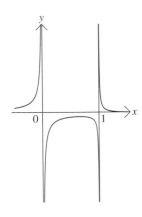

has the domain $\mathbb{R}$, excluding the points 0 and 1. The lines $x = 0$ and 1 are vertical asymptotes of its graph.

Let $\mathbb{Z}$ denote the set of integers, namely, $0, \pm 1, \pm 2, \pm 3, \ldots$. The definitions for other **trigonometric functions** may be expressed in terms of the sine and cosine functions as below:

Note that the domains of the functions defined exclude points at which the denominators on the right are zero.

$$\tan x = \frac{\sin x}{\cos x} \qquad \sec x = \frac{1}{\cos x} \qquad x \neq \frac{(2n+1)\pi}{2} \quad (n \in \mathbb{Z})$$

$$\cot x = \frac{\cos x}{\sin x} \qquad \operatorname{cosec} x = \frac{1}{\sin x} \qquad x \neq n\pi \quad (n \in \mathbb{Z})$$

The **tangent** function is the most important of these further functions. Its graph, which has vertical asymptotes $x = \dfrac{(2n+1)\pi}{2}$ $(n \in \mathbb{Z})$ is illustrated next:

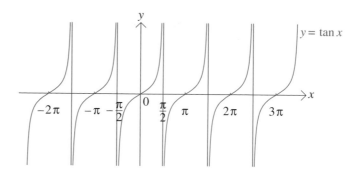

2.5 Inverse functions◇

Let X and Y be sets of real numbers, and suppose that $f: X \to Y$ is a function. This means that for each $x \in X$, there is a unique $y \in Y$ such that $y = f(x)$. If $g: Y \to X$ is another function and has the property that

$$y = f(x) \text{ if and only if } x = g(y)$$

then we call g the **inverse function** to f. The notation used for g is f^{-1}, but we shall see that there is not always a unique inverse function. Writing $g = f^{-1}$ we have

$$y = f(x) \text{ if and only if } x = f^{-1}(y)$$

Observe that $x = f^{-1}(y)$ is what we obtain by solving the equation $y = f(x)$ for x in terms of y. It must not be confused with the reciprocal function $\dfrac{1}{f(y)}$.

For example, if

$$y = \frac{2x - 1}{x + 1} = f(x) \qquad (x \neq -1)$$

then

$$x = \frac{y + 1}{2 - y} = f^{-1}(y) \qquad (y \neq 2)$$

Interchanging x and y,

$$f^{-1}(x) = \frac{x + 1}{2 - x}$$

The reciprocal function $\dfrac{1}{f}(x) = \dfrac{1}{f(x)} = \dfrac{x + 1}{2x - 1} \neq f^{-1}(x)$.

In the general case, the equation $y = f(x)$ may have no solutions at all, or else may have many solutions, as shown in the following diagram of a function.

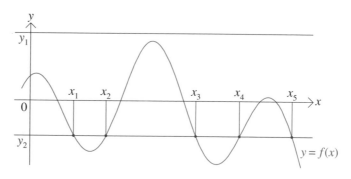

57

The equation $y_1 = f(x)$ has no solutions, while the equation $y_2 = f(x)$ has five solutions (namely x_1, x_2, x_3, x_4 and x_5).

But in order that there exists a *function* $f^{-1}: Y \to X$ which is an inverse function of $f: X \to Y$ it is necessary that there is a *unique* $x \in X$ for *each* $y \in Y$, such that $x = f^{-1}(y)$ or $y = f(x)$. For such a function, we deduce that, for each $x \in X$ and for each $y \in Y$,

$$f(f^{-1}(y)) = f(x) = y \quad \text{and} \quad f^{-1}(f(x)) = f^{-1}(y) = x$$

Let x_1 and x_2 be points of an interval I on which f is defined.

(1) If $f(x_1) < f(x_2)$ whenever $x_1 < x_2$, we say that f is **increasing** on I.

(2) If $f(x_1) > f(x_2)$ whenever $x_1 < x_2$, we say that f is **decreasing** on I.

The following diagrams illustrate two functions, an increasing and a decreasing function $f: I \to J$ which admit inverse functions.

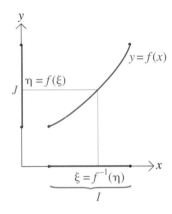
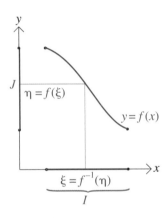

Example 1

The function $f: \mathbb{R} \to \mathbb{R}$ defined by $y = f(x) = x^2$ has *no* inverse function $f^{-1}: \mathbb{R} \to \mathbb{R}$. Observe that the equation

$$-1 = x^2$$

has *no* (real) solution, while the equation

$$1 = x^2$$

has *two* solutions.

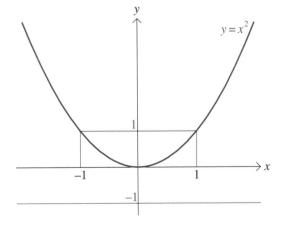

In this example the equation $y = f(x)$ fails to have a solution for x in terms of y or else has too many solutions. One can usually get round this problem by restricting the sets of values of x and y which one takes into account. This leads to the notion of a 'local inverse', which is discussed in detail in Chapter 7.

2.6 Inverses of the elementary functions$^\diamond$

2.6.1 Root functions$^\diamond$

Consider the function $f : [0, \infty) \to [0, \infty)$ defined by

$$y = f(x) = x^n \qquad (n \in \mathbb{N})$$

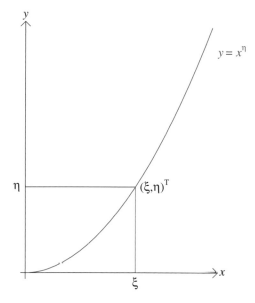

Note that we are here deliberately excluding consideration of negative real numbers and restricting attention to *nonnegative* values of x and y.

For each $y \geq 0$, the equation $y = x^n$ has a unique solution $x \geq 0$ and hence the function $f : [0, \infty) \to [0, \infty)$ has an inverse function $f^{-1} : [0, \infty) \to [0, \infty)$ at any $x > 0$. We use the notation

$$x = f^{-1}(y) = y^{1/n}$$

Thus, if $x \geq 0$ and $y \geq 0$,

$$x = y^{1/n} \text{ if and only if } y = x^n$$

It is instructive to observe that the graph of $x = y^{1/n}$ is just the same as the graph of $y = x^n$ but looked at from a different viewpoint. The graph of $x = y^{1/n}$ is what you would see if the graph of $y = x^n$ were drawn on a piece of glass and you viewed it sideways from behind.

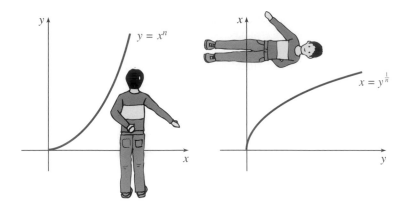

Let us now make x the independent variable in the inverse function so that $y = x^{1/n}$. Then the graphs of the function and its inverse can both be drawn on the same diagram. Now, an interchange of x and y is equivalent to a reflection in the line $y = x$. It is easy to see this from simple geometry, since $(y, x)^{\mathrm{T}}$ is the reflection of $(x, y)^{\mathrm{T}}$ in this line, as shown in the figure on the left below. Hence the curve $y = x^{1/n}$ can be obtained by reflecting the curve $y = x^n$ in the line $y = x$, as shown in the figure on the right below.

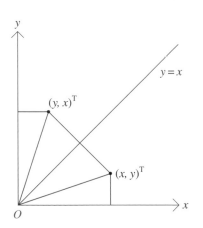
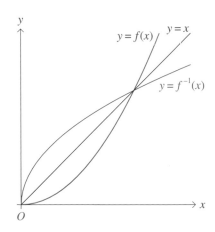

The same principles, of course, apply in respect of the graph of *any* inverse function.

If $m \in \mathbb{Z}$ and $x \geq 0$, we define $x^{m/n}$ by

$$x^{m/n} = (x^m)^{1/n}$$

In particular, we obtain

$$(x^n)^{1/n} = x$$

2.6.2 Exponential and logarithmic functions$^{\diamond}$

For the function $f: \mathbb{R} \to (0, \infty)$ defined by

$$y = f(x) = a^x \qquad (a > 0)$$

we may deduce the existence of an inverse function $f^{-1}\colon (0, \infty) \to \mathbb{R}$. This inverse function is used to define the **logarithm to base** a. We have

$$x = \log_a y = f^{-1}(y) \qquad (y > 0)$$

Thus $\quad x = \log_a y$ if and only if $y = a^x$

Note that $\log_a y$ is defined only for positive values of y because a^x takes only positive values. Also $\log_a a^x = x$ and $a^{\log_a x} = x$.

Let $y_1 = a^{x_1}$ and $y_2 = a^{x_2}$. Then

$$\log_a(y_1 y_2) = \log_a(a^{x_1} a^{x_2}) = \log_a(a^{x_1 + x_2}) = x_1 + x_2 = \log_a y_1 + \log_a y_2$$

$$\log_a(y_1^b) = \log_a(a^{x_1})^b = \log_a(a^{x_1 b}) = x_1 b = b \log_a y_1$$

The number $e \approx 2.7183$ has the special property that the slope of the tangent to $y = e^x$ when $x = 0$ is equal to one. The function $f\colon \mathbb{R} \to (0, \infty)$ defined by $y = f(x) = e^x$ is called the **exponential function** and often written as 'exp(x)'. Its inverse function $f^{-1}\colon (0, \infty) \to \mathbb{R}$ is the logarithm to base e. We call this function the **natural logarithm** and write $\log_e y = \ln y$.

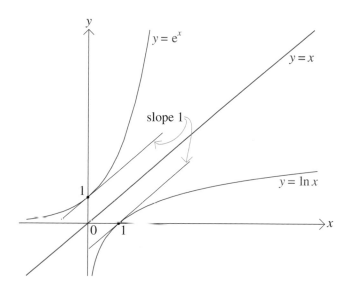

Finally, since e^x ranges over all positive values, all powers of positive numbers can be expressed in terms of the exponential and natural logarithm functions. We have

$$e^{b \ln a} = e^{\ln(a^b)} = a^b \qquad (a > 0)$$

2.7 Derivatives$^\diamond$

The concept of the derivative is one of the most fundamental in calculus. The derivative is concerned with the responses of dependent variables to changes in independent variables. We begin by revising the idea of the derivative of a real valued function of one real variable.

Denote the **derivative** of the function f evaluated at the point ξ by $f'(\xi)$.

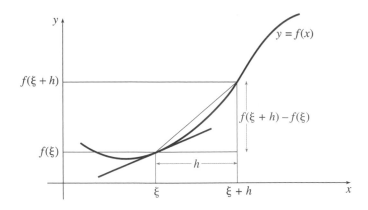

We have

$$f'(\xi) = \lim_{h \to 0} \left\{ \frac{f(\xi + h) - f(\xi)}{h} \right\}$$

when this limit exists as a finite real number. In this case we say that f has a derivative or is **differentiable** at ξ. We will also use the notation $\mathrm{D}f(x)$ for the derivative of f at a general point x. *Note that $\mathrm{D}f$ is itself a function defined by*

$$\mathrm{D}f(x) = \frac{\mathrm{d}}{\mathrm{d}x}\{f(x)\} = f'(x)$$

Geometrically, $f'(\xi)$ is the *slope* or *gradient* of the tangent line to the graph $y = f(x)$ at the point $(\xi, f(\xi))^{\mathrm{T}}$.

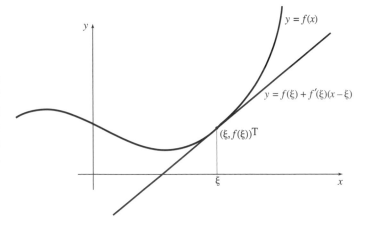

Since the tangent line has slope $f'(\xi)$ and passes through the point $(\xi, f(\xi))^{\mathrm{T}}$, it has equation

$$y - f(\xi) = f'(\xi)(x - \xi)$$

A function is differentiable at ξ if it has a nonvertical tangent line at ξ. It is differentiable in an interval if it is differentiable at every point of the interval.

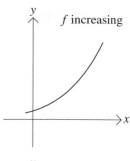

f increasing

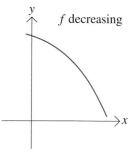

f decreasing

It is easily seen that

(1) f is increasing on I if $f'(x) > 0$.

(2) f is decreasing on I if $f'(x) < 0$.

As x increases from ξ to $\xi + h$ the expression

$$\frac{f(\xi + h) - f(\xi)}{h}$$

is the **average rate of increase** of $f(x)$ with respect to x. As h tends to 0, we obtain $f'(\xi)$ as the **instantaneous rate of increase** of $f(x)$ with respect to x at the point ξ. Economists call $f'(\xi)$ the **marginal** rate of increase.

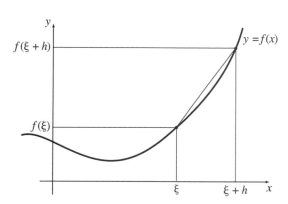

2.8 Existence of derivatives◇

Commonly occurring instances when the derivative fails to exist are shown below.

(1) Write $h \to 0^+$ for h tends to 0 from the right (through positive values) and $h \to 0^-$ for h tends to 0 from the left (through negative values).

The above limit exists only if the **derivative from the right** is equal to the **derivative from the left**:

$$\lim_{h \to 0^+} \left\{ \frac{f(\xi + h) - f(\xi)}{h} \right\} = \lim_{h \to 0^-} \left\{ \frac{f(\xi + h) - f(\xi)}{h} \right\}$$

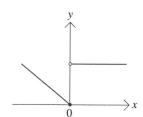

$y = |x|$

For example, this is not the case for the function $f(x) = |x|$ when $x = 0$, since

$$\lim_{x \to 0^+} \frac{|x| - 0}{x} = 1 \quad \text{and} \quad \lim_{x \to 0^-} \frac{|x| - 0}{x} = -1$$

Hence $f(x) = |x|$ is not differentiable when $x = 0$.

(2) The limit does not exist for functions which are *discontinuous* at ξ, like the function

$$f(x) = \begin{cases} -x & x \le 0 \\ 1 & x > 0 \end{cases}$$

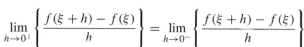

In this case, the derivative from the right does not exist since as $h \to 0^+$

$$\frac{f(\xi + h) - f(\xi)}{h} = \frac{1 - 0}{h} \to \infty$$

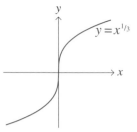

(3) Lastly, if the function is continuous and the limits from the left and the right do not exist at ξ, the function has a vertical tangent line and is not differentiable at ξ.

For example, the function

$$f(x) = x^{\frac{1}{3}}$$

has a vertical tangent line and is not differentiable when $x = 0$.

2.9 Derivatives of inverse functions$^\diamond$

The diagrams below illustrate a differentiable function $f: I \to J$ which has a differentiable inverse function $f^{-1}: J \to I$.

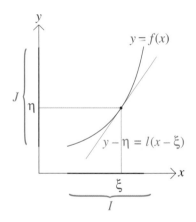
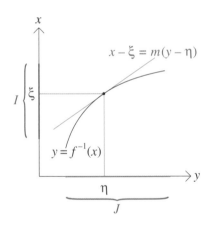

We know that the second diagram is really the same as the first but looked at from a different viewpoint. In particular, the lines $y - \eta = l(x - \xi)$ and $x - \xi = m(y - \eta)$ are really the *same* lines. Thus $m = l^{-1}$. But $l = f'(\xi)$ and $m = \left(f^{-1}\right)'(\eta)$. This gives the formula

$$\frac{dx}{dy} = \left(\frac{dy}{dx}\right)^{-1} \qquad (1)$$

This clearly indicates that special thought has to be given to what happens when the derivative of f is zero.

2.10 Calculation of derivatives

2.10.1 Derivatives of elementary functions and their inverses

Derivatives of power functions

We start by calculating the derivative of the function $f(x) = x^2$. By the definition of the derivative

$$f'(\xi) = \lim_{h \to 0} \left\{ \frac{f(\xi + h) - f(\xi)}{h} \right\} = \lim_{h \to 0} \frac{(\xi + h)^2 - \xi^2}{h}$$

$$= \lim_{h \to 0} \frac{\xi^2 + 2\xi h + h^2 - \xi^2}{h} = \lim_{h \to 0} (2\xi + h) = 2\xi$$

Hence, if $y = x^2$, then $\dfrac{dy}{dx} = 2x$.

Using expansions by the binomial theorem for integral powers, the derivatives of x^n where $n \in \mathbb{Z}$ may be calculated by the same method, yielding the result

$$D y^n = n y^{n-1} \qquad \text{where } n \in \mathbb{Z}$$

Derivatives of root functions

If $x = y^{1/n}$, then $y = x^n$. Hence

$$D y^{1/n} = \frac{dx}{dy} = \left(\frac{dy}{dx} \right)^{-1} = \frac{1}{nx^{n-1}} = \frac{1}{n} \left(y^{1/n} \right)^{1-n} = \frac{1}{n} y^{1/n-1}$$

Derivatives of exponential and logarithmic functions

Using the definition of the derivative

$$\frac{d}{dx} \left(e^x \right) = \lim_{h \to 0} \left(\frac{e^{x+h} - e^x}{h} \right) = e^x \lim_{h \to 0} \left(\frac{e^h - e^0}{h} \right)$$

But the second limit is the slope of the exponential function at $x = 0$ and e was chosen to make this equal to one. So e^x *is its own derivative*, i.e.

$$\frac{d}{dx} \left(e^x \right) = e^x$$

Recall that $y = e^x$ if and only if $x = \ln y$.

It follows that

$$\frac{d}{dy} (\ln y) = \frac{dx}{dy} = \left(\frac{dy}{dx} \right)^{-1} = \frac{1}{e^x} = \frac{1}{y}$$

That is

$$\frac{d}{dy} (\ln y) = \frac{1}{y} \qquad (y > 0)$$

Derivatives of the sine and cosine functions

$$\frac{\mathrm{d}}{\mathrm{d}x}(\sin x) = \cos x \qquad \frac{\mathrm{d}}{\mathrm{d}x}(\cos x) = -\sin x$$

A justification for these formulas may be based on the diagram below:

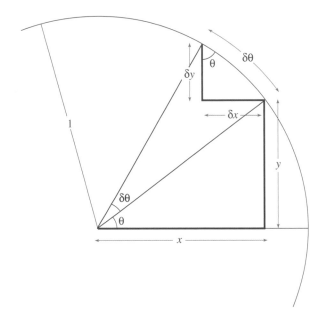

From the larger right angled triangle, we have $x = \cos\theta$ and $y = \sin\theta$. For small incremental angle $\delta\theta$ the arc of length $\delta\theta$ approximates to the hypotenuse of the smaller right angled 'triangle', which is tangential to the radius in the limit as $\delta\theta \to 0$. Then, the vertical angle is θ in the smaller 'triangle', and so we have

$$\cos\theta \approx \frac{\delta y}{\delta\theta}, \qquad \sin\theta \approx -\frac{\delta x}{\delta\theta}$$

Taking limits as $\delta\theta \to 0$ we obtain

$$\frac{\mathrm{d}}{\mathrm{d}\theta}(\sin\theta) = \cos\theta, \qquad \frac{\mathrm{d}}{\mathrm{d}\theta}(\cos\theta) = -\sin\theta$$

2.10.2 Derivatives of combinations of functions◊

There are rules for finding the derivatives of combinations of functions which are listed on the following page and are worth memorising.

Let f and g be functions of x and α and β be constants $\in \mathbb{R}$.

(I) $\dfrac{d}{dx}(\alpha f + \beta g) = \alpha \dfrac{df}{dx} + \beta \dfrac{dg}{dx}$

(II) $\dfrac{d}{dx}(fg) = f\dfrac{dg}{dx} + g\dfrac{df}{dx}$ Product rule

(III) $\dfrac{d}{dx}\left(\dfrac{f}{g}\right) = \dfrac{g\dfrac{df}{dx} - f\dfrac{dg}{dx}}{g^2}$ Quotient rule

If $z = f(y)$ and $y = g(x)$, then

(IV) $\dfrac{dz}{dx} = \dfrac{dz}{dy}\dfrac{dy}{dx}$ Chain rule

or

$$\frac{d}{dx}f(g(x)) = f'(y)g'(x) = f'((g(x))g'(x)$$

We can use the chain rule to obtain the derivative of any rational power of x:
If $z = y^{m/n}$ and $t = y^m$ where $m \in \mathbb{Z}$. Then $z = t^{1/n}$ and so

$$Dy^{m/n} = \frac{dz}{dy} = \frac{dz}{dt}\frac{dt}{dy} = \frac{1}{n}t^{1/n-1}my^{m-1} = \frac{m}{n}y^{m/n-1}$$

This justifies the formula

$$Dy^\alpha = \alpha y^{\alpha-1} \qquad (\alpha \in \mathbb{Q})$$

Using the formula for differentiating a quotient, we obtain

$$\frac{d}{dx}(\tan x) = \sec^2 x$$

Example 2 _____

(i) To find $\dfrac{d}{dx}(\cos(x^2 + 2x + 1))$ let $y = x^2 + 2x + 1$ and $z = \cos y$. Then

$$\frac{dy}{dx} = 2x + 2 \qquad \frac{dz}{dy} = -\sin y$$

Hence by the chain rule

$$\frac{dz}{dx} = \frac{dz}{dy}\frac{dy}{dx} = -(\sin y)(2x + 2) = -(2x + 2)\sin(x^2 + 2x + 1)$$

(ii)

$$\frac{d}{dx}(\ln(f(x))) = \frac{f'(x)}{f(x)}$$

Let $y = f(x)$ and $z = \ln y$. Then

$$\frac{dy}{dx} = f'(x), \qquad \frac{dz}{dy} = \frac{1}{y}$$

Hence by the chain rule

$$\frac{dz}{dx} = \frac{dz}{dy}\frac{dy}{dx} = \frac{1}{y}f'(x) = \frac{f'(x)}{f(x)}$$

Example 3

An economics application

The **rate of growth** of a function $y = f(t)$ in economics is defined by

$$\frac{f'(t)}{f(t)} = \frac{d}{dt}(\ln f(t))$$

See Example 2(ii).

It tells us how fast a function increases in percentage terms.

For example, suppose that national investment $I(t)$ increases by 0.5% per year and population $P(t)$ by 1.25% per year. Then the investment per head of population (the **per capita investment**) is

$$\text{PCI}(t) = \frac{I(t)}{P(t)}$$

On taking logarithms,

$$\ln \text{PCI}(t) = \ln I(t) - \ln P(t)$$

Hence, the rate of growth of per capita investment $\text{PCI}(t)$ is

$$\frac{d}{dt}(\ln \text{PCI}(t)) = \frac{d}{dt}(\ln I(t)) - \frac{d}{dt}(\ln P(t))$$

$$= \text{rate of growth of } I(t) - \text{rate of growth of } P(t)$$

$$= (0.5 - 1.25)\% = -0.75\%$$

We can conclude that per capita investment falls by 0.75% per year.

Example 4

Another economics application

A revenue function $R(x)$ of output $x = x(l)$, itself a function of labour input l, is indirectly a function of l. If an employer in a small cottage industry, producing a small number of goods, wishes to know if it is worth increasing labour from 10 to 11 (measured by the number of working hours per day), he can estimate the increase in revenue measured in dollars by calculating dR/dl. The chain rule states that

$$\frac{dR}{dl} = \frac{dR}{dx}\frac{dx}{dl}$$

This expresses, in mathematical terms, the statement in economics:

$$\begin{array}{c}\text{marginal revenue} \\ \text{product of labour}\end{array} = \text{marginal revenue} \times \begin{array}{c}\text{marginal physical} \\ \text{product of labour}\end{array}$$

If the production function has a Cobb–Douglas form $x = \frac{1}{100}l^3$ and the revenue function $R(x) = 22x - x^2$, then

$$R(x(l)) = \frac{22}{100}l^3 - \frac{1}{10\,000}l^6$$

We evaluate the marginal revenue when $l = 10$.

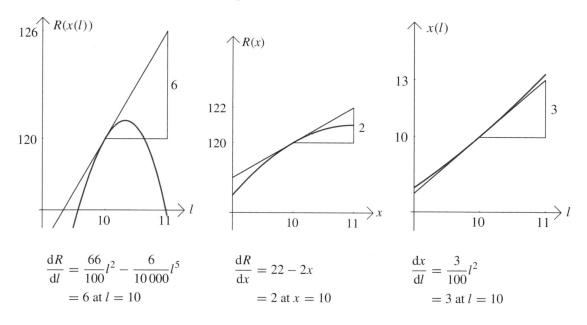

$$\frac{dR}{dl} = \frac{66}{100}l^2 - \frac{6}{10\,000}l^5 \qquad \frac{dR}{dx} = 22 - 2x \qquad \frac{dx}{dl} = \frac{3}{100}l^2$$
$$= 6 \text{ at } l = 10 \qquad\qquad = 2 \text{ at } x = 10 \qquad\qquad = 3 \text{ at } l = 10$$

Hence, increasing the number of working hours from 10 to 11 increases revenue by only \$6.[†] This is not worth it for the employer since wages would be more than \$6 an hour.

[†] This is only an approximation to the increase in revenue as the increase in l is not small.

Often, as in the above instance, the derivative may be calculated directly from the composite function, $R(x(l))$, but this is by no means always the case.

2.11 Exercises

1. Calculate $f'(1)$ using the definition of the derivative when

$$f(x) = x^3 + 2x + 2.$$

2. Differentiate the following expressions:

    (i) $x^4 - 3x^2 + 1$ (ii) $\sqrt{x}$ (iii)* $\sqrt{(x^2 + 1)}$

    (iv)* $(x + 1)^{3/2}$ (v) $x \sin x$ (vi)* $\dfrac{\tan x}{x}$

3. (i) Find $\dfrac{dy}{dx}$ in the following cases:

    (a)* $y = \ln\left(\dfrac{x + \sqrt{1 + x^2}}{1 + \sqrt{2}}\right)$ (b) $y = e^x \sin(x^3 - 1)$

    (c)* $y = \dfrac{x^2 - 1}{x^2 + x - 1}$ (d) $y = \dfrac{-x}{x^2 + x - 1}$

    Explain the relationship between the results for (c) and (d).

    (ii) Find the equation of the tangent line at the point $(1, 0)^{\mathsf{T}}$ to the curves given by (a)*, (b) and (c).

4.* Obtain the following results where a and b are constants:

    (i) $\dfrac{d}{dx} e^{(ax+b)} = ae^{(ax+b)}$ (ii) $\dfrac{d}{dx}(e^{-x}) = -e^{-x}$

    (iii) $\dfrac{d}{dx}(\ln(ax + b)) = \dfrac{a}{ax + b}$ (iv) $\dfrac{d}{dx}(\cos(ax + b)) = -a \sin(ax + b)$

5.* Money obtained ($\$y$) is related to amount sold (x kg) by

    $$y = ((x - 1)^2 + 1)^{1/2}$$

    What is the marginal price per kg when the amount sold is $x = 2$?

6.♣ Differentiate the following expressions:

    (i)* $\ln(\ln x)$ $(x > 1)$ (ii) $(\ln(x + 1))^2$ $(x > -1)$

    (iii) $e^{(\ln x)^{1/2}}$ $(x > 1)$ (iv) 2^x

7.♣ Differentiate the following expressions:

    (i)* $\operatorname{cosec} x$ $(x \neq n\pi)$ (ii) $\cot\left(\tfrac{1}{2}x\right)$ $(x \neq 2n\pi)$

    (iii)* $\ln(\sin x)$ $(0 < x < \pi)$ (iv) $\ln\left(\cot \tfrac{1}{2}x\right)$ $(0 < x < \pi)$.

8.* Differentiate the following expressions:

    (i) $x^{2/3} \tan((x + 1)^{1/3})$ (ii) $\exp(x\sqrt{4 - x^2})$ (iii) $\sqrt{\dfrac{x - 2}{5x + 1}}$

9. Find the equations of the tangent lines to the curve

    $$y = \dfrac{x}{1 - x}$$

    which are parallel to the line $4x - y = 7$.

10* Show that, for each real number $t \in$ the interval $(0, 1]$, the curve given in Exercise 3(i)(a) by

$$y = \ln\left(\frac{x + \sqrt{1 + x^2}}{1 + \sqrt{2}}\right)$$

has a tangent line with slope t. Find the points on the curve at which the tangent line has slope $2/3$.

11*♣ Given that there are distinct points on the curve

$$y = 3\sin x - x$$

which have a common tangent, find two such tangents, their equations and points of contact. Are there any more tangents with this property?

12* Find the number of tangent lines to the curve

$$y = \frac{3x}{x - 2}$$

which pass through the point $(-1, 9)^{\mathrm{T}}$. Find also the points of contact of these tangent lines with the curve.

13* Given a function $f(x)$ in economics, its **average function**, $Af(x)$, is defined by $Af(x) = \dfrac{f(x)}{x}$.

Prove the following results relating the average and marginal functions, $Af(x)$ and $Mf(x)$ respectively:

$$Af(x) < Mf(x) \quad \text{if and only if} \quad Af(x) \quad \text{is increasing.}$$

$$Af(x) > Mf(x) \quad \text{if and only if} \quad Af(x) \quad \text{is decreasing.}$$

$$Af(x) = Mf(x) \quad \text{if and only if} \quad Af(x) \quad \text{is stationary.}$$

14* The price of a type of computer scanner is dropping by 15% each year. The quantity of scanners sold is increasing by 25% each year. Find the annual rate of growth of the revenue derived from this type of scanner.[†]

† See Example 3.

15. For each of the following cost functions, find the largest possible domain and the marginal cost function. Compare the marginal cost at the production level of 100 units with the extra cost of producing one more unit.

$$(i)^* \ C(x) = \frac{2x^3 + 3000x}{x^2 + 2000} \qquad (ii) \ C(x) = 2000x - x^2\sqrt{200 - x}$$

2.12 Higher order derivatives

The second derivative of a function $f\colon \mathbb{R} \to \mathbb{R}$ at the point ξ is simply the derivative of the function f' evaluated at the point ξ. It is denoted by $f''(\xi)$ or by $\mathrm{D}^2 f(\xi)$. If $y = f(x)$ we also use the notation

$$\frac{\mathrm{d}^2 y}{\mathrm{d}x^2} = \frac{\mathrm{d}}{\mathrm{d}x}\left(\frac{\mathrm{d}y}{\mathrm{d}x}\right)$$

The second derivative can be interpreted geometrically as the slope of the tangent

$$y = f'(\xi) + f''(\xi)(x - \xi)$$

For instance, the second derivative is the slope of a marginal function in economics.

to the curve $y = f'(x)$ at the point $x = \xi$.

Higher order derivatives can be defined in a similar way. Like the first derivative, the derivative of any order is a function of x. We denote the nth order derivative of $f: \mathbb{R} \to \mathbb{R}$ at the point ξ by $f^{(n)}(\xi)$ or by $D^n f(\xi)$. Alternatively, we may write

$$\frac{d^n y}{dx^n} = \frac{d}{dx}\left(\frac{d^{n-1} y}{dx^{n-1}}\right)$$

indicating that $\dfrac{d^n y}{dx^n}$ is the derivative or rate of change of $\dfrac{d^{n-1} y}{dx^{n-1}}$ with respect to x.

Example 5

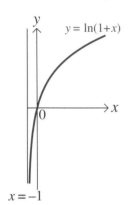

$y = \ln(1+x)$

$x = -1$

Let $y = \ln(1 + x)$. Then

$$\frac{dy}{dx} = \frac{1}{1 + x}$$

$$\frac{d^2 y}{dx^2} = -\frac{1}{(1 + x)^2}$$

$$\frac{d^3 y}{dx^3} = \frac{2}{(1 + x)^3}$$

and, in general,

$$\frac{d^n y}{dx^n} = \frac{(-1)^{n-1}(n - 1)!}{(1 + x)^n}$$

We will restrict our study to 'well behaved' functions, whose derivatives of all orders exist, and begin by approximating a general function of this type by one of the simplest kind – a polynomial function.

2.13 Taylor series for functions of one variable

As polynomials are simple functions, which are easily known, in order to study the behaviour of a function near a point, it is useful to find a polynomial approximation of it near the point. If we are investigating the function at points x *near* a point $x = \xi$, we consider a polynomial in powers of $x - \xi$. A useful approximation is the Taylor polynomial for a function $f: \mathbb{R} \to \mathbb{R}$ near a point ξ given by

$$P_n(x) = f(\xi) + (x - \xi)f'(\xi) + \frac{(x - \xi)^2}{2!}f''(\xi) + \cdots + \frac{(x - \xi)^n}{n!}f^{(n)}(\xi)$$

The polynomial closely resembles $f(x)$ near ξ, since not only does it have the same value as $f(x)$ at ξ, but also its first derivative and all higher order derivatives up to order n match those of $f(x)$ at ξ.

To see this differentiate both sides of the equation:

$$P_n'(x) = f'(\xi) + (x - \xi)f''(\xi) + \frac{3(x - \xi)^2}{3!} f'''(\xi) + \cdots + \frac{n(x - \xi)^{n-1}}{n!} f^{(n)}(\xi)$$

At $x = \xi$ both sides of the equation are equal to $f'(\xi)$.
Differentiating a second time

$$P_n''(x) = f''(\xi) + (x - \xi)f'''(\xi) + \cdots + \frac{n(n - 1)(x - \xi)^{n-2}}{n!} f^{(n)}(\xi)$$

At $x = \xi$ both sides of the equation are equal to $f''(\xi)$.
Similarly

$$P_n'(\xi) = f'(\xi) \quad \ldots \quad P_n^{(n)}(\xi) = f^{(n)}(\xi)$$

When the function has derivatives of all orders at $x = \xi$ we define the infinite Taylor series about the point $x = \xi$:

$$f(x) = f(\xi) + (x - \xi)f'(\xi) + \frac{(x - \xi)^2}{2!} f''(\xi) + \frac{(x - \xi)^3}{3!} f'''(\xi) + \cdots$$

This formula is valid for some set of values of x including the point ξ. When the series does not converge to $f(x)$, this set contains only the point ξ (in which case the formula is pretty useless). In exceptional cases, it is even possible for the series to converge to a function other than $f(x)$. On other occasions the formula is valid for *all* values of x. Usually, however, the formula is valid for values of x close to ξ and invalid for values of x not close to ξ. The question of convergence will be briefly discussed in §10.11.

Example 6

We shall find the Taylor polynomials for the function $\sin x$ about the point $x = 0$. Let $f(x) = \sin x$. Then

$$f(x) = \sin x \qquad f(0) = 0$$

$$f'(x) = \cos x \qquad f'(0) = 1$$

$$f''(x) = -\sin x \qquad f''(0) = 0$$

$$f'''(x) = -\cos x \qquad f'''(0) = -1$$

$$f^{iv}(x) = \sin x \qquad f^{iv}(0) = 0$$

As we have arrived back at the function $\sin x$, the pattern of values for the derivatives will repeat itself. Since the derivatives of even order are zero at the point $x = 0$, only derivatives of odd order, and so only odd powers, will appear in the Taylor polynomials. Hence the polynomials of even power degenerate to the polynomial of the previous odd power.

$$P_1(x) = P_2(x) = 0 + x = x$$

$$P_3(x) = P_4(x) = 0 + x - 0 - \frac{x^3}{3!} = x - \frac{x^3}{3!}$$

$$P_5(x) = P_6(x) = x - \frac{x^3}{3!} + \frac{x^5}{5!}$$

$$\cdots$$

In general, the Taylor polynomial for $\sin x$ about $x = 0$ is

$$P_{2n-1}(x) = P_{2n}(x) = x - \frac{x^3}{3!} + \frac{x^5}{5!} - \frac{x^7}{7!} + \cdots + (-1)^n \frac{x^{2n-1}}{(2n-1)!} + \cdots$$

The figure below shows the graph of $\sin x$ together with its approximations $P_1(x)$, $P_3(x)$ and $P_5(x)$ at $x = 0$. The polynomial $P_1(x)$ is the tangent at the point. The graphs show that the approximations deteriorate as we move away from the point. Also, the approximation improves as the degree of the polynomial increases.

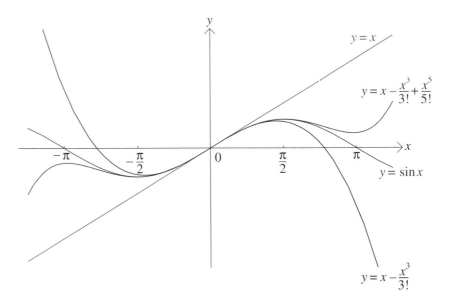

The Taylor series for $\sin x$ is

$$\sin x = x - \frac{x^3}{3!} + \frac{x^5}{5!} - \frac{x^7}{7!} + \cdots$$

Example 7

Suppose that $f(x) = \ln(1 + x)$. As we know from Example 5

$$f(0) = 0 \quad f'(0) = 1 \quad f''(0) = -1 \quad f'''(0) = 2 \ldots.$$

The Taylor series for $\ln(1 + x)$ about the point 0 is therefore

$$\ln(1 + x) = 0 + \frac{x}{1!} + \frac{x^2(-1)}{2!} + \frac{x^3(2)}{3!} + \frac{x^4(-6)}{4!} + \cdots$$

$$= x - \frac{x^2}{2} + \frac{x^3}{3} - \frac{x^4}{4} + \cdots$$

The range of validity of the formula is $-1 < x \le 1$.

You may find it instructive to calculate the sum of the first n terms of the expansion, using a calculator or a computer, when $x = -1$, $x = 1$ and $x = 1.01$.

The Taylor series about $\xi = 0$ is also known as the **Maclaurin series** and takes the simpler form

$$f(x) = f(0) + xf'(0) + \frac{x^2}{2!} f''(0) + \frac{x^3}{3!} f'''(0) + \cdots$$

The following series can be verified as demonstrated in Examples 6 and 7:

$$e^x = 1 + x + \frac{x^2}{2!} + \frac{x^3}{3!} + \cdots$$

$$\cos x = 1 - \frac{x^2}{2!} + \frac{x^4}{4!} - \frac{x^6}{6!} + \cdots$$

$$(1 + x)^\alpha = 1 + \alpha x + \frac{\alpha(\alpha - 1)}{2!} x^2 + \cdots \quad (-1 < x < 1)$$

Example 8

For many purposes one only needs terms up to the second order in the Taylor expansion of a function. When the terms of third order and higher have been discarded, the polynomial $P_2(x)$ which remains is known as the **quadratic Taylor approximation to f about ξ.**

The next figure illustrates how a quadratic approximation about a point indicates the shape of the graph near the point. It shows the graph of the function $y = \sin x$ together with the graphs of quadratic Taylor approximations $P_2(x)$ to the function about the points $x = \frac{\pi}{2}$ and $\frac{3\pi}{2}$ and also, their common tangents at these points.

If $f(x) = \sin x$ then $f'(x) = \cos x$ and $f''(x) = -\sin x$.

$$f\left(\frac{\pi}{2}\right) = \sin\left(\frac{\pi}{2}\right) = 1 \quad f'\left(\frac{\pi}{2}\right) = \cos\left(\frac{\pi}{2}\right) = 0 \quad f''\left(\frac{\pi}{2}\right) = -\sin\left(\frac{\pi}{2}\right) = -1$$

Hence near $\pi/2$,

$$P_2(x) = 1 - \frac{1}{2}\left(x - \frac{\pi}{2}\right)^2$$

Similarly, near $3\pi/2$

$$P_2(x) = -1 + \frac{1}{2}\left(x - \frac{3\pi}{2}\right)^2$$

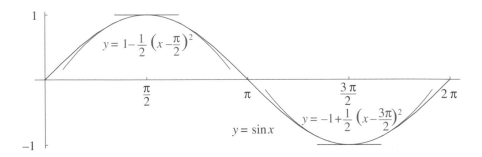

The negative quadratic term of the Taylor polynomial at $\pi/2$ shows that the graph is below the tangent line there. The positive quadratic term at $3\pi/2$ shows that it is above the tangent line there. Again, the graphs indicate that the approximations deteriorate as we move away from the points.

The quadratic Taylor approximation degenerates at π to the tangent approximation. There is no quadratic term near π since the graph crosses the tangent line there, and so cannot be approximated by a quadratic. Example 6 shows that this is also the case at $x = 0$.

2.14 Conic sections

We have defined functions which are expressed explicitly in terms of an independent variable. Functions may also be defined *implicitly* by a relation between x and y, which may be solved to give y as one or more explicit functions of x. This will be studied in Chapter 8. For the moment, we conclude this chapter with a study of the simplest type of implicit relation, which is a second degree equation in x and y

$$Ax^2 + Bxy + Cy^2 + Dx + Ey + F = 0 \qquad (A, B, C, D, E, F \in \mathbb{R})$$

This always represents a **conic section** or **conic**, which is the intersection of a double-napped cone and a plane, as the following illustrations show.

We shall see that conics play a central role in our study of the behaviour of functions of one and two variables. This is due to the crucial part played by the quadratic term of the Taylor approximation to a function at a point, in determining the shape of the function near the point.

The figures below show that when the plane does not pass through the vertex of the cone there are three basic **nondegenerate** conics: the parabola, the ellipse and the hyperbola together with the circle, which can be regarded as a special type of ellipse.

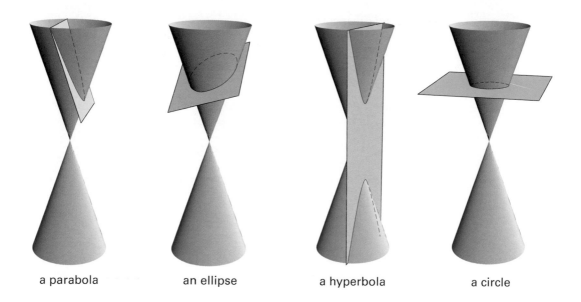

a parabola   an ellipse   a hyperbola   a circle

When the plane passes through the vertex of the cone, the conic assumes one of the **degenerate** forms below: a point, a line or a pair of intersecting lines.

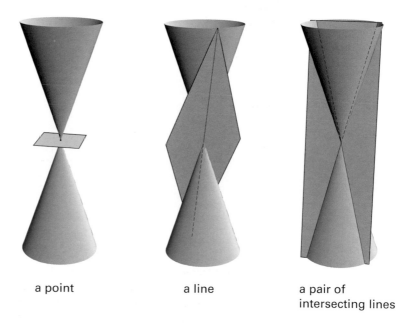

a point   a line   a pair of
intersecting lines

The equation can be transformed into one of three 'standard forms', considered below, in the following way. First, a rotation of the coordinate axes will eliminate the 'xy' term. This procedure will be described in §6.1 using the methods of linear algebra.

When neither coefficient of x^2 or y^2 is zero, a translation of the axes can then eliminate the linear terms in x and y to give one of the forms

$$\frac{x^2}{a^2} + \frac{y^2}{b^2} = 1 \qquad \text{and} \qquad \frac{x^2}{a^2} - \frac{y^2}{b^2} = 1$$

When one of these coefficients, say that of y^2, is zero, the linear term in the other variable x can be eliminated by a translation to give the form

$$x^2 = 4ay$$

These procedures will be described in Exercise 6.5.14. Each of the three basic conics may be characterised geometrically by the path of a point P which moves so that

$$\frac{\text{PF}}{\text{P}d} = e$$

where F is a fixed point called the **focus** and d is a fixed line called the **directrix**, with Pd denoting the distance of P from d. The constant e is known as the **eccentricity** of the conic.

By choosing the positions of the focus and the directrix conveniently, we can obtain the conics in standard position.

(i) $e = 1$. Choose d to be the line $y = -a$ and F to be the point $(0, a)$.

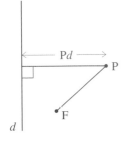

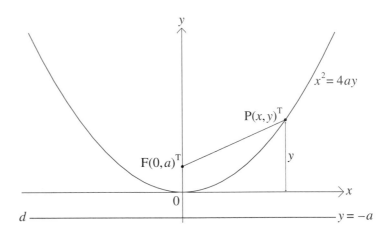

$$\begin{aligned} \text{PF}^2 &= \text{P}d^2 \\ x^2 + (y-a)^2 &= (y+a)^2 \\ x^2 + y^2 - 2ay + a^2 &= y^2 + 2ay + a^2 \end{aligned}$$

therefore

$$x^2 = 4ay$$

This is the equation of a **parabola**.

(ii) $0 < e < 1$. Choose d to be the line $x = \dfrac{a}{e}$ and F to be the point $(ae, 0)$.

$$\text{P}d = \frac{a}{e} - x \qquad \text{PF} = \sqrt{(x - ae)^2 + y^2}$$

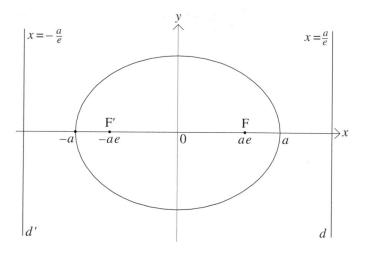

$$PF^2 = e^2 Pd^2$$

$$x^2 - 2aex + a^2e^2 + y^2 = e^2\left(x^2 - 2\frac{a}{e}x + \frac{a^2}{e^2}\right)$$

$$= e^2x^2 - 2aex + a^2$$

therefore

$$x^2(1 - e^2) + y^2 = a^2(1 - e^2)$$

or $$\frac{x^2}{a^2} + \frac{y^2}{a^2(1 - e^2)} = 1$$

Since $e < 1$, $b^2 = a^2(1 - e^2) > 0$ so the equation may be written as

$$\frac{x^2}{a^2} + \frac{y^2}{b^2} = 1$$

This is the equation of an ellipse. By symmetry there is an alternative focus–directrix pair F′, d′, as shown in the figure.

As $e \to 0$ the ellipse tends to the circle

$$x^2 + y^2 = a^2$$

Hence the circle may be thought of as a conic with '$e = 0$'.

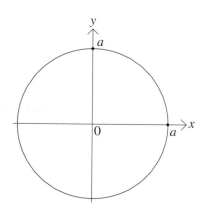

(iii) $e > 1$. Choosing d and F as in the previous case we obtain the identical equation

$$\frac{x^2}{a^2} + \frac{y^2}{a^2(1 - e^2)} = 1$$

Here, since $e > 1$, $a^2(e^2 - 1) > 0$ so we can write $b^2 = a^2(e^2 - 1)$, and the equation becomes the **hyperbola**

$$\frac{x^2}{a^2} - \frac{y^2}{b^2} = 1$$

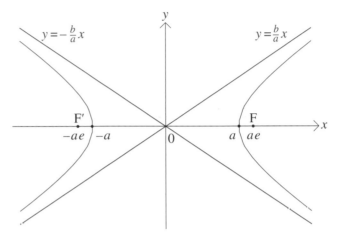

Rearranging the equation

$$\frac{y^2}{x^2} - \frac{b^2}{a^2} = -\frac{b^2}{x^2}$$

It follows that the hyperbola is approximately the same as

$$\frac{y^2}{x^2} - \frac{b^2}{a^2} = \left(\frac{y}{x} - \frac{b}{a}\right)\left(\frac{y}{x} + \frac{b}{a}\right) = 0 \qquad (1)$$

when x is large enough. But the graph of (1) consists of two straight lines

$$\frac{y}{x} = \frac{b}{a} \quad \text{and} \quad \frac{y}{x} = -\frac{b}{a}$$

These lines are said to be **slant asymptotes** for the hyperbola.

2.15 Exercises

1. Find the quadratic Taylor polynomial approximation about the point $x = \xi$ for the following functions:

   (i) e^x; $\xi = 1$

   (ii)* $e^x \sin x$; $\xi = 2$

   (iii) $\sqrt{1 + x}$; $\xi = 1$

(iv)* $\tan x$; $\quad \xi = \frac{\pi}{4}$ and $\xi = -\frac{\pi}{4}$

(v) $\sqrt{x}e^{-x}$; $\quad \xi = \frac{1}{2}$

(vi)* $\sqrt{x} - \ln x$; $\quad \xi = 4$

In each case, use a computer to plot the graph of the function, together with the graph of its Taylor approximation near the chosen point to see the resemblance between the two curves.

2.♣ Find the Taylor series about the point $x = \xi$ for the following functions indicating the form of the general terms:

(i)* e^x; $\quad \xi = 1$ (ii) $\cos x$; $\quad \xi = \frac{\pi}{4}$

(iii)* $\sqrt{x}$; $\quad \xi = 1$ (iv) $(x+1)^{-1}$; $\quad \xi = 2$

(v)* e^{-x}; $\quad \xi = \frac{1}{2}$ (vi) $\ln x$; $\quad \xi = 4$

3. Use a computer to plot the graphs of the following functions, in an interval centred at the following points, together with the approximating polynomials $P_1(x)$, $P_2(x)$, $P_3(x)$ at the points:

(i)* e^x; $\quad \xi = 1$ (ii) $2\sin 3x$; $\quad \xi = \pi/6$

(iii) $(1+x)^{-2}$; $\quad \xi = 1$ (iv) xe^{-x}; $\quad \xi = 1$

(v) $\cos x$; $\quad \xi = \frac{\pi}{4}$ (vi) $-\cos 2x$; $\quad \xi = \pi/2$

Observe that the accuracy of the approximation increases as the degree of the polynomial increases.

4. Find the Maclaurin series for the following functions:

(i)* e^x (ii) $\cos x$ (iii)* $(1+x)^\alpha$ ($\alpha \in \mathbb{N}$)

(iv) $\dfrac{1}{(1+x)}$ (v)* $\dfrac{1}{1-x}$ (vi) $\dfrac{1}{(1-x)^2}$

5. Find the Maclaurin series for the following functions using the various series given in the text and found in the previous exercise.

(i)* e^{-x} (ii) $\cos^2 x$ (iii)* $(1+x)^{-2}$

(iv) $e^{\frac{x^3}{2}}$ (v)* $\dfrac{\sin x \cos x}{x}$ (vi) $\ln\left(\dfrac{1+x}{1-x}\right)$

6.* Assuming that e^x is equal to its Maclaurin series expansion, prove that, for any fixed n,

$$x^{n+1}e^{-x} < (n+1)! \quad (x > 0)$$

Deduce that $x^n e^{-x} \to 0$ as $x \to \infty$.

7. Sketch the following conics, giving their intersections with the x and y axes. In the case of a hyperbola, also sketch its asymptotes.

(i)* $9y + 4x^2 = 0$

(ii) $3x^2 + 3y^2 = 14$

(iii) $2x = 5y^2$

(iv)* $2x = -5y^2$

(v) $x^2 = 3y$

(vi)* $\dfrac{4x^2}{25} + 9y^2 = 4$

(vii)* $36x^2 - 25y^2 = 1$

(viii) $\dfrac{y^2}{49} - x^2 = 9$

(ix) $25y^2 + 2x = 0$

8.*♣ The conics have properties which are exploited in a variety of situations. Parabolic shapes are used in the design of suspension bridges and parabolas have a reflective property that has wide applications.

This is due to the fact that in the figure on the right

$$\angle QPT = \angle FPT'$$

as shown, which you may prove as follows:

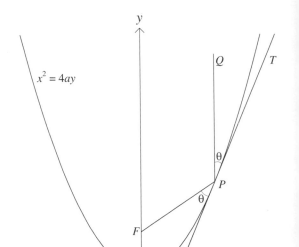

(i) Differentiate the equation $y = x^2/4a$ and deduce that a tangent vector $\overrightarrow{PT}$ is $(2a, x)^{\mathrm{T}}$.

(ii) Show that the cosine of the angle QPT between the vectors $\overrightarrow{PT}$ and $\overrightarrow{PQ} = (0, 1)^{\mathrm{T}}$ is equal to the cosine of the angle between the vectors

$$\overrightarrow{PF} = -\begin{pmatrix} x \\ y - a \end{pmatrix} \quad \text{and} \quad \overrightarrow{PT'} = -\begin{pmatrix} 2a \\ x \end{pmatrix}$$

Using $x^2 = 4ay$ you can show that they are both equal to $x/\sqrt{x^2 + 4a^2}$.

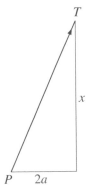

This property is exploited in parabolic reflectors in several ways.

Light beams parallel to the axis can be concentrated at the focus. On the other hand, a light source at the focus is reflected parallel to the axis, which intensifies the beam. Parabolic reflectors have a multitude of other uses for sound, heat and radar rays.

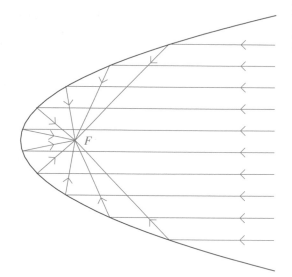

9.♣ The reflective property of ellipses is based on the fact that in the figure on the right

$$\angle FPT = \angle F'PT'$$

as shown, which you may prove as follows:

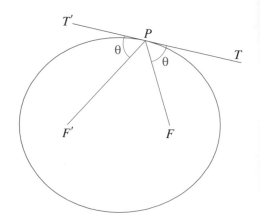

(i)* Differentiate the equation

$$\frac{x^2}{a^2} + \frac{y^2}{a^2(1 - e^2)} = 1$$

and deduce that a tangent vector $\overrightarrow{PT}$ is $(y, -x(1 - e^2))^{\mathsf{T}}$.

(ii)* Show that the cosine of the angle FPT between the vectors $\overrightarrow{PT}$ and $\overrightarrow{PF} = -(x - ae, y)^{\mathsf{T}}$ is equal to the cosine of the angle between the vectors

$$\overrightarrow{PF'} = -\begin{pmatrix} x + ae \\ y \end{pmatrix} \quad \text{and} \quad \overrightarrow{PT'} = \begin{pmatrix} -y \\ x(1 - e^2) \end{pmatrix}$$

This involves some manipulation using the fact that

$$x^2 + y^2 = x^2 e^2 + a^2(1 - e^2)$$

This reflective property is used in whispering galleries, where sound emanating at one focus is reflected towards the other focus. In lithotripsy, shockwaves from a transducer at one focus are reflected towards a patient at the other focus, in order to pulverise kidney stones.

(iii) The reflective property of hyperbolas is similarly due to the fact that in the figure on the right

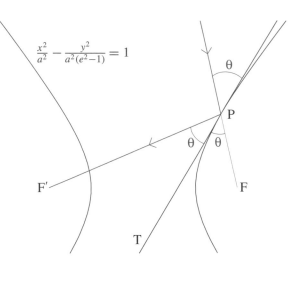

$$\frac{x^2}{a^2} - \frac{y^2}{a^2(e^2-1)} = 1$$

$$\angle FPT = \angle F'PT$$

as shown. Prove it in an almost identical way to the case of the ellipse, with $a^2(1-e^2)$ replaced by $a^2(e^2 - 1)$. As shown in the figure, rays directed towards one focus are reflected towards the other focus.

This, again, has a variety of uses and parabolic and hyperbolic mirrors are used together in telescopes.

10.♣ Suppose that $R(x) = P(x)/Q(x)$ is a rational function for which the degree of $P(x)$ is greater by one than the degree of $Q(x)$, then on dividing $P(x)$ by $Q(x)$, $R(x)$ may be written in the form

$$R(x) = ax + b + \frac{r(x)}{Q(x)}$$

where the 'remainder' r has degree less than $Q(x)$. Hence

$$R(x) - ax - b = \frac{r(x)}{Q(x)} \to 0 \quad \text{as} \quad x \to \infty \quad \text{or} \quad -\infty$$

We say that the line $y = ax + b$ is a **slant asymptote** of $R(x)$.

E.g. the line $y = 3x - 5$ is a slant asymptote of the function

$$R(x) = 3x - 5 + \frac{2}{x+4}$$

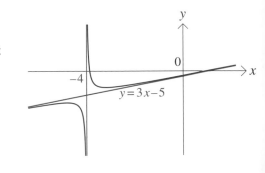

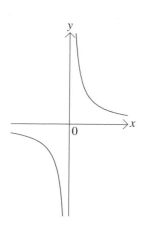

Suppose that $R(x) = P(x)/Q(x)$ is a rational function for which the degree of $P(x)$ is less than or equal to the degree of $Q(x)$. Then

$$R(x) \to c \quad \text{where} \quad c \in \mathbb{R} \quad \text{as} \quad x \to \infty \quad \text{or} \quad -\infty$$

We say that the line $y = c$ is a **horizontal asymptote** of $R(x)$.

E.g. The lines $y = 0$ and $y = 4$ are respectively horizontal asymptotes of the graphs

$$y = \frac{1}{x} \quad \text{and} \quad y = \frac{4x - 1}{x + 2}$$

Find all asymptotes of the following graphs $y = f(x)$ and match each function $f(x)$ with one of the graphs shown.

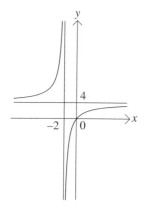

(i)* $f(x) = -\dfrac{1}{x - 2}$

(ii) $f(x) = \dfrac{x^2}{x + 1}$

(iii)* $f(x) = \dfrac{x^3}{(x + 1)(x - 1)}$

(iv) $f(x) = \dfrac{1}{(x + 2)^2}$

(v) $f(x) = -\dfrac{x^2}{(x - 2)(x + 1)}$

(vi) $f(x) = -\dfrac{x^3}{x^2 - 1}$

(vii)* $f(x) = \dfrac{x}{x + 1}$

(viii) $f(x) = \dfrac{-x^2(x - 2)}{x^2 - 1}$

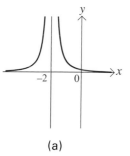

(a)

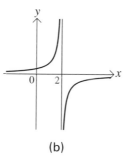

(b)

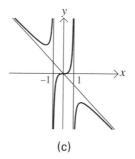

(c)

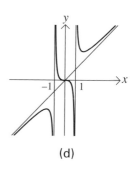

(d)

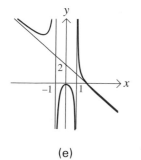

(e)

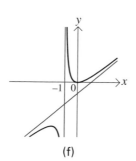

(f)

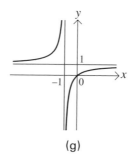

(g)

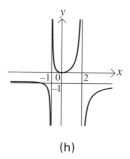

(h)

11.♣ Find all asymptotes of the graphs of the following functions:

$$(i)^* \quad f(x) = \frac{2}{(x-3)^3} \qquad\qquad (ii) \quad f(x) = \frac{x^2+4}{x+2}$$

$$(iii) \quad f(x) = \frac{4x^3+5x^2}{x^2-3x+1} \qquad (iv)^* \quad f(x) = -\frac{x^3+7}{x^2+1}$$

$$(v) \quad f(x) = \frac{5}{x^2-x-12} \qquad (vi) \quad f(x) = \frac{5x^2-1}{x^2+2}$$

$$(vii)^* \quad f(x) = \frac{(x+2)(x-1)}{x^3(x+1)} \qquad (viii) \quad f(x) = \frac{x^3(x-3)}{(x-1)^2(2x+1)}$$

Three	**Functions of several variables**

In this chapter we shall consider functions of two or more variables and extend the ideas developed for functions of one variable to such functions. We initially consider functions of two variables whose graphs can be visualised, before generalising to functions of more than two variables.

We define the concept of a partial derivative with respect to one of the variables and use it to find the equation of the tangent plane to a function at a point. The concept of the derivative of a function of several variables is then considered and a notation for the derivative introduced which accords with the rest of the theory. We also explain the role that partial derivatives play in finding the rate of change of a function in a given direction, called the 'directional derivative'.

This chapter, like the last, is vital for nearly everything which follows in this book.

3.1 Real valued functions of two variables

We denote the set of all two dimensional vectors by $\mathbb{R}^2$. Given a set $D \subseteq \mathbb{R}^2$ a function $f : D \to \mathbb{R}$ is a rule which assigns to each point $(x, y)^\mathrm{T}$ in the set D a unique $z \in \mathbb{R}$ denoted by

$$z = f(x, y)$$

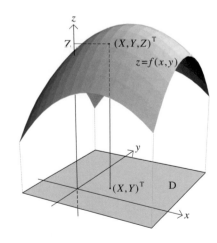

In this case, x and y are the independent variables and z is the dependent variable. The domain D is a **region** in the $(x, y)^\mathrm{T}$ plane and the graph is a **surface** in three dimensional space. To each point $(X, Y)^\mathrm{T}$ in D with $f(X, Y) = Z$ there corresponds a unique point $(X, Y, Z)^\mathrm{T}$ on the surface.

The surface is usually not easy to sketch. We therefore sometimes choose to think of it as a landscape. It is then natural to represent this landscape by drawing a map of horizontal curves at a fixed level called **contours** or **level curves**, along which the value of the function is constant. Each such contour corresponds to a horizontal section through the surface. Vertical sections also help describe it, by presenting side views. The mesh on computer plots of a surface corresponds to vertical sections taken through the surface in two orthogonal directions.

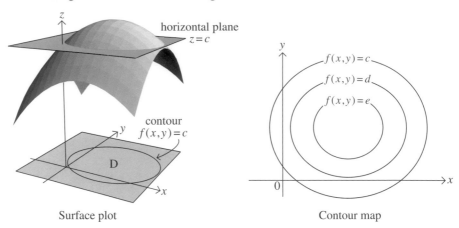

Surface plot   Contour map

3.1.1 Linear and affine functions

We shall begin by examining the simplest surfaces - firstly those for which $f(x, y)$ is an expression of degree one in x and y.

If M is a 1×2 matrix, (l, m), the function $L\colon \mathbb{R}^2 \to \mathbb{R}$ defined by

$$z = L(\mathbf{x}) = (l, m)\begin{pmatrix} x \\ y \end{pmatrix} = lx + my$$

is said to be **linear**. If c is a scalar, the function $A\colon \mathbb{R}^2 \to \mathbb{R}$ defined by

$$z = A(\mathbf{x}) = (l, m)\begin{pmatrix} x \\ y \end{pmatrix} + c = lx + my + c$$

is said to be **affine**.

The graph of a linear function and the graph of an affine function are each a *nonvertical* plane. We have studied these graphs in Chapter 1 in the context of vectors.

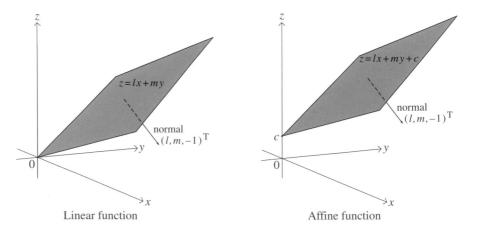

Linear function   Affine function

Example 1

The function $f\colon \mathbb{R}^2 \to \mathbb{R}$, $f(x, y) = 2 - 2x - y$ has graph the plane

$$z = 2 - 2x - y$$

which can also be represented by the vector equation

$$\langle \mathbf{x} - (0, 0, 2)^{\mathrm{T}}, (-2, -1, -1)^{\mathrm{T}} \rangle = 0$$

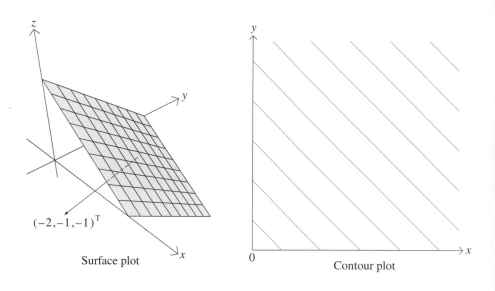

$(-2,-1,-1)^{\mathrm{T}}$

Surface plot

Contour plot

3.1.2 Quadric surfaces

We next examine surfaces of the form $z = f(x, y)$ where $f(x, y)$ is an expression of degree 2 in x and y. These belong to a class called **quadric surfaces**. Their contours are conics, which, as in Chapter 2, can be transformed by rotations and translations into conics in standard position. In the general standard position, the surfaces take the form

$$z = ax^2 + by^2 \qquad (a \text{ and } b \text{ not both zero, } a, b \in \mathbb{R})$$

Example 2

Let $f\colon \mathbb{R}^2 \to \mathbb{R}$ be defined by $f(x, y) = x^2 + 4y^2$.

The contours or level curves of the surface $z = x^2 + 4y^2$ are given by $x^2 + 4y^2 = c$, $c \in \mathbb{R}$. When $c > 0$ the level curves are the ellipses $x^2 + 4y^2 = c$ centred at the origin. When $c = 0$ the level curve reduces to the origin. There are no level curves for $c < 0$.

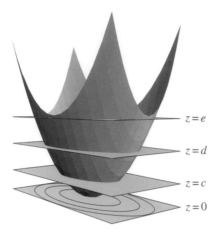

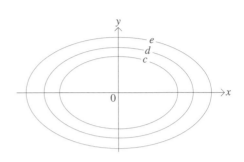

Surface with horizontal planes at
levels 0, c, d and e

Contours at levels c, d and e

The sections of vertical planes $y = c$, parallel to the $(x, z)^{\mathrm{T}}$ plane, with the surface $z = x^2 + 4y^2$ are parabolas $z = x^2 + 4c^2 = x^2 + c_1$. The sections of vertical planes $x = c$, parallel to the $(y, z)^{\mathrm{T}}$ plane, with $z = x^2 + 4y^2$ are parabolas $z = c^2 + 4y^2 = c_2 + 4y^2$.

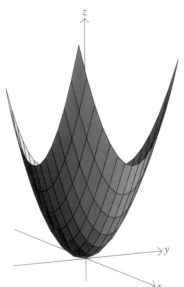

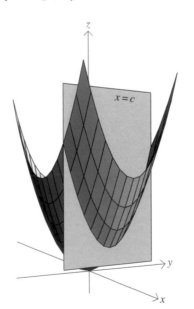

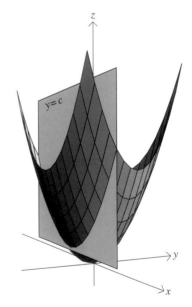

Surface with mesh

$z = x^2 + 4y^2$

Section parallel to $(y, z)^{\mathrm{T}}$ plane

$z = 4y^2 + c^2$

Section parallel to $(x, z)^{\mathrm{T}}$ plane

$z = x^2 + 4c^2$

Since the vertical sections are parabolas and the contours are ellipses, the graph of f, $z = x^2 + 4y^2$ is known as an **elliptic paraboloid**.

A general elliptic paraboloid in standard position has equation

$$z = \frac{x^2}{a^2} + \frac{y^2}{b^2} \qquad (a, b \in \mathbb{R})$$

Example 3

Let $f \colon \mathbb{R}^2 \to \mathbb{R}$ be defined by $f(x, y) = x^2 - y^2$. The contours are hyperbolas $x^2 - y^2 = c$.

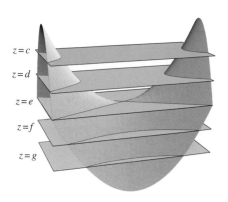

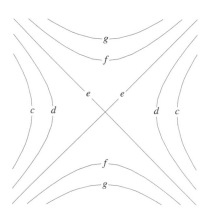

Level curves or contours on the surface with horizontal planes at levels c, d, e, f and g

Contours at levels c, d, e, f and g

The sections of vertical planes $y = c$, parallel to the $(x, z)^{\mathrm{T}}$ plane, with the surface $z = x^2 - y^2$ are parabolas $z = x^2 - c^2 = x^2 - c_1$. The sections of vertical planes $x = c$, parallel to the $(y, z)^{\mathrm{T}}$ plane, with $z = x^2 - y^2$ are parabolas $z = c^2 - y^2 = -y^2 - c_2$.

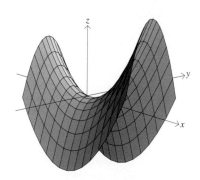

Surface with mesh

$z = x^2 - y^2$

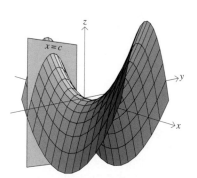

Section parallel to $(y, z)^{\mathrm{T}}$ plane

$z = -y^2 + c^2$

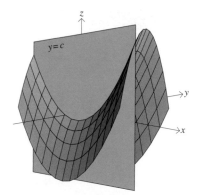

Section parallel to $(x, z)^{\mathrm{T}}$ plane

$z = x^2 - c^2$

91

Since the vertical sections are parabolas and the contours are hyperbolas (with asymptotes the lines $y = \pm x$ which are also contours of the graph), the surface is called a **hyperbolic paraboloid**, or more usually a **saddle**, because of its shape.

The general equation of a hyperbolic paraboloid in the standard position is

$$z = \frac{x^2}{a^2} - \frac{y^2}{b^2} \qquad (a, b \in \mathbb{R})$$

Example 4

Let $f: \mathbb{R}^2 \to \mathbb{R}$ be defined by $f(x, y) = 3x^2$. This degenerate function f has graph $z = 3x^2$, a parabolic 'cylinder'.

Vertical sections with planes $y = c$ are just the parabolas $z = 3x^2$. Vertical sections with planes $x = c$ are horizontal straight lines $z = 3c^2 = c_1$.

Contours $3x^2 = c$ are parallel lines $x = \pm\sqrt{c/3} = c_2$.

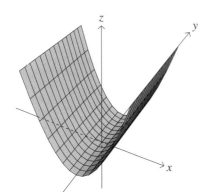

Surface with mesh

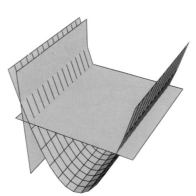

Sections parallel to $(x, y)^{\mathrm{T}}$ and $(y, z)^{\mathrm{T}}$ planes

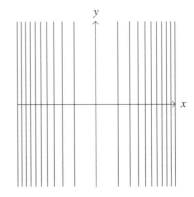

Contours $\pm x = \pm\sqrt{c/3} = c_2$

3.2 Partial derivatives

The concept of the derivative of a function of two variables is a more complex one than that of the derivative of a function of a single variable, since there are two variables which vary independently. We shall begin by considering how the function alters when only one of the variables changes. So we keep one of the variables constant and calculate the derivative of the function with respect to the other variable. This is called a **partial derivative** of the function.

The partial derivative of a function $f: \mathbb{R}^2 \to \mathbb{R}$ in the x direction with y kept constant is defined by

$$\frac{\partial f}{\partial x} = \lim_{h \to 0} \left\{ \frac{f(x+h, y) - f(x, y)}{h} \right\}$$

Similarly, the partial derivative in the y direction with x kept constant is defined by

$$\frac{\partial f}{\partial y} = \lim_{k \to 0} \left\{ \frac{f(x, y+k) - f(x, y)}{k} \right\}$$

The computation of partial derivatives is easy. Simply differentiate with respect to one of the variables treating the other variable as a constant.

It is important to distinguish between the symbol ∂ for a partial derivative, the Roman letter d for an ordinary derivative and the Greek letter δ for an increment.

An alternative notation for $\partial f / \partial x$ is f_x. This has the advantage that it allows one to distinguish between the variable with respect to which differentiation takes place and the point at which the partial derivative is evaluated. Thus

$$f_x(X, Y)$$

means 'the partial derivative with respect to x evaluated at the point $(X, Y)^{\mathrm{T}}$'.

Example 5

If $z = \sin(xy + y^2)$, then

$$\frac{\partial z}{\partial x} = y \cos(xy + y^2)$$

$$\frac{\partial z}{\partial y} = (x + 2y) \cos(xy + y^2).$$

Example 6

If $z = x^2 y + y^3 x$, then

$$\frac{\partial z}{\partial x} = 2xy + y^3$$

$$\frac{\partial z}{\partial y} = x^2 + 3y^2 x.$$

The following picture shows a surface $z = f(x, y)$ sliced by the plane $y = Y$. On the plane $y = Y$, one can see the graph of the function $z = f(x, Y)$. The partial derivative $f_x(X, Y)$ is simply the slope of this graph and of its tangent line at the point $x = X$.

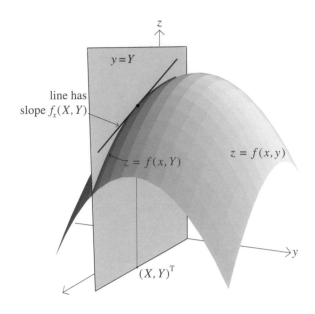

Geometrical interpretation of f_x

The plots below show the surface $z = x^2y + y^3x$ of Example 6, sliced first by the plane $y = 2$ with curve of intersection the parabola $z = 2x^2 + 8x$, whose slope is the partial derivative $f_x = 4x + 8$. Then the surface is sliced by the plane $x = 1$, with curve of intersection the cubic $z = y^3 + y$, whose slope is the partial derivative $f_y = 3y^2 + 1$.

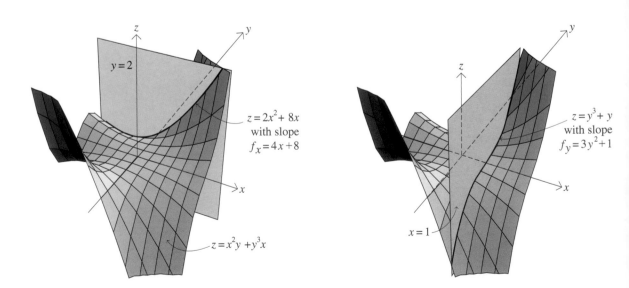

The following diagrams illustrate both partial derivatives simultaneously:

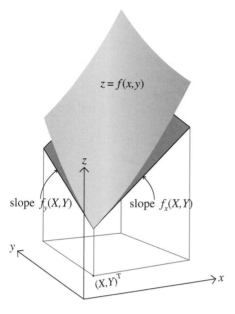

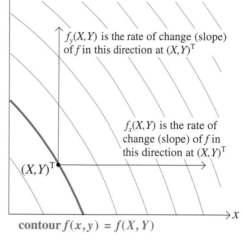

$f_y(X,Y)$ is the rate of change (slope) of f in this direction at $(X,Y)^{\mathrm{T}}$

$f_x(X,Y)$ is the rate of change (slope) of f in this direction at $(X,Y)^{\mathrm{T}}$

$(X,Y)^{\mathrm{T}}$

contour $f(x,y) = f(X,Y)$

Geometric interpretation of f_x and f_y Contour plot

As the diagram above implies, the tangent lines at $(x, y, z)^{\mathrm{T}} = (X, Y, Z)^{\mathrm{T}}$ in the x and y directions, determine a plane which is the tangent plane to the surface $z = f(x, y)$ at $(X, Y, Z)^{\mathrm{T}}$, apart from certain unusual cases.

Example 7

This example illustrates that it is possible for the partial derivatives in the x and y directions to exist at a point $(X, Y)^{\mathrm{T}}$, without the function having a tangent plane at the point $(X, Y, Z)^{\mathrm{T}}$.

$$\text{Let} \quad f(x, y) = 1 \quad \text{for} \quad x > 1 \text{ and } y > 1$$
$$= 0 \quad \text{otherwise}$$

Then $f_x(1, 1) = 0$ and $f_y(1, 1) = 0$ but f is discontinuous at $(x, y)^{\mathrm{T}} = (1, 1)^{\mathrm{T}}$ and its graph has no tangent plane at $(1, 1, 0)^{\mathrm{T}}$.

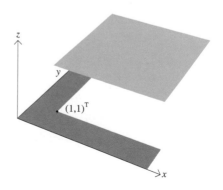

However, if a function is continuous and has continuous partial derivatives f_x and f_y at a point $\boldsymbol{\xi}$, then it can be shown that its graph has a nonvertical tangent plane at the corresponding point on the surface.

3.3 Tangent plane

If $Z = f(X, Y)$, then $(X, Y, Z)^{\mathrm{T}}$ is a point on the surface $z = f(x, y)$. If the surface admits a nonvertical tangent plane at $(X, Y, Z)^{\mathrm{T}}$, then we say that f is **differentiable** at $\boldsymbol{\xi} = (X, Y)^{\mathrm{T}}$.

The general equation of a plane through the point $(X, Y, Z)^{\mathrm{T}}$ with normal $(u, v, w)^{\mathrm{T}}$ is

$$\left\langle \begin{pmatrix} u \\ v \\ w \end{pmatrix}, \begin{pmatrix} x - X \\ y - Y \\ z - Z \end{pmatrix} \right\rangle = 0$$

See §1.7.

or

$$u(x - X) + v(y - Y) + w(z - Z) = 0$$

where $(u, v, w)^{\mathrm{T}}$ is normal to the plane. We assume that this plane is nonvertical, which means that its normal is not orthogonal to the z axis; that is, $w \neq 0$. Hence we can rewrite the equation in the form

$$z = A(x, y) = -\frac{u}{w}(x - X) - \frac{v}{w}(y - Y) + Z$$

For the plane to be tangent to the surface, our original function f and the affine function A must have the *same* partial derivatives at the point $(X, Y)^{\mathrm{T}}$. It follows that

$$\left. \begin{aligned} f_x(X, Y) &= A_x(X, Y) = -\frac{u}{w} \\[2mm] f_y(X, Y) &= A_y(X, Y) = -\frac{v}{w} \end{aligned} \right\}$$

Hence, if f is differentiable at $(X, Y)^{\mathrm{T}}$ its tangent plane must have equation

$$z - Z = f_x(X, Y)(x - X) + f_y(X, Y)(y - Y)$$

which can be written as

$$\left\langle \begin{pmatrix} x - X \\ y - Y \\ z - Z \end{pmatrix}, \begin{pmatrix} f_x(X, Y) \\ f_y(X, Y) \\ -1 \end{pmatrix} \right\rangle = 0$$

Thus

$\left(f_x(X, Y), f_y(X, Y), -1 \right)^{\mathrm{T}}$ is normal to the tangent plane and hence also to the surface at the point $(X, Y)^{\mathrm{T}}$

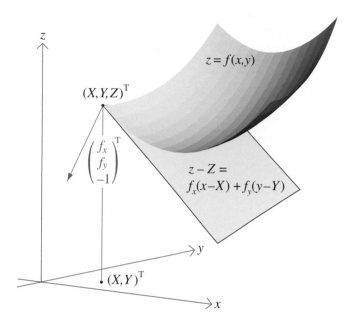

Although both $\left(f_x(X, Y), f_y(X, Y), -1\right)^{\mathrm{T}}$ and $\left(-f_x(X, Y), -f_y(X, Y), 1\right)^{\mathrm{T}}$ are normal to the surface, it is convenient to use the former vector for a reason which will soon be obvious.

Example 8

The equation of the tangent plane to the surface $z = x^2 y + y^3 x$ at the point $(x, y)^{\mathrm{T}} = (1, 2)^{\mathrm{T}}$ is

$$z - f(1, 2) = f_x(1, 2)(x - 1) + f_y(1, 2)(y - 2)$$

At this point $z = 2 + 8 = 10$ and

$$\left. \begin{array}{l} f_x(1, 2) = 2xy + y^3 = 2 \times 2 + 8 = 12 \\ f_y(1, 2) = x^2 + 3y^2 x = 1 \times 1 + 3 \times 4 \times 1 = 13 \end{array} \right\}$$

The tangent plane therefore has equation

$$z - 10 = 12(x - 1) + 13(y - 2)$$

A *normal* to the surface $z = x^2 y + y^3 x$ at $(1, 2, 10)^{\mathrm{T}}$ is therefore $(12, 13, -1)^{\mathrm{T}}$.

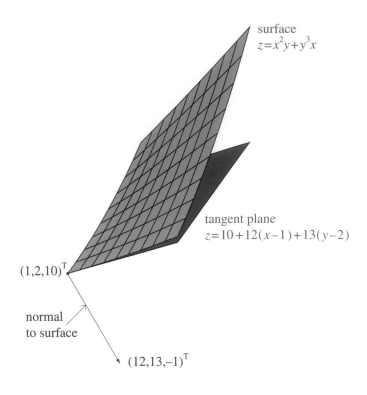

surface
$z=x^2y+y^3x$

tangent plane
$z=10+12(x-1)+13(y-2)$

$(1,2,10)^{\mathrm{T}}$

normal
to surface

$(12,13,-1)^{\mathrm{T}}$

3.4 Gradient

The tangent plane to the surface $z = f(x, y)$ at the point $(X, Y, Z)^{\mathrm{T}}$ has equation

$$z - Z = f_x(X, Y)(x - X) + f_y(X, Y)(y - Y)$$

If we slice the surface and its tangent plane by the horizontal plane $z = Z$, we find that the line

$$0 = f_x(X, Y)(x - X) + f_y(X, Y)(y - Y)$$

is tangent to the contour

$$Z = f(x, y)$$

at the point $(X, Y)^{\mathrm{T}}$.

The equation of the tangent line can be written as

$$\left\langle \begin{pmatrix} f_x \\ f_y \end{pmatrix}, \begin{pmatrix} x - X \\ y - Y \end{pmatrix} \right\rangle = 0$$

It follows immediately that the vector

$$\nabla f = \left(f_x(X, Y), f_y(X, Y) \right)^{\mathrm{T}}$$

is normal to the contour $f(x, y) = Z$ at the point $(X, Y)^T$.

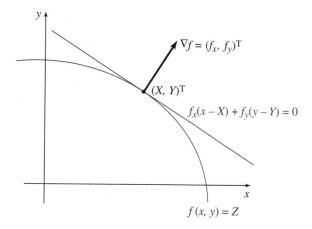

Contours at height $z = Z$ with gradient vector ∇f

The vector $\nabla f = \begin{pmatrix} f_x \\ f_y \end{pmatrix}$ is called the **gradient** of the function f.

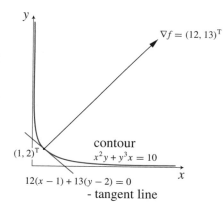

$(X, Y, Z)^T$ $\qquad \nabla f$

$(f_x, f_y, -1)^T$
normal to surface
in vertical plane

$\begin{pmatrix} 0 \\ 0 \\ -1 \end{pmatrix}$

Note that the components of the gradient vector are the scalar gradients of the tangent lines in the x and y directions. Observe also that the normal to the surface $\left(f_x(X, Y), f_y(X, Y), -1\right)^T$ lies in the vertical plane through the point $(X, Y)^T$ in the direction of the gradient vector.

Example 9

Let $f(x, y) = x^2 y + y^3 x$. Then

$$\nabla f = \left(f_x, f_y\right)^T = (2xy + y^3, x^2 + 3y^2 x)^T$$

The gradient vector $(12, 13)^T$ is normal to the contour $x^2 y + y^3 x = 10$ at the point $(1, 2)^T$. The accompanying diagram shows contours of surface and tangent plane at height $z = 10$ with gradient vector at the point $(1, 2)^T$.

$\nabla f = (12, 13)^T$

contour
$x^2 y + y^3 x = 10$

$(1, 2)^T$

$12(x - 1) + 13(y - 2) = 0$
- tangent line

3.5 Derivative

If a function f is differentiable at a point $\xi = (X, Y)^T$ and $Z = f(\xi)$, the equation of its tangent plane can be written as

$$z - Z = (f_x(\xi), f_y(\xi)) \begin{pmatrix} x - X \\ y - Y \end{pmatrix}$$

If $x = \begin{pmatrix} x \\ y \end{pmatrix}$, this suggests the definition of the **derivative** of f with respect to x at ξ, as

$$Df(\xi) = f'(\xi) = (f_x(\xi), f_y(\xi))$$

Then the equation of the tangent plane takes the same form as the equation of a tangent line:

$$z - f(\xi) = f'(\xi)(x - \xi)$$

Notice that the gradient of a real valued function is just the transpose of the derivative, i.e.

$$\nabla f = (Df)^T$$

3.6 Directional derivatives

Given a function $f: \mathbb{R}^2 \to \mathbb{R}$ the partial derivatives f_x and f_y at a point ξ give the rate of change of the function (which are the slopes of the tangent lines) in the x and y directions.

We now consider the derivative or rate of change of f (which is the slope of the tangent line) in *any* direction u where we assume that $u = (u_1, u_2)^T$ is a *unit* vector. We know from §1.6, that $x = \xi + tu$ is a point on the straight line through ξ in the direction u whose distance from ξ is t. We may then define the **derivative of f in the direction u** at the point ξ by

$$f_u(\xi) = \lim_{t \to 0} \left\{ \frac{f(\xi + tu) - f(\xi)}{t} \right\}$$

Using the following geometrical argument, we find an expression for the directional derivative in terms of the gradient vector, which dispenses with the need to calculate the above limit.

Since the derivative f_u can be interpreted as the slope of the tangent line in the direction u at ξ and $||u|| = 1$, the vector

$$v = (u_1, u_2, f_u)^T$$

lies in the tangent plane to the surface at the point $\xi = (X, Y, Z)^T$.

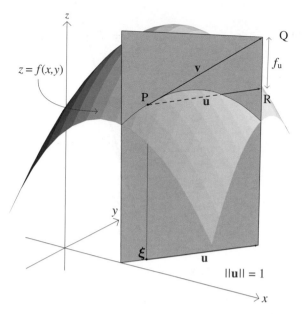

$f_{\mathbf{u}} = <\nabla f, \mathbf{u}>$ is the rate of increase of f at $\boldsymbol{\xi}$ in this direction

Surface plot with vertical plane in direction u, showing curve of intersection and tangent line with slope $f_{\mathbf{u}}$.

Contour map

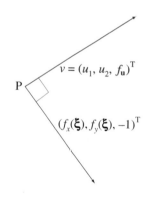

Hence $\mathbf{v}$ is orthogonal to the normal to the surface $\left(f_x(\boldsymbol{\xi}), f_y(\boldsymbol{\xi}), -1\right)^{\mathrm{T}}$ at this point,

i.e. the scalar product of the vectors is zero:

$$\left\langle \begin{pmatrix} u_1 \\ u_2 \\ f_{\mathbf{u}} \end{pmatrix}, \begin{pmatrix} f_x(\boldsymbol{\xi}) \\ f_y(\boldsymbol{\xi}) \\ -1 \end{pmatrix} \right\rangle = 0$$

or

$$u_1 f_x(\boldsymbol{\xi}) + u_2 f_y(\boldsymbol{\xi}) - f_{\mathbf{u}} = 0$$

Hence, rearranging this equation, we obtain

$$f_{\mathbf{u}} = \left\langle \begin{pmatrix} u_1 \\ u_2 \end{pmatrix}, \begin{pmatrix} f_x(\boldsymbol{\xi}) \\ f_y(\boldsymbol{\xi}) \end{pmatrix} \right\rangle$$

where the right hand side is the scalar product of vectors $\mathbf{u}$ and $\nabla f(\boldsymbol{\xi})$.

Hence the **directional derivative of** f **at the point** $\boldsymbol{\xi}$ **in the direction of the** *unit* **vector u** is given by

$$f_{\mathbf{u}} = \langle \nabla f(\boldsymbol{\xi}), \mathbf{u} \rangle$$

If θ is the angle between the vectors ∇f and $\mathbf{u}$, then, because $\|\mathbf{u}\| = 1$,

$$\langle \nabla f, \mathbf{u} \rangle = \|\nabla f\| \cos \theta$$

So the derivative in the direction $\mathbf{u}$ is the component of ∇f along $\mathbf{u}$. The diagram on the right shows the direction vectors ∇f and $\mathbf{u}$ in a horizontal plane.

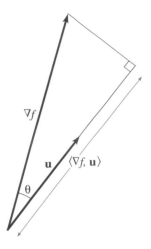

Now the maximum value of $\|\nabla f\| \cos \theta$ is attained when $\cos \theta = 1$ which occurs when $\theta = 0$. It follows that the direction in which ∇f points is the direction of maximum rate of increase of f at $\xi = (X, Y)^T$ and that this maximum rate of increase is equal to $\|\nabla f\|$. The minimum value of $\|\nabla f\| \cos \theta$ is attained when $\cos \theta = -1$ which occurs when $\theta = \pi$. It follows that the direction in which $-\nabla f$ points is the direction of maximum rate of decrease of f at $\xi = (X, Y)^T$ and that this maximum rate of decrease is equal to $-\|\nabla f\|$. We have $\|\nabla f\| \cos \theta = 0$ when $\theta = \dfrac{\pi}{2}$ and $\dfrac{3\pi}{2}$, which are the directions orthogonal to ∇f. These are along the directions of the contour or level curve of the surface that passes through $(X, Y, Z)^T$.

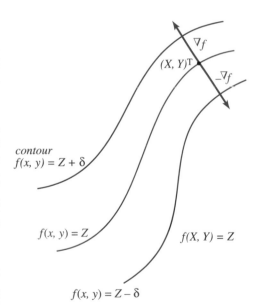

Example 10

Let $f(x, y) = x^2 y + y^3 x$. We know from the Example 9 that

$$\nabla f = (12, 13)^T$$

at the point $(1, 2)^T$. To find the rate of increase of f at $(1, \; 2)^T$ in the direction $(3, 4)^T$, we must first find a unit vector which points in the same direction. Since $3^2 + 4^2 = 25$, $\left(\frac{3}{5}, \frac{4}{5}\right)^T$ is the required vector. The rate of increase of f in this direction is therefore

$$\langle \nabla f, \mathbf{u} \rangle = 12 \cdot \frac{3}{5} + 13 \cdot \frac{4}{5} = \frac{88}{5} = 17.6$$

The direction of maximum increase of f at $(1, 2)^T$ is simply $\nabla f = (12, 13)^T$ and the rate of increase in this direction is

$$\|\nabla f\| = \{12^2 + 13^2\}^{1/2} = \sqrt{(313)} \approx 17.69$$

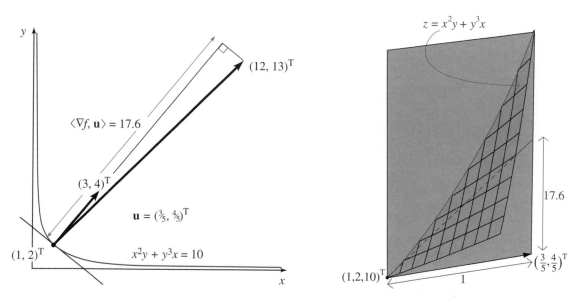

Contour and surface plots of function

Example 11

The function $f: \mathbb{R}^2 \to \mathbb{R}$ defined by

$$f(x, y) = 2400e^{1-10^{-6}x^2 - 6 \times 10^{-6}y^2}$$

represents the surface of a mountain.

$\|\nabla f\|$

∇f 1.44

$\leftarrow$ 1 $\rightarrow$

$(100, 100, f(100, 100))^{\mathrm{T}}$

If a climber at the point $(100, 100)^{\mathrm{T}}$, wishes to ascend at the greatest rate, she should move in the direction of

$$\nabla f = \begin{pmatrix} f_x \\ f_y \end{pmatrix} = 2400e^{1-10^{-6}x^2-6\times10^{-6}y^2} \begin{pmatrix} -2 \times 10^{-6}x \\ -12 \times 10^{-6}y \end{pmatrix}$$

Therefore her direction at $(100, 100)^{\mathrm{T}}$ is

$$\nabla f(100, 100) = 2400e^{1-10^{-2}-6\times10^{-2}} \begin{pmatrix} -2 \times 10^{-4} \\ -12 \times 10^{-4} \end{pmatrix} = 0.24e^{0.93} \begin{pmatrix} -2 \\ -12 \end{pmatrix}$$

Her *rate* of ascent would then be

$$\|\nabla f\| = 0.24e^{0.93}\sqrt{(-2)^2 + (-12)^2} \approx 7.400\,07$$

The angle that her path would make with the horizontal is $\arctan\|\nabla f\| \approx 1.44$ rad.

Suppose that a climber at the point $(10, 10)^{\mathrm{T}}$, sensing a thunderstorm looming, requires to descend as quickly as possible. He should take the direction of

$$-\nabla f(10, 10) = 0.024e^{0.9993} \begin{pmatrix} 2 \\ 12 \end{pmatrix}$$

as shown in the contour plot of the peak on the right. In this direction his rate of descent (the maximum possible) is

$$\|\nabla f\| \approx 0.793\,108$$

His path makes an angle $\arctan\|\nabla f\| \approx 0.67$ rad with the horizontal.

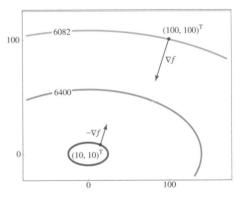

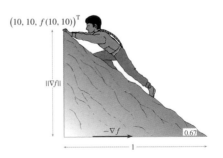

$(10, 10, f(10, 10))^{\mathrm{T}}$

$\|\nabla f\|$

$-\nabla f$ 0.67

$\leftarrow$ 1 $\rightarrow$

A walker at the point $(-1000, 1000)^{\mathrm{T}}$ considers a path in the direction $(-3, -4)^{\mathrm{T}}$. A unit vector in that direction is $\mathbf{u} = (-3/5, -4/5)^{\mathrm{T}}$. The rate of ascent of the path is the directional derivative

$$\langle \nabla f, \mathbf{u} \rangle = \left\langle 2400e^{1-1-6} \begin{pmatrix} 0.002 \\ -0.012 \end{pmatrix}, \begin{pmatrix} -\frac{3}{5} \\ -\frac{4}{5} \end{pmatrix} \right\rangle \approx 0.05$$

and so, initially, the path is almost horizontal.

Example 12

The temperature distribution of a lake with a hot spring centred at the origin is given by the function

$$\tau(x, y) = 32 - 3\ln(1 + x^2 + 4y^2)$$

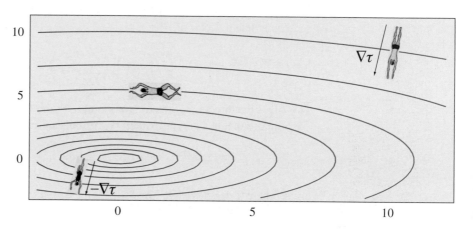

If a swimmer at the point $(x, y)^T = (10, 10)^T$ with a temperature of $\tau(10, 10) \approx 13.4\,°C$ is cold, he should swim in the direction of $\nabla\tau$, which is the direction of greatest increase in τ, to get warmer as quickly as possible. Now

$$\nabla\tau = \begin{pmatrix} \dfrac{-3 \times 2x}{(1 + x^2 + 4y^2)} \\ \dfrac{-3 \times 8y}{(1 + x^2 + 4y^2)} \end{pmatrix} \quad \text{which has direction} \quad \begin{pmatrix} -x \\ -4y \end{pmatrix}.$$

At the point $(x, y)^T = (10, 10)^T$ the gradient is

$$\nabla\tau = \begin{pmatrix} -3 \times \dfrac{20}{501} \\ -3 \times \dfrac{80}{501} \end{pmatrix} = \begin{pmatrix} -60/501 \\ -240/501 \end{pmatrix} \quad \text{which has direction} \quad \begin{pmatrix} -1 \\ -4 \end{pmatrix}.$$

This is clearly not directly towards the hottest point which is the origin, but rather is the direction of greatest increase in temperature *at the point*.

The rate at which the temperature increases at the point in this direction is

$$\|\nabla\tau\| = \sqrt{60^2 + 240^2}/501 \approx 0.49$$

If the swimmer at $(-2, -2)^T$ is too hot, she should swim in the direction of $-\nabla\tau = (x, 4y)^T$, which, at the point, has the direction $(-1, -4)^T$.

A swimmer who finds the temperature of the water comfortable should swim along the contours.

Example 13 _____

An economics example

A Cobb–Douglas production function is defined for $x, y > 0$ by the formula

$$P(x, y) = Ax^\alpha y^\beta$$

in which x and y are factors of production, with $A > 0$, $\alpha > 0$ and $\beta > 0$.
The gradient vector of $P(x, y)$ is

$$\nabla P(x, y) = \begin{pmatrix} P_x \\ P_y \end{pmatrix} = \begin{pmatrix} \alpha Ax^{\alpha-1} y^\beta \\ \beta Ax^\alpha y^{\beta-1} \end{pmatrix} = Ax^{\alpha-1} y^{\beta-1} \begin{pmatrix} \alpha y \\ \beta x \end{pmatrix}$$

This gives the direction of maximum increase of the function at $(x, y)^T$.
Suppose that at a particular factory the daily output is

$$P(x, y) = 100x^{\frac{1}{2}} y^{\frac{1}{5}}$$

units, where x denotes capital investment measured in units of \$1000 and y the total number of hours, in units of 10, for which the work force is employed each day.

If the manufacturer wishes to expand production when the capital investment is 30 and the total number of working hours is 24 each day, he wants to know the ratio in which he should increase capital and labour to obtain the maximum increase in output. This is given by the direction of

$$\nabla P(x, y)$$

which is along the direction $(\alpha y, \beta x)^T$. Hence at the point $(x, y)^T = (30, 24)^T$ the required ratio is

$$\frac{\alpha y}{\beta x} = \frac{\frac{1}{2} \times 24}{\frac{1}{5} \times 30} = \frac{2}{1}$$

Therefore, to maximise output, for each extra \$2000 he invests in capital, he should increase the total number of working hours by 10.

_____

3.7 Exercises

1.* Match the following functions (i)–(vi) with the contour plots (a)–(f) and the surface plots (g)–(l):

   (i) $(x + 2)^2 + 2(y - 3)^2$ (ii) $4x + 3y$ (iii) $10 - x^2 - (y - 2)^2$
   (iv) $(3y - 5x)^2$ (v) $5 - (y + 2)^2$ (vi) $4(x + 3)^2 - y^2$

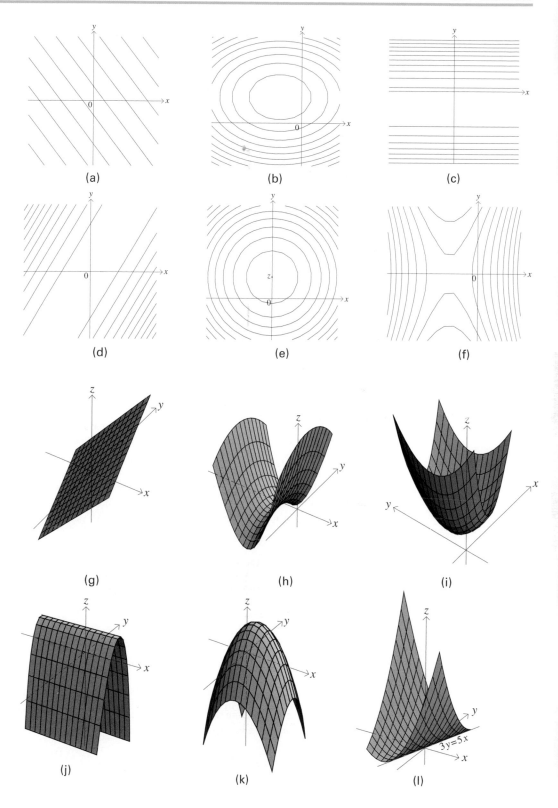

(a)

(b)

(c)

(d)

(e)

(f)

(g)

(h)

(i)

(j)

(k)

(l)

2. Prove that for any function f

$$\frac{\partial f}{\partial x} = \langle \nabla f, (1, 0)^{\mathrm{T}} \rangle \quad \frac{\partial f}{\partial y} = \langle \nabla f, (0, 1)^{\mathrm{T}} \rangle.$$

* Interpret these results.

3* The functions $f \colon \mathbb{R}^2 \to \mathbb{R}$ are defined in this and each of the following exercises.

Calculate

$$\frac{\partial f}{\partial x} \quad \text{and} \quad \frac{\partial f}{\partial y}$$

in the following cases:

(i) $f(x, y) = x^2 + 3y^2$ (ii) $f(x, y) = \dfrac{x - y}{x + y}$

(iii) $f(x, y) = e^{xy^2}$ (iv) $f(x, y) = x^2 y^5$

4. Calculate

$$\frac{\partial f}{\partial x} \quad \text{and} \quad \frac{\partial f}{\partial y}$$

in the following cases:

(i) $f(x, y) = \sin(x + 2y)$ (ii)* $f(x, y) = \cos(xy^2)$

(iii)* $f(x, y) = \sin\left(\dfrac{x}{y}\right)$ (iv) $f(x, y) = \ln(y \tan x)$.

5* For each of the functions of Exercise 3, find the equation of the tangent plane to the surface $z = f(x, y)$ where $x = 1$ and $y = 1$. Write down in each case a vector which is normal to the contour $f(x, y) = f(1, 1)$ at the point $(1, 1)^{\mathrm{T}}$.

6. For each of the functions of Exercise 4, find the equation of the tangent plane to the surface $z = f(x, y)$ where $x = \pi/4$ and $y = 1$. Write down in each case a vector which is normal to the contour $f(x, y) = f(\pi/4, 1)$ at the point $(\pi/4, 1)^{\mathrm{T}}$.

7* For each of the functions of Exercise 3, find

(1) the rate of increase at $(1, 1)^{\mathrm{T}}$ in the direction $(-3, 4)^{\mathrm{T}}$,

(2) the direction of maximum rate of increase at $(1, 1)^{\mathrm{T}}$. What is the rate of increase in this direction?

8. For each of the functions of Exercise 4, find

(a) the rate of increase at $(\pi/4, 1)^{\mathrm{T}}$ in the direction $(-1, 3)^{\mathrm{T}}$,

(b) the direction of maximum rate of increase at $(\pi/4, 1)^{\mathrm{T}}$. What is the rate of increase in this direction?

9. A function $f \colon \mathbb{R}^2 \to \mathbb{R}$ is defined by

$$f(x, y) = 5y^2 - x^2$$

(i) Find its gradient vector and evaluate it at the point $(x, y)^T = (1, 1)^T$. Find the rate of change of the function in the direction $(2, 1)^T$ at the point $(1, 1)^T$.

(ii) In what direction is the rate of change of the function at the point $(x, y)^T = (2, 0)^T$ equal to -2? Is there a direction for which the rate of change at the point $(x, y)^T = (2, 0)^T$ is equal to -5? Find the greatest rate of decrease of the function at this point.

10. The function P is defined for $x > 0$ and $y > 0$ in each of the following cases. In each case calculate

$$\frac{\partial P}{\partial x} \quad \text{and} \quad \frac{\partial P}{\partial y}$$

and find the equation of the tangent plane to the surface $z = P(x, y)$ where $(x, y)^T = (1, 1)^T$:

(i) $P(x, y) = 10x^{2/3} y^{1/3}$

(ii)* $P(x, y) = 15x^{4/5} y^{7/8}$

(iii) $P(x, y) = 5(\frac{2}{5}x^{1/2} + \frac{3}{5}y^{1/2})^2$

(iv)* $P(x, y) = 20(\frac{1}{3}x^{-1/4} + \frac{2}{3}y^{-1/4})^{-4}$

(v) $P(x, y) = 20(\frac{1}{3}x^{-4} + \frac{2}{3}y^{-4})^{-1/4}$

(vi) A general Cobb–Douglas production function which is defined by the formula

$$P(x, y) = Ax^\alpha y^\beta$$

where x and y are the factors of production, and $A > 0, 0 < \alpha < 1$ and $0 < \beta < 1$.

(vii) A general constant elasticity of substitution (CES) production function which is defined by

$$P(x, y) = A[\alpha x^\beta + (1 - \alpha) y^\beta]^{\frac{1}{\beta}}$$

where x and y are the factors of production and $A > 0, 0 < \alpha < 1$ and $\beta < 1$.

11. Find the direction of fastest growth in output for the CES production function

$$P(x, y) = 50 \left(\frac{2}{5} x^{-\frac{1}{3}} + \frac{3}{5} y^{-\frac{1}{3}} \right)^{-3}$$

at the point $(x, y)^T = (1, 1)^T$. Find the magnitude of growth in this direction. A contour $P(x, y) = $ constant is known as an **isoquant**, since the quantity (output) is the same at each point of the contour. Find the tangent line at the point $(x, y)^T = (1, 1)^T$ to the isoquant $P(x, y) = 50$. Verify that the tangent line is orthogonal to the direction of fastest growth of $P(x, y)$ at the point.

12.*♣ Let

$$f(x, y) = \sqrt{x^2 + y^2 + 2y + 1}$$

(i)   Find the equation of the tangent plane to the surface

$$z = \sqrt{x^2 + y^2 + 2y + 1}$$

at the point $(x, y, z)^{\mathrm{T}} = (a, b, c)^{\mathrm{T}}$.

(ii)   Write down a normal vector to the surface at this point.

(iii) It is known that all the tangent planes to the given surface intersect in a unique point. Find this point of intersection.

(iv) Verify that every tangent plane to the surface passes through this point.

3.8 Functions of more than two variables

We denote the set of all n dimensional vectors by $\mathbb{R}^n$. We now consider functions

$$f \colon \mathbb{R}^n \to \mathbb{R}$$

– i.e. functions which assign to each $\mathbf{x} = (x_1, x_2, \ldots, x_n)^{\mathrm{T}}$ a unique real number $f(\mathbf{x}) = f(x_1, x_2, \ldots, x_n)$.

All the above work applies equally well in the case of such functions. We cannot draw pictures when $n > 2$, but we continue to use geometrical language. When $n = 3$, the **contours** or **level surfaces** can be drawn in $\mathbb{R}^3$. They are not level in the sense of being horizontal, but are surfaces on which all values of f are the same.

Example 14 _____

The function

$$f \colon \mathbb{R}^3 \to \mathbb{R}$$

is defined by

$$f(x, y, z) = \sqrt{x^2 + y^2 + z^2}.$$

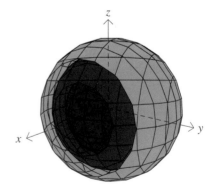

Its contours are spheres of radius c centred at the origin, which exist only for $c \geq 0$.

_____

The simplest nonlinear contours of functions $f \colon \mathbb{R}^3 \to \mathbb{R}$ are those for which f is an expression of degree 2 in x, y and z. These contours are known as **quadric surfaces**. Together with the quadric surfaces of the form $z = f(x, y)$ studied in §3.1.2 they represent the class of quadric surfaces in three dimensional space.

As in the case of two variables, functions of degree 2 can be transformed to the standard form

$$f(x, y, z) = Ax^2 + By^2 + Cz^2 \qquad (A, B, C \in \mathbb{R})$$

Those where one or two of A, B and C are zero have been considered before. So we assume that $A, B, C \neq 0$. There are three main types which are illustrated in the following diagrams and given together with their equations:

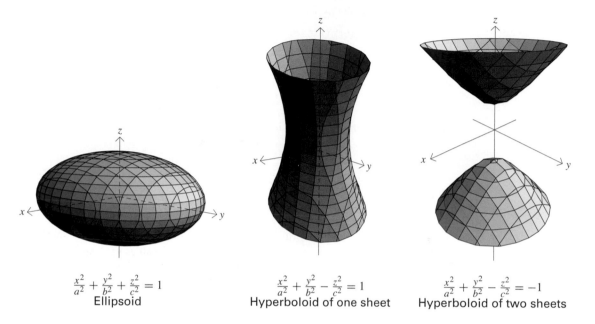

$\frac{x^2}{a^2} + \frac{y^2}{b^2} + \frac{z^2}{c^2} = 1$	$\frac{x^2}{a^2} + \frac{y^2}{b^2} - \frac{z^2}{c^2} = 1$	$\frac{x^2}{a^2} + \frac{y^2}{b^2} - \frac{z^2}{c^2} = -1$
Ellipsoid	Hyperboloid of one sheet	Hyperboloid of two sheets

3.8.1 Tangent hyperplanes

The graph of a function $f \colon \mathbb{R}^n \to \mathbb{R}$ is the 'hypersurface' in $\mathbb{R}^{n+1}$, $y = f(\xi_1, \xi_2, \ldots, \xi_n)$. If $Y = f(x_1, x_2, \ldots, x_n)$ then by identical reasoning to that of the case $n = 2$, if $\boldsymbol{\xi} = (\xi_1, \xi_2, \ldots, \xi_n)^{\mathrm{T}}$ the equation

$$y - Y = f_{x_1}(\boldsymbol{\xi})(x_1 - \xi_1) + f_{x_2}(\boldsymbol{\xi})(x_2 - \xi_2) + \cdots + f_{x_n}(\boldsymbol{\xi})(x_n - \xi_n)$$

is the tangent hyperplane to the hypersurface $y = f(x_1, x_2, \ldots, x_n)$ at the point $(\xi_1, \xi_2, \ldots, \xi_n, Y)^{\mathrm{T}}$ provided that $Y = f(\boldsymbol{\xi})$. Rearranging this as

$$\left\langle \begin{pmatrix} f_{x_1}(\boldsymbol{\xi}) \\ f_{x_2}(\boldsymbol{\xi}) \\ \vdots \\ f_{x_n}(\boldsymbol{\xi}) \\ -1 \end{pmatrix}, \begin{pmatrix} x_1 - \xi_1 \\ x_2 - \xi_2 \\ \vdots \\ x_n - \xi_n \\ y - Y \end{pmatrix} \right\rangle = 0$$

where the scalar product is in $\mathbb{R}^{n+1}$, shows that

the normal to the hypersurface is $\left(f_{x_1}(\boldsymbol{\xi}), f_{x_2}(\boldsymbol{\xi}), \ldots, f_{x_n}(\boldsymbol{\xi}), -1 \right)^{\mathrm{T}}$

The tangent hyperplane may be written as

$$y - Y = \left\langle \begin{pmatrix} f_{x_1}(\xi) \\ f_{x_2}(\xi) \\ \vdots \\ f_{x_n}(\xi) \end{pmatrix}, \begin{pmatrix} x_1 - \xi_1 \\ x_2 - \xi_2 \\ \vdots \\ x_n - \xi_n \end{pmatrix} \right\rangle$$

with the scalar product in $\mathbb{R}^n$.

This intersects the hyperplane $y = Y$ where

$$0 = \left\langle \begin{pmatrix} f_{x_1}(\xi) \\ f_{x_2}(\xi) \\ \vdots \\ f_{x_n}(\xi) \end{pmatrix}, \begin{pmatrix} x_1 - \xi_1 \\ x_2 - \xi_2 \\ \vdots \\ x_n - \xi_n \end{pmatrix} \right\rangle$$

which is the tangent hyperplane in $\mathbb{R}^n$ to the contour

$$Y = f(x_1, x_2, \ldots, x_n)$$

Hence

$$\nabla f = \left(\frac{\partial f}{\partial x_1}, \frac{\partial f}{\partial x_2}, \ldots, \frac{\partial f}{\partial x_n} \right)^{\mathrm{T}}$$

evaluated at ξ is the normal to the tangent hyperplane and so to the contour $Y = f(x_1, x_2, \ldots, x_n)$ at ξ. This can be visualised when $n = 3$.

The graph of $u = f(x, y, z)$ involves four variables, so cannot be drawn in three dimensional space

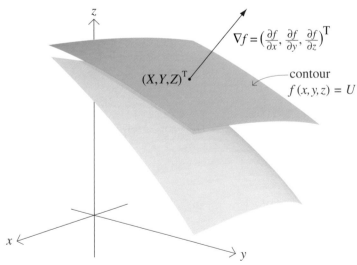

Two contours of $f(x,y,z)$ with gradient vector

Example 15

To find the tangent hyperplane to the hypersurface
$u = x^2 + y^2 + z^2$ at $(3, -1, 1)^{\mathrm{T}}$:

$$u = 3^2 + (-1)^2 + 1^2 = 11$$

$$\left(\frac{\partial f}{\partial x}, \frac{\partial f}{\partial y}, \frac{\partial f}{\partial z} \right)^{\mathrm{T}} = (2x, 2y, 2z)^{\mathrm{T}} = (6, -2, 2)^{\mathrm{T}} \text{ at the point.}$$

The required equation is

$$u - 11 = (6, -2, 2) \begin{pmatrix} x - 3 \\ y + 1 \\ z - 1 \end{pmatrix}$$

The normal to the hypersurface is $(6, -2, 2, -1)^{\mathrm{T}}$.
When $u = 11$,

$$6(x - 3) - 2(y + 1) + 2(z - 1) = 0$$

or

$$3x - y + z = 11$$

is a tangent plane to the contour $x^2 + y^2 + z^2 = 11$ at $(3, -1, 1)^{\mathrm{T}}$, with normal $\nabla f = (6, -2, 2)^{\mathrm{T}}$.

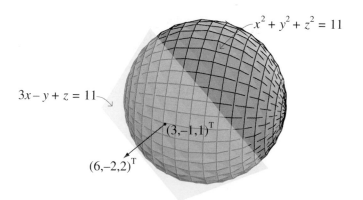

It is not surprising that the direction of ∇f is that of $(3, -1, 1)^{\mathrm{T}}$ since the normal to a sphere at a point is in the direction of the radius vector through the point.

Example 16

This example illustrates that we can find tangent planes to surfaces that are described by *implicit* equations of the form $f(x, y, z) = c$, by considering the equation $f(x, y, z) = c$ as a level surface for the hypersurface in $\mathbb{R}^4$, $u = f(x, y, z)$.
We can find the equation of the plane that is tangent to the ellipsoid

$$x^2 + 4y^2 + z^2 = 18$$

at the point $(1, 2, -1)^T$, by recognising that the ellipsoid is a level surface of the hypersurface given by

$$u = x^2 + 4y^2 + z^2$$

A normal to the ellipsoid is
$\nabla f = (2x, 8y, 2z)^T$ which is $(2, 16, -2)^T$ at the point $(1, 2, -1)^T$.
The equation of the tangent plane at $(1, 2, -1)^T$ is

$$\langle \nabla f, \mathbf{x} - \boldsymbol{\xi} \rangle = 0$$

or

$$2(x - 1) + 16(y - 2) - 2(z + 1) = 0$$

or

$$x + 8y - z = 18$$

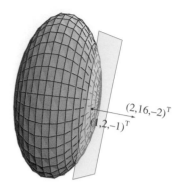

A function $f: \mathbb{R}^n \rightarrow \mathbb{R}$ is **differentiable** at the point $\boldsymbol{\xi}$ if there is a tangent hyperplane to the hypersurface $u = f(\mathbf{x})$ where $\mathbf{x} = \boldsymbol{\xi}$. Analogous to the two variable case we have the following definition:

> The **derivative of a scalar function** f with respect to $\mathbf{x}$ is the row matrix $\left(\dfrac{\partial f}{\partial x_1}, \dfrac{\partial f}{\partial x_2}, \ldots, \dfrac{\partial f}{\partial x_n} \right)$

This enables the equation of the tangent hyperplane to assume the usual pattern

$$z - f(\boldsymbol{\xi}) = f'(\boldsymbol{\xi})(\mathbf{x} - \boldsymbol{\xi})$$

Again, the **gradient** ∇f of a real valued function is just the *transpose* of the *derivative Df* – i.e.

$$\nabla f = (Df)^T$$

3.8.2 Directional derivatives

Finally, note that, if $\mathbf{u}$ is a unit vector, then

$$\langle \nabla f(\boldsymbol{\xi}), \mathbf{u} \rangle$$

remains the **directional derivative of f in the direction u at the point $\boldsymbol{\xi}$**.

The previous geometrical argument works just as well when extended to n dimensions. For a function $f: \mathbb{R}^n \rightarrow \mathbb{R}$, if $f_{\mathbf{u}}$ is its rate of increase at the point $\boldsymbol{\xi} = (x_1, x_2, \ldots, x_n)^T$ in the direction of the unit vector $\mathbf{u} = (u_1, u_2, \ldots, u_n)^T$, then the vector

$$\mathbf{v} = (u_1, u_2, \ldots, u_n, f_{\mathbf{u}})^T$$

lies in the tangent hyperplane to the hypersurface $y = f(x_1, x_2, \ldots, x_n)$ at the point $(x_1, x_2, \ldots, x_n, f(\boldsymbol{\xi}))^{\mathrm{T}}$. So $\mathbf{v}$ is orthogonal to the normal to the hypersurface $(f_{x_1}(\boldsymbol{\xi}), f_{x_2}(\boldsymbol{\xi}), \ldots, f_{x_n}(\boldsymbol{\xi}), -1)^{\mathrm{T}}$ at this point. That is, the scalar product of the two vectors is zero.

$$\langle (u_1, u_2, \ldots, u_n, f_{\mathbf{u}})^{\mathrm{T}}, (f_{x_1}(\boldsymbol{\xi}), f_{x_2}(\boldsymbol{\xi}), \ldots, f_{x_n}(\boldsymbol{\xi}), -1)^{\mathrm{T}} \rangle = 0$$

or

$$u_1 f_{x_1}(\boldsymbol{\xi}) + u_2 f_{x_2}(\boldsymbol{\xi}) + \cdots + u_n f_{x_n}(\boldsymbol{\xi}) - f_{\mathbf{u}} = 0$$

Hence

$$f_{\mathbf{u}} = \langle (u_1, u_2, \ldots, u_n)^{\mathrm{T}}, (f_{x_1}(\boldsymbol{\xi}), f_{x_2}(\boldsymbol{\xi}), \ldots, f_{x_n}(\boldsymbol{\xi}))^{\mathrm{T}} \rangle$$

$$= \langle \nabla f(\boldsymbol{\xi}), \mathbf{u} \rangle$$

where the scalar product is in $\mathbb{R}^n$.

This result can also be derived using the chain rule for functions of several variables, which will be introduced in Chapter 5.

In particular, the maximum rate of increase at the point $\boldsymbol{\xi}$ is $\|\nabla f(\boldsymbol{\xi})\|$ in the direction of $\nabla f(\boldsymbol{\xi})$ and its maximum rate of decrease is $-\|\nabla f(\boldsymbol{\xi})\|$ in the direction $-\nabla f(\boldsymbol{\xi})$.

Example 17

The derivative of the function $f(x, y, z) = x^2 + y^2 + z^2$ in the direction $(1, 1, 1)^{\mathrm{T}}$ at the point $(3, -1, 1)^{\mathrm{T}}$ is

$$\left\langle \nabla f, \left(\frac{1}{\sqrt{3}}, \frac{1}{\sqrt{3}}, \frac{1}{\sqrt{3}} \right)^{\mathrm{T}} \right\rangle$$

$$= \left\langle (6, -2, 2)^{\mathrm{T}}, \left(\frac{1}{\sqrt{3}}, \frac{1}{\sqrt{3}}, \frac{1}{\sqrt{3}} \right)^{\mathrm{T}} \right\rangle = 2\sqrt{3}$$

The maximum rate of increase of f at this point is

$$\|\nabla f\| = \|(6, -2, 2)^{\mathrm{T}}\| = 2\sqrt{11}$$

in the direction $(6, -2, 2)^{\mathrm{T}}$.

3.9 Exercises

The functions $f \colon \mathbb{R}^3 \to \mathbb{R}$ are defined in Exercises 1 to 6.

1.* Let $f(x, y, z) = x^2 + yz$.

    (i)   Find

$$\frac{\partial f}{\partial x}, \quad \frac{\partial f}{\partial y} \quad \text{and} \quad \frac{\partial f}{\partial z}$$

(ii) Find the tangent hyperplane to the hypersurface in $\mathbb{R}^4$, $u = x^2 + yz$ where $(x, y, z)^T = (1, 0, 2)^T$.

(iii) Find the normal and the tangent plane to the contour $1 = x^2 + yz$ at $(1, 0, 2)^T$.

(iv) What is the rate of increase of f in the direction $(2, -1, 2)^T$ at the point $(1, 0, 2)^T$?

2. Let $f(x, y, z) = xy^2 z^3$.

(i) Find the gradient of f at the point $(x, y, z)^T$.

(ii) Find the tangent hyperplane to the hypersurface $u = xy^2 z^3$ where $(x, y, z)^T = (1, 1, 1)^T$.

(iii) Find the normal and the tangent plane to the contour $1 = xy^2 z^3$ at $(1, 1, 1)^T$.

(iv) What is the rate of increase of f in the direction $(1, 2, 3)^T$ at the point $(1, 1, 1)^T$?

3. Let $f(x, y, z) = \sin(\pi + x + yz^2)$.

(i) Find the gradient of f at the point $(x, y, z)^T$.

(ii) Find the tangent hyperplane to the hypersurface

$$u = \sin(\pi + x + yz^2) \quad \text{where} \quad (x, y, z)^T = (0, 0, 0)^T$$

(iii) Find the normal and the tangent plane to the contour

$$\sin(\pi + x + yz^2) = 0 \quad \text{at} \quad (0, 0, 0)^T$$

(iv) What is the rate of increase of f in the direction $(2,1,2)^T$ at the point $(0,0,0)^T$?

4.* Let $f(x, y, z) = y^2 \sin(x^2 z)$.

(i) Find the gradient of f at the point $(x, y, z)^T$.

(ii) Find the tangent hyperplane to the hypersurface $u = y^2 \sin(x^2 z)$ where $(x, y, z, u)^T = (1, 2, \pi, 0)^T$.

(iii) Find the normal and the tangent plane to the contour

$$y^2 \sin(x^2 z) = 0 \quad \text{at} \quad (1, 2, \pi)^T$$

(iv) What is the rate of increase of f in the direction $(-3, 2, 6\pi)^T$ at the point $(1, 2, \pi)^T$? Is this the minimum rate? Justify your answer.

5.* Let $f(x, y, z) = \ln(x^2 + y^2 + z^2)$.

(i) Find the gradient of f at the point $(x, y, z)^T$.

(ii) Find the tangent hyperplane to the hypersurface

$$u = \ln(x^2 + y^2 + z^2)$$

where $(x, y, z, u)^T = (1/\sqrt{3}, 1/\sqrt{3}, 1/\sqrt{3}, 0)^T$.

(iii) Find the normal and the tangent plane to the contour

$$\ln(x^2 + y^z + z^2) = 0$$

at $(1/\sqrt{3}, 1/\sqrt{3}, 1/\sqrt{3})^{\mathrm{T}}$.

(iv) Find the equation of the surface on which lie all the points $(x, y, z)^{\mathrm{T}}$ at which the rate of increase of f in the direction $(x, y, z)^{\mathrm{T}}$ is equal to 1.

6.   Let $f(x, y, z) = \ln(xy + z)$.

    (i)   Find the gradient of f at the point $(x, y, z)^{\mathrm{T}}$.

    (ii)   Find the tangent hyperplane to the hypersurface

$$u = \ln(xy + z) \text{ where } (x, y, z, u)^{\mathrm{T}} = (1, 1, 0, 0)^{\mathrm{T}}.$$

    (iii)   Find the normal and the tangent plane at $(1, 1, 0)^{\mathrm{T}}$ to the contour

$$\ln(xy + z) = 0.$$

    (iv)   Prove that all the points $(x, y, z)^{\mathrm{T}}$ at which the rate of increase of f in the direction $(x/2, y/2, z)^{\mathrm{T}}$ is equal to 2 lie on the surface with equation

$$x^2 + y^2 + 4z^2 = 1.$$

7.*   At a certain factory the joint cost function for producing quantities x, y, z of three goods is given by

$$C(x, y) = 120x^2 + 120xy + 200y^2 + 60xz + 80z^2$$

Currently the quantities, in hundreds, of the three goods produced are 12, 15 and 20. If the manufacturer wishes to cut down on production to reduce costs, find the ratio in which he should cut production of the three goods to make the greatest savings.

8.   (i)   Find a normal vector to the surface

$$z = x^2 - 6xy + y^2$$

        and a normal vector to the sphere

$$g(x, y, z) = x^2 + y^2 + z^2 - 6y - 3z + 1 = 0$$

        at the point $(a, b, c)^{\mathrm{T}}$.

    (ii)   Verify that the point $(1, 0, 1)^{\mathrm{T}}$ lies on both surfaces and deduce from (i) that the surfaces touch tangentially at this point.
Write down the equation of their common tangent plane.

    (iii)   Write down, also, the equation of the line that is normal to both surfaces at this point.

(iv) Using the normal line, or otherwise, find the point on the sphere at which the tangent plane is parallel to the above common tangent plane. Show that this second tangent plane has equation

$$2x - 6y - z + 40 = 0$$

9.* (i) Find a normal vector to the surface

$$z = 3x^2 + 2xy + 3y^2$$

and the gradient vector to the ellipsoid

$$x^2 + y^2 + 2z^2 = 34$$

at the point $(x_0, y_0, z_0)^T$.

(ii) Deduce that the surfaces always intersect orthogonally.

(iii) Show that $(1, -1, 4)^T$ is one point of intersection of the two surfaces.

(iv) Find the tangent plane to the ellipsoid at $(1, -1, 4)^T$.

(v) Using symmetry, or otherwise, write down another point on the ellipsoid at which the tangent plane is parallel to the tangent plane at $(1, -1, 4)^T$.

10.* Functions $f, g : \mathbb{R}^3 \to \mathbb{R}$ are defined by

$$f(x, y, z) = x^2 + 4y^2 + 2z^2$$
$$g(x, y, z) = x^2 + y^2 - 2z^2$$

(i) Find the gradients of f and g at the point $(x, y, z)^T$.

(ii) Show that the point $(3, -2, 1)^T$ lies on both of the following level surfaces and hence on their curve of intersection:

$$f(x, y, z) = 27 \quad \text{and} \quad g(x, y, z) = 11$$

Describe these surfaces, sketching their intersections with the coordinate planes.

(iii) Find tangent planes and normal vectors to both surfaces at $(3, -2, 1)^T$. Do the surfaces touch tangentially at this point?

(iv) Use the normal vectors found in (iii) to find the direction of the tangent line to the curve of intersection of the level surfaces at $(3, -2, 1)^T$. Then write down equations for this tangent line.

(v) Find the rate of increase of f at $(3, -2, 1)^T$ in the direction $(2, 1, 1)^T$. What does your answer imply?

11.*♣ Suppose that the temperature distribution in space is given by

$$\tau(x, y, z) = 3x^2 + yz - 4xz + z^2$$

If an insect flies along the curve of intersection of the two surfaces given in Exercise 10, find the rate of change in temperature that it experiences as it flies through the point $(x, y, z)^T = (3, -2, 1)^T$.

12. Functions $f, g : \mathbb{R}^3 \to \mathbb{R}$ are defined by

$$f(x, y, z) = 3x^2 + 2y^2 + z^2 - 9$$

and

$$g(x, y, z) = x^2 + y^2 + z^2 - 8x - 6y - 8z + 24.$$

(i) Find the equation of the tangent plane to the ellipsoid

$$3x^2 + 2y^2 + z^2 = 9$$

at the point $(x, y, z)^{\mathrm{T}} = (a, b, c)^{\mathrm{T}}$.

(ii) Find the equation of the tangent plane to the sphere

$$x^2 + y^2 + z^2 - 8x - 6y - 8z + 24 = 0$$

at the point $(x, y, z)^{\mathrm{T}} = (\alpha, \beta, \gamma)^{\mathrm{T}}$.

(iii)* Given that the surfaces touch tangentially at the point $(x, y, z)^{\mathrm{T} } = (1, b, c)^{\mathrm{T}}$, find the values of b and c.

(iv)* The ellipsoid

$$3x^2 + 2y^2 + z^2 = 9$$

intersects the surface

$$-3x^2 + dy^2 + ez^2 = 4$$

orthogonally at the point $(x, y, z)^{\mathrm{T}} = (-1, -1, 2)^{\mathrm{T}}$. Find the values of d and e and describe the surface.

(v)* Write down the direction of maximum increase of f at $(x, y, z)^{\mathrm{T}} = (-1, -1, 2)^{\mathrm{T}}$ and also the maximum rate of increase.

(vi)* Calculate the value of the directional derivative of f in the direction $(1, 2, 2)^{\mathrm{T}}$ at the point $(x, y, z)^{\mathrm{T}} = (-1, -1, 2)^{\mathrm{T}}$.

3.10 Applications (optional)

3.10.1 Indifference curves

It is often convenient to describe the preferences of a consumer over a set of commodity bundles (§1.12.1) with a real valued utility function u. We then interpret

$$u(\mathbf{x}) < u(\mathbf{y})$$

as meaning that commodity bundle $\mathbf{y}$ is preferred to commodity bundle $\mathbf{x}$.

The contours of such a utility function are called **indifference curves** because the consumer is indifferent between any two bundles on the same contour.

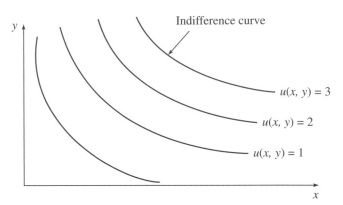

If the utility function is differentiable, its gradient indicates the direction in which utility increases fastest – i.e. the direction in which the consumer would most like to go. However, the consumer will be happy to move in any direction $\mathbf{v}$ for which $\langle \nabla u, \mathbf{v} \rangle \geq 0$.

In our diagram, this means that points above an indifference curve are preferred to points on an indifference curve.

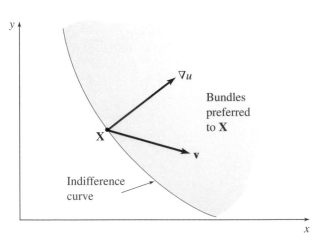

It is not unusual for a consumer to be constrained in his choice of a commodity bundle by the amount c of money in his possession. If the relevant price vector is $\mathbf{p}$, he will only be able to purchase those bundles $\mathbf{x}$ for which

$$\langle \mathbf{p}, \mathbf{x} \rangle \leq c$$

(see §1.12.3). We call this inequality the consumer's *budget constraint* and the set of **x** which satisfies this constraint is the consumer's budget set. What **X** in the budget set maximises the consumer's utility?

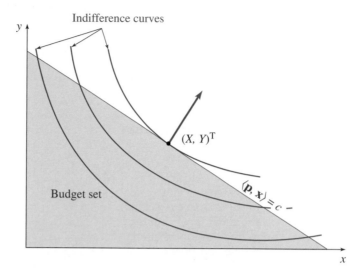

The only point in the budget set where the consumer cannot improve his position without going outside the budget set is $(X, Y)^{\mathrm{T}}$. Observe that, at this point, the gradient ∇u points in the same direction as the normal **p** to the hyperplane $\langle \mathbf{p}, \mathbf{x} \rangle = c$. This means that, for some scalar k,

$$\nabla u = k\mathbf{p}$$

In the two dimensional case, this can be written as

$$\left.\begin{array}{l} u_x(X, Y) = kp_1 \\ u_y(X, Y) = kp_2 \end{array}\right\}$$

Eliminating k, we obtain

$$\frac{u_x(X, Y)}{u_x(X, Y)} = \frac{p_1}{p_2}$$

In the special case of a utility Cobb–Douglas function when $u(x, y) = Ax^\alpha y^\beta$, this equation reduces to

$$\frac{\alpha X^{\alpha-1} Y^\beta}{\beta X^\alpha Y^{\beta-1}} = \frac{p_1}{p_2}$$

i.e.

$$p_2 Y = \frac{\beta}{\alpha} p_1 X$$

We also have that $(X, Y)^{\mathrm{T}}$ lies on $\langle \mathbf{p}, \mathbf{x} \rangle = c$ – i.e. $p_1 x + p_2 y = c$. Thus

$$c = p_1 X + p_2 Y = p_1 X + \frac{\beta}{\alpha} p_1 X$$

and solving for X and Y we obtain

$$\left.\begin{array}{l} X = \dfrac{\alpha c}{(\alpha + \beta)p_1} \\[3mm] Y = \dfrac{\beta c}{(\alpha + \beta)p_2} \end{array}\right\}$$

3.10.2 Profit maximisation

Suppose that a producer can costlessly produce any bundle in the region shaded. If the producer is a profit maximiser, he will choose the **X** in the production set which maximises

$$u(\mathbf{x}) = \langle \mathbf{p}, \mathbf{x} \rangle = c.$$

The point **X** will therefore be the point at which $\langle \mathbf{p}, \mathbf{x} \rangle$ is tangent to the curve bounding the production set.

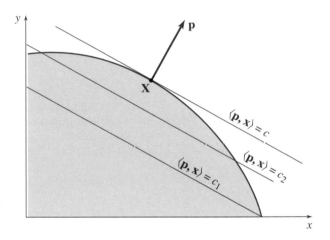

3.10.3 Contract curve

Robinson Crusoe and Man Friday wish to trade in wheat and fish. Robinson Crusoe begins with a commodity bundle $(W, 0)^{\mathrm{T}}$ representing a quantity W of wheat and no fish. Man Friday begins with a commodity bundle $(0, F)^{\mathrm{T}}$ representing a quantity F of fish and no wheat. We call these bundles their respective endowments.

We denote Robinson's utility function by u and Friday's by v. If the result of the trading is that Robinson receives commodity bundle $(w, f)^{\mathrm{T}}$, then Friday will receive $(W - w, F - f)^{\mathrm{T}}$. From the trade in which Robinson receives $(w, f)^{\mathrm{T}}$, it follows that Robinson will derive utility $u(w, f)$ and Friday will derive utility $V(w, f) = v(W - w, F - f)$.

The diagram drawn below is called the **Edgeworth box**. A point $(w, f)^{\mathrm{T}}$ in this box represents a trade in which Robinson receives the bundle $(w, f)^{\mathrm{T}}$ and Friday receives $(W - w, F - f)^{\mathrm{T}}$. The red curves represent Robinson's indifference curves (i.e. $u(w, f) = $ constant) and the blue curves represent Friday's indifference curves (i.e. $V(w, f) = $ constant).

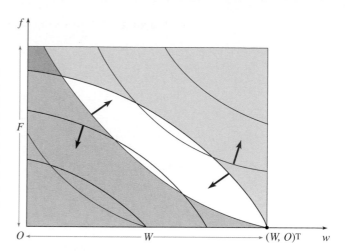

The points in the shaded regions represent trades on which it is impossible that they will agree because one or the other receives less than the utility he enjoys at the 'no trade' point.

'no trade' point, Robinson receives $(W, O)^T$
and Friday receives $(O, F)^T$

The traders will also not agree on the trades P or P' in the accompanying diagram because *both* traders prefer the trade Q.

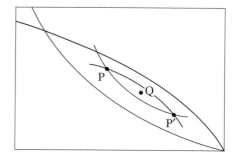

The only trades not ruled out by these considerations are those in the diagram. The curve on which these lie is called the *contract curve*.

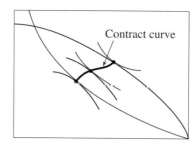

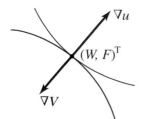

At points $(W, F)^T$ on the contract curve, the indifference curves touch and their normals point in opposite directions. Thus, if ∇u and ∇V are evaluated at $(W, F)^T$, then

$$\nabla V = -\alpha \nabla u$$

for some positive scalar α. We can use this fact to work out the equation of the contract curve. Consider, for example, the case when

$$\left.\begin{array}{l} u(w, f) = wf^2 \\ \\ v(w, f) = w^2 f \end{array}\right\}$$

and

Let $W = 2$ and $F = 1$. Then

$$V(w, f) = v(2 - w, 1 - f) = (2 - w)^2(1 - f)$$

123

We have

$$\nabla u = (f^2, 2wf)^{\mathrm{T}}$$
$$\nabla V = (-2(2-w)(1-f), -(2-w)^2)^{\mathrm{T}}$$

Since $\nabla V = -\alpha \nabla u$ at the point $(W, F)^{\mathrm{T}}$, we obtain

$$\left. \begin{array}{r} -2(2-W)(1-F) = -\alpha F^2 \\ -(2-W)^2 = -\alpha 2WF \end{array} \right\}$$

Hence

$$\frac{2(1-F)}{(2-W)} = \frac{F}{2W}$$
$$4W - 4WF = 2F - FW$$

and so the contract curve is a segment of the curve with equation

$$3wf - 4w + 2f = 0.$$

Four   Stationary points

In this chapter we consider the use of first order partial derivatives in the first stage of optimisation. The one variable case (studied in §4.1, 4.2 and §4.3) is worth some careful attention since the discussion makes it clear that there is more to optimisation than simply setting derivatives equal to zero.

In the one variable case the study realises its conclusion with the optimisation of functions, including optimisation under constraints. In the case of several variables, this must be deferred till vector functions have been introduced in Chapter 5. These enable Taylor's theorem to be extended to functions of several variables, paving the way for the optimisation of such functions in Chapter 6.

4.1   Stationary points for functions of one variable$^\diamond$

Consider the Taylor expansion of a function $f\colon \mathbb{R} \to \mathbb{R}$ about a point $x = \xi$. For points x sufficiently close to ξ, $|x-\xi|$ is small, so powers of $(x-\xi)$ get progressively smaller in magnitude. Hence the third and higher order terms in the expansion are negligible compared with the quadratic term

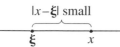

$$\frac{(x - \xi)^2}{2!} f''(\xi)$$

We may then approximate the function by its quadratic Taylor polynomial

$$f(x) \approx f(\xi) + (x - \xi)f'(\xi) + \frac{(x - \xi)^2}{2} f''(\xi) \tag{1}$$

A **stationary point** of f is a point $x = \xi$, for which $f'(\xi) = 0$. Therefore at a stationary point $x = \xi$, the tangent line

$$y = f(\xi) + f'(\xi)(x - \xi) \tag{2}$$

becomes the horizontal line $y = f(\xi)$.

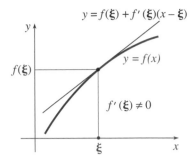
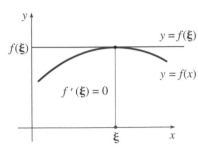

The word 'stationary' arises from the fact that, if a tiny ball bearing were to be balanced on the graph of $y = f(x)$ at any point where the tangent line is horizontal, then the ball bearing would not move. At any other point on the graph it would begin to roll down.

Also when $f'(\xi) = 0$

$$f(x) \approx f(\xi) + \frac{(x - \xi)^2}{2} f''(\xi)$$

The sign of $f(x) - f(\xi)$ is therefore the same as that of $f''(\xi)$. This plays a crucial part in the shape of the function.

(i) If $f''(\xi) > 0$, $f(x) > f(\xi)$ for all x near ξ, and so ξ is a **local minimum**.

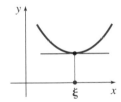

(ii) If $f''(\xi) < 0$, $f(x) < f(\xi)$ for all x near ξ, and so ξ is a **local maximum**.

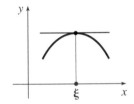

When $f''(x) > 0$ on an interval, we say f is **convex** on the interval. Comparing (1) and (2) we can deduce that the graph of a convex function lies above all its tangent lines.

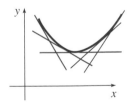

When $f''(x) < 0$ on an interval, we say f is **concave** on the interval. The graph of a concave function lies below all its tangent lines.

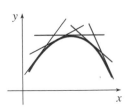

If $f(x)$ changes from convex to concave or vice versa at ξ then ξ is called a **point of inflection**.

(iii) Hence, if $f''(\xi) = 0$ and $f''(x)$ changes sign at ξ, then ξ is a point of inflection. The tangent line at a point of inflection ξ cuts across the graph of the function.

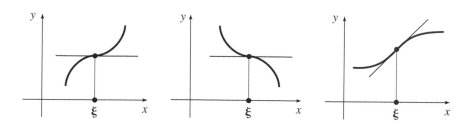

Observe that ξ need not be a stationary point to be a point of inflection. Points of inflection are a very common feature of curves.

Also, if $f''(\xi) = 0$ but $f''(x)$ does not change sign at ξ then ξ is not a point of inflection. For example, if $f(x) = x^4$, then $f''(0) = 0$, but f has no point of inflection at ξ.

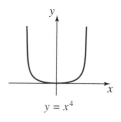

$y = x^4$

To summarise:

A stationary point ξ of a twice differentiable function f can be classified as a

(i) local maximum if $f''(\xi) < 0$,

(ii) local minimum if $f''(\xi) > 0$, or a

(iii) point of inflection if $f''(\xi) = 0$ and $f''(x)$ changes sign at ξ;

ξ is a point of inflection if the conditions in (iii) are satisfied even when it is not a stationary point. The function $y = x^4$ illustrates that condition (ii) does not necessarily hold for a local minimum. A similar remark applies to condition (i).

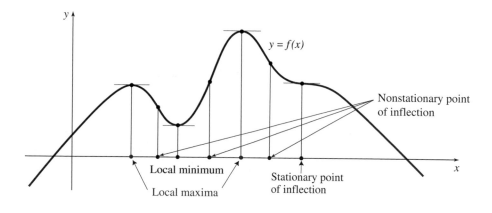

4.2 Optimisation

A common problem is to find a point ξ at which $f(x)$ achieves its maximum value – i.e.

$$f(\xi) = \max_x f(x)$$

Such a point ξ is called a **global maximum** for f. A function need not have a global maximum. However, if f does have a global maximum and f is differentiable, then the global maximum must be one of the *local* maxima.

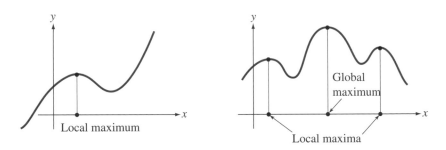

A helpful way to start finding a global maximum of a function is therefore to find the stationary points of the function. Amongst these stationary points are the local maxima and hence the global maximum, when this exists.

Assuming that a global maximum exists, how does one decide which of the stationary points it is? A simple minded, but not to be despised, method is to evaluate $f(x)$ at each stationary point and to see at which of these points $f(x)$ is largest. One can often avoid some or all of these calculations by sketching a graph of $y = f(x)$. Amongst other things, it helps avoid the error of identifying a global maximum when none exists.

Clearly, if a differentiable function of one variable has a *unique* stationary point which is a local maximum, then we are forced to conclude that this must be the global maximum of the function.

We shall see later that this is not necessarily the case for functions of more than one variable.

It is vital to realise that our reasoning relies on the condition of differentiability that we had imposed on the function. The following figures show that a *nondifferentiable* function may attain a maximum at an interior nonstationary point.

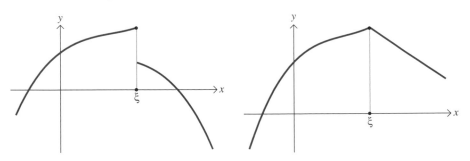

All the above remarks apply equally well to **global minima**.

Example 1

Let $f: \mathbb{R} \to \mathbb{R}$ be defined by

$$f(x) = 2 + 4x^3 - 6x^2 + 3x$$

Then

$$f'(x) = 12x^2 - 12x + 3 = 3(4x^2 - 4x + 1) = 3(2x - 1)^2$$

The stationary points are therefore found by solving the equation

$$3(2x - 1)^2 = 0$$

i.e.

$$x = \frac{1}{2}$$

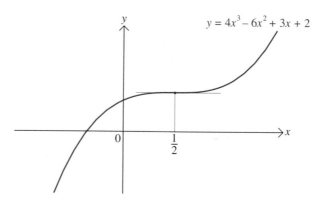

$$f''\left(\tfrac{1}{2}\right) = 0 \text{ and } f''(x) \text{ changes sign at } x = \frac{1}{2}.$$
Hence $x = 1/2$ is a point of inflection.

Note that this function has *no* global maximum and *no* global minimum.

Example 2

Let $f: \mathbb{R} \to \mathbb{R}$ be defined by

$$f(x) = x^3 - 3x^2 + 2x$$

Then

$$f'(x) = 3x^2 - 6x + 2$$

The stationary points are therefore found by solving the equation

$$3x^2 - 6x + 2 = 0$$

i.e.

$$x = \frac{6 \pm \sqrt{(36 - 24)}}{6} = 1 \pm \frac{1}{6}\sqrt{(12)}$$

$$x = 1 \pm \frac{1}{\sqrt{3}}$$

Because $x^3 - 3x^2 + 2x = x(x^2 - 3x + 2) = x(x-1)(x-2)$, it is fairly easy to sketch the graph of $y = f(x)$. From this graph it is apparent that $\{1 - (1/\sqrt{3})\}$ is a *local maximum* and that $\{1 + (1/\sqrt{3})\}$ is a *local minimum*.

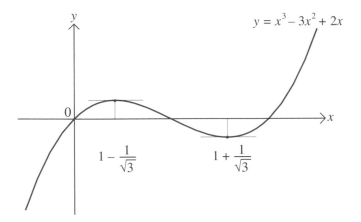

It is important to note that this function has *no* global maximum. There is *no* largest value of $f(x)$. The fact that a function has just one *local* maximum therefore does *not* guarantee that this local maximum is a global maximum, since this is not the only stationary point. Precisely the same remarks apply in the case of minima.

4.3 Constrained optimisation

If a differentiable function $f: \mathbb{R} \to \mathbb{R}$ has a global maximum, it attains this maximum at a stationary point.

In many cases it is required to find the maximum of $f(x)$ *over a restricted range of values of x*; that is x must belong to an interval or union of intervals. So x must satisfy constraints such as

$$a \leq x \leq b \text{ or } a \leq x \text{ or } x \leq b$$

where any of the inequalities may be strict.

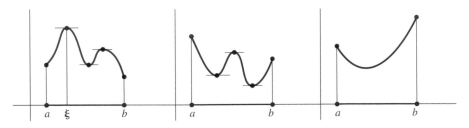

For example, suppose that $a \leq x \leq b$. In this case the maximum must be attained at an interior point ξ, which must be a stationary point, or at one of the boundary points a or b. Hence the maximum can be found by evaluating $f(a)$, $f(b)$ and $f(\xi)$ for each local maximum ξ satisfying $a < \xi < b$. The maximum of f will be the largest of these numbers. Analogous results hold for a constrained minimum.

Example 3

A producer is assumed to seek to maximise profit. If a quantity x of his product is sold at price p (per unit quantity) his *profit* is given by

$$\pi(x) = R(x) - C(x)$$

where $R(x) = px$ is the *revenue* obtained from selling x and $C(x)$ is the *cost* of producing x.

Stationary points are obtained by solving the equation

$$\pi'(x) = R'(x) - C'(x) = 0$$

i.e.

$$R'(x) = C'(x)$$

Recalling that economists use the word 'marginal', to describe the derivative of a quantity, we obtain the 'economic law' that profit is maximised when

$$\text{marginal revenue} = \text{marginal cost.}$$

Note, however, that a capitalist who blindly sets marginal revenue equal to marginal cost could equally well be *minimising* profit. The 'law' must therefore be applied with some caution.

Consider the case of 'perfect competition'. Here the producer is assumed to have such a small share of the market that variations in his output have a negligible effect on the price p of his product – i.e. p may be regarded as constant.

As an example, we take $p = \frac{4}{3}$ and $C(x) = x^3 - 3x^2 + 3x$. Thus

$$\pi(x) = \tfrac{4}{3}x - x^3 + 3x^2 - 3x$$

Then

$$\pi'(x) = -\tfrac{5}{3} - 3x^2 + 6x$$

This is actually a *constrained* optimisation problem because it is impossible to produce a *negative* value of x. We are therefore really interested in

$$\max_{x \geq 0} \pi(x)$$

where $\pi(x) = R(x) - C(x) = \frac{4}{3}x - x^3 + 3x^2 - 3x$. The stationary points are found by solving

$$9x^2 - 18x + 5 = 0$$

i.e.

$$x = 1 \pm 18^{-1}\sqrt{(18^2 - 9 \times 20)} = 1 \pm \tfrac{2}{3}$$

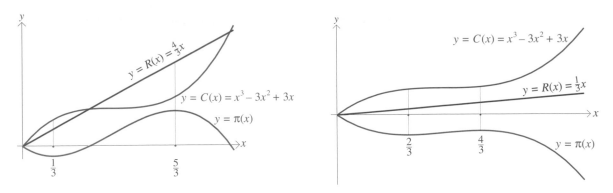

Cost, revenue and profit functions when $p = \frac{4}{3}$ and $p = \frac{1}{3}$

An examination of the diagram on the left shows that it is the larger of these stationary points (i.e. $x = \frac{5}{3}$) which yields the required maximum – i.e. profit is maximised at $x = \frac{5}{3}$. However, marginal revenue = marginal cost also at the point $x = \frac{1}{3}$ where profit has a local minimum.

Now consider the same example again but with $p = \frac{1}{3}$, illustrated in the diagram on the right.

The stationary points are found by solving

$$\pi'(x) - \tfrac{1}{3} - 3x^2 + 6x - 3 = 0$$

i.e.

$$x^2 - 2x + \left(1 - \tfrac{1}{9}\right) = 0$$

i.e.

$$x = 1 \pm \sqrt{(1 - 8/9)} = 1 \pm \tfrac{1}{3}$$

The stationary points are therefore $x = \frac{2}{3}$ and $x = \frac{4}{3}$. Observe that $\pi(\frac{2}{3}) = -\frac{20}{27}$ and $\pi(\frac{4}{3}) = -\frac{16}{27}$. But one *cannot* deduce that the maximum is achieved at $x = \frac{4}{3}$. We also need to note the fact that $\pi(0) = 0$. It follows that the maximum is achieved at $x = 0$ – i.e. no production at all is the optimal course of action.

4.4   The use of computer systems

The advances in the computational and graphic facilities of computers and calculators enable a very much wider range of functions to be studied than in the past. Computational systems are capable of providing derivatives of functions and solutions of algebraic equations that are either unobtainable or prohibitively laborious by elementary techniques. Also, graphs of functions or parts of functions can be instantly obtained. The systems, then, can enhance understanding of the fundamental concepts.

However, without an understanding of what the computer is doing, one can easily be led astray by the results it produces. One should therefore think of modern computational techniques as a companion to a standard mathematical analysis, rather than as a substitute.

We illustrate this point with a study of the function $f: \mathbb{R} \to \mathbb{R}$ defined by

$$f(x) = \frac{2x^4 - x^3 + 4x + 5}{x^3 - 11x^2 + 4x - 11}$$

This is a reasonably simple rational function, but the computation involved in carrying out the analysis would be considerable if attempted without a computer or calculator. Using the graphic facilities we can plot a graph of the function. It must be appreciated that we can plot only a finite part of the graph. In the absence of any information on the function, we plot a graph of f over an arbitrarily chosen interval, centred at the origin for convenience – say, $[-10, 10]$.

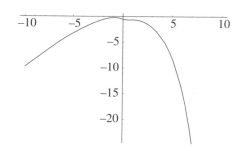

There is a subtle indication of critical points near the origin, but the picture appears far from complete. We need to include *all* critical points of f to observe its significant features.

See §13.6.

Also, the denominator is a third degree polynomial which must vanish for at least one point, because polynomials of odd degree must have at least one real root. So f has vertical asymptotes at one or more points, which must be included as well. Hence we use basic calculus methods to find the necessary interval for x.

We can use a computer system, first to locate vertical asymptotes by solving $x^3 - 11x^2 + x - 11 = 0$ which gives one real root $x \approx 10.7226$, and then to differentiate f, giving

$$f'(x) = \frac{2x^6 - 44x^5 + 35x^4 - 104x^3 + 62x^2 + 110x - 64}{(x^3 - 11x^2 + 4x - 11)^2}$$

The computer will solve the equation $f'(x) = 0$ yielding four real roots, the stationary points $x \approx -0.873\,123, 0.561\,651, 0.977\,985$ and 21.2892. A further differentiation gives

$$f''(x) = \frac{2(223x^6 - 174x^5 + 654x^4 - 571x^3 + 3723x^2 - 2310x - 349)}{(x^3 - 11x^2 + 4x - 11)^3}$$

Solving the equation $f''(x) = 0$ results in two real roots, $x \approx -0.125\,243$ and $0.756\,555$ which are possible points of inflection. The graph will indicate the nature of the stationary points and whether the above nonstationary points are inflection points.

An overall view of the graph is provided by a plot over an interval which contains all the points of interest, namely $[-25, 25]$.

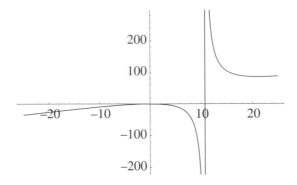

Unfortunately, because f takes very large values near the vertical asymptote, finer details are lost and we get only a very rough impression. To explore more refined aspects we need to consider neighbourhoods of critical points. For the stationary point at $x \approx 21.2892$, we consider the graph over the range $x \in [20, 23]$.

A local minimum at the point is clearly visible on this plot. The picture bears little resemblance to the previous picture near this point. This is because the total variation in the function here is just 0.4 compared with a total variation in function values of nearly 500 there.

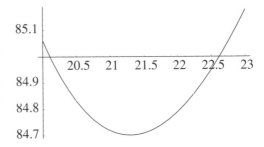

Next, to examine the critical points near the origin, look at the range $x \in [-4, 4]$.

Our new picture distinctly shows two local maxima near $-0.873\,123$ and $0.977\,985$, a local minimum near $0.561\,651$ and two points of inflection near $-0.125\,243$ and $0.756\,555$. Since the total variation in the function here is just 1.2 these subtle features are completely obscured in the overall plot.

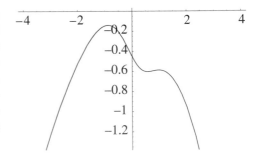

Finally, since the degree of the numerator exceeds that of the denominator, by long division or by computer algebra we obtain

$$y = 2x + 21 + \frac{223x^2 - 58x + 236}{x^3 - 11x^2 + 4x - 11}$$

showing that the graph has the slant asymptote $y = 2x + 21$. This is shown together with the graph in the final plot.

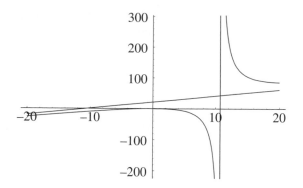

The previous discussion shows how a combination of analysis and computation can be used to tease out the properties of a function without back breaking labour.

There are definite limitations, however, to computer systems. A function like the one above belongs to the class of functions which are called **algebraic**. These are functions which can be obtained by combinations of power and root functions. Equations involving nonalgebraic or transcendental functions, like $\sin x = 0$, may have an infinite number of solutions which cannot feasibly be investigated computationally. However, in this case, given the rough location of the root, which can be ascertained from a graph, in most situations the systems provide individual numerical solutions to the degree of accuracy of the computer.

4.5 Exercises

1.* Find and classify the stationary points of the function $f : \mathbb{R} \to \mathbb{R}$ defined by

$$f(x) = 6 - x - x^2$$

Sketch the graph of the function and compute the following quantities:

(i) $\max\limits_{-3 \leq x \leq 2} f(x)$ (ii) $\min\limits_{-3 \leq x \leq 2} f(x)$

(iii) $\max\limits_{1 \leq x \leq 3} f(x)$ (iv) $\min\limits_{1 \leq x \leq 3} f(x)$

2. Let $f : \mathbb{R} \to \mathbb{R}$ be defined by

$$f(x) = x^3 - 3x^2 + 2x$$

Calculate the following quantities.

(i) $\max\limits_{0 \leq x \leq 1} f(x)$ (ii) $\min\limits_{0 \leq x \leq 1} f(x)$

(iii) $\max\limits_{0 \leq x \leq 2} f(x)$ (iv) $\min\limits_{0 \leq x \leq 2} f(x)$

(v) $\max\limits_{0 \leq x \leq 3} f(x)$ (vi) $\min\limits_{0 \leq x \leq 3} f(x)$

3. Find the domains of the following functions, investigate their behaviour and sketch their graphs:

$$\text{(i) } f(x) = \frac{x}{(x-4)^2} \quad \text{(ii)}^* \ f(x) = x\sqrt{3-x}$$

4. Find the stationary points of the functions $f: \mathbb{R} \to \mathbb{R}$ defined by

$$\text{(i)}^* \ f(x) = 5x^3 - 3x^5 \qquad \text{(ii) } f(x) = x(x+2)^3$$

$$\text{(iii)}^* \ f(x) = \frac{x^2 + x - 1}{x^2 + 1} \qquad \text{(iv) } f(x) = x^3 e^{-x^2}$$

Sketch the graphs and hence classify the stationary points.

5.* (i) Show that $\ln x - x$ has a global maximum and find the value of $x > 0$ which maximises it.

   (ii) Do the same for $\ln x - x^n$ where $n \in \mathbb{N}$.

6.* A quantity x of a product realises revenue $R(x) = px$ but costs $C(x) = x^3 - 3x^2 + 3x$ to produce. The profit $\pi(x)$ is given by $\pi(x) = R(x) - C(x)$. Prove that

   (i) If $0 \le p \le \frac{3}{4}$, then

   $$\max_{x \ge 0} \pi(x) = \pi(0) = 0$$

   (ii) If $p \ge \frac{3}{4}$, then

   $$\max_{x \ge 0} \pi(x) = \pi(\xi)$$

   where $\xi = 1 + \sqrt{p/3}$.

7. A quantity x of a product realises revenue $R(x) = px$ but costs $C(x)$ to produce. The profit $\pi(x)$ is given by $\pi(x) = R(x) - C(x)$. For each $p \ge 0$, discuss the problem of finding the value of $x \ge 0$ which maximises profit in the two cases

   (i) $C(x) = x^2$ (ii) $C(x) = \sqrt{x}$

8.* Find the local maxima and minima of the profit function $\pi : \mathbb{R} \to \mathbb{R}$ defined by

$$\pi(x) = x^3 - 24x^2 + 117x + 784$$

and sketch it. Find and sketch both the marginal profit function $\pi'(x)$ and the average profit function $A\pi(x) = \pi(x)/x$ and investigate the relationship between them.[†]

9. If p is the price per unit required to sell x units of a commodity, verify that each of the demand functions

$$(i)^* \quad x = 85e^{-6p} \qquad \text{and} \qquad (ii) \quad x = 90 - 2p$$

is invertible, and sketch the graphs of the functions and their inverses. Find the revenue functions $R(x) = px$ and the outputs x and prices p which maximise them.

The **price elasticity of demand** measures the percentage change in quantity associated with a percentage change in price. It is given by

$$\epsilon = \frac{\mathrm{d}x/x}{\mathrm{d}p/p} \quad \text{which can be written as} \quad \frac{\mathrm{d}x}{\mathrm{d}p}\frac{p}{x}$$

Show that $\epsilon = -1$ where the revenue function is maximised.

For the demand function (i), express the price elasticity of demand as a function of x and verify that $\epsilon = -1$ when the revenue function is maximised.

10. If the average cost function for a commodity x is

$$AC(x) = \frac{C(x)}{x} = x^2 - 8x + 157 + \frac{2}{x}$$

find the output x which minimises marginal costs.

11. If the average cost function for a commodity x is

$$AC(x) = \frac{C(x)}{x} = 3x + 12 + \frac{12}{x}$$

and its demand function is $x = 60 - 4p$, find the output x which maximises a monopolistic profit.

Also, find the output x which maximises profit in the cases when the following are introduced:

(i) a lump sum tax of 15

(ii) a tax of 5 per unit of output

(iii) a subsidy of 2 per unit of output.

12.* Suppose that an electricity generating company meets all demand made on it for the supply of electricity (i.e. no 'brown-outs') but is able to control demand by fixing the price. Suppose that demand x is related to price p by the formula $xp^2 = 1$. If the cost of meeting demand x is $C(x) = 2x^2$, what price maximises profit $\pi(x) = px - C(x)$ and what will be the demand at this price?

13. If the company need not meet all demand in the previous problem, we have to maximise $\pi(x)$ subject to $x \geq 0$, $p \geq 0$, $xp^2 \leq 1$. What is the result?

14.* Prove that a polynomial of degree 3

$$P_3(x) = ax^3 + bx^2 + cx + d \qquad (a, b, c, d \in \mathbb{R}, a \neq 0)$$

has between zero and two stationary points and at most one point of inflection. Find conditions on a, b, c, d for each number of stationary points. Sketch the shape of the curve in each case.

Deduce conditions on a, b, c, d for the polynomial to be always (i) increasing and (ii) decreasing.

15. Prove that a polynomial of degree n

$$P_n(x) = a_n x^n + a_{n-1} x^{n-1} + \cdots + a_0 \qquad (a_i \in \mathbb{R})$$

has at most $(n-1)$ stationary points and at most $(n-2)$ points of inflection.

16.* Prove that if n is even, then P_n has at least one stationary point but may not have any points of inflection and if n is odd, it may not have any stationary points but has at least one point of inflection.

(i) Sketch a graph of P_4 with no inflection point and a graph of P_5 with no stationary point.

(ii) Sketch a graph of P_4 with three stationary points and two inflection points and a graph of P_5 with four stationary points and three inflection points.

17. For each of the following polynomials, divide $\mathbb{R}$ into the intervals on which the polynomial is convex and the intervals on which the polynomial is concave:

(i)* $P_3(x) = x^3 - 5x^2 + 7x + 2$

(ii) $P_4(x) = x^4 + 5x^3 - 9x^2 + 2x - 3$

18. Use your knowledge of trigonometric functions to show that the following functions have an infinite number of stationary points:

(i)* $f(x) = x \sin x$ (ii) $f(x) = x^2 \sin x$ (iii) $f(x) = e^{-x} \sin x$

In each case, discuss the nature of the largest negative stationary point. Use a computer to graph each function and to locate the largest negative stationary point to approximately five decimal places.

19. With the help of a computer find the significant features of the function $f: \mathbb{R} \to \mathbb{R}$ defined by
$$f(x) = \frac{x(x-4)}{(x-1)^2(x-3)}$$
locating all the critical points and asymptotes to five decimal places. Obtain graph plots that display all these features.

4.6 Stationary points for functions of two variables

For functions $f: \mathbb{R}^2 \to \mathbb{R}$, $\boldsymbol{\xi}$ is a stationary point if and only if $z = f(x, y)$ has a *horizontal* tangent plane where $(x, y)^{\mathrm{T}} = \boldsymbol{\xi}$.

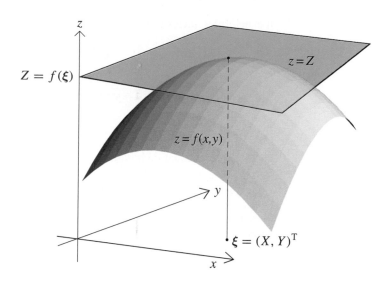

If $\boldsymbol{\xi} = (X, Y)^{\mathrm{T}}$ and $Z = f(X, Y)$, then we know from §3.3 that a differentiable function has tangent plane

$$z - Z = f_x(X, Y)(x - X) + f_y(X, Y)(y - Y)$$

at the point in question. For this to be horizontal, it must take the form $z = Z$. Thus the condition for $\boldsymbol{\xi} = (X, Y)^{\mathrm{T}}$ to be a stationary point is that

$$f_x(X, Y) = 0 \qquad \text{and} \qquad f_y(X, Y) = 0$$

To find the stationary points of a function $f: \mathbb{R}^2 \to \mathbb{R}$, one must therefore find *all* solutions of the simultaneous equations

$$\frac{\partial f}{\partial x} = 0 \qquad \text{and} \qquad \frac{\partial f}{\partial y} = 0.$$

Example 4

Let $f: \mathbb{R}^2 \to \mathbb{R}$ be defined by

$$f(x, y) = x^2 y + y^3 x - xy$$

Then

$$\frac{\partial f}{\partial x} = 2xy + y^3 - y \qquad \text{and} \qquad \frac{\partial f}{\partial y} = x^2 + 3y^2 x - x$$

The stationary points are therefore found by solving the *simultaneous* equations

$$\left.\begin{array}{c} 2xy + y^3 - y = 0 \\ x^2 + 3y^2 x - x = 0 \end{array}\right\}$$

It is helpful to factorise these equations as below:

$$\left.\begin{array}{c} y(2x + y^2 - 1) = 0 \\ x(x + 3y^2 - 1) = 0 \end{array}\right\}$$

It is then clear that the first equation holds if and only if

$$\text{(a) } y = 0 \quad or \quad \text{(b) } 2x + y^2 - 1 = 0$$

and the second equation holds if and only if

$$\text{(c) } x = 0 \quad or \quad \text{(d) } x + 3y^2 - 1 = 0$$

All solutions to the simultaneous equations are found by taking these latter equations in pairs – i.e.

$$\left.\begin{array}{ll} \text{(i)} & \text{(a) and (c)} \\ \text{(ii)} & \text{(a) and (d)} \\ \text{(iii)} & \text{(b) and (c)} \\ \text{(iv)} & \text{(b) and (d)} \end{array}\right\}$$

Notice how easy it would be to lose one or more of the solutions by being careless at this stage. A systematic approach is therefore essential.

(i)   (a) and (c)

$$y = 0 \quad \text{and} \quad x = 0$$

These equations have the unique solution $(0, 0)^{\mathrm{T}}$.

(ii) (a) and (d)

$$y = 0 \quad \text{and} \quad x + 3y^2 - 1 = 0$$

Substituting $y = 0$ in $x + 3y^2 - 1 = 0$ yields $x = 1$. The equations therefore have the unique solution $(1, 0)^{\mathrm{T}}$.

(iii) (b) and (c)

$$2x + y^2 - 1 = 0 \quad \text{and} \quad x = 0$$

Substituting $x = 0$ in $2x + y^2 - 1 = 0$ yields $y^2 = 1$ – i.e. $y = \pm 1$. The equations therefore have two solutions, namely, $(0, 1)^{\mathrm{T}}$ and $(0, -1)^{\mathrm{T}}$.

(iv) (b) and (d)

$$2x + y^2 - 1 = 0 \qquad \text{and} \qquad x + 3y^2 - 1 = 0$$

Multiply the first equation by 3 and then subtract the second equation. This yields $5x - 2 = 0$ – i.e. $x = \frac{2}{5}$. Substitute this result in the first equation. This gives $y^2 = \frac{1}{5}$ – i.e. $y = \pm 1/\sqrt{5}$. The equations therefore have two solutions, namely, $\left(\frac{2}{5}, 1/\sqrt{5}\right)^{\mathrm{T}}$ and $\left(\frac{2}{5}, -1/\sqrt{5}\right)^{\mathrm{T}}$.

The full list of stationary points is

$$(0,0)^{\mathrm{T}}, (1,0)^{\mathrm{T}}, (0,1)^{\mathrm{T}}, (0-1)^{\mathrm{T}}, \left(\frac{2}{5}, \frac{1}{\sqrt{5}}\right)^{\mathrm{T}}, \left(\frac{2}{5}, \frac{-1}{\sqrt{5}}\right)^{\mathrm{T}}$$

Examine the contour lines near the six stationary points on the contour plot below and compare with the surface plot. The third plot shows the surface, together with the horizontal plane $z = 0$, which is tangent to the surface at the first four points, each of which is a saddle point (see page 142).

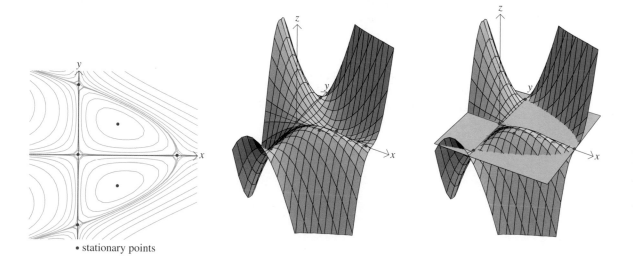

• stationary points

In general, the stationary points of 'well behaved' functions $f : \mathbb{R}^2 \to \mathbb{R}$ fall into three classes:

(i)   local maxima;

(ii) local minima;

(iii) saddle points.

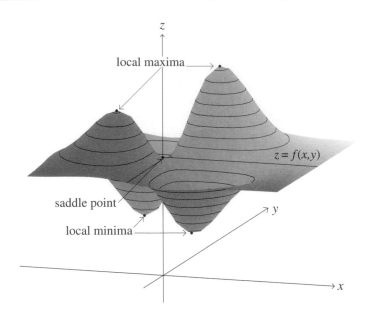

Observe again that the stationary points of f are those where a ball bearing can be balanced on the surface. Saddle points are so called because the surface looks like a saddle at such a point. It is important to note that saddle points are a very common feature of a surface $z = f(x, y)$.

In the contour map drawn below it is not hard to see that something peculiar is going on at the stationary points. As in the case of the preceding example, closed contours surround individual local maxima and minima and contours at the height of a saddle point intersect at the point.

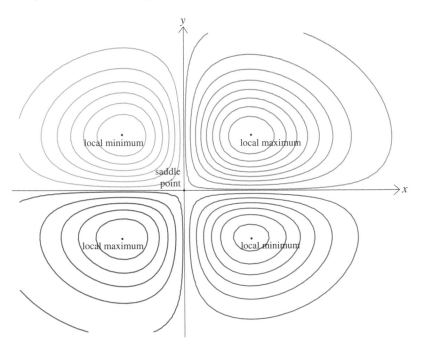

When there is an *infinite* set of stationary points lying along a line or a curve, these cannot be detected from a contour map. For instance, the parabolic cylinder $z = 3x^2$ of Example 3.4 has an infinite set of stationary points along the line $z = x = 0$, which is indistinguishable from the other contour lines.

4.7 Gradient and stationary points

A stationary point of a function $f : \mathbb{R}^2 \to \mathbb{R}$ occurs when

$$\frac{\partial f}{\partial x} = 0 \quad \text{and} \quad \frac{\partial f}{\partial y} = 0$$

An alternative way of expressing this is by saying that, at a stationary point,

$$\nabla f = \left(\frac{\partial f}{\partial x}, \frac{\partial f}{\partial y} \right)^{\mathrm{T}} = \mathbf{0}$$

We have seen in §3.6 that the derivative in a direction $\mathbf{u}$ where $\|\mathbf{u}\| = 1$ is $\langle \nabla f, \mathbf{u} \rangle$. It follows that, at a stationary point, the rate of change of f in *all* directions is zero.

We shall discuss how one goes about classifying the stationary points of a function $f : \mathbb{R}^2 \to \mathbb{R}$ in Chapter 6. For the moment, we shall only observe that, if one is seeking to find a global maximum or minimum of a function $f : \mathbb{R}^2 \to \mathbb{R}$, then it is a good idea to begin by finding the stationary points of f.

4.8 Stationary points for functions of more than two variables

Given a function $f : \mathbb{R}^n \to \mathbb{R}$, one cannot draw pictures of tangent hyperplanes when $n > 2$. But one can still define a stationary point to be a point at which the rate of change of f in all directions is zero – i.e. a point at which

$$\nabla f = \mathbf{0}$$

To find the stationary points of a function $f : \mathbb{R}^n \to \mathbb{R}$, one must therefore solve the *simultaneous* equations

$$\frac{\partial f}{\partial x_1} = 0 \quad \frac{\partial f}{\partial x_2} = 0 \quad \dots \quad \frac{\partial f}{\partial x_n} = 0$$

Example 5 _____

Let $f : \mathbb{R}^3 \to \mathbb{R}$ be defined by

$$f(x, y, z) = x^2 + yz + 3z^3 + y^4$$

Then

$$\frac{\partial f}{\partial x} = 2x \qquad \frac{\partial f}{\partial y} = z + 4y^3 \qquad \frac{\partial f}{\partial z} = y + 9z^2$$

The stationary points are therefore found by solving the simultaneous equations

$$\left. \begin{array}{r} 2x = 0 \\ z + 4y^3 = 0 \\ y + 9z^2 = 0 \end{array} \right\}$$

Substituting from the third equation into the second yields that

$$z + 4(-9z^2)^3 = 0$$

i.e.

$$z = 0 \quad \text{or} \quad 1 - 4 \times 9^3 z^5 = 0$$

$$z^5 = \frac{1}{4 \times 9^3} = \frac{1}{12 \times 3^5}$$

$$z = \left\{ \frac{1}{12} \right\}^{1/5} \frac{1}{3}$$

We therefore obtain two stationary points, namely $(0, 0, 0)^{\mathrm{T}}$ and

$$\left(0, -\left\{ \frac{1}{12} \right\}^{2/5}, \frac{1}{3} \left\{ \frac{1}{12} \right\}^{1/5} \right)^{\mathrm{T}}$$

The Mathematica plot below shows three contours of this function, together with the gradient vector $(2, 0, 0)^{\mathrm{T}}$ which is normal to the contour $f(\mathbf{x}) = 1$ at the point $(1, 0, 0)^{\mathrm{T}}$.

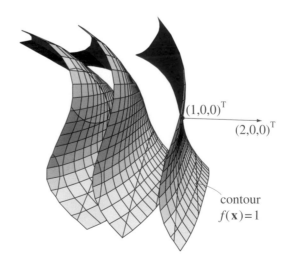

$(1,0,0)^{\mathrm{T}}$

$(2,0,0)^{\mathrm{T}}$

contour
$f(\mathbf{x})=1$

4.9 Exercises

1. Find all stationary points of the function $f: \mathbb{R}^2 \to \mathbb{R}$ defined by

(i)* $\quad f(x, y) = x^3 + y^3 - 3xy^2$

(ii) $\quad f(x, y) = 5x^2y - y^2 - 6x^2 - 4x^4$

(iii)* $\quad f(x, y) = 2x^2 - 8xy + y^2$

(iv) $\quad f(x, y) = x^3 - 2x^2 - 4x - y^2 - 6y$

(v)* $\quad f(x, y) = xy^2 + 2y - x$

(vi) $\quad f(x, y) = 8y^3 + x^4 - 2x^2$

(vii)* $\quad f(x, y) = 5xy^2 - 20x - 3y^2$

(viii) $\quad f(x, y) = 2x^2y + 2y - 5x^2$

(ix)* $\quad f(x, y) = 16x^4 - 8x^2 + y^3 - 3y$

(x) $\quad f(x, y) = \dfrac{4}{x} - \dfrac{2x}{y} + y$ where $x, y \neq 0$

(xi)* $\quad f(x, y) = 2y^2x + 3x^2y - 4xy$

(xii) $\quad f(x, y) = (4x^2 + y^2)e^{1 - x^2 - 4y^2}$

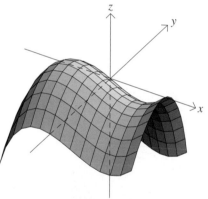

Surface plot (iv)

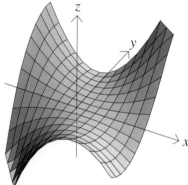

Surface plot (xi)

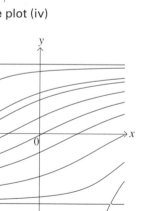

Contour plot (ii)

Contour plot (v)

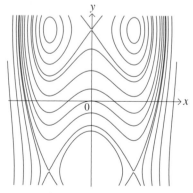

Contour plot (ix)

2. Investigate the following functions $f: \mathbb{R}_+^2 \to \mathbb{R}$ for stationary points:

(i)* $f(x, y) = Ax^\alpha y^\beta$ where A, α, β are constants and $0 < \alpha < 1, 0 < \beta < 1$

(ii)* $f(x, y) = [\alpha x^\beta + (1 - \alpha) y^\beta]^{\frac{1}{\beta}}$ where $0 < \alpha < 1$ and $\beta < 1$

(iii) $f(x, y) = \ln[(1 + x)^\alpha (1 + y)^\beta]$ where $\alpha, \beta \in \mathbb{N}$

3.* (i) If $\mathbf{x} \in \mathbb{R}^n$ explain why the functions $f(\mathbf{x})$ and $e^{f(\mathbf{x})}$ have the same stationary points.

(ii) If $f(\mathbf{x}) > 0$, explain why the functions $f(\mathbf{x})$, $\ln f(\mathbf{x})$ and $(f(\mathbf{x}))^\alpha$ where $\alpha \neq 0, \alpha \in \mathbb{Q}$ have the same stationary points.

(iii) In addition explain why the contours $f(\mathbf{x}) = c$ of the function $f(\mathbf{x})$ are the contours (of different values) of the functions $e^{f(\mathbf{x})}$, $\ln f(\mathbf{x})$ and $(f(\mathbf{x}))^\alpha$ where $\alpha \neq 0, \alpha \in \mathbb{Q}$.

(iv)♣ If $\mathbf{x} \in \mathbb{R}^n$, investigate whether the functions $f(\mathbf{x}), \sin(f(\mathbf{x}))$ and $\cos(f(\mathbf{x}))$ have the same stationary points.

Also, investigate whether the contours $f(\mathbf{x}) = c$ of the function $f(\mathbf{x})$ are the contours (of different values) of the functions $\sin(f(\mathbf{x}))$ and $\cos(f(\mathbf{x}))$.

4.*♣ Match the following functions (i)–(vi) with their contour plots (a)–(f) and their surface plots (g)–(l):

(i) $\cos(x^2 + y^2)$ (ii) $\exp(-x^2 - y^2)$ (iii) $\ln(1 + 4x^2 + y^2)$
(iv) $\sqrt{x^2 + 4y^2}$ (v) $\sin(x^2 - y^2)$ (vi) $(x^2 + 2y^2)^3$

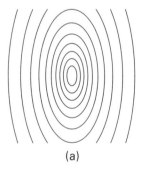

(a)

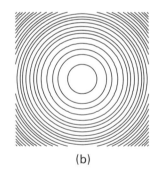

(b)

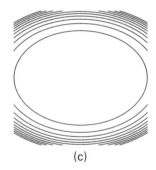

(c)

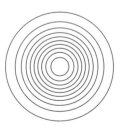

(d)

(e)

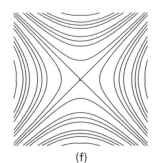

(f)

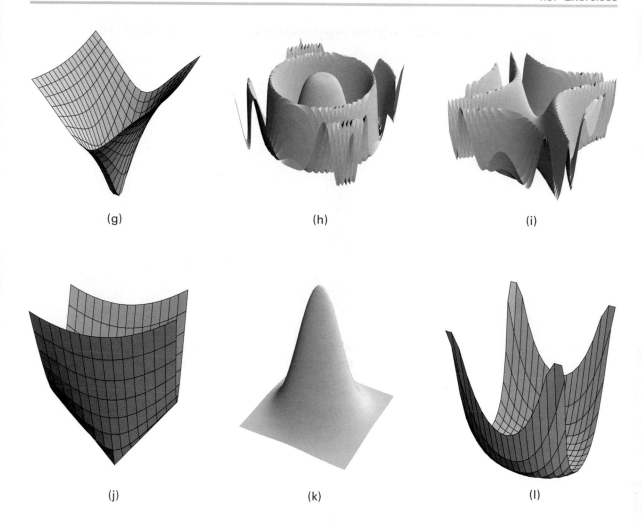

(g)

(h)

(i)

(j)

(k)

(l)

Use Exercise 3, but the case when $f(\mathbf{x}) = 0$ in (iv) must be treated separately. It would be helpful to consider vertical sections of the surface with the $(x, z)^{\mathrm{T}}$ and $(y, z)^{\mathrm{T}}$ planes.

5. Find all stationary points of the function $f \colon \mathbb{R}^3 \to \mathbb{R}$ defined by

   (i)* $f(x, y, z) = 4x^2 z - 2xy - 4x^2 - z^2 + y$

   (ii) $f(x, y, z) = 6xy + z^2 - 18x^2 - 4y^2 z$

   (iii) $f(x, y, z) = x^2 y + 2y^2 z - 4z - 2x$

   (iv)* $f(x, y, z) = 3(x^2 + y^2 + z^2) - (x + y + z)^2$

   (v) $f(x, y, z) = \ln((x^3 - 2y^3 - 3z^3 - 6xyz + 4)^{\frac{1}{2}})$

   where $(x^3 - 2y^3 - 3z^3 - 6xyz + 4) > 0$

6.   Find all stationary points of the function $f: \mathbb{R}^2 \to \mathbb{R}$ defined by

(i)* $f(x, y) = \exp(8x + 12y - x^2 - y^3)$

(ii) $f(x, y) = \dfrac{1}{x^2 + 9y^2 + 6xy + 75}$

(iii)* $f(x, y) = (6 + x^2 + y^2 - 2x + 4y)^3$

(iv) $f(x, y) = \ln(10 - 4x - 4y^2 + x^2 + y^4)$

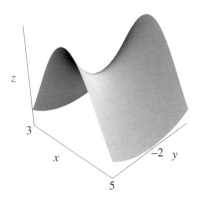 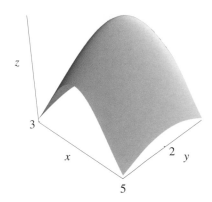

Two stationary points of (i) $f(x, y) = \exp(8x + 12y - x^2 - y^3)$.

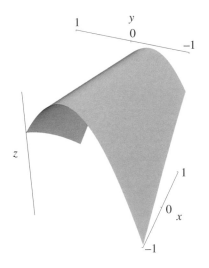 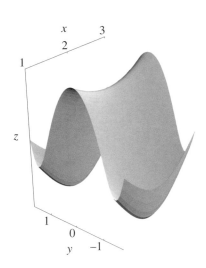

(ii) $f(x, y) = \dfrac{1}{x^2 + 9y^2 + 6xy + 75}$ (iv) $f(x, y) = \ln(10 - 4x - 4y^2 + x^2 + y^4)$

Five   **Vector functions**

In this chapter we introduce the idea of a vector function and examine the instances in which it can be visualised. We have seen in Chapter 3 how partial derivatives are used to generalise the idea of the derivative of a real valued function of one real variable to that of a total derivative of a function of several variables. We extend this further to the idea of the total derivative of a vector function. This object is defined to be a matrix, and an understanding of the material covered in this chapter requires a good knowledge of the matrix algebra surveyed in Chapter 1. (Note especially §1.10.) Although it is quite demanding at first to think systematically in terms of matrices and vectors, the payoff is considerable. The theory is elegant, the formulas are easily remembered, and the amount that has to be written down is very much less than would otherwise be necessary. Some effort with this chapter will therefore be amply repaid.

5.1   **Vector valued functions**

In this chapter we shall study functions

$$\mathbf{f}\colon \mathbb{R}^n \to \mathbb{R}^m$$

Such a function assigns to each $\mathbf{x} \in \mathbb{R}^n$ a unique vector $\mathbf{y} \in \mathbb{R}^m$. We write

$$\mathbf{y} = \mathbf{f}(\mathbf{x})$$

Suppose that

$$\mathbf{x} = \begin{pmatrix} x_1 \\ x_2 \\ \vdots \\ x_n \end{pmatrix} \quad \text{and} \quad \mathbf{y} = \begin{pmatrix} y_1 \\ y_2 \\ \vdots \\ y_m \end{pmatrix}$$

Then the vector function $\mathbf{f}$ assigns to each set of values of the n independent variables $x_1, x_2, \ldots, x_n$ a unique set of values of the m dependent variables $y_1, y_2, \ldots, y_m$. The vector equation $\mathbf{y} = \mathbf{f}(\mathbf{x})$ can be written out as a list of m *real* equations as below.

$$\left. \begin{array}{l} y_1 = f_1(x_1, x_2, \ldots, x_n) \\ y_2 = f_2(x_1, x_2, \ldots, x_n) \\ \quad \cdots \\ y_m = f_m(x_1, x_2, \ldots, x_n) \end{array} \right\}$$

A vector valued function can therefore be studied by looking at the *m real* valued functions $f_1, f_2, \ldots, f_m$. We shall say that $f_1, f_2, \ldots, f_m$ are the **component** functions of **f**.

We shall concentrate mainly on the case $\mathbf{f}: \mathbb{R}^2 \to \mathbb{R}^2$, which represents the mapping of planes. We denote a mapping of the $(x, y)^{\mathrm{T}}$ plane to the $(u, v)^{\mathrm{T}}$ plane by

$$\begin{pmatrix} u \\ v \end{pmatrix} = \mathbf{f}(x, y) = \begin{pmatrix} f_1(x, y) \\ f_2(x, y) \end{pmatrix}$$

There are four variables in total, x and y independent and u and v dependent; and since space is three dimensional, there is no point of course in trying to draw the *graph* of such a function. We illustrate the idea with a diagram.

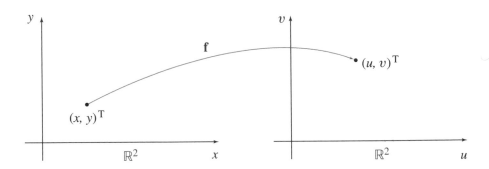

Example 1

The equations

$$\left. \begin{array}{l} u = x^2 + y^2 \\ v = x^2 - y^2 \end{array} \right\}$$

define a function $\mathbf{f}: \mathbb{R}^2 \to \mathbb{R}^2$. We may write

$$\begin{pmatrix} u \\ v \end{pmatrix} = \mathbf{f}(x, y) = \begin{pmatrix} x^2 + y^2 \\ x^2 - y^2 \end{pmatrix}$$

The component functions of **f** are defined by

$$\left. \begin{array}{l} u = f_1(x, y) = x^2 + y^2 \\ v = f_2(x, y) = x^2 - y^2 \end{array} \right\}$$

To illustrate a function $\mathbf{f}: \mathbb{R}^2 \to \mathbb{R}^2$ one can draw contour maps of the real valued component functions $f_1: \mathbb{R}^2 \to \mathbb{R}$ and $f_2: \mathbb{R}^2 \to \mathbb{R}$.

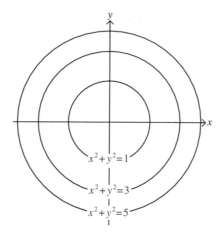

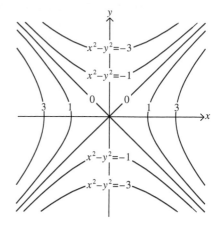

Contour map of $f_1 : \mathbb{R}^2 \to \mathbb{R}$ Contour map of $f_2 : \mathbb{R}^2 \to \mathbb{R}$

These contour maps can then be superimposed producing the pattern indicated in the diagram below on the left. The image of this pattern under the function $\mathbf{f} : \mathbb{R}^2 \to \mathbb{R}^2$ is then the grid indicated in the diagram on the right.

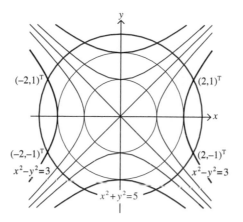

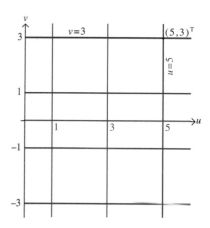

As the diagram indicates,

$$\begin{pmatrix} 5 \\ 3 \end{pmatrix} = \begin{pmatrix} 2^2 + (-1)^2 \\ 2^2 - (-1)^2 \end{pmatrix} = \mathbf{f}(2, -1)$$

Thus the function $\mathbf{f}$ assigns the point $(5, 3)^{\mathrm{T}}$ to the point $(2, -1)^{\mathrm{T}}$. It is also true, however, that

$$\begin{pmatrix} 5 \\ 3 \end{pmatrix} = \mathbf{f}(2, -1) = \mathbf{f}(2, 1) = \mathbf{f}(-2, 1) = \mathbf{f}(-2, -1)$$

Hence $(5, 3)^{\mathrm{T}}$ is assigned to all four of the points $(2, 1)^{\mathrm{T}}$, $(2, -1)^{\mathrm{T}}$, $(-2, 1)^{\mathrm{T}}$ and $(-2, -1)^{\mathrm{T}}$.

Example 2

(i) The equations

$$\left. \begin{array}{l} x = \cos t \\ y = \sin t \end{array} \right\}$$

define a function $\mathbf{f}: \mathbb{R}^1 \rightarrow \mathbb{R}^2$ which maps the independent variable t to the dependent variables x and y. We may write

$$\left(\begin{array}{c} x \\ y \end{array} \right) = \mathbf{f}(t) = \left(\begin{array}{c} f_1(t) \\ f_2(t) \end{array} \right) = \left(\begin{array}{c} \cos t \\ \sin t \end{array} \right)$$

In this instance, the graph of $\mathbf{f}$ may be plotted in three dimensional $(x, y, t)^{\mathrm{T}}$ space. We first plot the component functions of $\mathbf{f}$ in $(x, y, t)^{\mathrm{T}}$ space, separately and together.

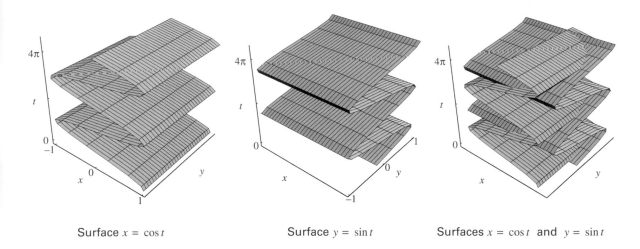

Surface $x = \cos t$ Surface $y = \sin t$ Surfaces $x = \cos t$ and $y = \sin t$

The two components intersect in a graph in space which is a circular helix, as shown in the third plot.

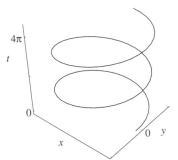

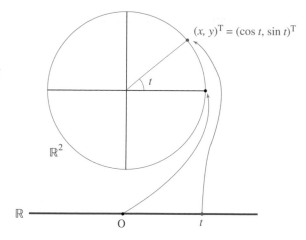

As t varies over the set of real numbers, the point

$$(x, y)^{\mathrm{T}} = (\cos t, \sin t)^{\mathrm{T}}$$

describes a circle in the $(x, y)^{\mathrm{T}}$ plane.

In a similar way, any function $\mathbf{g} : \mathbb{R}^1 \to \mathbb{R}^m$ defines a curve in $\mathbb{R}^m$.

5.2 Affine functions and flats

If M is an $m \times n$ matrix and $\mathbf{c}$ is an $m \times 1$ vector, then the function $\mathbf{A} : \mathbb{R}^n \to \mathbb{R}^m$ defined by

$$\mathbf{y} = \mathbf{A}(\mathbf{x}) = M\mathbf{x} + \mathbf{c}$$

[†] See §2.4.

is called an **affine** function.[†] Observe to begin with that the graph of this affine function is a set of pairs

$$(\mathbf{x}, \mathbf{y})^{\mathrm{T}} = (x_1, x_2, \ldots, x_n, y_1, y_2, \ldots, y_m)^{\mathrm{T}}$$

and hence defines a region in $(n + m)$ dimensional space, $\mathbb{R}^{m+n}$.

The equations

$$\mathbf{y} = M\mathbf{x} + \mathbf{c}$$

may be written as m component linear equations in the variables $x_1, x_2, \ldots, x_n, y_1, y_2, \ldots, y_m$. As each of these represents a hyperplane in $\mathbb{R}^{n+m}$, the graph of $\mathbf{y} = M\mathbf{x} + \mathbf{c}$ is simply a flat in $\mathbb{R}^{n+m}$.

If this flat passes through the point $(\boldsymbol{\xi}, \boldsymbol{\eta})^{\mathrm{T}}$, then its equation may be rewritten in the form

$$\mathbf{y} - \boldsymbol{\eta} = M(\mathbf{x} - \boldsymbol{\xi})$$

Example 3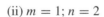

(i) $m = 1; n = 1$

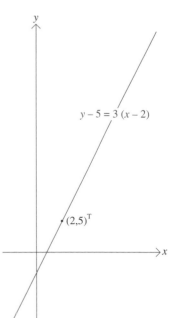

$y - 5 = 3 (x - 2)$

If $(\xi, \eta)^T = (2, 5)^T$, and M is the 1×1 matrix (3), the equation takes the form

$$y - 5 = 3(x - 2).$$

$(2,5)^T$

(ii) $m = 1; n = 2$

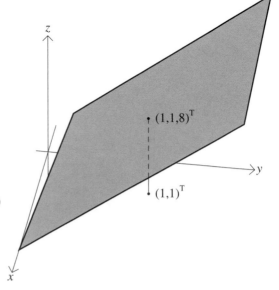

If $(\xi_1, \xi_2, \eta)^T = (1, 1, 8)^T$ and M is the 1×2 matrix

$$M = (-2, 5)$$

the equation takes the form

$$z - 8 = (-2, 5) \begin{pmatrix} x - 1 \\ y - 1 \end{pmatrix}$$
$$= -2(x - 1) + 5(y - 1)$$

$(1,1,8)^T$

$(1,1)^T$

(iii) $m = 2; n = 1$

If $(\xi, \eta_1, \eta_2)^T = (1, 3, -1)^T$, and M is the 2×1 matrix

$$M = \begin{pmatrix} 2 \\ 3 \end{pmatrix}$$

the equation takes the form

$$\begin{pmatrix} x - 3 \\ y + 1 \end{pmatrix} = \begin{pmatrix} 2 \\ 3 \end{pmatrix} (t - 1)$$

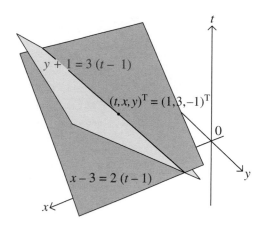

i.e.

$$\left. \begin{array}{c} x - 3 = 2(t - 1) \\ y + 1 = 3(t - 1) \end{array} \right\}$$

Thus the flat is the line of intersection of the two planes $x - 3 = 2(t - 1)$ and $y + 1 = 3(t - 1)$.

5.3 Derivatives of vector functions

A function $\mathbf{f} \colon \mathbb{R}^n \to \mathbb{R}^m$ is **differentiable** at the point $\boldsymbol{\xi}$ if and only if there is a flat

$$\mathbf{y} - \boldsymbol{\eta} = M(\mathbf{x} - \boldsymbol{\xi})$$

which is *tangent* to $\mathbf{y} = \mathbf{f}(\mathbf{x})$ where $\mathbf{x} = \boldsymbol{\xi}$. We must then have $\boldsymbol{\eta} = \mathbf{f}(\boldsymbol{\xi})$.

If $\mathbf{f} \colon \mathbb{R}^n \to \mathbb{R}^m$ is differentiable at $\boldsymbol{\xi}$, we define its **derivative**

$$D\mathbf{f}(\boldsymbol{\xi}) = \mathbf{f}'(\boldsymbol{\xi})$$

to be the *matrix M*. The tangent to the graph $\mathbf{y} = \mathbf{f}(\mathbf{x})$ at the point $(\boldsymbol{\xi}, f(\boldsymbol{\xi}))^T$ then has equation

$$\mathbf{y} - \mathbf{f}(\boldsymbol{\xi}) = \mathbf{f}'(\boldsymbol{\xi})(\mathbf{x} - \boldsymbol{\xi})$$

Example 4

(i) Let $f \colon \mathbb{R}^1 \to \mathbb{R}^1$ be defined by

$$f(x) = x^3 - 9x + 15$$

So

$$f'(x) = 3x^2 - 9$$

155

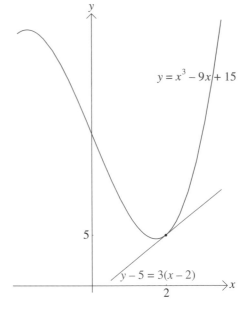

$$y = x^3 - 9x + 15$$

If $\xi = 2$, $f(\xi) = 5$, and the derivative $f'(\xi)$ is the 1×1 matrix (3). In this case, $f'(\xi)$ is therefore just a number. The tangent equation is

$$y - 5 = 3(x - 2)$$

$$y - 5 = 3(x - 2)$$

(ii) Let $f : \mathbb{R}^2 \to \mathbb{R}^1$ be defined by

$$f(\mathbf{x}) = 2x^2 + y^2 + 4$$

So

$$f_x = 4x \qquad f_y = 2y$$

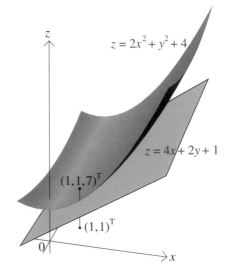

$$z = 2x^2 + y^2 + 4$$

$$z = 4x + 2y + 1$$

If $\boldsymbol{\xi} = (1, 1)^{\mathrm{T}}$, $f(\boldsymbol{\xi}) = 7$, and the derivative at $\boldsymbol{\xi}$ is the 1×2 matrix, $f'(\boldsymbol{\xi}) = (4, 2)$. The tangent equation

$$\mathbf{y} = f(\boldsymbol{\xi}) + f'(\boldsymbol{\xi})(\mathbf{x} - \boldsymbol{\xi})$$

assumes the form

$$z = 7 + (4, 2) \begin{pmatrix} x - 1 \\ y - 1 \end{pmatrix}$$
$$= 4x + 2y + 1$$

$(1,1,7)^{\mathrm{T}}$

$(1,1)^{\mathrm{T}}$

(iii) Consider $\mathbf{f} : \mathbb{R}^1 \to \mathbb{R}^2$ of Example 2, with $\xi = \pi/4$. The value of $\mathbf{f}$ at ξ is a 2×1 matrix – i.e.

$$\mathbf{f}(\pi/4) = \begin{pmatrix} 1/\sqrt{2} \\ 1/\sqrt{2} \end{pmatrix}$$

– and the derivative is a 2×1 matrix:

$$\mathbf{f}'(t) = \begin{pmatrix} -\sin t \\ \cos t \end{pmatrix} = \begin{pmatrix} -1/\sqrt{2} \\ 1/\sqrt{2} \end{pmatrix} \text{ at } t = \pi/4$$

The tangent equation

$$\mathbf{x} = \mathbf{f}(\pi/4) + \mathbf{f}'(\pi/4)(t - \pi/4)$$

assumes the form

$$\begin{pmatrix} x \\ y \end{pmatrix} - \begin{pmatrix} 1/\sqrt{2} \\ 1/\sqrt{2} \end{pmatrix} = \begin{pmatrix} -1/\sqrt{2} \\ 1/\sqrt{2} \end{pmatrix}(t - \pi/4)$$

This flat of intersection of the tangent planes is the tangent line to the curve of intersection of the two surfaces at the point $t = \pi/4$ shown below.

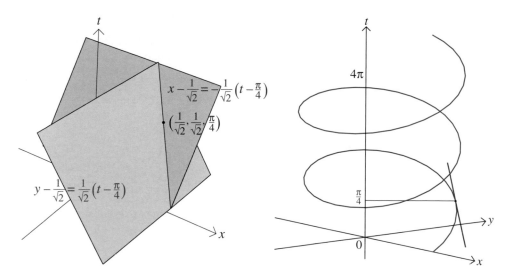

Intersecting tangent planes   Helix of intersection and tangent line

Writing out the equation $\mathbf{y} = \mathbf{f}(\mathbf{x})$ in terms of its component functions, we obtain

$$\left. \begin{array}{l} y_1 = f_1(x_1, x_2, \ldots, x_n) \\ y_2 = f_2(x_1, x_2, \ldots, x_n) \\ \cdots \\ y_m = f_m(x_1, x_2, \ldots, x_n) \end{array} \right\}$$

If the flat $\mathbf{y} - \boldsymbol{\eta} = M(\mathbf{x} - \boldsymbol{\xi})$ is tangent to $\mathbf{y} = \mathbf{f}(\mathbf{x})$, then it consists of m hyperplanes each of which is tangent to one of the m hypersurfaces given above. But this observation means that the ith hyperplane has the same partial derivatives at the point $\boldsymbol{\xi} = (\xi_1, \xi_2, \ldots, \xi_n)^{\mathrm{T}}$ as the hypersurface $y_i = f_i(\mathbf{x})$ for $i = 1, 2, \ldots, m$. So its equation takes the form derived in § 3.8.1:

$$y_i - f_i(\xi) = \left(\frac{\partial f_i}{\partial x_1}, \frac{\partial f_i}{\partial x_2}, \ldots, \frac{\partial f_i}{\partial x_n} \right) \begin{pmatrix} x_1 - \xi_1 \\ x_2 - \xi_2 \\ \vdots \\ x_n - \xi_n \end{pmatrix} \qquad i = 1, 2, \ldots, m$$

where the partial derivatives are evaluated at the point ξ.

The equation of the tangent flat can be written as

$$\begin{pmatrix} y_1 - f_1(\xi) \\ y_2 - f_2(\xi) \\ \vdots \\ y_m - f_m(\xi) \end{pmatrix} = \begin{pmatrix} \dfrac{\partial f_1}{\partial x_1} & \dfrac{\partial f_1}{\partial x_2} & \cdots & \dfrac{\partial f_1}{\partial x_n} \\ \dfrac{\partial f_2}{\partial x_1} & \dfrac{\partial f_2}{\partial x_2} & \cdots & \dfrac{\partial f_2}{\partial x_n} \\ \vdots & \vdots & & \vdots \\ \dfrac{\partial f_m}{\partial x_1} & \dfrac{\partial f_m}{\partial x_2} & \cdots & \dfrac{\partial f_m}{\partial x_n} \end{pmatrix} \begin{pmatrix} x_1 - \xi_1 \\ x_2 - \xi_2 \\ \vdots \\ x_n - \xi_n \end{pmatrix}$$

where the partial derivatives are evaluated at the point ξ.

Thus, if $\mathbf{f} \colon \mathbb{R}^m \to \mathbb{R}^n$ is differentiable at the point ξ, its derivative is

$$D\mathbf{f}(\xi) = \mathbf{f}'(\xi) = \begin{pmatrix} \dfrac{\partial f_1}{\partial x_1} & \dfrac{\partial f_1}{\partial x_2} & \cdots & \dfrac{\partial f_1}{\partial x_n} \\ \dfrac{\partial f_2}{\partial x_1} & \dfrac{\partial f_2}{\partial x_2} & \cdots & \dfrac{\partial f_2}{\partial x_n} \\ \vdots & \vdots & & \vdots \\ \dfrac{\partial f_m}{\partial x_1} & \dfrac{\partial f_m}{\partial x_2} & \cdots & \dfrac{\partial f_m}{\partial x_n} \end{pmatrix}$$

where the partial derivatives are evaluated at the point ξ.

If $\mathbf{y} = \mathbf{f}(\mathbf{x})$, we shall also find it convenient to use the notation

$$\frac{d\mathbf{y}}{d\mathbf{x}} = \begin{pmatrix} \dfrac{\partial y_1}{\partial x_1} & \dfrac{\partial y_1}{\partial x_2} & \cdots & \dfrac{\partial y_1}{\partial x_n} \\ \dfrac{\partial y_2}{\partial x_1} & \dfrac{\partial y_2}{\partial x_2} & \cdots & \dfrac{\partial y_2}{\partial x_n} \\ \vdots & \vdots & & \vdots \\ \dfrac{\partial y_m}{\partial x_1} & \dfrac{\partial y_m}{\partial x_2} & \cdots & \dfrac{\partial y_m}{\partial x_n} \end{pmatrix}$$

Example 5

Consider the function $f \colon \mathbb{R}^3 \to \mathbb{R}^1$ defined by

$$v = f(\mathbf{u}) = xy^2 + yz^3 + zx^4$$

where

$$\mathbf{u} = \begin{pmatrix} x \\ y \\ z \end{pmatrix}$$

We have

$$\frac{dv}{d\mathbf{u}} = \left(\frac{\partial v}{\partial x}, \frac{\partial v}{\partial y}, \frac{\partial v}{\partial z} \right) = (y^2 + 4zx^3, 2xy + z^3, 3yz^2 + x^4)$$

It follows that the tangent hyperplane to the 'hypersurface' $v = xy^2 + yz^3 + zx^4$ at $(x, y, z, v)^T = (1, 1, 1, 3)^T$ has equation

$$v - 3 = (5, 3, 4) \begin{pmatrix} x - 1 \\ y - 1 \\ z - 1 \end{pmatrix} = 5(x - 1) + 3(y - 1) + 4(z - 1)$$

Example 6

Consider the function $\mathbf{f} \colon \mathbb{R}^1 \to \mathbb{R}^3$ defined by

$$\begin{pmatrix} x \\ y \\ z \end{pmatrix} = \mathbf{u} = \mathbf{f}(t) = \begin{pmatrix} t \\ t^2 \\ t^3 \end{pmatrix}$$

The function is therefore determined by the equations

$$\left. \begin{array}{l} x = t \\ y = t^2 \\ z = t^3 \end{array} \right\}$$

Thus

$$\frac{d\mathbf{u}}{dt} = \begin{pmatrix} \dfrac{\partial x}{\partial t} \\[2mm] \dfrac{\partial y}{\partial t} \\[2mm] \dfrac{\partial z}{\partial t} \end{pmatrix} = \begin{pmatrix} 1 \\ 2t \\ 3t^2 \end{pmatrix}$$

The tangent at $(t, x, y, z)^T = (2, 2, 4, 8)^T$ is the line

$$\begin{pmatrix} x - 2 \\ y - 4 \\ z - 8 \end{pmatrix} = \begin{pmatrix} 1 \\ 4 \\ 12 \end{pmatrix} (t - 2)$$

i.e.

$$\left. \begin{array}{l} x - 2 = (t - 2) \\ y - 4 = 4(t - 2) \\ z - 8 = 12(t - 2) \end{array} \right\}$$

which we can also write as

$$t - 2 = x - 2 = \frac{y - 4}{4} = \frac{z - 8}{12}$$

Example 7

Consider the function $\mathbf{f} \colon \mathbb{R}^3 \to \mathbb{R}^2$ defined by

$$\begin{pmatrix} s \\ t \end{pmatrix} = \mathbf{v} = f(\mathbf{u}) = \begin{pmatrix} xy \\ xyz \end{pmatrix}$$

where

$$\mathbf{u} = \begin{pmatrix} x \\ y \\ z \end{pmatrix}$$

The function is determined by the equations

$$s = xy$$
$$t = xyz$$

Thus

$$\frac{d\mathbf{v}}{d\mathbf{u}} = \begin{pmatrix} \dfrac{\partial s}{\partial x} & \dfrac{\partial s}{\partial y} & \dfrac{\partial s}{\partial z} \\[2mm] \dfrac{\partial t}{\partial x} & \dfrac{\partial t}{\partial y} & \dfrac{\partial t}{\partial z} \end{pmatrix} = \begin{pmatrix} y & x & 0 \\ yz & xz & xy \end{pmatrix}$$

The tangent at $(x, y, z, s, t)^{\mathrm{T}} = (1, 2, 3, 2, 6)^{\mathrm{T}}$ is therefore the flat

$$\begin{pmatrix} s - 2 \\ t - 6 \end{pmatrix} = \begin{pmatrix} 2 & 1 & 0 \\ 6 & 3 & 2 \end{pmatrix} \begin{pmatrix} x - 1 \\ y - 2 \\ z - 3 \end{pmatrix}$$

The affine function $\mathbf{A} \colon \mathbb{R}^n \to \mathbb{R}^m$ defined by

$$\mathbf{y} = \mathbf{A}(\mathbf{x}) = \mathbf{f}(\boldsymbol{\xi}) + \mathbf{f}'(\boldsymbol{\xi})(\mathbf{x} - \boldsymbol{\xi})$$

is tangent to $\mathbf{f} \colon \mathbb{R}^n \to \mathbb{R}^m$ at the point $(\boldsymbol{\xi}, \boldsymbol{\eta})^{\mathrm{T}}$ provided $\boldsymbol{\eta} = \mathbf{f}(\boldsymbol{\xi})$. This tangent flat is the intersection of tangent hyperplanes to the component functions of $\mathbf{f}$. It follows that the contours of this affine function are tangent to the corresponding contours of the function $\mathbf{f}$ at the point.

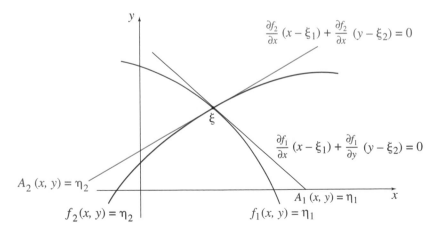

Contours of **f** and its tangent at the point $\boldsymbol{\xi}$

Example 8

Consider the function $\mathbf{f} : \mathbb{R}^2 \to \mathbb{R}^2$ of Example 1. This is defined by the equations

$$u = f_1(x, y) = x^2 + y^2$$
$$v = f_2(x, y) = x^2 - y^2$$

Since

$$\begin{pmatrix} \dfrac{\partial u}{\partial x} & \dfrac{\partial u}{\partial y} \\[2ex] \dfrac{\partial v}{\partial x} & \dfrac{\partial v}{\partial y} \end{pmatrix} = \begin{pmatrix} 2x & 2y \\ 2x & -2y \end{pmatrix}$$

we have

$$\mathbf{f}'(2, 1) = \begin{pmatrix} 4 & 2 \\ 4 & -2 \end{pmatrix}$$

Also,

$$\mathbf{f}(2, 1) = \begin{pmatrix} 5 \\ 3 \end{pmatrix}$$

Hence the equation of the tangent to the function where $(x, y)^{\mathrm{T}} = (2, 1)^{\mathrm{T}}$ is

$$\begin{pmatrix} u \\ v \end{pmatrix} = A(x, y) = \begin{pmatrix} 5 \\ 3 \end{pmatrix} + \begin{pmatrix} 4 & 2 \\ 4 & -2 \end{pmatrix} \begin{pmatrix} x - 2 \\ y - 1 \end{pmatrix}$$

i.e.

$$\left.\begin{array}{l} u = A_1(x, y) = 5 + 4(x - 2) + 2(y - 1) \\ v = A_2(x, y) = 3 + 4(x - 2) - 2(y - 1) \end{array}\right\}$$

The accompanying figure shows that the contour $A_1(x, y) = 5$ is tangent to the contour $f_1(x, y) = 5$ at the point $(x, y)^{\mathrm{T}} = (2, 1)^{\mathrm{T}}$ – i.e. $4(x - 2) + 2(y - 1) = 0$ is tangent to $x^2 + y^2 = 5$ at $(2, 1)^{\mathrm{T}}$.

Also, the contour $A_2(x, y) = 3$ is tangent to the contour $f_2(x, y) = 3$ at the point $(x, y)^{\mathrm{T}} = (2, 1)^{\mathrm{T}}$ – i.e. $4(x - 2) - 2(y - 1) = 0$ is tangent to $x^2 - y^2 = 3$ at $(2, 1)^{\mathrm{T}}$.

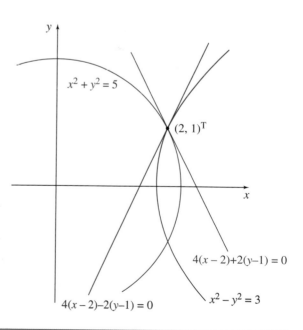

5.4 Manipulation of vector derivatives

In §2.10.2 we gave a list of rules for manipulating derivatives. It is a very satisfactory fact that these rules remain valid for vector derivatives *provided* that they make sense in this broader context. We list below some of the more useful versions of these rules:

$$\text{(I)} \quad \frac{d}{dx}(A\mathbf{y} + B\mathbf{z}) = A\frac{d\mathbf{y}}{dx} + B\frac{d\mathbf{z}}{dx}$$

(where A and B are constant matrices).

$$\text{(II)} \quad \frac{d}{dx}(\mathbf{y}^T\mathbf{z}) = \mathbf{z}^T\frac{d\mathbf{y}}{dx} + \mathbf{y}^T\frac{d\mathbf{z}}{dx}$$

$$\text{(III)} \quad \frac{d\mathbf{z}}{dx} = \frac{d\mathbf{z}}{d\mathbf{y}}\frac{d\mathbf{y}}{dx}$$

$$\text{(IV)} \quad \frac{d\mathbf{x}}{d\mathbf{y}} = \left(\frac{d\mathbf{y}}{d\mathbf{x}}\right)^{-1}$$

Notice the necessity of introducing the transpose signs in rule (II), in order that the matrix multiplication may be defined. The object $\mathbf{y}^T\mathbf{z}$ is then a scalar provided $\mathbf{y}$ and $\mathbf{z}$ are both $m \times 1$ column vectors. Notice also that the product in rule (III) is a *matrix* product – i.e. the two derivatives are matrices and they are multiplied using matrix multiplication.

Also useful are the following formulas:

$$\text{(V)} \quad \frac{d}{dx}(A\mathbf{x} + \mathbf{b}) = A$$

$$\text{(VI)} \quad \frac{d}{dx}(\mathbf{x}^T A\mathbf{x}) = \mathbf{x}^T(A^T + A)$$

The first of these is a consequence of the fact that an affine function is obviously tangent to itself. Note incidentally the special case

$$\frac{d\mathbf{x}}{d\mathbf{x}} = I$$

where I is the identity matrix. Formula (VI) is a generalization of the familiar result that

$$\frac{d}{dx}(ax^2) = 2ax$$

This is more apparent in the case when A is a *symmetric* matrix (i.e. $A^T = A$). In this special case

$$\frac{d}{dx}(\mathbf{x}^T A\mathbf{x}) = 2\mathbf{x}^T A$$

In order for (VI) to make sense, it is necessary that $\mathbf{x}$ be an $n \times 1$ column vector and that A be an $n \times n$ square matrix. In this case $\mathbf{x}^T A\mathbf{x}$ is a scalar.

It is instructive to see how rule (VI) may be deduced from rules (II) and (V). We have

$$\frac{d}{d\mathbf{x}}\{\mathbf{x}^{\mathsf{T}}A\mathbf{x}\} = \frac{d}{d\mathbf{x}}\{\mathbf{x}^{\mathsf{T}}(A\mathbf{x})\}$$

$$= (A\mathbf{x})^{\mathsf{T}}\frac{d\mathbf{x}}{d\mathbf{x}} + \mathbf{x}^{\mathsf{T}}\frac{d}{d\mathbf{x}}(A\mathbf{x})$$

$$= \mathbf{x}^{\mathsf{T}}A^{\mathsf{T}}I + \mathbf{x}^{\mathsf{T}}A$$

$$= \mathbf{x}^{\mathsf{T}}(A^{\mathsf{T}} + A)$$

Example 9

Suppose that

$$z = \mathbf{u}^{\mathsf{T}}A\mathbf{u} = (x, y)\begin{pmatrix} 1 & 2 \\ 2 & 4 \end{pmatrix}\begin{pmatrix} x \\ y \end{pmatrix} = x^2 + 4xy + 4y^2$$

Then

$$\frac{\partial z}{\partial x} = 2x + 4y \qquad \frac{\partial z}{\partial y} = 4x + 8y$$

Thus

$$\frac{dz}{d\mathbf{u}} = \left(\frac{\partial z}{\partial x}, \frac{\partial z}{\partial y}\right) = (2x + 4y, 4x + 8y).$$

Alternatively, using rule (VI),

$$\frac{dz}{d\mathbf{u}} = \frac{d}{d\mathbf{u}}(\mathbf{u}^{\mathsf{T}}A\mathbf{u}) = 2\mathbf{u}^{\mathsf{T}}A$$

$$= 2(x, y)\begin{pmatrix} 1 & 2 \\ 2 & 4 \end{pmatrix}$$

$$= (2x + 4y, 4x + 8y)$$

5.5 Chain rule

We begin with the simplest form of the chain rule involving the composition of a scalar function $f : \mathbb{R}^n \to \mathbb{R}$ and a vector one variable function $\mathbf{g}: \mathbb{R} \to \mathbb{R}^n$. We shall write

$$y = f(\mathbf{x}) \text{ and } \mathbf{x} = \mathbf{g}(t)$$

Then

$$\frac{dy}{d\mathbf{x}} = \left(\frac{\partial y}{\partial x_1}, \frac{\partial y}{\partial x_2}, \dots, \frac{\partial y}{\partial x_n}\right)$$

Note that the derivatives with respect to t are ordinary derivatives since t is the only independent variable.

$$\frac{d\mathbf{x}}{dt} = \begin{pmatrix} \dfrac{dx_1}{dt} \\ \dfrac{dx_2}{dt} \\ \vdots \\ \dfrac{dx_n}{dt} \end{pmatrix}$$

163

Let $h: \mathbb{R} \to \mathbb{R}$ be the function defined by $y = h(t) = f(\mathbf{g}(t))$. Then the chain rule **(III)** asserts that

$$\frac{dy}{dt} = \frac{dy}{d\mathbf{x}}\frac{d\mathbf{x}}{dt}$$

$$= \left(\frac{\partial y}{\partial x_1}, \frac{\partial y}{\partial x_2}, \ldots, \frac{\partial y}{\partial x_n}\right) \begin{pmatrix} \dfrac{dx_1}{dt} \\[2mm] \dfrac{dx_2}{dt} \\[1mm] \vdots \\[1mm] \dfrac{dx_n}{dt} \end{pmatrix}$$

Thus

$$\frac{dy}{dt} = \frac{\partial y}{\partial x_1}\frac{dx_1}{dt} + \frac{\partial y}{\partial x_2}\frac{dx_2}{dt} + \cdots + \frac{\partial y}{\partial x_n}\frac{dx_n}{dt}$$

It can be helpful to represent this by a diagram:

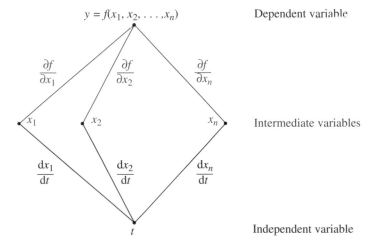

$y = f(x_1, x_2, \ldots, x_n)$ — Dependent variable

x_1, x_2, x_n — Intermediate variables

t — Independent variable

More generally, consider the composition of a scalar function $f: \mathbb{R}^m \to \mathbb{R}$ with an n variable vector function $\mathbf{g}: \mathbb{R}^n \to \mathbb{R}^m$. We may then write

$$y = f(\mathbf{x}) \quad \text{and} \quad \mathbf{x} = \mathbf{g}(\mathbf{u})$$

We have

$$\frac{dy}{d\mathbf{x}} = \left(\frac{\partial y}{\partial x_1}, \frac{\partial y}{\partial x_2}, \ldots, \frac{\partial y}{\partial x_m}\right)$$

$$\frac{d\mathbf{x}}{d\mathbf{u}} = \begin{pmatrix} \dfrac{\partial x_1}{\partial u_1} & \dfrac{\partial x_1}{\partial u_2} & \cdots & \dfrac{\partial x_1}{\partial u_n} \\[2mm] \dfrac{\partial x_2}{\partial u_1} & \dfrac{\partial x_2}{\partial u_2} & \cdots & \dfrac{\partial x_2}{\partial u_n} \\[2mm] \vdots & \vdots & & \vdots \\[2mm] \dfrac{\partial x_m}{\partial u_1} & \dfrac{\partial x_m}{\partial u_2} & \cdots & \dfrac{\partial x_m}{\partial u_n} \end{pmatrix}$$

Thus, by the chain rule (III),

$$\left(\frac{\partial y}{\partial u_1}, \frac{\partial y}{\partial u_2}, \ldots, \frac{\partial y}{\partial u_n} \right) = \frac{dy}{d\mathbf{u}} = \frac{dy}{d\mathbf{x}} \frac{d\mathbf{x}}{d\mathbf{u}}$$

$$= \left(\frac{\partial y}{\partial x_1}, \ldots, \frac{\partial y}{\partial x_m} \right) \begin{pmatrix} \dfrac{\partial x_1}{\partial u_1} & \cdots & \dfrac{\partial x_1}{\partial u_n} \\[2mm] \vdots & & \vdots \\[2mm] \dfrac{\partial x_m}{\partial u_1} & \cdots & \dfrac{\partial x_m}{\partial u_n} \end{pmatrix}$$

Writing this out in full,

$$\frac{\partial y}{\partial u_1} = \frac{\partial y}{\partial x_1} \frac{\partial x_1}{\partial u_1} + \frac{\partial y}{\partial x_2} \frac{\partial x_2}{\partial u_1} + \cdots + \frac{\partial y}{\partial x_m} \frac{\partial x_m}{\partial u_1}$$

$$\frac{\partial y}{\partial u_2} = \frac{\partial y}{\partial x_1} \frac{\partial x_1}{\partial u_2} + \frac{\partial y}{\partial x_2} \frac{\partial x_2}{\partial u_2} + \cdots + \frac{\partial y}{\partial x_m} \frac{\partial x_m}{\partial u_2}$$

$$\cdots$$

$$\frac{\partial y}{\partial u_n} = \frac{\partial y}{\partial x_1} \frac{\partial x_1}{\partial u_n} + \frac{\partial y}{\partial x_2} \frac{\partial x_2}{\partial u_n} + \cdots + \frac{\partial y}{\partial x_m} \frac{\partial x_m}{\partial u_n}$$

When a number of variables are in play, it is not always obvious which are to be held constant when differentiating partially. But it is of the *greatest* importance to be clear on this point. Where necessary, use the notation

$$\left(\frac{\partial f}{\partial x} \right)_y$$

to mean 'the partial derivative with respect to x keeping y constant'.

Example 10 _____

Consider the equations

$$\left. \begin{array}{l} u = x + y \\ v = x - y \end{array} \right\}$$

From the first equation

$$\frac{\partial u}{\partial x} = 1$$

If the equations are solved for x and y in terms of u and v, we obtain

$$x = \tfrac{1}{2}(u + v) \ \left.\vphantom{\begin{matrix}a\\b\end{matrix}}\right\}$$
$$y = \tfrac{1}{2}(u - v)$$

and therefore

$$\frac{\partial x}{\partial u} = \frac{1}{2}$$

Note that

$$\frac{\partial u}{\partial x} \neq \left(\frac{\partial x}{\partial u}\right)^{-1}$$

This is not in the least surprising since *different* variables were held constant during the two differentiations. What has been shown is that

$$\left(\frac{\partial u}{\partial x}\right)_y = 1 \qquad \left(\frac{\partial x}{\partial u}\right)_v = \frac{1}{2}$$

It is instructive to calculate

$$\left(\frac{\partial u}{\partial x}\right)_v$$

Eliminate y from the first set of equations. We obtain

$$u = x + (x - v) = 2x - v$$

and therefore

$$\left(\frac{\partial u}{\partial x}\right)_v = 2 = \left(\frac{\partial x}{\partial u}\right)_v^{-1}$$

as one would expect.

Example 11

Suppose that $z = xy^2$ where

$$x = \tfrac{1}{2}(s^2 - t^2) \ \left.\vphantom{\begin{matrix}a\\b\end{matrix}}\right\}$$
$$y = st$$

Then

$$\frac{\partial z}{\partial s} = \frac{\partial z}{\partial x}\frac{\partial x}{\partial s} + \frac{\partial z}{\partial y}\frac{\partial y}{\partial s}$$

In these partial derivatives it is, as always, important to be quite clear about what variables are being held constant. This is made clear in the equation below:

$$\left(\frac{\partial z}{\partial s}\right)_t = \left(\frac{\partial z}{\partial x}\right)_y \left(\frac{\partial x}{\partial s}\right)_t + \left(\frac{\partial z}{\partial y}\right)_x \left(\frac{\partial y}{\partial s}\right)_t$$

We obtain

$$\left(\frac{\partial z}{\partial s}\right)_t = y^2 s + 2xyt$$
$$= s^3 t^2 + (s^2 - t^2)st^2$$
$$= 2s^3 t^2 - st^4$$

Similarly,

$$\left(\frac{\partial z}{\partial t}\right)_s = \left(\frac{\partial z}{\partial x}\right)_y \left(\frac{\partial x}{\partial t}\right)_s + \left(\frac{\partial z}{\partial y}\right)_x \left(\frac{\partial y}{\partial t}\right)_s$$
$$= -y^2 t + 2yxs$$
$$= -s^2 t^3 + (s^2 - t^2)s^2 t$$
$$= -2s^2 t^3 + s^4 t$$

Example 12 _____

An economics application

Consider a Cobb–Douglas production function $P \colon \mathbb{R}_+^3 \to \mathbb{R}$ defined by

$$P(A, x, y) = Ax^\alpha y^\beta$$

where $0 < \alpha < 1$ and $0 < \beta < 1$ and where A, x and y are respectively the available levels of technology, capital and labour. If A, x and y are functions of time, we need to consider

$$P(A(t), x(t), y(t)) = A(t)x(t)^\alpha y(t)^\beta$$

By the chain rule

$$\frac{\mathrm{d}P}{\mathrm{d}t} = \frac{\partial P}{\partial A}\frac{\mathrm{d}A}{\mathrm{d}t} + \frac{\partial P}{\partial x}\frac{\mathrm{d}x}{\mathrm{d}t} + \frac{\partial P}{\partial y}\frac{\mathrm{d}y}{\mathrm{d}t}$$

$$= x^\alpha y^\beta \frac{\mathrm{d}A}{\mathrm{d}t} + \alpha Ax^{\alpha-1}y^\beta \frac{\mathrm{d}x}{\mathrm{d}t} + \beta Ax^\alpha y^{\beta-1}\frac{\mathrm{d}y}{\mathrm{d}t}$$

Dividing throughout by $P = Ax^\alpha y^\beta$ we obtain

$$\frac{1}{P}\frac{\mathrm{d}P}{\mathrm{d}t} = \frac{1}{A}\frac{\mathrm{d}A}{\mathrm{d}t} + \alpha\frac{1}{x}\frac{\mathrm{d}x}{\mathrm{d}t} + \beta\frac{1}{y}\frac{\mathrm{d}y}{\mathrm{d}t}$$

This attributes the rate of growth in output to the rate of growth of the input factors in the following proportions:

rate of growth in output	=	rate of growth of technology	+	$\alpha \times$ rate of growth of capital	+	$\beta \times$ rate of growth of labour

5.6 Second derivatives

Consider a function $f: \mathbb{R}^2 \to \mathbb{R}$ and write

$$z = f(x, y)$$

From such a function one can form four **second order partial derivatives**. These are defined by

$$\frac{\partial^2 z}{\partial x^2} = \frac{\partial}{\partial x}\left(\frac{\partial z}{\partial x}\right) \qquad \frac{\partial^2 z}{\partial y\, \partial x} = \frac{\partial}{\partial y}\left(\frac{\partial z}{\partial x}\right)$$

$$\frac{\partial^2 z}{\partial x\, \partial y} = \frac{\partial}{\partial x}\left(\frac{\partial z}{\partial y}\right) \qquad \frac{\partial^2 z}{\partial y^2} = \frac{\partial}{\partial y}\left(\frac{\partial z}{\partial y}\right)$$

The **second order partial derivatives** of a function of n variables $f: \mathbb{R}^n \to \mathbb{R}$ are defined in a similar way. Thus, for example,

$$\frac{\partial^2 f}{\partial x_2 \partial x_5} = \frac{\partial}{\partial x_2}\left(\frac{\partial f}{\partial x_5}\right)$$

Example 13

Suppose that $z = x^2 y + xy^3$. Then

$$\frac{\partial z}{\partial x} = 2xy + y^3 \qquad \frac{\partial z}{\partial y} = x^2 + 3xy^2$$

Hence

$$\frac{\partial^2 z}{\partial x^2} = \frac{\partial}{\partial x}(2xy + y^3) = 2y$$

$$\frac{\partial^2 z}{\partial y\, \partial x} = \frac{\partial}{\partial y}(2xy + y^3) = 2x + 3y^2$$

$$\frac{\partial^2 z}{\partial x\, \partial y} = \frac{\partial}{\partial x}(x^2 + 3xy^2) = 2x + 3y^2$$

$$\frac{\partial^2 z}{\partial y^2} = \frac{\partial}{\partial y}(x^2 + 3xy^2) = 6xy$$

Note in the above example that

$$\frac{\partial^2 z}{\partial y\, \partial x} = \frac{\partial^2 z}{\partial x\, \partial y}.$$

This is a special case of the general rule for a function $f: \mathbb{R}^n \to \mathbb{R}$:

$$\frac{\partial^2 f}{\partial x_i \partial x_j} = \frac{\partial^2 f}{\partial x_j \partial x_i}$$

There are exceptions to this rule, but the exceptional functions are rather peculiar and we therefore exclude them from our study.

Having defined the second order *partial* derivatives of $f: \mathbb{R}^n \to \mathbb{R}$, we can now consider the **second order derivative** of f. The first order derivative

$$\frac{\mathrm{d}f}{\mathrm{d}\mathbf{x}} = \left(\frac{\partial f}{\partial x_1}, \frac{\partial f}{\partial x_2}, \ldots, \frac{\partial f}{\partial x_n} \right)$$

is a $1 \times n$ *row* vector. In order to define its derivative we need to take its transpose, which is a column vector. So we may define the second derivative of f with respect to x by

$$\frac{\mathrm{d}^2 f}{\mathrm{d}\mathbf{x}^2} = \frac{\mathrm{d}}{\mathrm{d}\mathbf{x}} \left(\frac{\mathrm{d}f}{\mathrm{d}\mathbf{x}} \right)^{\mathrm{T}} = \begin{pmatrix} \dfrac{\partial^2 f}{\partial x_1^2} & \dfrac{\partial^2 f}{\partial x_2 \partial x_1} & \cdots & \dfrac{\partial^2 f}{\partial x_n \partial x_1} \\[2mm] \dfrac{\partial^2 f}{\partial x_1 \partial x_2} & \dfrac{\partial^2 f}{\partial x_2^2} & \cdots & \dfrac{\partial^2 f}{\partial x_n \partial x_2} \\[2mm] \vdots & \vdots & & \vdots \\[2mm] \dfrac{\partial^2 f}{\partial x_1 \partial x_n} & \dfrac{\partial^2 f}{\partial x_2 \partial x_n} & \cdots & \dfrac{\partial^2 f}{\partial x_n^2} \end{pmatrix}$$

The second derivative of a real valued function $f: \mathbb{R}^n \to \mathbb{R}$ is therefore an $n \times n$ matrix whose entries are the second order partial derivatives of f. Observe that this matrix is *symmetric*.

The geometric interpretation is that the flat

$$\mathbf{y} = f'(\boldsymbol{\xi})^{\mathrm{T}} + f''(\boldsymbol{\xi})(\mathbf{x} - \boldsymbol{\xi})$$

is tangent to $\mathbf{y} = f'(\mathbf{x})^{\mathrm{T}}$ at the point where $\mathbf{x} = \boldsymbol{\xi}$.

Example 14 _____

Let $f: \mathbb{R}^2 \to \mathbb{R}$ be defined by

$$z = f(x, y) = x^2 y + xy^3$$

The second derivative is

$$\begin{pmatrix} \dfrac{\partial^2 z}{\partial x^2} & \dfrac{\partial^2 z}{\partial x\, \partial y} \\[3mm] \dfrac{\partial^2 z}{\partial y\, \partial x} & \dfrac{\partial^2 z}{\partial y^2} \end{pmatrix} = \begin{pmatrix} 2y & 2x + 3y^2 \\[2mm] 2x + 3y^2 & 6xy \end{pmatrix}$$

5.7 Taylor series for scalar valued functions of n variables

We have studied Taylor series expansions of a function $f: \mathbb{R} \to \mathbb{R}$.

We now turn to the Taylor series expansion about the point $\boldsymbol{\xi}$ of a scalar valued function of n variables $f: \mathbb{R}^n \to \mathbb{R}$. We shall employ a cunning trick to transform this into a function $F(t)$ of one variable, so that we can exploit the results that we have already obtained on the Taylor series expansions of functions of one variable.

Let $\mathbf{X}$ and $\boldsymbol{\xi}$ be constant vectors in $\mathbb{R}^n$ and write

$$\mathbf{x} = \boldsymbol{\xi} + t(\mathbf{X} - \boldsymbol{\xi}) \tag{1}$$

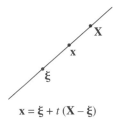

$\mathbf{x} = \boldsymbol{\xi} + t\,(\mathbf{X} - \boldsymbol{\xi})$

This means that we restrict $\mathbf{x}$ to lie on the line through the fixed points $\mathbf{X}$ and $\boldsymbol{\xi}$, so that its position is determined by the single variable t.

Define $F: \mathbb{R} \to \mathbb{R}$ by

$$F(t) = f(\mathbf{x}) = f(\boldsymbol{\xi} + t(\mathbf{X} - \boldsymbol{\xi}))$$

To expand F by its Taylor series, we must first find its derivatives. Using the chain rule, we obtain

$$\frac{\mathrm{d}F}{\mathrm{d}t} = \frac{\mathrm{d}f}{\mathrm{d}\mathbf{x}} \frac{\mathrm{d}\mathbf{x}}{\mathrm{d}t} = \frac{\mathrm{d}f}{\mathrm{d}\mathbf{x}}(\mathbf{X} - \boldsymbol{\xi}) = (\mathbf{X} - \boldsymbol{\xi})^{\mathrm{T}} \left(\frac{\mathrm{d}f}{\mathrm{d}\mathbf{x}} \right)^{\mathrm{T}} \tag{2}$$

For the last step note that we are dealing with a scalar quantity which must be equal to its transpose.

$$\frac{\mathrm{d}^2 F}{\mathrm{d}t^2} = \frac{\mathrm{d}}{\mathrm{d}t}\left(\frac{\mathrm{d}F}{\mathrm{d}t} \right) = (\mathbf{X} - \boldsymbol{\xi})^{\mathrm{T}} \frac{\mathrm{d}}{\mathrm{d}t} \left(\frac{\mathrm{d}f}{\mathrm{d}\mathbf{x}} \right)^{\mathrm{T}}$$

$$= (\mathbf{X} - \boldsymbol{\xi})^{\mathrm{T}} \frac{\mathrm{d}}{\mathrm{d}\mathbf{x}} \left(\frac{\mathrm{d}f}{\mathrm{d}\mathbf{x}} \right)^{\mathrm{T}} \frac{\mathrm{d}\mathbf{x}}{\mathrm{d}t}$$

$$= (\mathbf{X} - \boldsymbol{\xi})^{\mathrm{T}} \frac{\mathrm{d}^2 f}{\mathrm{d}\mathbf{x}^2}(\mathbf{X} - \boldsymbol{\xi}) \tag{3}$$

The Taylor expansion for $F(t)$ about $t = 0$ is

$$F(t) = F(0) + \frac{t}{1!}F'(0) + \frac{t^2}{2!}F''(0) + \cdots$$

so

$$F(1) = F(0) + \frac{1}{1!}F'(0) + \frac{1}{2!}F''(0) + \cdots$$

From (1) we deduce that

$$\begin{cases} \mathbf{x} & = \boldsymbol{\xi} \ \text{ when } \ t = 0 \\ \mathbf{x} & = \mathbf{X} \ \text{ when } \ t = 1 \end{cases}$$

Hence, $F(0) = f(\boldsymbol{\xi})$ and $F(1) = f(\mathbf{X})$.

Also

$$\begin{cases} F'(0) & = f'(\boldsymbol{\xi})(\mathbf{X} - \boldsymbol{\xi}) \ \text{ from (2)} \\ F''(0) & = (\mathbf{X} - \boldsymbol{\xi})^{\mathrm{T}} f''(\boldsymbol{\xi})(\mathbf{X} - \boldsymbol{\xi}) \ \text{ from (3)} \end{cases}$$

It follows that

$$f(\mathbf{X}) = f(\xi) + \frac{1}{1!}f'(\xi)(\mathbf{X} - \xi) + \frac{1}{2!}(\mathbf{X} - \xi)^{\mathrm{T}}f''(\xi)(\mathbf{X} - \xi) + \cdots$$

Relaxing the assumption that $\mathbf{X}$ is fixed and replacing it by the general point $\mathbf{x}$, which can be *anywhere* near ξ, we obtain

$$f(\mathbf{x}) = f(\xi) + \frac{1}{1!}f'(\xi)(\mathbf{x} - \xi) + \frac{1}{2!}(\mathbf{x} - \xi)^{\mathrm{T}}f''(\xi)(\mathbf{x} - \xi) + \cdots$$

Suppose that $f: \mathbb{R}^2 \to \mathbb{R}$. Then

$$f'(\xi, \eta) = \left(\frac{\partial f}{\partial x}, \frac{\partial f}{\partial y} \right) = (f_x, f_y)$$

$$f''(\xi, \eta) = \begin{pmatrix} \dfrac{\partial^2 f}{\partial x^2} & \dfrac{\partial^2 f}{\partial x\,\partial y} \\[2mm] \dfrac{\partial^2 f}{\partial y\,\partial x} & \dfrac{\partial^2 f}{\partial y^2} \end{pmatrix} = \begin{pmatrix} f_{xx} & f_{xy} \\ f_{yx} & f_{yy} \end{pmatrix}$$

provided the partial derivatives are evaluated at the point $(\xi, \eta)^{\mathrm{T}}$.

The Taylor expansion of f about $(\xi, \eta)^{\mathrm{T}}$ is

$$f(\xi + h, \eta + k) = f + (f_x, f_y) \begin{pmatrix} h \\ k \end{pmatrix} + \frac{1}{2}(h, k) \begin{pmatrix} f_{xx} & f_{xy} \\ f_{yx} & f_{yy} \end{pmatrix} \begin{pmatrix} h \\ k \end{pmatrix} + \cdots$$

$$= f + \{hf_x + kf_y\} + \frac{1}{2}\{h^2 f_{xx} + 2hkf_{xy} + k^2 f_{yy}\} + \cdots$$

As in the one variable case, we discard the terms of third order and higher to obtain the **quadratic Taylor approximation** *to f about ξ.*

Example 15

Let $f: \mathbb{R}^2 \to \mathbb{R}$ be defined by

$$z = f(x, y) = x^2 y + xy^3 - xy$$

We shall find the quadratic Taylor approximation to this function about the point $(1, 0)^{\mathrm{T}}$.

$$f_x = 2xy + y^3 - y \qquad f_y = x^2 + 3xy^2 - x$$

At this point $f = f_x = f_y = 0$ and the second derivative is

$$\begin{pmatrix} 2y & 2x + 3y^2 - 1 \\ 2x + 3y^2 - 1 & 6xy \end{pmatrix}$$

which is

$$\begin{pmatrix} 0 & 1 \\ 1 & 0 \end{pmatrix}$$

at the point $(1, 0)^T$. Hence the quadratic Taylor approximation of f about $(1, 0)^T$ is $\frac{1}{2}(2(x-1)y)$ or

$$xy - y$$

This quadric displays the essential features of the graph of f at this point. This is illustrated by the surface plots of the function and its Taylor approximation below which have similar saddle shapes.

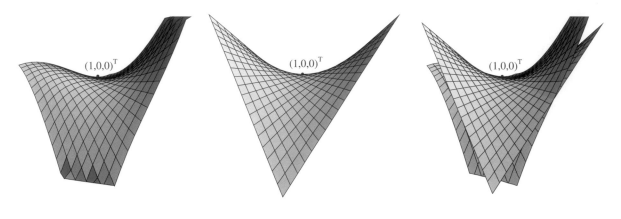

Plot of surface near $(1, 0)^T$ Plot of $z = xy - y$ near $(1, 0)^T$ Plot of superimposed surfaces

The resemblance between the saddle shapes of the surface and its Taylor approximation near $(1, 0)^T$ is also clearly shown by their contour plots.

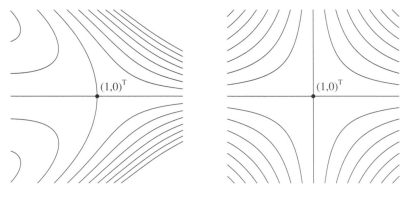

Contours of $z = x^2y + y^3x - xy$ Contours of $z = xy - y$ near $(1, 0)^T$
near $(1, 0)^T$

We now consider the quadratic Taylor approximation to this function about the point $(\frac{1}{2}, -\frac{1}{2})^T$. At this point the second derivative can be calculated to be

$$\begin{pmatrix} -1 & \frac{3}{4} \\ \frac{3}{4} & -\frac{3}{2} \end{pmatrix}$$

Hence the quadratic term is

$$-\left(x - \frac{1}{2}\right)^2 - \frac{3}{2}\left(y + \frac{1}{2}\right)^2 + \frac{3}{2}\left(x - \frac{1}{2}\right)\left(y + \frac{1}{2}\right)$$

This time the quadratic Taylor approximation is a paraboloid pointing downwards. The surface plot near the point shows a similar shape.

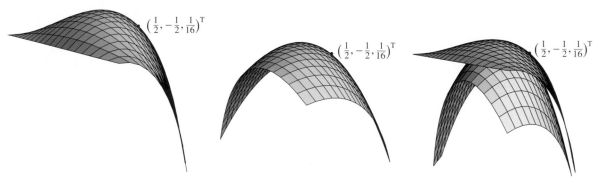

$\left(\frac{1}{2}, -\frac{1}{2}, \frac{1}{16}\right)^T$

Plot of $z = x^2y + y^3x - xy$ near $\left(\frac{1}{2}, -\frac{1}{2}\right)^T$

Plot of its Taylor approximation about $\left(\frac{1}{2}, -\frac{1}{2}\right)^T$

Plot of superimposed surfaces

The contour plot of the surface near $\left(\frac{1}{2}, -\frac{1}{2}\right)^T$, shown next, again resembles that of the paraboloid which is its Taylor approximation about the point.

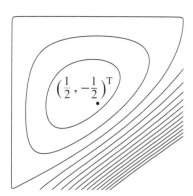

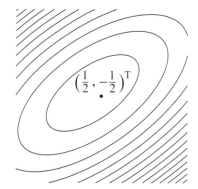

Contours of $z = x^2y + y^3x - xy$ near $\left(\frac{1}{2}, -\frac{1}{2}\right)^T$

Contours of the Taylor approximation about $\left(\frac{1}{2}, -\frac{1}{2}\right)^T$

Example 16

The quadratic Taylor approximation provides the quadric surface which most closely resembles the surface at the point. However, sometimes the quadric surface becomes degenerate, and one must then look at third or higher order terms in the Taylor expansion to learn what the graph of f looks like near $\boldsymbol{\xi}$. This is the case for the function defined next.

The surface and contour plots of the function

$$f(x, y) = x^2 y + y^3 x$$

near the point $(0, 0)^{\mathrm{T}}$ indicate that it cannot be adequately represented by a quadric surface at this point. This reflects the fact that all second order partial derivatives of f are zero at the point, so that the quadratic term in the Taylor approximation vanishes. We use the term 'saddle point' to include all such stationary points which are neither local maxima nor local minima.

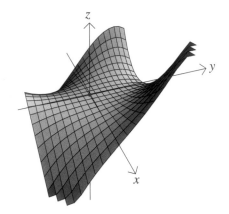

Surface plot of $z = x^2 y + y^3 x$
near the origin

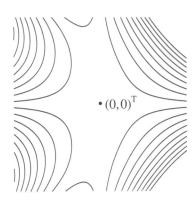

Contours of $z = x^2 y + y^3 x$ near
the origin

5.8 Exercises

1* The equations

$$\left. \begin{array}{l} u = f_1(x, y) = \tfrac{1}{2}(x^2 - y^2) \\ v = f_2(x, y) = xy \end{array} \right\}$$

define a function $\mathbf{f}: \mathbb{R}^2 \rightarrow \mathbb{R}^2$ whose component functions are f_1 and f_2.

Sketch the contours $f_1(x, y) = 6$ and $f_2(x, y) = 8$ on the same diagram. Indicate on the diagram the values of $(x, y)^{\mathrm{T}}$ for which

$$\mathbf{f}(x, y) = (6, 8)^{\mathrm{T}}.$$

2. The equations

$$\left. \begin{array}{l} u = f_1(x, y) = x^2 + y^2 \\ v = f_2(x, y) = x^2 - y^2 \end{array} \right\}$$

define a function $\mathbf{f}: \mathbb{R}^2 \rightarrow \mathbb{R}^2$ whose component functions are f_1 and f_2.

Sketch the contours $f_1(x, y) = 2$, $f_1(x, y) = 8$, $f_2(x, y) = -3$ and $f_2(x, y) = 4$ on the same diagram. Indicate on the diagram the set of $(x, y)^T$ which correspond to values of $(u, v)^T$ for which $2 \leq u \leq 8$ and $-3 \leq v \leq 4$.

3. The function $\mathbf{f} \colon \mathbb{R}^2 \to \mathbb{R}^2$ is defined by

$$\begin{pmatrix} u \\ v \end{pmatrix} = \mathbf{f}(x, y) = \begin{pmatrix} xy \\ f(x, y) \end{pmatrix}$$

where $\qquad f(x, y) = $ (i) $y \quad$ (ii) $y + 2x \quad$ and (iii) $x^2 + y^2$.

In each case, sketch the contours $xy = 1$, $xy = 2$, $f(x, y) = 3$ and $f(x, y) = 4$ on the same diagram. Indicate on the diagram, if possible, the set of $(x, y)^T$ which correspond to values of $(u, v)^T$ for which $1 \leq u \leq 2$ and $3 \leq v \leq 4$.

4.* Let S be the set of vectors $(x, y, z)^T$ in $\mathbb{R}^3$ for which $y \neq 0$ and $z \neq 0$. Find the derivative at the point $(x, y, z)^T = (1, 2, 3)^T$ of the function $\mathbf{f} \colon S \to \mathbb{R}^2$ defined by the equations

$$u = \frac{x}{y}$$
$$v = \frac{y}{z}$$

Write down the equations for the tangent flat at the point where $(x, y, z)^T = (1, 2, 3)^T$.

5. Find the derivative at the given point of the functions defined by the following sets of equations. Write down the equations for the corresponding tangent flats in each case.

(i) $\quad z = x^2 + 2xy + y^3 \qquad$ at $(x, y)^T = (1, 2)^T$

(ii) $\quad \begin{cases} z_1 = \cos(xy) \\ z_2 = \sin \dfrac{x}{y} \end{cases} \qquad$ at $(x, y)^T = (\pi, 1)^T$

(iii)* $\quad \begin{cases} r = \{x^2 + y^2\}^{1/2} \\ \theta = \arctan \dfrac{y}{x} \end{cases} \qquad$ at $(x, y)^T = (1, 0)^T$

(iv) $\quad \begin{cases} u = x \\ v = x + y \\ w = xy \end{cases} \qquad$ at $(x, y)^T = (1, 1)^T$

(v) $\quad \begin{cases} u = 1 \\ v = t \\ w = t^2 \end{cases} \qquad$ at $t = 0$

(vi) $\quad \begin{cases} u = xy \\ v = yz \end{cases} \qquad$ at $(x, y, z)^T = (1, 1, 1)^T$

6.* Show that $z = f(x^2 + y^2)$ satisfies the partial differential equation

$$y \frac{\partial z}{\partial x} = x \frac{\partial z}{\partial y}$$

for any differentiable function $f: \mathbb{R} \to \mathbb{R}$.

7. Show that $z = f(y \ln x)$ satisfies the partial differential equation

$$x\frac{\partial z}{\partial x} + y\frac{\partial z}{\partial y} = xy\frac{\partial^2 z}{\partial x\,\partial y} - x^2 \ln x\,\frac{\partial^2 z}{\partial x^2}$$

for any differentiable function $f: \mathbb{R} \to \mathbb{R}$.

8.* Let the function $P: \mathbb{R}_+^2 \to \mathbb{R}$ be defined by

$$P(x, y) = Ax^{\frac{1}{3}}y^{\frac{2}{3}}$$

where $x = 2t^2$ and $y = t^3 + 1$.

Verify the chain rule for the composite function $P(x(t), y(t))$.

9.* Use the chain rule to calculate dz/dt in the following cases:

(i) $z = \cos(x^2 y)$ and let $x = t^3$ and $y = t^2$

(ii) $z = x^2 y^5$ and let $x = \cos t$ and $y = \sin t$

10.* Find the second derivative of the real valued function $f: \mathbb{R}^2 \to \mathbb{R}$ defined by

$$f(x, y) = xe^{-y}$$

Write down the quadratic Taylor approximation about the point $(x, y)^{\mathrm{T}} = (1, 0)^{\mathrm{T}}$.

Use a computer to obtain surface and contour plots of the function and its quadratic Taylor approximation near $(1, 0)^{\mathrm{T}}$, and observe the resemblance between them.

11. Find the second derivative of the real valued function $P: \mathbb{R}_+^2 \to \mathbb{R}$ defined by

$$P(x, y) = x^{\frac{1}{2}}y^{\frac{1}{2}}$$

Write down the quadratic Taylor approximation about the point $(x, y)^{\mathrm{T}} = (1, 1)^{\mathrm{T}}$.

Use a computer to obtain surface and contour plots of the function and its quadratic Taylor approximation near $(1, 0)^{\mathrm{T}}$, and observe the resemblance between them.

12. Let S be the set of all vectors $(x, y, z)^{\mathrm{T}}$ in $\mathbb{R}^3$ for which $x > 0$, $y > 0$ and $z > 0$. Find the second derivative of the real valued function $f: S \to \mathbb{R}$ defined by

$$u = f(x, y, z) = \ln(x + yz^{-1})$$

Write down the quadratic Taylor expansion about the point $(x, y, z)^{\mathrm{T}} = (1, 1, 1)^{\mathrm{T}}$.

13.* Let $f: \mathbb{R}^2 \to \mathbb{R}$ be homogeneous of degree n. This means that, for all x, y and λ,

$$f(\lambda x, \lambda y) = \lambda^n f(x, y)$$

Let $X = \lambda x$, $Y = \lambda y$ where x and y are constant. Apply the chain rule to the function $f(X, Y)$ to prove that

$$n\lambda^{n-1} f(x, y) = x f_X(X, Y) + y f_Y(X, Y)$$

Let $\lambda = 1$ to deduce **Euler's theorem** for a homogeneous function of degree n:

$$x \frac{\partial f}{\partial x} + y \frac{\partial f}{\partial y} = nf$$

14. (i) Show that the Cobb–Douglas function $P: \mathbb{R}_+^2 \to \mathbb{R}$ defined by

$$P(x, y) = x^\alpha y^\beta \quad \text{where } 0 < \alpha < 1, 0 < \beta < 1$$

is homogeneous of degree $\alpha + \beta$.

(ii) Show that the constant elasticity of substitution (CES) function $P: \mathbb{R}_+^2 \to \mathbb{R}$ defined by

$$P(x, y) = [\alpha x^\beta + (1 - \alpha) y^\beta]^{\frac{1}{\beta}} \quad \text{where } 0 < \alpha < 1, \ \beta < 1$$

is homogeneous of degree one.

Verify Euler's theorem for each of these functions.

Six   Optimisation of scalar valued functions

The technique in classifying the stationary points of a real valued function of one real variable is to examine the sign of the second derivative. The same technique works in the case of a real valued function of a vector variable except that, in this case, the second derivative is a matrix. Instead of asking whether this second derivative is positive or negative we ask instead whether it is 'positive definite' or 'negative definite'. The chapter begins with a section on linear algebra which explains these notions, although it is possible for most purposes to manage just by remembering statements (I) and (II) of §6.2.

Only the linear algebra that is required to use the technique is supplied. This is only a survey of the appropriate ideas and is not an adequate introduction for those to whom these ideas are entirely new.

6.1   Change of basis in quadratic forms$^\diamond$

A set of n vectors $\mathbf{v}_1, \mathbf{v}_2, \ldots, \mathbf{v}_n$ is **linearly independent** if the equation

$$c_1\mathbf{v}_1 + c_2\mathbf{v}_2 + \cdots + c_n\mathbf{v}_n = \mathbf{0}$$

holds if and only if $c_1 = c_2 = \cdots = c_n = 0$.

A set of n vectors $\mathbf{b}_1, \mathbf{b}_2, \ldots, \mathbf{b}_n$ is a **basis** for $\mathbb{R}^n$ if and only if each vector $\mathbf{x}$ in $\mathbb{R}^n$ can be expressed uniquely in the form

$$\mathbf{x} = X_1\mathbf{b}_1 + X_2\mathbf{b}_2 + \cdots + X_n\mathbf{b}_n \tag{1}$$

Any set of n linearly independent vectors in $\mathbb{R}^n$ is a basis for $\mathbb{R}^n$ and any set of basis vectors is linearly independent. The entries in the column vector

$$\mathbf{X} = \begin{pmatrix} X_1 \\ X_2 \\ \vdots \\ X_n \end{pmatrix}$$

are then called the **coordinates** of the vector $\mathbf{x}$ in (1) with respect to the basis $\mathbf{b}_1$, $\mathbf{b}_2, \ldots, \mathbf{b}_n$.

The **natural basis** for $\mathbb{R}^n$ is the set of vectors

$$
\mathbf{e}_1 = \begin{pmatrix} 1 \\ 0 \\ 0 \\ \vdots \\ 0 \end{pmatrix} \quad
\mathbf{e}_2 = \begin{pmatrix} 0 \\ 1 \\ 0 \\ \vdots \\ 0 \end{pmatrix} \quad \cdots \quad
\mathbf{e}_n = \begin{pmatrix} 0 \\ 0 \\ \vdots \\ 0 \\ 1 \end{pmatrix}
$$

The reason for calling these vectors the natural basis is that

$$
\mathbf{x} = \begin{pmatrix} x_1 \\ x_2 \\ \vdots \\ x_n \end{pmatrix} = x_1\mathbf{e}_1 + x_2\mathbf{e}_2 + \cdots + x_n\mathbf{e}_n
$$

Given $x_1, x_2, \ldots, x_n$, one can find $X_1, X_2, \ldots, X_n$, the coordinates of $\mathbf{x}$ with respect to the new basis $\mathbf{b}_1, \mathbf{b}_2, \ldots, \mathbf{b}_n$ as follows.

Observe from (1) that

$$
\mathbf{x} = X_1\mathbf{b}_1 + X_2\mathbf{b}_2 + \cdots + X_n\mathbf{b}_n
$$

$$
= X_1 \begin{pmatrix} b_{11} \\ b_{21} \\ \vdots \\ b_{n1} \end{pmatrix}
+ X_2 \begin{pmatrix} b_{12} \\ b_{22} \\ \vdots \\ b_{n2} \end{pmatrix}
+ \cdots + X_n \begin{pmatrix} b_{1n} \\ b_{2n} \\ \vdots \\ b_{nn} \end{pmatrix}
$$

$$
= \begin{pmatrix} b_{11}X_1 + b_{12}X_2 + \cdots + b_{1n}X_n \\ b_{21}X_1 + b_{22}X_2 + \cdots + b_{2n}X_n \\ \vdots \\ b_{n1}X_1 + b_{n2}X_2 + \cdots + b_{nn}X_n \end{pmatrix}
= \begin{pmatrix} b_{11} & b_{12} & \cdots & b_{1n} \\ b_{21} & b_{22} & \cdots & b_{2n} \\ \vdots & \vdots & & \vdots \\ b_{n1} & b_{n2} & \cdots & b_{nn} \end{pmatrix} \begin{pmatrix} X_1 \\ X_2 \\ \vdots \\ X_n \end{pmatrix}
$$

$$
= P\mathbf{X}
$$

where

$$
P = \begin{pmatrix} | & | & & | \\ \mathbf{b}_1 & \mathbf{b}_2 & \cdots & \mathbf{b}_n \\ | & | & & | \end{pmatrix}
$$

The matrix P is invertible because its columns are a basis and so are linearly independent. We can therefore multiply $\mathbf{x} = P\mathbf{X}$ by P^{-1} to obtain

$$
\mathbf{X} = P^{-1}\mathbf{x}
$$

A basis is **orthonormal** if its constituent vectors are mutually orthogonal and of unit length. The natural basis is orthonormal. If a *new* basis is created by a rotation of the natural basis vectors about an arbitrary axis, the new basis vectors will then also be orthonormal. The vectors $\mathbf{b}_1, \mathbf{b}_2, \ldots, \mathbf{b}_n$ are orthonormal if and only if

$$P^{\mathrm{T}}P = \begin{pmatrix} \underline{\quad} & \mathbf{b}_1^{\mathrm{T}} & \underline{\quad} \\ \underline{\quad} & \mathbf{b}_2^{\mathrm{T}} & \underline{\quad} \\ & \vdots & \\ \underline{\quad} & \mathbf{b}_n^{\mathrm{T}} & \underline{\quad} \end{pmatrix} \begin{pmatrix} | & | & & | \\ \mathbf{b}_1 & \mathbf{b}_2 & \cdots & \mathbf{b}_n \\ | & | & & | \end{pmatrix}$$

$$= \begin{pmatrix} \mathbf{b}_1^{\mathrm{T}}\mathbf{b}_1 & \mathbf{b}_1^{\mathrm{T}}\mathbf{b}_2 & \cdots & \mathbf{b}_1^{\mathrm{T}}\mathbf{b}_n \\ \mathbf{b}_2^{\mathrm{T}}\mathbf{b}_1 & \mathbf{b}_2^{\mathrm{T}}\mathbf{b}_2 & \cdots & \mathbf{b}_2^{\mathrm{T}}\mathbf{b}_n \\ \vdots & \vdots & & \vdots \\ \mathbf{b}_n^{\mathrm{T}}\mathbf{b}_1 & \mathbf{b}_n^{\mathrm{T}}\mathbf{b}_2 & \cdots & \mathbf{b}_n^{\mathrm{T}}\mathbf{b}_n \end{pmatrix}$$

$$= \begin{pmatrix} \|\mathbf{b}_1\|^2 & \langle\mathbf{b}_1,\mathbf{b}_2\rangle & \cdots & \langle\mathbf{b}_1,\mathbf{b}_n\rangle \\ \langle\mathbf{b}_2,\mathbf{b}_1\rangle & \|\mathbf{b}_2\|^2 & \cdots & \langle\mathbf{b}_2,\mathbf{b}_n\rangle \\ \vdots & \vdots & & \vdots \\ \langle\mathbf{b}_n,\mathbf{b}_1\rangle & \langle\mathbf{b}_n,\mathbf{b}_2\rangle & \cdots & \|\mathbf{b}_n\|^2 \end{pmatrix} = \begin{pmatrix} 1 & 0 & \cdots & 0 \\ 0 & 1 & \cdots & 0 \\ \vdots & \vdots & & \vdots \\ 0 & 0 & \cdots & 1 \end{pmatrix} = I$$

A matrix P satisfying $P^{\mathrm{T}}P = I$ is called **orthogonal**. As we have seen, a matrix P is orthogonal if and only if its columns are orthonormal. It can save a lot of calculation to note that an orthogonal matrix is invertible and that its inverse is simply its transpose – i.e.

$$P^{-1} = P^{\mathrm{T}}.$$

A change of basis can be used to transform a general quadratic expression into a 'standard form' which involves only the squares of the variables. This method entails first expressing the second degree terms in the form

$$z = \mathbf{x}^{\mathrm{T}}A\mathbf{x}$$

where $\mathbf{x}$ is an $n \times 1$ column vector and A is an $n \times n$ *symmetric* matrix. We call such an expression a **quadratic form**. A quadratic form can be reduced to a sum of squares by an orthogonal change of variable that transforms A into a diagonal matrix. When A is symmetric there always exists an orthogonal matrix P, for which

$$\mathbf{x} = P\mathbf{X}$$

$$P^{\mathrm{T}}AP = D$$

where D is a diagonal matrix. Changing the basis to the columns of P therefore transforms a quadratic form in the following way:

$$z = \mathbf{x}^{\mathrm{T}}A\mathbf{x} = (P\mathbf{X})^{\mathrm{T}}AP\mathbf{X} = \mathbf{X}^{\mathrm{T}}(P^{\mathrm{T}}AP)\mathbf{X} = \mathbf{X}^{\mathrm{T}}D\mathbf{X}$$

It follows that

$$z = (X_1, X_2, \ldots, X_n) \begin{pmatrix} \lambda_1 & 0 & \cdots & 0 \\ 0 & \lambda_2 & \cdots & 0 \\ \vdots & \vdots & & \vdots \\ 0 & 0 & \cdots & \lambda_n \end{pmatrix} \begin{pmatrix} X_1 \\ X_2 \\ \vdots \\ X_n \end{pmatrix}$$

i.e.

$$z = \lambda_1 X_1^2 + \lambda_2 X_2^2 + \cdots + \lambda_n X_n^2$$

The change of variable $\mathbf{X} = P^{\mathrm{T}}\mathbf{x}$ from the natural basis to the basis $\mathbf{b}_1, \mathbf{b}_2, \ldots, \mathbf{b}_n$ therefore reduces the quadratic form to a simple sum of squares.

How do we work out $\lambda_1, \lambda_2, \ldots, \lambda_n$?[†] They satisfy the equation

$$\det(A - \lambda I) = 0$$

[†] These numbers are called the **eigenvalues** of the matrix A.

To see this we begin by observing that

$$P^{\mathrm{T}}AP = D$$

implies

$$AP = PD$$

because $P^{-1} = P^{\mathrm{T}}$ when P is orthogonal. The ith column of P is $\mathbf{b}_i$ and so the ith column of AP is $A\mathbf{b}_i$ for each $i = 1, 2, \ldots, n$. Since $AP = PD$, $A\mathbf{b}_i$ must be equal to the ith column of PD. But

$$PD = \begin{pmatrix} | & | & & | \\ \mathbf{b}_1 & \mathbf{b}_2 & \cdots & \mathbf{b}_n \\ | & | & & | \end{pmatrix} \begin{pmatrix} \lambda_1 & 0 & \cdots & 0 \\ 0 & \lambda_2 & \cdots & 0 \\ \vdots & \vdots & & \vdots \\ 0 & 0 & \cdots & \lambda_n \end{pmatrix}$$

and so

$$A\mathbf{b}_i = \lambda_i \mathbf{b}_i \quad \text{for each } i = 1, 2, \ldots, n$$

[‡] The vectors $\mathbf{b}_i$ are called the **eigenvectors** of the matrix A and the scalars λ_i are the corresponding eigenvalues.

[‡] Since $\lambda_i \mathbf{b}_i = \lambda_i I \mathbf{b}_i$, we have

$$(A - \lambda_i I)\mathbf{b}_i = \mathbf{0} \quad \text{for each } i = 1, 2, \ldots, n \quad (2)$$

The matrix $(A - \lambda_i I)$ cannot therefore be invertible, otherwise

$$\mathbf{b}_i = (A - \lambda_i I)^{-1}\mathbf{0} = \mathbf{0}$$

No $\mathbf{b}_i = \mathbf{0}$ since $\mathbf{b}_1, \mathbf{b}_2, \ldots, \mathbf{b}_n$ are linearly independent. The determinant of each matrix $(A - \lambda_i I)$ must therefore be zero. Hence the eigenvalues are the roots of the equation

$$\det(A - \lambda I) = 0$$

which is called the **characteristic equation** of A. This equation is a polynomial of degree n in λ and hence has n roots, namely $\lambda_1, \lambda_2, \ldots, \lambda_n$.

In general, the eigenvalues of a matrix are complex numbers. However, when A is *symmetric* the eigenvalues are always *real* numbers. For each eigenvalue the corresponding eigenvectors can be calculated from equation (2). We need to restrict our attention to eigenvectors of unit length – i.e. satisfying $\|\mathbf{b}\| = 1$.

If A is symmetric and its eigenvalues $\lambda_1, \lambda_2, \ldots, \lambda_n$ are *distinct*, then the corresponding unit eigenvectors $\mathbf{b}_1, \mathbf{b}_2, \ldots, \mathbf{b}_n$ are mutually orthogonal and so form an orthonormal set. The matrix P whose columns are $\mathbf{b}_1, \mathbf{b}_2, \ldots, \mathbf{b}_n$ is then an *orthogonal* matrix – i.e. $P^{\mathrm{T}}P = I$.

If the eigenvalues of A are not distinct – i.e. the equation $\det(A - \lambda I) = 0$ has multiple roots – then the discussion above remains valid except that several of the eigenvectors $\mathbf{b}_1, \mathbf{b}_2, \ldots, \mathbf{b}_n$ will be derived from the *same* eigenvalue.

Considering the case when $n = 2$, suppose that the basis in $\mathbb{R}^2$ is changed from the natural basis

$$\mathbf{e}_1 = \begin{pmatrix} 1 \\ 0 \end{pmatrix} \quad \mathbf{e}_2 = \begin{pmatrix} 0 \\ 1 \end{pmatrix}$$

to the basis

$$\mathbf{b} = \begin{pmatrix} \cos\theta \\ \sin\theta \end{pmatrix} \quad \tilde{\mathbf{b}} = \begin{pmatrix} -\sin\theta \\ \cos\theta \end{pmatrix}$$

This is a rotation which is an orthogonal transformation with

$$P = \begin{pmatrix} \cos\theta & -\sin\theta \\ \sin\theta & \cos\theta \end{pmatrix}$$

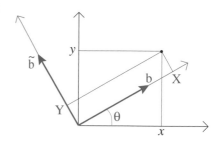

and so

$$P^{-1} = P^{\mathrm{T}} = \begin{pmatrix} \cos\theta & \sin\theta \\ -\sin\theta & \cos\theta \end{pmatrix}$$

Thus

$$\begin{pmatrix} X \\ Y \end{pmatrix} = \begin{pmatrix} \cos\theta & \sin\theta \\ -\sin\theta & \cos\theta \end{pmatrix} \begin{pmatrix} x \\ y \end{pmatrix}$$

Example 1

Consider the quadratic form

$$z = 5x^2 - 6xy + 5y^2$$

We write this in the form

$$z = (x, y) \begin{pmatrix} 5 & -3 \\ -3 & 5 \end{pmatrix} \begin{pmatrix} x \\ y \end{pmatrix}$$

The characteristic equation of the matrix is

$$\left| \begin{pmatrix} 5 & -3 \\ -3 & 5 \end{pmatrix} - \lambda \begin{pmatrix} 1 & 0 \\ 0 & 1 \end{pmatrix} \right| = \left| \begin{matrix} 5-\lambda & -3 \\ -3 & 5-\lambda \end{matrix} \right| = 0$$

i.e.

$$(5 - \lambda)^2 - 9 = 0$$
$$5 - \lambda = \pm 3$$
$$\lambda = 2 \quad \text{or} \quad \lambda = 8$$

The eigenvalues of the matrix are therefore $\lambda_1 = 2$ and $\lambda_2 = 8$.

We can change the variables from $(x, y)^{\mathrm{T}}$ to $(X, Y)^{\mathrm{T}}$ in such a way that the quadratic form reduces to

$$z = 2X^2 + 8Y^2$$

This makes it obvious that z has a minimum at the origin, and that the contour $5x^2 - 6xy + 5y^2 = c$ is an ellipse:

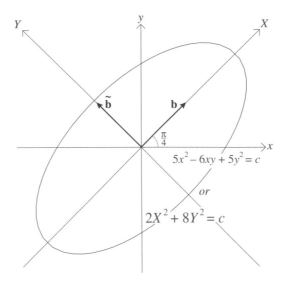

The new X and Y axes point in the directions of the eigenvectors $\mathbf{b}$ and $\tilde{\mathbf{b}}$. We can find $\mathbf{b}$ by solving

$$\begin{pmatrix} 5 & -3 \\ -3 & 5 \end{pmatrix} \begin{pmatrix} b_1 \\ b_2 \end{pmatrix} = 2 \begin{pmatrix} b_1 \\ b_2 \end{pmatrix} - \text{i.e.} \quad \left. \begin{array}{c} 5b_1 - 3b_2 = 2b_1 \\ -3b_1 + 5b_2 = 2b_2 \end{array} \right\}$$

Both equations yield $b_1 = b_2$. We want an eigenvector of unit length and so we take $b_1 = b_2 = 1/\sqrt{2}$. Thus

$$\mathbf{b} = \begin{pmatrix} \frac{1}{\sqrt{2}} \\ \frac{1}{\sqrt{2}} \end{pmatrix}$$

We find $\tilde{\mathbf{b}}$ by solving

$$\begin{pmatrix} 5 & -3 \\ -3 & 5 \end{pmatrix} \begin{pmatrix} \tilde{b}_1 \\ \tilde{b}_2 \end{pmatrix} = 8 \begin{pmatrix} \tilde{b}_1 \\ \tilde{b}_2 \end{pmatrix} - \text{i.e.} \quad \left. \begin{array}{c} 5\tilde{b}_1 - 3\tilde{b}_2 = 8\tilde{b}_1 \\ -3\tilde{b}_1 + 5\tilde{b}_2 = 8\tilde{b}_2 \end{array} \right\}$$

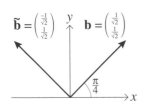

Both equations yield $\tilde{b}_1 = -\tilde{b}_2$. Again we want a unit eigenvector and we take $\tilde{b}_1 = -\tilde{b}_2 = -\frac{1}{\sqrt{2}}$. Thus

$$\tilde{\mathbf{b}} = \begin{pmatrix} -\frac{1}{\sqrt{2}} \\ \frac{1}{\sqrt{2}} \end{pmatrix}$$

The orthogonal matrix P is given by

$$P = \begin{pmatrix} | & | \\ \mathbf{b} & \tilde{\mathbf{b}} \\ | & | \end{pmatrix} = \begin{pmatrix} \dfrac{1}{\sqrt{2}} & -\dfrac{1}{\sqrt{2}} \\ \dfrac{1}{\sqrt{2}} & \dfrac{1}{\sqrt{2}} \end{pmatrix}$$

Thus the new coordinates $(X, Y)^T$ are related to the old coordinates $(x, y)^T$ by the formulas

$$\begin{pmatrix} x \\ y \end{pmatrix} = \begin{pmatrix} \dfrac{1}{\sqrt{2}} & -\dfrac{1}{\sqrt{2}} \\ \dfrac{1}{\sqrt{2}} & \dfrac{1}{\sqrt{2}} \end{pmatrix} \begin{pmatrix} X \\ Y \end{pmatrix}, \qquad \begin{pmatrix} X \\ Y \end{pmatrix} = \begin{pmatrix} \dfrac{1}{\sqrt{2}} & \dfrac{1}{\sqrt{2}} \\ -\dfrac{1}{\sqrt{2}} & \dfrac{1}{\sqrt{2}} \end{pmatrix} \begin{pmatrix} x \\ y \end{pmatrix}$$

Finally, it is of some interest to note that

$$P = \begin{pmatrix} \dfrac{1}{\sqrt{2}} & -\dfrac{1}{\sqrt{2}} \\ \dfrac{1}{\sqrt{2}} & \dfrac{1}{\sqrt{2}} \end{pmatrix} = \begin{pmatrix} \cos\frac{\pi}{4} & -\sin\frac{\pi}{4} \\ \sin\frac{\pi}{4} & \cos\frac{\pi}{4} \end{pmatrix}$$

Hence we see that the new axes are obtained from the old axes by an anticlockwise rotation through $\frac{\pi}{4}$.

In fact, it is generally true, that whenever a quadratic form has the coefficient of x^2 equal to the coefficient of y^2, the xy term can be eliminated by an anticlockwise rotation through $\frac{\pi}{4}$.

See Exercise 6.5.12.

6.2 Positive and negative definite

A symmetric matrix is called **positive definite** if *all* its eigenvalues are positive. It is called **negative definite** if *all* its eigenvalues are negative.

As explained in §6.1,

$$\mathbf{x}^T A \mathbf{x} = \lambda_1 X_1^2 + \lambda_2 X_2^2 + \cdots + \lambda_n X_n^2$$

and hence A is positive definite if and only if

$$z = \mathbf{x}^T A \mathbf{x} > 0$$

unless $\mathbf{x} = \mathbf{0}$. Similarly, A is negative definite if and only if

$$z = \mathbf{x}^T A \mathbf{x} < 0$$

unless $\mathbf{x} = \mathbf{0}$.

When $\mathbf{x}$ is a two dimensional column vector, a quadratic form $z = \mathbf{x}^T A \mathbf{x}$ is a surface in three dimensional space. We can figure out the type of surface by introducing an orthogonal change of variable from $\mathbf{x}$ to $\mathbf{X}$ that reduces the quadratic form to a sum of squares:

$$z = \lambda_1 X^2 + \lambda_2 Y^2$$

The possible shapes that can arise are the following:

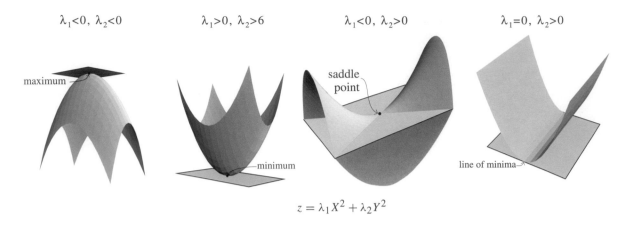

$$z = \lambda_1 X^2 + \lambda_2 Y^2$$

Fortunately, it is not necessary to calculate the eigenvalues of A to determine whether or not A is positive definite or negative definite. Simpler criteria exist for this. First define the **principal minors** of a symmetric matrix

$$A = \begin{pmatrix} a_{11} & a_{12} & \dots & a_{1n} \\ a_{21} & a_{22} & \dots & a_{2n} \\ \vdots & \vdots & & \vdots \\ a_{n1} & a_{n2} & \dots & a_{nn} \end{pmatrix}$$

as the determinants

$$a_{11} \qquad \begin{vmatrix} a_{11} & a_{12} \\ a_{21} & a_{22} \end{vmatrix} \qquad \begin{vmatrix} a_{11} & a_{12} & a_{13} \\ a_{21} & a_{22} & a_{23} \\ a_{31} & a_{32} & a_{33} \end{vmatrix} \qquad \dots \qquad \begin{vmatrix} a_{11} & a_{12} & \dots & a_{1n} \\ a_{21} & a_{22} & \dots & a_{2n} \\ \vdots & \vdots & & \vdots \\ a_{n1} & a_{n2} & \dots & a_{nn} \end{vmatrix}$$

We then have

(I)   A symmetric matrix is positive definite if and only if *all* of its principal minors are *positive*.

(II)   A symmetric matrix is negative definite if and only if all of its principal minors of *even* order are *positive* and all its principal minors of *odd* order are *negative*.

This result can be proved quite easily for the case of a 2×2 symmetric matrix

$$A = \begin{pmatrix} \alpha & \gamma \\ \gamma & \beta \end{pmatrix}$$

The characteristic equation is

$$\begin{vmatrix} \alpha - \lambda & \gamma \\ \gamma & \beta - \lambda \end{vmatrix} = (\alpha - \lambda)(\beta - \lambda) - \gamma^2$$

$$= \lambda^2 - \lambda(\alpha + \beta) + (\alpha\beta - \gamma^2)$$

If the eigenvalues are λ_1 and λ_2, we therefore have

$$(\lambda - \lambda_1)(\lambda - \lambda_2) = \lambda^2 - \lambda(\alpha + \beta) + (\alpha\beta - \gamma^2)$$

and so

$$\begin{cases} \lambda_1 + \lambda_2 = \alpha + \beta \\ \lambda_1\lambda_2 = \alpha\beta - \gamma^2 = \begin{vmatrix} \alpha & \gamma \\ \gamma & \beta \end{vmatrix} = \det(A) \end{cases}$$

The roots λ_1 and λ_2 have the same sign if and only if $\lambda_1\lambda_2 > 0$ – i.e.

$$\det(A) = \alpha\beta - \gamma^2 = \lambda_1\lambda_2 > 0$$

But $\alpha\beta > \gamma^2$ implies that α and β have the same sign. If $\lambda_1 > 0$ and $\lambda_2 > 0$, then $\alpha + \beta = \lambda_1 + \lambda_2 > 0$ and hence $\alpha > 0$. Similarly, if $\lambda_1 < 0$ and $\lambda_2 < 0$, then $\alpha + \beta = \lambda_1 + \lambda_2 < 0$ and hence $\alpha < 0$.

These results, incidentally, are special cases of more general results. For any $n \times n$ matrix A,

$$\lambda_1 + \lambda_2 + \cdots + \lambda_n = a_{11} + a_{22} + \cdots + a_{nn}$$

$$\lambda_1\lambda_2 \ldots \lambda_n = \det(A)$$

Example 2

(i) Consider the matrix

$$A = \begin{pmatrix} 5 & -3 \\ -3 & 5 \end{pmatrix}$$

We know that this matrix has eigenvalues $\lambda_1 = 2$ and $\lambda_2 = 8$ and hence is positive definite (see Example 1). The matrix can therefore be used to check our criterion (I) concerning the principal minors of a matrix. Observe that

$$(a) \; \begin{vmatrix} 5 & -3 \\ -3 & 5 \end{vmatrix} = 25 - 9 = 16 > 0 \quad \text{and} \quad (b) \; 5 > 0$$

Thus criterion (I) is satisfied.

(ii) Consider the matrix

$$B = \begin{pmatrix} 1 & 6 \\ 6 & 4 \end{pmatrix}$$

Calculating the principal minors, we obtain

$$(a) \; \begin{vmatrix} 1 & 6 \\ 6 & 4 \end{vmatrix} = 4 - 36 = -32 < 0 \quad \text{and} \quad (b) \; 1 > 0$$

Thus the matrix is neither positive definite nor negative definite because neither criterion (I) nor criterion (II) is satisfied.[†]

[†] Note that (b) is unnecessary for this conclusion.

(iii) Consider the matrix

$$C = \begin{pmatrix} -2 & 1 \\ 1 & -2 \end{pmatrix}$$

Calculating the principal minors, we obtain

$$\text{(a)} \quad \begin{vmatrix} -2 & 1 \\ 1 & -2 \end{vmatrix} = 4 - 1 = 3 > 0 \quad \text{and} \quad \text{(b)} \quad -2 < 0$$

Hence criterion (II) is satisfied and so C is negative definite.

Since $\lambda_1 \lambda_2 \ldots \lambda_n = \det(A)$, it follows that $\det(A) = 0$ implies that at least one of the eigenvalues of A is zero. In view of this result, it is tempting to suppose that if some of the principal minors of A are zero and the rest are positive, then the matrix A will necessarily have nonnegative eigenvalues. But this is *false* as the next example shows. When some of the principal minors are zero, criteria (I) and (II) have nothing to say about the signs of the eigenvalues of the matrix A.

Example 3

Consider the matrix

$$A = \begin{pmatrix} 1 & 1 & 0 \\ 1 & 1 & 0 \\ 0 & 0 & t \end{pmatrix}$$

The principal minors are all nonnegative because

$$\text{(a)} \quad \begin{vmatrix} 1 & 1 & 0 \\ 1 & 1 & 0 \\ 0 & 0 & t \end{vmatrix} = t - t = 0, \quad \text{(b)} \quad \begin{vmatrix} 1 & 1 \\ 1 & 1 \end{vmatrix} = 1 - 1 = 0 \text{ and (c) } 1 > 0.$$

But the characteristic equation of A is

$$\begin{vmatrix} 1 - \lambda & 1 & 0 \\ 1 & 1 - \lambda & 0 \\ 0 & 0 & t - \lambda \end{vmatrix} = (t - \lambda)\{(1 - \lambda)^2 - 1\} = 0$$

and hence the eigenvalues are $\lambda_1 = t$, $\lambda_2 = 0$, $\lambda_3 = 2$. But t may be positive or it may be negative. The values of the principal minors are therefore not much use in this case.

6.3 Maxima and minima

We have seen that the stationary points for a function $f : \mathbb{R}^n \to \mathbb{R}$ are those at which all the partial derivatives are zero. That is

$$\frac{dy}{dx} = \left(\frac{\partial y}{\partial x_1}, \frac{\partial y}{\partial x_2}, \ldots, \frac{\partial y}{\partial x_n} \right) = (0, 0, \ldots, 0) = \mathbf{0}$$

Thus ξ is a stationary point if and only if

$$f'(\xi) = \mathbf{0}$$

How does one determine whether a stationary point ξ is a local maximum, a local minimum or a saddle point?

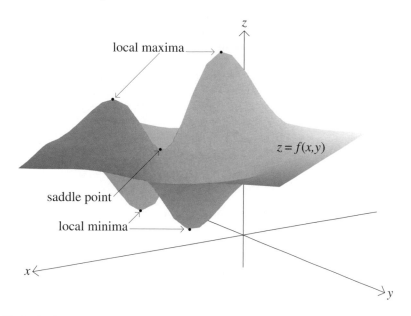

Taylor's theorem asserts that

$$f(\mathbf{x}) = f(\xi) + \frac{1}{1!}f'(\xi)(\mathbf{x} - \xi) + \frac{1}{2!}(\mathbf{x} - \xi)^{\mathrm{T}}f''(\xi)(\mathbf{x} - \xi) + \cdots$$

At a stationary point, $f'(\xi) = \mathbf{0}$ and so

$$f(\mathbf{x}) - f(\xi) = \frac{1}{2!}(\mathbf{x} - \xi)^{\mathrm{T}}f''(\xi)(\mathbf{x} - \xi) + \cdots$$

Provided that $\mathbf{x}$ is sufficiently close to ξ, the third and higher order terms in this expression are negligible compared with

$$\frac{1}{2!}(\mathbf{x} - \xi)^{\mathrm{T}}f''(\xi)(\mathbf{x} - \xi)$$

(unless this term happens to be zero). Recall that $f''(\xi)$ is an $n \times n$ symmetric matrix. If we write $\mathbf{u} = \mathbf{x} - \xi$, we therefore have to deal with a quadratic form

$$\frac{1}{2}\mathbf{u}^{\mathrm{T}}f''(\xi)\mathbf{u}$$

If this quadratic form is always positive (unless $\mathbf{u} = \mathbf{0}$), then $f(\mathbf{x}) > f(\xi)$ for all $\mathbf{x}$ close to ξ (except $\mathbf{x} = \xi$). If the quadratic form is always negative (unless $\mathbf{u} = \mathbf{0}$), then $f(\mathbf{x}) < f(\xi)$ for all $\mathbf{x}$ close to ξ (except $\mathbf{x} = \xi$).

Note that a stationary point ξ may well be a local maximum or minimum even though $f''(\xi)$ is neither positive definite nor negative definite. In particular, if

$\det f''(\xi) = 0$, then at least one of the eigenvalues of $f''(\xi)$ is zero. Hence near ξ, $\mathbf{u}^T f''(\xi)\mathbf{u}$ is *zero* either along some line through ξ or everywhere near ξ. Then either along this line or everywhere near ξ the third and higher order terms of the Taylor approximation are *not* negligible and the question as to whether ξ has a local maximum or minimum at ξ will depend on their nature.

If $f''(\xi)$ has some positive eigenvalues and some negative eigenvalues, then the function has a **saddle point** at ξ.

We summarise these results:

Let $f'(\xi) = \mathbf{0}$.

(I) If $f''(\xi)$ is positive definite, then ξ is a local minimum.

(II) If $f''(\xi)$ is negative definite, then ξ is a local maximum.

(III) If $\det f''(\xi) \neq 0$ but $f''(\xi)$ is neither positive definite nor negative definite, then ξ is a saddle point.

In terms of the eigenvalues:

- $f''(\xi)$ *positive definite means that* $\lambda_1 > 0, \lambda_2 > 0, \ldots, \lambda_n > 0$ *and* ξ *is a local minimum;*

- $f''(\xi)$ *negative definite means that* $\lambda_1 < 0, \lambda_2 < 0, \ldots, \lambda_n < 0$ *and* ξ *is a local maximum;*

- $f''(\xi)$ *invertible and neither positive definite nor negative definite means that the λs differ in sign, so* ξ *is a saddle point.*

In the case of a function $f \colon \mathbb{R}^2 \to \mathbb{R}$, it is best to begin the process of classifying a stationary point by calculating the determinant Δ called the **Hessian** defined by

$$\Delta = \det f''(\xi) = \begin{vmatrix} f_{xx} & f_{xy} \\ f_{xy} & f_{yy} \end{vmatrix} = f_{xx}f_{yy} - f_{xy}^2$$

There are then three possibilities:

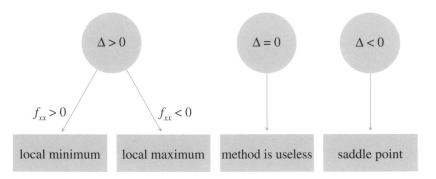

Example 4

Consider the function $f: \mathbb{R}^2 \to \mathbb{R}$ defined by

$$f(x, y) = x^3 - 3xy^2 + y^4$$

The stationary points are found by solving

$$\left. \begin{array}{l} \dfrac{\partial f}{\partial x} = 3x^2 - 3y^2 = 0 \\[2mm] \dfrac{\partial f}{\partial y} = -6xy + 4y^3 = 0 \end{array} \right\}$$

We obtain

$$x = y \quad \text{or} \quad x = -y$$

and

$$y = 0 \quad \text{or} \quad y^2 = \frac{3}{2}x$$

The stationary points are therefore $(0, 0)^{\mathrm{T}}$, $\left(\frac{3}{2}, \frac{3}{2}\right)^{\mathrm{T}}$ and $\left(\frac{3}{2}, \frac{-3}{2}\right)^{\mathrm{T}}$.

The second derivative is

$$f''(x, y) = \begin{pmatrix} \dfrac{\partial^2 f}{\partial x^2} & \dfrac{\partial^2 f}{\partial x \, \partial y} \\[3mm] \dfrac{\partial^2 f}{\partial x \, \partial y} & \dfrac{\partial^2 f}{\partial y^2} \end{pmatrix} = \begin{pmatrix} 6x & -6y \\ -6y & -6x + 12y^2 \end{pmatrix}$$

and so the principal minors are δ and Δ where

$$\Delta = \det f''(x, y) = \begin{vmatrix} 6x & -6y \\ -6y & -6x + 12y^2 \end{vmatrix}$$
$$= 36(-x^2 + 2xy^2 - y^2)$$

and

$$\delta = f_{xx}(x, y) = 6x$$

(i) At the point $\left(\frac{3}{2}, \frac{3}{2}\right)^{\mathrm{T}}$, we have

$$\Delta = 36\left(-\frac{9}{4} + 2 \cdot \frac{3}{2} \cdot \frac{9}{4} - \frac{9}{4}\right) = 81 > 0$$
$$\delta = 6 \cdot \frac{3}{2} = 9 > 0$$

and hence this point is a local minimum.

(ii) At the point $\left(\frac{3}{2}, -\frac{3}{2}\right)^{\mathrm{T}}$, we have

$$\Delta = 81 > 0$$
$$\delta = 9 > 0$$

as before. Hence this point is also a local minimum.

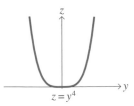

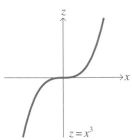

(iii) At the point $(0, 0)^T$, $\Delta = 0$ and $\delta = 0$. Hence the method is useless. Observe, however, that

$$z = f(0, y) = y^4$$
$$z = f(x, 0) = x^3$$

Thus the function certainly does not have a local maximum or minimum, but a saddle point at $(0, 0)^T$.

The surface plot and contour plot below illustrate the three stationary points of the function.

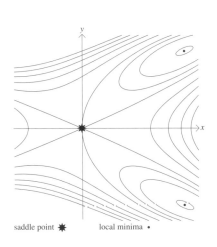

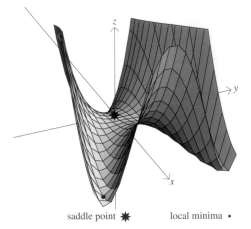

saddle point ✳   local minima •

Example 5

Consider the function $f: \mathbb{R}^3 \to \mathbb{R}$ defined by

$$f(x, y, z) = xy + yz + zx$$

The stationary points are found by solving the equations

$$\frac{\partial f}{\partial x} = y + z = 0 \qquad \frac{\partial f}{\partial y} = x + z = 0 \qquad \frac{\partial f}{\partial z} = y + x = 0$$

and hence $(0, 0, 0)^T$ is the only stationary point. The second derivative is

$$\begin{pmatrix} f_{xx} & f_{xy} & f_{xz} \\ f_{yx} & f_{yy} & f_{yz} \\ f_{zx} & f_{zy} & f_{zz} \end{pmatrix} = \begin{pmatrix} 0 & 1 & 1 \\ 1 & 0 & 1 \\ 1 & 1 & 0 \end{pmatrix}$$

The principal minors are

$$\begin{vmatrix} 0 & 1 & 1 \\ 1 & 0 & 1 \\ 1 & 1 & 0 \end{vmatrix} = 2 < 0 \qquad \begin{vmatrix} 0 & 1 \\ 1 & 0 \end{vmatrix} = -1 < 0 \qquad 0 = 0$$

Since $\det f''(0,0,0) \neq 0$ and $f''(0,0,0)$ is neither positive definite nor negative definite, $(0,0,0)^T$ is a saddle point.

6.4 Convex and concave functions

The discussion in this chapter so far has centred on the identification of *local* maxima and minima. But, as noted in §4.2 for functions of one variable, the identification of local maxima and minima is usually only a step on the way to finding *global* maxima or minima for an optimisation problem. However, in the case of a *concave* function, a local maximum is automatically a global maximum. Similarly, in the case of a *convex* function, a local minimum is automatically a global minimum. This observation is of some importance since concave and convex functions appear quite frequently in applications.

A **convex set** in $\mathbb{R}^n$ is a set S with the property that the line segment joining any two points of S lies entirely inside S.

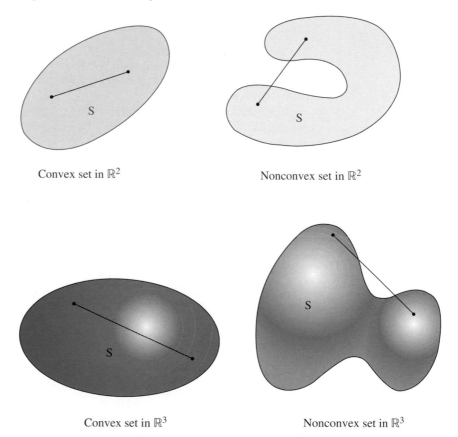

Convex set in $\mathbb{R}^2$ Nonconvex set in $\mathbb{R}^2$

Convex set in $\mathbb{R}^3$ Nonconvex set in $\mathbb{R}^3$

A **convex function** $f\colon \mathbb{R}^n \to \mathbb{R}$ is a function with the property that the set of points in $\mathbb{R}^{n+1}$ which lie *above* the graph $y = f(\mathbf{x})$ is convex. This accords with the concept of a function of one variable being convex, which was briefly mentioned in §4.1. A

concave function $f : \mathbb{R}^n \to \mathbb{R}$ is a function with the property that the set of points which lie *below* the graph $y = f(\mathbf{x})$ is convex.

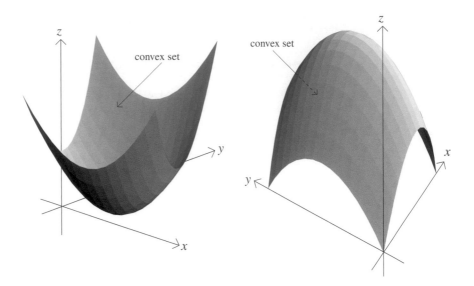

The diagrams above make it clear why a local minimum for a convex function is necessarily a global minimum and why a local maximum for a concave function is necessarily a global maximum.

If a function $f : \mathbb{R}^n \to \mathbb{R}$ is twice differentiable, it is possible to determine whether or not it is convex or concave by examining the second derivative. We consider the case of a concave function.

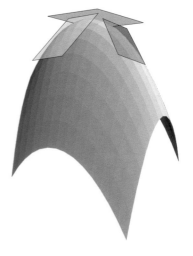

As in §4.1 in the case $n = 1$, the tangent hyperplanes to the graph of a concave function $f : \mathbb{R}^n \to \mathbb{R}$ all lie on or above the graph.

As we know from §5.3, the equation of the tangent hyperplane where $\mathbf{x} = \boldsymbol{\xi}$ is given by

$$y = f(\boldsymbol{\xi}) + f'(\boldsymbol{\xi})(\mathbf{x} - \boldsymbol{\xi})$$

To say that this lies on or above the graph $y = f(\mathbf{x})$ is the same as asserting that

$$f(\mathbf{x}) \le f(\boldsymbol{\xi}) + f'(\boldsymbol{\xi})(\mathbf{x} - \boldsymbol{\xi})$$

for all values of $\mathbf{x}$. Applying Taylor's theorem as in §6.3, we obtain for all $\boldsymbol{\xi}$

$$f(\boldsymbol{\xi}) + f'(\boldsymbol{\xi})(\mathbf{x} - \boldsymbol{\xi}) + \frac{1}{2}(\mathbf{x} - \boldsymbol{\xi})^{\mathrm{T}} f''(\boldsymbol{\xi})(\mathbf{x} - \boldsymbol{\xi}) \le f(\boldsymbol{\xi}) + f'(\boldsymbol{\xi})(\mathbf{x} - \boldsymbol{\xi})$$

or

$$(\mathbf{x} - \boldsymbol{\xi})^{\mathrm{T}} f''(\boldsymbol{\xi})(\mathbf{x} - \boldsymbol{\xi}) \le 0$$

Equivalently, for all $\boldsymbol{\xi}$, all the eigenvalues of $f''(\boldsymbol{\xi})$ must be nonpositive in which case the matrix is called **nonpositive definite**. Similarly, if f is convex, for all $\boldsymbol{\xi}$ all the eigenvalues of $f''(\boldsymbol{\xi})$ must be nonnegative, in which case the matrix is called **nonnegative definite**.

The general result is stated below.

(I)   A function f is convex if and only if $f''(\boldsymbol{\xi})$ is nonnegative definite *for all $\boldsymbol{\xi}$*.

(II)   A function f is concave if and only if $f''(\boldsymbol{\xi})$ is nonpositive definite *for all $\boldsymbol{\xi}$*.

Example 6

Consider the function $f: \mathbb{R}^2 \to \mathbb{R}$ defined by

$$f(x, y) = e^{x^2 + y^2}$$

The partial derivatives are

$$f_x = 2x e^{x^2 + y^2} \qquad f_y = 2y e^{x^2 + y^2}$$

$$f_{xx} = (4x^2 + 2)e^{x^2 + y^2} \quad f_{xy} = 4xy e^{x^2 + y^2} \quad f_{yy} = (4y^2 + 2)e^{x^2 + y^2}$$

The second derivative is

$$f''(x, y) = e^{x^2 + y^2} \begin{pmatrix} 4x^2 + 2 & 4xy \\ 4xy & 4y^2 + 2 \end{pmatrix}$$

Now

$$\Delta = e^{2(x^2 + y^2)} \begin{vmatrix} 4x^2 + 2 & 4xy \\ 4xy & 4y^2 + 2 \end{vmatrix} = e^{2(x^2 + y^2)} 4(2x^2 + 2y^2 + 1) > 0$$

and $f_{xx} = (4x^2 + 2)e^{x^2 + y^2} > 0$ *for all* $(x, y)^{\mathrm{T}}$ so the second derivative is positive definite *for all* $\boldsymbol{\xi} = (x, y)^{\mathrm{T}}$ and hence nonnegative definite. The function is therefore convex.

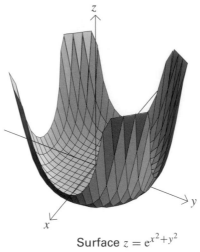

Surface $z = e^{x^2+y^2}$

Example 7

Consider the function $f\colon \mathbb{R}^2 \to \mathbb{R}$ defined by

$$f(x, y) = 2xy - x^2 - y^2$$

The second derivative is

$$f''(x, y) = \begin{pmatrix} -2 & 2 \\ 2 & -2 \end{pmatrix}$$

The eigenvalues are found by solving $(-2-\lambda)^2 = 4$ for which the roots are $\lambda_1 = -4$ and $\lambda_2 = 0$. Thus the second derivative is always nonpositive definite and so the function is concave.

Alternatively, $-(x - y)^2 \le 0$, which shows that the function is nonpositive definite and so is concave.

See Example 3.4.

The surface is a concave parabolic cylinder with the quadratic term equal to zero along the line $x = y$. This is reflected in the fact that the Hessian

$$\Delta = \begin{vmatrix} -2 & 2 \\ 2 & -2 \end{vmatrix} = 0$$

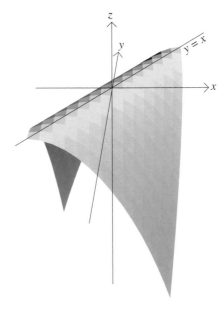

Example 8

The saddle function $f: \mathbb{R}^2 \to \mathbb{R}$ is defined by

$$f(x, y) = -x^2 + y^2$$

We investigate the sign of the quadratic term, which in this case is the whole function.

We have

$$y^2 - x^2 \geq 0$$

when $|y| \geq |x|$ in which case the function is convex, and

$$y^2 - x^2 \leq 0$$

for the remaining values of x and y, in which case the function is concave. Hence the function is neither convex, nor concave overall.

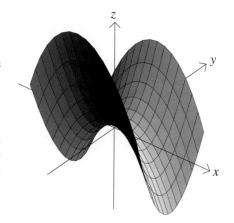

Example 9

The function $f: \mathbb{R}^2 \to \mathbb{R}$ is defined by

$$f(x, y) = x^3 + y^2$$

with second derivative

$$f''(x, y) = \begin{pmatrix} 6x & 0 \\ 0 & 2 \end{pmatrix}$$

On the half plane $x \geq 0$ this is nonnegative definite, so f is convex; f is neither convex nor concave *anywhere* on the half plane $x < 0$, since $f''(x)$ is *never* either nonpositive definite or nonnegative definite at any point. This means that at every point for which $x < 0$, the tangent plane cuts across the surface.

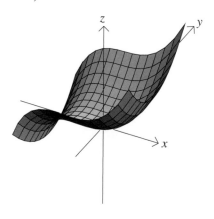

See Exercise 4.5.17.

Observe how this differs from the situation for one variable. Any interval I of $\mathbb{R}$ can be divided into intervals on which a twice differentiable function $f: I \to \mathbb{R}$ has $f''(x) \geq 0$ and intervals on which $f''(x) \leq 0$. That is, I can be divided into intervals on which $f(x)$ is convex, and intervals on which $f(x)$ is concave.

Example 10

See Exercise 6.5.4.

By evaluating the Hessian Δ and considering the sign of P_{xx}, it can be shown that the graph of the Cobb–Douglas function $P: \mathbb{R}_+^2 \to \mathbb{R}$ defined by

$$P(x, y) = x^\alpha y^\beta \quad \text{where } 0 < \alpha < 1, 0 < \beta < 1$$

is a concave surface when $\alpha + \beta \leq 1$ and neither convex nor concave when $\alpha + \beta > 1$.

The surface plots of the Cobb–Douglas functions for different values of α and β shown below endorse this statement. The surface is concave when $\alpha + \beta \leq 1$. The surface $z = x^{9/10} y^{9/10}$ where $\alpha + \beta = \frac{9}{10} + \frac{9}{10} > 1$ is neither convex nor concave, which is shown by its tangent plane at $(x, y)^{\mathrm{T}} = (1, 1)^{\mathrm{T}}$ cutting across the surface.

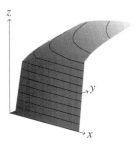

$\alpha = 1/8, \beta = 1/10,$
$\alpha + \beta < 1$

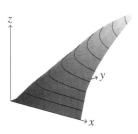

$\alpha = \beta = 1/2, \; \alpha + \beta = 1$

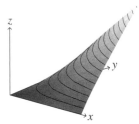

$\alpha = \beta = \frac{9}{10}, \; \alpha + \beta > 1$

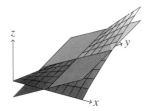

$\alpha = \beta = \frac{9}{10}$, surface and
tangent plane

Example 11

The constant elasticity of substitution (CES) function $P: \mathbb{R}_+^2 \to \mathbb{R}$ is defined by

$$P(x, y) = [\alpha x^\beta + (1 - \alpha) y^\beta]^{\frac{1}{\beta}} \quad \text{where } 0 < \alpha < 1, \beta < 1$$

See Exercise 6.5.5.

Showing the value of the Hessian Δ to be 0 and so to be nonpositive and considering the sign of P_{xx}, it can be deduced that the graph of P is a concave surface, as illustrated in the surface plots below, where $\alpha = 1/4$.

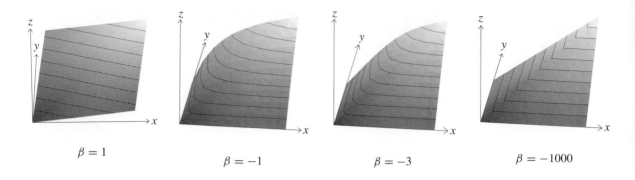

$\beta = 1$

$\beta = -1$

$\beta = -3$

$\beta = -1000$

Example 12

We have noted in §4.2 that when a differentiable one variable function has a *unique* stationary point which is a local maximum, this must be a global maximum of the function. However, this example illustrates that it would be a mistake to conclude that this is also the case for functions of more than one variable.

The function $f \colon \mathbb{R}^2 \to \mathbb{R}$ defined below has a single stationary point which is a local maximum, but it has no global maximum.

$$f(x, y) = 6x^2 e^y - 4x^3 - e^{6y}$$

Its stationary points are given by

$$f_x = 12xe^y - 12x^2 = 0$$
$$f_y = 6x^2 e^y - 6e^{6y} = 0$$

or

$$x(e^y - x) = 0$$
$$e^y(x^2 - e^{5y}) = 0$$

So $x^2 = e^{5y} = e^{2y}$ which implies that $e^{2y}(e^{3y} - 1) = 0$. Since $e^{2y} \neq 0$ we must have $e^{3y} = 1$. We are forced to conclude that $y = 0$; in which case, $x = e^y = e^0 = 1$. So there is just one stationary point, namely $(1, 0)^{\mathrm{T}}$.

To find the nature of the point, calculate the second order partial derivatives.

$$f_{xx} = 12e^y - 24x \qquad f_{yy} = 6x^2 e^y - 36e^{6y} \qquad f_{xy} = 12xe^y$$

The Hessian is then

$$\Delta = \begin{vmatrix} 12e^y - 24x & 12xe^y \\ 12xe^y & 6x^2 e^y - 36e^{6y} \end{vmatrix}$$

At the point $(x, y)^{\mathrm{T}} = (1, 0)^{\mathrm{T}}$, $\Delta = \begin{vmatrix} -12 & 12 \\ 12 & -30 \end{vmatrix} > 0$ and $f_{xx} = -12 < 0$.

local maximum
$(1,0,1)^{\mathrm{T}}$

Hence the point is a local maximum. But

$$f(1,0) = 6 - 4 - 1 = 1$$

and, for instance,

$$f(-1,0) = 6 + 4 - 1 = 9 > f(1,0)$$

So the point is *not* a global maximum.

In the special case when a differentiable function is known to be concave, the unique stationary point which is a local maximum must also be the global maximum.

6.5 Exercises

1. Classify the stationary points of the functions $f: \mathbb{R}^2 \to \mathbb{R}$ defined by

(i)* $f(x, y) = (x - 2)^2 + (y - 3)^2$

(ii) $f(x, y) = 5x^2y - y^2 - 6x^2 - 4x^4$

(iii)* $f(x, y) = 2x^2 - 8xy + y^2$

(iv) $f(x, y) = x^3 - 2x^2 - 4x - y^2 - 6y$

(v)* $f(x, y) = xy^2 + 2y - x$

(vi) $f(x, y) = 8y^3 + x^4 - 2x^2$

(vii)* $f(x, y) = 5xy^2 - 20x - 3y^2$

(viii) $f(x, y) = 2x^2y + 2y - 5x^2$

(ix)* $f(x, y) = 16x^4 - 8x^2 + y^3 - 3y$

(x) $f(x, y) = \dfrac{4}{x} - \dfrac{2x}{y} + y$ where $x, y \neq 0$

(xi)* $f(x, y) = 2y^2x + 3x^2y - 4xy$

(xii) $f(x, y) = (4x^2 + y^2)e^{1-x^2-4y^2}$

(xiii)* $f(x, y) = x^3 + y^3 - 3xy^2$

(xiv)* $f(x, y) = x^2y + y^3x - xy$

(xv) $f(x, y) = x^2y + y^3x - xy^2$

(xvi) $f(x, y) = e^{x+y}(x^2 - 2xy + 3y^2)$

(xvii)* $f(x, y) = y^3 + 3x^2y - 3x^2 - 3y^2 + 2$

(xviii) $f(x, y) = 8x^2y - x^3y - 5y^3x$

2.   Classify the stationary points of the function $f: \mathbb{R}^3 \to \mathbb{R}$ defined by

(i)* $f(x, y, z) = x^2y + y^2z + z^2x$

(ii) $f(x, y, z) = 4x^2z - 2xy - 4x^2 - z^2 + y$

(iii) $f(x, y, z) = 6xy + z^2 - 18x^2 - 4y^2z$

(iv)* $f(x, y, z) = x^2y + 2y^2z - 4z - 2x$

(v)♣ $f(x, y, z) = 3(x^2 + y^2 + z^2) - (x + y + z)^2$

(vi)* $f(x, y, z) = \ln((x^3 - 2y^3 - 3z^3 - 6xy + 4)^{\frac{1}{2}})$
       where $(x^3 - 2y^3 - 3z^3 - 6xy + 4) > 0$.

3.   Classify the stationary points of the function $f: \mathbb{R}^2 \to \mathbb{R}$ defined by

(i)* $f(x, y) = \exp(8x + 12y - x^2 - y^3)$

(ii) $f(x, y) = \dfrac{1}{x^2 + 9y^2 + 6xy + 75}$

(iii)* $f(x, y) = \ln(10 - 4x - 4y^2 + x^2 + y^4)$

(iv) $f(x, y) = (6 + x^2 + y^2 - 2x + 4y)^3$

4.* The Cobb–Douglas function $P: \mathbb{R}_+^2 \to \mathbb{R}$ is defined by

$$P(x, y) = x^\alpha y^\beta \quad \text{where } 0 < \alpha < 1, 0 < \beta < 1$$

Evaluate the Hessian Δ and use the sign of P_{xx} to show that the graph of the function is a concave surface when $\alpha + \beta \leq 1$ and neither convex nor concave when $\alpha + \beta > 1$.

5.*♣ The constant elasticity of substitution (CES) function $P: \mathbb{R}_+^2 \to \mathbb{R}$ is defined by

$$P(x, y) = [\alpha x^\beta + (1 - \alpha)y^\beta]^{\frac{1}{\beta}} \quad \text{where } 0 < \alpha < 1, \ \beta < 1$$

Show that

$$P_{xx} = -\alpha(1 - \alpha)(1 - \beta)x^{\beta-2}y^\beta[\alpha x^\beta + (1 - \alpha)y^\beta]^{1/\beta-2}$$

Find similar expressions for P_{yy} and P_{xy} and deduce that the value of the Hessian Δ is 0 and so is nonpositive. Use the sign of P_{xx} to deduce that the graph of P is a concave surface.

6.*♣ Find the stationary points of the function $f: \mathbb{R}^3 \to \mathbb{R}$ defined by

$$f(x, y, z) = 5x^2 + 5y^2 + 9z^2 - 6xz - 12yz$$

and determine their nature.
Show that the function is convex. What is a global minimum for this function?

7.♣ Find the stationary points of the function $f: \mathbb{R}^6 \to \mathbb{R}$ defined by

$$f(u, v, w, x, y, z) = (u^3 - 3uv^2 + v^4) + (w^3 - 3wx^2 + x^4) + (y^3 - 3yz^2 + z^4)$$

and determine their nature.

8.*♣ Reduce $z = xy$ to the form

$$z = \lambda_1 X^2 + \lambda_2 Y^2$$

by an orthogonal transformation of the variables. Draw a diagram indicating the curve $xy = 1$ together with the new X and Y axes. Discuss the behaviour of the function $f: \mathbb{R}^2 \to \mathbb{R}$ defined by $f(x, y) = xy$ at the point $(0, 0)^{\mathrm{T}}$.

9.♣ Reduce $u = xy + yz + zx$ to the form

$$u = \lambda_1 X^2 + \lambda_2 Y^2 + \lambda_3 Z^2$$

by an orthogonal transformation of the variables. Use your result to discuss the behaviour of the function $f: \mathbb{R}^3 \to \mathbb{R}$ defined by $f(x, y, z) = xy + yz + zx$ at the point $(0, 0, 0)^{\mathrm{T}}$.

10.* Let x, y, z denote quantities and p_1, p_2, p_3 respectively denote prices of three related goods produced by a monopolistic firm. Its demand functions are

$$p_1 = 290 - 2x - 7y - 3z$$
$$p_2 = 312 - 3x - 5y - 4z$$
$$p_3 = 245 - 5x - 2y - 2z$$

and its joint cost function is

$$C(x, y, z) = x^2 + 2y^2 + z^2 + xz$$

The revenue function is

$$R(x, y, z) = p_1 x + p_2 y + p_3 z$$

and the profit function is

$$\pi(x, y, z) = R(x, y, z) - C(x, y, z)$$

Find output levels for the goods for which the profits are maximised.

11. Investigate whether the utility function $U: \mathbb{R}_+^2 \to \mathbb{R}$ defined by

$$U(x, y) = \ln[(1 + x)^\alpha (1 + y)^\beta] \quad \text{where } 0 < \alpha < 1, 0 < \beta < 1$$

is convex or concave or neither.

12.* Show that if a conic has an equation with the coefficient of x^2 equal to the coefficient of y^2, that is

$$Ax^2 + Bxy + Ay^2 + Dx + Ey + F = 0 \text{ where } A \neq 0$$

the xy term can be eliminated by an anticlockwise rotation through $\pi/4$.

13.**♣** The ideas developed in §6.1 can be used to bring *any* conic to standard form. To convert the conic

$$5x^2 - 8xy + 5y^2 - 10x + 8y = 1$$

into standard form, begin with a rotation to eliminate the xy term. Since the coefficients of x and y are equal, an anticlockwise rotation through $\pi/4$ will achieve this, as stated in Exercise 12.

Show that the eigenvalues are 1 and 9, with corresponding eigenvectors $(1/\sqrt{2}, 1/\sqrt{2})^T$ and $(1/\sqrt{2}, -1/\sqrt{2})^T$ respectively. Use these as the columns of the required rotation matrix P, and write $\mathbf{x} = P\mathbf{X}$ in the equation of the conic, to obtain

$$(P\mathbf{X})^T \begin{pmatrix} 5 & -4 \\ -4 & 5 \end{pmatrix} P\mathbf{X} + (-10, \ 8)P\mathbf{X} = 1$$

Using the method of §6.1 write the equation as

$$(X, \ Y) \begin{pmatrix} 1 & 0 \\ 0 & 9 \end{pmatrix} \begin{pmatrix} X \\ Y \end{pmatrix} + (-10, \ 8) \begin{pmatrix} 1/\sqrt{2} & -1/\sqrt{2} \\ 1/\sqrt{2} & 1/\sqrt{2} \end{pmatrix} \begin{pmatrix} X \\ Y \end{pmatrix} = 1$$

Show that this gives

$$X^2 + 9Y^2 - \sqrt{2}X + 9\sqrt{2}Y = 1$$

See §10.7.

Complete the squares to obtain

$$(X - 1/\sqrt{2})^2 + 9(Y + 1/\sqrt{2})^2 = 6$$

Finally, the translation $X' = X - 1/\sqrt{2}$ and $Y' = Y + 1/\sqrt{2}$ gives the standard form

$$(X')^2 + 9(Y')^2 = 6$$

Sketch the conic, finding its intersections with the original axes, $y = 0$ and $x = 0$.

14.**♣** Use the procedure outlined in the previous exercise to convert the following conics, which may be degenerate, to standard form:

(i) $x^2 + 2xy + y^2 - x + y = 0$

(ii) $2x^2 + 6xy + 2y^2 - 3\sqrt{2}x + 3\sqrt{2}y = 5$

(iii) $xy + y - x = 1$

6.6 Constrained optimisation

Suppose we wish to maximise a function $f: \mathbb{R}^2 \to \mathbb{R}$ subject to the constraint that $(x, y)^T$ lies in some specified region D in $\mathbb{R}^2$. Assuming the maximum exists, it is attained *either* at an interior point of D *or* at a boundary point of D. If the maximum is attained at an interior point ξ, then ξ is a stationary point of f, but this need *not* be true if ξ is a boundary point of D.

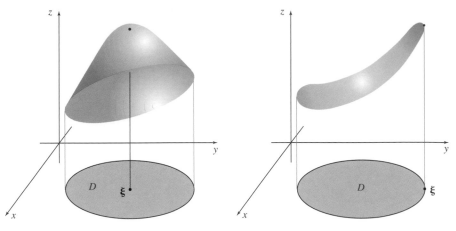

Maximum attained at interior point $\boldsymbol{\xi}$ Maximum attained at boundary point $\boldsymbol{\xi}$

Suppose that $f \colon \mathbb{R}^2 \to \mathbb{R}$ and $g \colon \mathbb{R}^2 \to \mathbb{R}$ and we are seeking to evaluate

$$\max f(x, y)$$

subject to the constraint

$$g(x, y) = 0$$

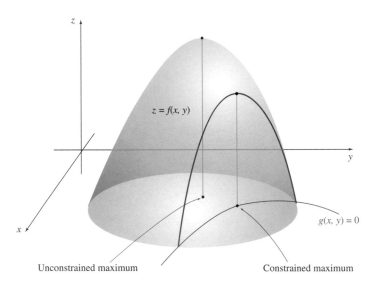

Unconstrained maximum   Constrained maximum

It may be, for example, that $g(x, y) = 0$ is the equation of the curve which bounds the region D in $\mathbb{R}^2$. The solution to our optimisation problem will then provide the maximum value of $f(x, y)$ for $(x, y)^{\mathrm{T}}$ on the boundary of D.

The most straightforward way of tackling this problem is to begin by solving the equation $g(x, y) = 0$ to obtain y in terms of x. Suppose that the formula which results from the solution of $g(x, y) = 0$ is

$$y = h(x)$$

Then we know that the points which lie on the curve $g(x, y) = 0$ are those of the form $(x, h(x))^T$.

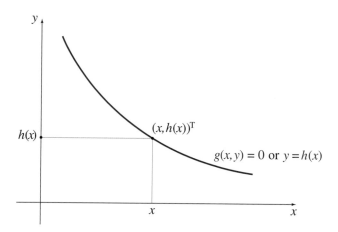

Our constrained optimisation problem therefore reduces to finding the maximum of

$$\phi(x) = f(x, h(x))$$

for all x. But this is a one variable *unconstrained* problem. If ξ is a stationary point for ϕ, we shall call $(\xi, h(\xi))^T$ a constrained stationary point for f.

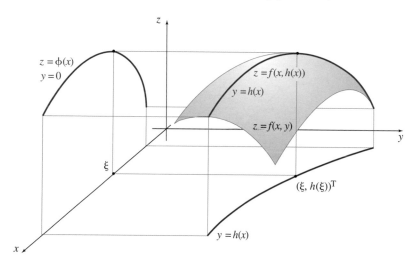

The same method can be used when $f: \mathbb{R}^3 \to \mathbb{R}$, $g: \mathbb{R}^3 \to \mathbb{R}$ and we are seeking to find

$$\max f(x, y, z)$$

subject to the constraint

$$g(x, y, z) = 0$$

One can solve $g(x, y, z) = 0$ to obtain $z = h(x, y)$ and then examine the stationary points of

$$\phi(x, y) = f(x, y, h(x, y))$$

The method can also be applied when there are several constraints. Suppose, for example, that $f: \mathbb{R}^3 \to \mathbb{R}$, $g_1: \mathbb{R}^3 \to \mathbb{R}$, $g_2: \mathbb{R}^3 \to \mathbb{R}$ and we are seeking to find

$$\max f(x, y, z)$$

subject to the constraints

$$\left.\begin{array}{c} g_1(x, y, z) = 0 \\ g_2(x, y, z) = 0 \end{array}\right\}$$

One can then solve the equations $g_1(x, y, z) = 0$ and $g_2(x, y, z) = 0$ simultaneously to obtain y and z as functions of x. Suppose that $y = h_1(x)$ and $z = h_2(x)$. The next step is then to examine the stationary points of

$$\phi(x) = f(x, h_1(x), h_2(x))$$

Example 13

Find

$$\min(3x^2 - 2y^2)$$

subject to the constraints

$$\left.\begin{array}{c} x + y \le 2 \\ x - y \le 0 \end{array}\right\}$$

The region D in $\mathbb{R}^2$ defined by these constraints is illustrated below:

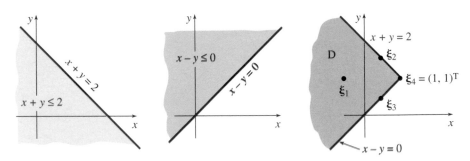

Assuming that the minimum exists, there are four different types of point in D at which the minimum could possibly be achieved.

(i) *Minimum achieved at a point like ξ_1.* Then ξ_1 will be an unconstrained stationary point of f. Such points are found by solving the simultaneous equations

$$\left.\begin{array}{l} \dfrac{\partial f}{\partial x} = 6x = 0 \\[2mm] \dfrac{\partial f}{\partial y} = -4y = 0 \end{array}\right\}$$

The only possibility is therefore $(0, 0)^T$. But this is *not* an interior point of D and so this case may be eliminated.

(ii) *Minimum achieved at a point like* ξ_2. Then ξ_2 will be a stationary point of f subject to the constraint $x + y = 2$. To find these stationary points, we can solve $x + y = 2$ for y in terms of x to obtain $y = 2 - x$. Then substitute the result in $f(x, y) = 3x^2 - 2y^2$. This yields

$$\phi(x) = 3x^2 - 2(2 - x)^2$$
$$= x^2 + 8x - 8.$$

But

$$\phi'(x) = 2x + 8$$

and hence ϕ has the single stationary point $x = -4$. The corresponding value of y is $y = 2 - (-4) = 6$ and so we obtain the constrained stationary point $(-4, 6)^T$ as a possibility for the point at which the minimum is achieved.

(iii) *Minimum achieved at a point like* ξ_3. Then ξ_3 will be a stationary point of f subject to the constraint $x - y = 0$. To find these stationary points, we can solve $x - y = 0$ for y in terms of x to obtain $y = x$. Then substitute the result in $f(x, y) = 3x^2 - 2y^2$. This yields

$$\phi(x) = 3x^2 - 2x^2 = x^2$$
$$\phi'(x) = 2x$$

and hence ϕ has the single stationary point $x = 0$. The corresponding value of y is $y = 0$, and so we obtain the constrained stationary point $(0, 0)^T$ as a possibility for the point at which the minimum is achieved. (It follows, in fact, from (i) that $(0, 0)^T$ would be one of the constrained stationary points in this case.)

(iv) *Minimum achieved at* $\xi_4 = (1, 1)^T$. This possibility for the point at which the minimum is achieved has already been included in the sets (ii) and (iii), so does not need to be considered.

We next observe that

$$\left. \begin{array}{c} f(-4, 6) = 48 - 72 = -24 \\ f(0, 0) = 0 \\ f(1, 1) = 1 \end{array} \right\}$$

The minimum we are seeking is therefore -24 and this is achieved at the point $(-4, 6)^T$.

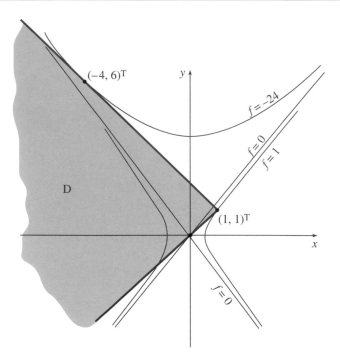

The contour map reveals a number of instructive facts:

1. At the minimising point $(-4, 6)^T$, the contour $3x^2 - 2y^2 = -24$ is tangent to the line $x + y = 2$.

2. The point $(1, 1)^T$ is only a *local* maximum for the problem. The function has *no* maximum on the whole set D.

3. The point $(0, 0)^T$ corresponds to a saddle point.

6.7   Constraints and gradients

Suppose that $f: \mathbb{R}^2 \to \mathbb{R}$ and $g: \mathbb{R}^2 \to \mathbb{R}$ and that we are seeking to evaluate

$$\max f(x, y)$$

subject to the constraint

$$g(x, y) = 0$$

The contour map that follows makes it clear that, if the maximum value of f subject to the constraint is M, then the contour $f(x, y) = M$ will be tangent to $g(x, y) = 0$ and the maximum will be achieved at the point $\boldsymbol{\xi}$ of tangency.

We encountered this phenomenon in Example 13 when considering the minimum of $3x^2 - 2y^2$ subject to the constraint $x + y = 2$. A minimum of -24 is achieved at $(-4, 6)^T$ and $3x^2 - 2y^2 = -24$ is tangent to $x + y = 2$ at the point $(-4, 6)^T$.

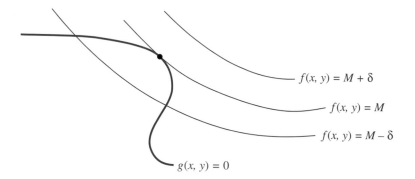

$f(x, y) = M + \delta$

$f(x, y) = M$

$f(x, y) = M - \delta$

$g(x, y) = 0$

One can exploit this phenomenon using the idea of the gradient, introduced in §3.4. We know that ∇f is a normal to the contour $f(x, y) = M$ at the point $\boldsymbol{\xi}$. Similarly, ∇g is a normal to the contour $g(x, y) = 0$ at the point $\boldsymbol{\xi}$.

Since $f(x, y) = M$ touches $g(x, y) = 0$, it follows that ∇f and ∇g must point in the same (or opposite) directions.

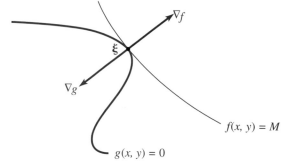

∇f

$\boldsymbol{\xi}$

∇g

$f(x, y) = M$

$g(x, y) = 0$

We conclude that ∇f must be a scalar multiple of ∇g at the point $\boldsymbol{\xi}$ and thus

$$\nabla f + \lambda \nabla g = 0$$

for some scalar λ. It follows that

$$\left. \begin{aligned} \frac{\partial f}{\partial x} + \lambda \frac{\partial g}{\partial x} = 0 \\ \frac{\partial f}{\partial y} + \lambda \frac{\partial g}{\partial y} = 0 \end{aligned} \right\}$$

If we apply this result in the case when $f(x, y) = 3x^2 - 2y^2$ and $g(x, y) = x + y - 2$, we obtain that

$$6x + \lambda = 0$$
$$-4y + \lambda = 0$$

Subtracting these equations yields

$$6x + 4y = 0.$$

In addition, we have the equation

$$x + y = 2$$

since ξ must satisfy the constraint $g(x, y) = 0$. Solving the simultaneous equations

$$\left.\begin{array}{r} 6x + 4y = 0 \\ x + y = 2 \end{array}\right\}$$

yields the result $(x, y)^{\mathrm{T}} = (-4, 6)^{\mathrm{T}}$ as obtained in Example 13.

Lagrange's method, which we describe in the next section, is a simple generalization of this technique.

6.8 Lagrange's method – optimisation with one constraint

We begin with the simplest case. Suppose that given $f: \mathbb{R}^2 \to \mathbb{R}$ and $g: \mathbb{R}^2 \to \mathbb{R}$ we have to consider the problem of evaluating the maximum or minimum of $f(x, y)$ subject to the constraint $g(x, y) = 0$. In §6.6, we saw that one approach was to solve the constraint equation for y in terms of x, to obtain $y = h(x)$ and substitute in $f(x, y)$. This reduces to the *unconstrained* problem of finding the maximum or minimum of the single variable function $f(x, h(x))$.

This method is not to be despised but it needs little imagination to see that it is not always possible to solve for y in terms of x. In §6.7, we examined an alternative approach. Lagrange's method is the general version of this. It may seem somewhat perverse, but we begin by introducing a *new* variable λ. We then form the **Lagrangian** $L: \mathbb{R}^3 \to \mathbb{R}$ defined by

$$L(x, y, \lambda) = f(x, y) + \lambda g(x, y)$$

and calculate the *stationary points* of L. If

$$(\tilde{x}, \tilde{y})^{\mathrm{T}}$$

is a constrained stationary point for f, then

$$(\tilde{x}, \tilde{y}, \tilde{\lambda})^{\mathrm{T}}$$

is an unconstrained stationary point for L (for some $\tilde{\lambda}$).

Example 14

Find the maximum of the function

$$f(x, y) = xy$$

subject to the constraint $x^2 + 4y^2 = 1$.

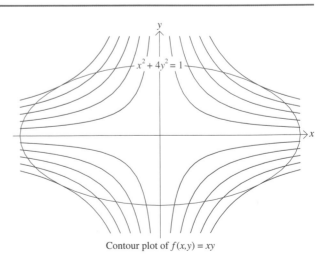

Contour plot of $f(x,y) = xy$

The Lagrangian is

$$L = xy + \lambda(x^2 + 4y^2 - 1)$$

and

$$\frac{\partial L}{\partial x} = y + 2\lambda x = 0 \qquad -2\lambda = \frac{y}{x} \tag{1}$$

$$\frac{\partial L}{\partial y} = x + 8\lambda y = 0 \qquad -\frac{1}{8\lambda} = \frac{y}{x} \tag{2}$$

$$\frac{\partial L}{\partial \lambda} = x^2 + 4y^2 - 1 = 0 \tag{3}$$

From equations (1) and (2) we obtain

$$\lambda^2 = \frac{1}{16}$$

$$\lambda = \pm\frac{1}{4}$$

It follows that

$$y = \pm\frac{1}{2}x$$

We substitute this result in equation (3) (the constraint equation) which has not yet been used. Then

$$x^2 + 4\left(\frac{1}{2}x\right)^2 = 1$$

$$x = \pm\frac{1}{\sqrt{2}}$$

We therefore obtain four constrained stationary points, namely

$$\left(\frac{1}{\sqrt{2}}, \frac{1}{2\sqrt{2}}\right)^{\mathrm{T}} \quad \left(-\frac{1}{\sqrt{2}}, \frac{1}{2\sqrt{2}}\right)^{\mathrm{T}} \quad \left(\frac{1}{\sqrt{2}}, -\frac{1}{2\sqrt{2}}\right)^{\mathrm{T}} \quad \left(-\frac{1}{\sqrt{2}}, -\frac{1}{2\sqrt{2}}\right)^{\mathrm{T}}$$

The value of xy at the first two of these points is $1/4$ and the value of xy at the second two is $-1/4$. The required maximum is therefore $1/4$.

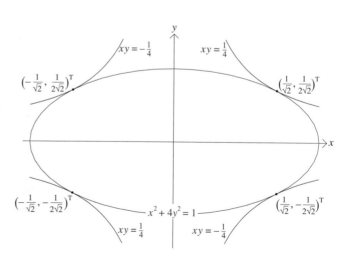

Example 15

An application to economics

The Cobb–Douglas production function $P: \mathbb{R}_+^2 \to \mathbb{R}$ for a particular manufacturer is given by

$$P(x, y) = 250x^{\frac{1}{4}} y^{\frac{3}{4}}$$

where x represents the units of labour at \$120 per unit, and y represents the units of capital at \$270 per unit.

We shall use Lagrange's method to

(i) find the maximum production level when the total cost of labour and capital is limited to \$30 000

(ii) find the minimum cost of producing 10 000 units of the product.

(i) We want to maximise the function $P(x, y)$ under the constraint

$$120x + 270y \leq 30\,000$$

or

$$4x + 9y \leq 1000$$

See Example 8.2.

A sketch of the contours of $P(x, y)$ together with the constraining line

$$4x + 9y = 1000$$

is shown on the right. The maximum is clearly attained at a point where the constraint is tangent to a contour of $P(x, y)$.

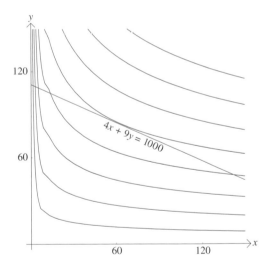

This point is a stationary point of the Lagrangian function

$$L = 250x^{\frac{1}{4}} y^{\frac{3}{4}} + \lambda(4x + 9y - 1000)$$

At stationary points

$$L_x = 250 \times \tfrac{1}{4}x^{-\frac{3}{4}} y^{\frac{3}{4}} + 4\lambda = 0 \tag{4}$$

$$L_y = 250 \times \tfrac{3}{4}x^{\frac{1}{4}} y^{-\frac{1}{4}} + 9\lambda = 0 \tag{5}$$

$$L_\lambda = \quad 4x + 9y - 1000 \quad = 0 \tag{6}$$

Hence

$$-\lambda = 250 \times \frac{1}{16} x^{-\frac{3}{4}} y^{\frac{3}{4}} = 250 \times \frac{1}{12} x^{\frac{1}{4}} y^{-\frac{1}{4}}$$

or

$$\frac{1}{4} y^{\frac{3}{4}+\frac{1}{4}} = \frac{1}{3} x^{\frac{1}{4}+\frac{3}{4}}$$

That is,

$$3y = 4x$$

Substituting in

$$4x + 9y = 1000$$
$$\tilde{y}_1 = \frac{1000}{12} = \frac{250}{3} = 83\frac{1}{3}$$
$$\tilde{x}_1 = \frac{3}{4} y = \frac{125}{2} = 62\frac{1}{2}$$

The maximum production level is

$$P\left(\frac{125}{2}, \frac{250}{3}\right) = 250 \times \left(\frac{125}{2}\right)^{\frac{1}{4}} \left(\frac{250}{3}\right)^{\frac{3}{4}} \approx 19\,388 \text{ product units}$$

(ii) We next want to minimise the cost function $C(x, y) = 120x + 270y$ where $x > 0$ and $y > 0$, subject to the constraint

$$250x^{\frac{1}{4}} y^{\frac{3}{4}} = 10\,000$$

or

$$x^{\frac{1}{4}} y^{\frac{3}{4}} = 40$$

A contour map of the function $C(x, y)$ together with the constraint curve is shown on the right. This indicates that the minimum occurs at a stationary point of the following Lagrangian function.

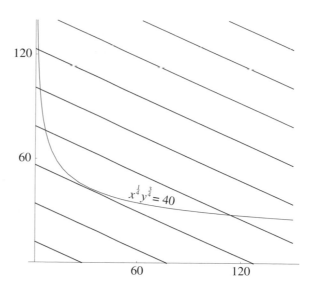

$$L = 120x + 270y + \lambda(x^{\frac{1}{4}} y^{\frac{3}{4}} - 40)$$

213

At stationary points

$$L_x = 120 + \tfrac{1}{4}\lambda x^{-\frac{3}{4}} y^{\frac{3}{4}} = 0 \tag{7}$$

$$L_y = 270 + \tfrac{3}{4}\lambda x^{\frac{1}{4}} y^{-\frac{1}{4}} = 0 \tag{8}$$

$$L_\lambda = \quad x^{\frac{1}{4}} y^{\frac{3}{4}} - 40 \quad = 0 \tag{9}$$

Hence

$$-\lambda = 4 \times 120 x^{\frac{3}{4}} y^{-\frac{3}{4}} = \frac{4 \times 270}{3} x^{-\frac{1}{4}} y^{\frac{1}{4}}$$

or

$$4 x^{\frac{3}{4}+\frac{1}{4}} = 3 y^{\frac{1}{4}+\frac{3}{4}}$$

That is,

$$4x = 3y$$

Substituting in the constraint equation

$$x^{\frac{1}{4}} y^{\frac{3}{4}} = 40$$

$$\left(\tfrac{3}{4}\right)^{\frac{1}{4}} y = 40$$

Hence

$$\tilde{y}_2 = 40\left(\tfrac{4}{3}\right)^{\frac{1}{4}} \approx 42.98$$

$$\tilde{x}_2 = \quad \tfrac{3}{4} y \approx 32.24$$

The minimum cost is $120 \times 30\left(\tfrac{4}{3}\right)^{\frac{1}{4}} + 270 \times 40\left(\tfrac{4}{3}\right)^{\frac{1}{4}} \approx 15\,474$

Observe that the two problems in (i) and (ii) are *dual* to each other. In (i), $P(x, y)$ is maximised along an isocost line $C(x, y) = $ constant and in (ii) $C(x, y)$ is minimised along an isoquant $P(x, y) = $ constant. The stationary equations (4) and (5) of problem (i), and (7) and (8) of problem (ii), each yield the same line $3y = 4x$, on which an optimal solution $(\tilde{x}, \tilde{y})^T$ lies. The line is known as the **expansion line** since it comprises all optimal solutions of the dual problems.

In the first problem, the solution is the intersection point of this line with the isocost line $C(x, y) = 30\,000$. In the second problem the solution is the intersection of the line and the isoquant $P(x, y) = 10\,000$.

The slope of the isocost lines is $-4/9$. On eliminating λ from equations (4) and (5) we can see that also

$$-\frac{P_x}{P_y} = -\frac{4}{9}$$

This is the slope of the tangent lines to the isoquant curves *at optimal points*, since the tangent lines have equations

$$P_x(x - \tilde{x}) + P_y(y - \tilde{y}) = 0$$

The following figure shows the expansion line containing the optimal solutions to the dual problems (i) and (ii).

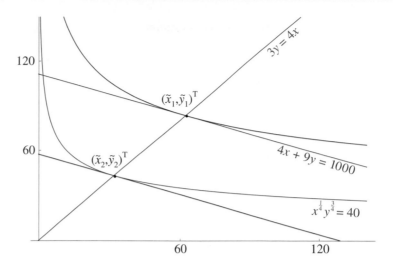

Example 16

In this example, instead of a two variable function, we minimise a three variable function, subject to a constraint. The same methods work here. We shall calculate the minimum distance from the point $(0, 0, 0)^T$ to the plane $2x - y - 2z = 15$. The vector $(2, -1, -2)^T$ is normal to the plane. Its length is

$$\sqrt{2^2 + (-1)^2 + (-2)^2} = 3$$

That is, $(\frac{2}{3}, -\frac{1}{3}, -\frac{2}{3})^T$ is a unit vector. If we write the equation of the plane as

$$\frac{2}{3}x - \frac{1}{3}y - \frac{2}{3}z = 5$$

we then know from §1.7, that the distance of the plane from the origin should be 5.

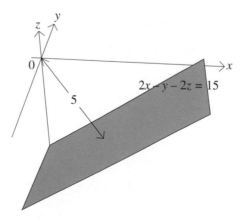

In analytical terms, the problem is to minimise

$$f(x, y, z) = x^2 + y^2 + z^2$$

215

subject to the constraint

$$2x - y - 2z = 15$$

Method 1

We use the constraint to eliminate z. Then

$$\phi(x, y) = f\left(x, y, \frac{15 - 2x + y}{(-2)}\right) = x^2 + y^2 + \frac{1}{4}(15 - 2x + y)^2$$

We then seek the unconstrained stationary points of this expression. Since

$$\left.\begin{array}{l} \phi_x = 2x + \dfrac{2}{4}(15 - 2x + y)(-2) \\[3mm] \phi_y = 2y + \dfrac{2}{4}(15 - 2x + y) \end{array}\right\}$$

the stationary points are found by solving

$$\left.\begin{array}{l} 2x - 15 + 2x - y = 0 \\[2mm] 4y + 15 - 2x + y = 0 \end{array}\right\}$$

These simplify to

$$\begin{array}{rcl} 4x - y & = & 15 \\ -2x + 5y & = & -15 \end{array}$$

and so

$$\left(\begin{array}{c} x \\ y \end{array}\right) = \left(\begin{array}{c} \frac{10}{3} \\ -\frac{5}{3} \end{array}\right)$$

Substituting these values in $z = (15 - 2x + y)/(-2)$, we obtain the corresponding value of z as $-\frac{10}{3}$. Thus there is a single constrained stationary point, namely $(\frac{10}{3}, -\frac{5}{3}, -\frac{10}{3})^T$. Since the geometry guarantees the existence of a minimum, this minimum must therefore be attained at the point and the required minimum value is

$$\sqrt{f} = \left[\left(\frac{10}{3}\right)^2 + \left(-\frac{5}{3}\right)^2 + \left(\frac{10}{3}\right)^2\right]^{1/2} = 5$$

Method 2

We use Lagrange's method. The contours of the function

$$u = x^2 + y^2 + z^2$$

are spheres centred at the origin. The minimum distance occurs where a contour is tangent to the given plane. So we may define the Lagrangian $L: \mathbb{R}^4 \to \mathbb{R}$ by

$$L(x, y, z, \lambda) = x^2 + y^2 + z^2 + \lambda(2x - y - 2z - 15)$$

To find the stationary points, we have to solve

$$L_x = 2x + 2\lambda = 0 \quad \text{i.e. } x = -\lambda \tag{10}$$

$$L_y = 2y - \lambda = 0 \quad \text{i.e. } y = \frac{1}{2}\lambda \tag{11}$$

$$L_z = 2z - 2\lambda = 0 \quad \text{i.e. } z = \lambda \tag{12}$$

$$L_\lambda = 2x - y - 2z - 15 = 0 \tag{13}$$

Note that the final equation is just the constraint equation again. Substitute $x = -\lambda$, $y = \frac{1}{2}\lambda$ and $z = \lambda$ in equation (13). We obtain

$$-\lambda\left(2 + \frac{1}{2} + 2\right) = 15$$

i.e.

$$\lambda = -\frac{10}{3}$$

It follows that

$$x = \frac{10}{3}, \quad y = -\frac{5}{3}, \quad z = -\frac{10}{3}$$

and hence the result.

6.9 Lagrange's method – general case♣

Suppose that $f: \mathbb{R}^n \to \mathbb{R}$ and that $g_1: \mathbb{R}^n \to \mathbb{R}, g_2: \mathbb{R}^n \to \mathbb{R}, \ldots, g_m: \mathbb{R}^n \to \mathbb{R}$. Write $\mathbf{x} = (x_1, x_2, \ldots, x_n)^T$. We now consider the problem of evaluating the maximum or minimum of the function of n variables

$$f(x_1, x_2, \ldots, x_n) \text{ or } f(\mathbf{x})$$

subject to the m constraints

$$\left.\begin{array}{c} g_1(x_1, x_2, \ldots, x_n) = 0 \\ g_2(x_1, x_2, \ldots, x_n) = 0 \\ \cdots \\ g_m(x_1, x_2, \ldots, x_n) = 0 \end{array}\right\} \quad \text{or} \quad \left.\begin{array}{c} g_1(\mathbf{x}) = 0 \\ g_2(\mathbf{x}) = 0 \\ \cdots \\ g_m(\mathbf{x}) = 0 \end{array}\right\}$$

If $\mathbf{g}: \mathbb{R}^n \to \mathbb{R}^m$ is the vector function $\mathbf{g}(\mathbf{x}) = (g_1(\mathbf{x}), g_2(\mathbf{x}), \ldots, g_m(\mathbf{x}))^T$ the problem, expressed in vector notation, is to find the stationary points of $y = f(\mathbf{x})$ subject to the constraint $\mathbf{g}(\mathbf{x}) = \mathbf{0}$. The Lagrangian for this problem is

$$L: \mathbb{R}^{n+m} \to \mathbb{R}$$

defined by

$$L(\mathbf{x}, \boldsymbol{\lambda}) = f(\mathbf{x}) + \boldsymbol{\lambda}^T \mathbf{g}(\mathbf{x}) = (f(\mathbf{x}))^T + (\mathbf{g}(\mathbf{x}))^T \boldsymbol{\lambda}$$

Each term $f(\mathbf{x})$ and $\boldsymbol{\lambda}^T\mathbf{g}(\mathbf{x})$ is a scalar and so is equal to its transpose.

where $\boldsymbol{\lambda} = (\lambda_1, \lambda_2, \ldots, \lambda_m)^T$. The stationary points of L are found by solving the simultaneous equations

$$\left\{\begin{array}{l} \dfrac{\partial L}{\partial \mathbf{x}} = \mathbf{0} - \text{i.e. } f'(\mathbf{x}) + \boldsymbol{\lambda}^T \mathbf{g}'(\mathbf{x}) = \mathbf{0} \\[2ex] \dfrac{\partial L}{\partial \boldsymbol{\lambda}} = \mathbf{0} - \text{i.e. } (\mathbf{g}(\mathbf{x}))^T = \mathbf{0} \quad \text{or} \quad \mathbf{g}(\mathbf{x}) = \mathbf{0} \end{array}\right.$$

Example 17♣

Suppose that A is an invertible $n \times n$ symmetric matrix and that $\mathbf{m}$ is a $1 \times n$ row vector. Prove that a maximum or minimum value of

$$y = \mathbf{mx}$$

subject to the constraint $\mathbf{x}^T A \mathbf{x} = 1$ must satisfy

$$y^2 = \mathbf{m} A^{-1} \mathbf{m}^T$$

We form the Lagrangian

$$L = \mathbf{m}\mathbf{x} + \lambda(\mathbf{x}^T A \mathbf{x} - 1) = 0$$

The stationary points are found by solving

$$\left.\begin{aligned} \frac{\partial L}{\partial \mathbf{x}} &= \mathbf{m} + 2\lambda \mathbf{x}^T A = \mathbf{0} \\[2mm] \frac{\partial L}{\partial \lambda} &= \mathbf{x}^T A \mathbf{x} - 1 = 0 \end{aligned}\right\} \qquad\begin{aligned} &(1)\\[4mm] &(2)\end{aligned}$$

Multiplying (1) through by A^{-1} we obtain

$$\mathbf{m} A^{-1} + 2\lambda \mathbf{x}^T = \mathbf{0}$$

$$\mathbf{x}^T = -\frac{1}{2\lambda}\mathbf{m} A^{-1}$$

$$\mathbf{x} = -\frac{1}{2\lambda}(\mathbf{m} A^{-1})^T$$

$$= -\frac{1}{2\lambda} A^{-1}\mathbf{m}^T$$

Recall that $A^T = A$.   Substituting in (2) yields

$$\left(-\frac{1}{2\lambda}\mathbf{m} A^{-1}\right) A \left(-\frac{1}{2\lambda} A^{-1}\mathbf{m}^T\right) = 1$$

$$(2\lambda)^2 = \mathbf{m} A^{-1}\mathbf{m}^T$$

But, multiplying (1) through by $\mathbf{x}$,

$$\mathbf{m}\mathbf{x} + 2\lambda\mathbf{x}^T A \mathbf{x} = 0$$
$$y + 2\lambda = 0$$
$$y^2 = (2\lambda)^2 = \mathbf{m} A^{-1}\mathbf{m}^T$$

as required.

For instance, a maximum or minimum height of the plane

$$z = x + y \qquad \text{(where } \mathbf{m} = (1, 1)^{\mathrm{T}})$$

restricted to the conic

$$2x^2 + 2xy + 2y^2 = 1$$

or

$$(x, y) \begin{pmatrix} 2 & 1 \\ 1 & 2 \end{pmatrix} \begin{pmatrix} x \\ y \end{pmatrix} = 1$$

This result can be verified using Lagrange's method.

is given by $z = \pm\sqrt{(1, 1) \begin{pmatrix} 2 & 1 \\ 1 & 2 \end{pmatrix}^{-1} \begin{pmatrix} 1 \\ 1 \end{pmatrix}} = \pm\sqrt{\tfrac{2}{3}}$.

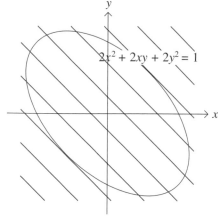

Contour plot of $z = x + y$

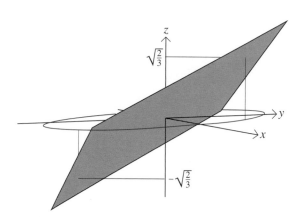

Plot of plane showing maximum constrained height and depth

6.10 Constrained optimisation – analytic criteria♣

Suppose that $f: \mathbb{R}^n \rightarrow \mathbb{R}$ and that $\mathbf{g}: \mathbb{R}^n \rightarrow \mathbb{R}^m$. The stationary points of $f(\mathbf{x})$ subject to the constraint $\mathbf{g}(\mathbf{x}) = \mathbf{0}$ are found by calculating the stationary points of the Lagrangian

$$L(\mathbf{x}, \boldsymbol{\lambda}) = f(\mathbf{x}) + \boldsymbol{\lambda}^{\mathrm{T}}\mathbf{g}(\mathbf{x})$$

We have classified these constrained stationary points by examining the contours of the functions. It is desirable to have analytic criteria for deciding the nature of the stationary points. When $n = 2$ and $m = 1$, the condition for a local maximum is that the determinant

$$\begin{vmatrix} 0 & g_x & g_y \\ g_x & L_{xx} & L_{xy} \\ g_y & L_{xy} & L_{yy} \end{vmatrix}$$

See Mathematics in Economics by A. Ostaszewski.

known as the **bordered Hessian** is positive at a stationary point of f.

In the general case, the condition for a local maximum is that the final $n - m$ principal minors of the following matrix should alternate in sign with the sign of the first being $(-1)^{m+1}$:

$$
\begin{pmatrix}
0 & \cdots & 0 & \dfrac{\partial g_1}{\partial x_1} & \cdots & \dfrac{\partial g_1}{\partial x_n} \\
\vdots & & \vdots & \vdots & & \vdots \\
0 & & 0 & \dfrac{\partial g_m}{\partial x_1} & \cdots & \dfrac{\partial g_m}{\partial x_n} \\
\dfrac{\partial g_1}{\partial x_1} & \cdots & \dfrac{\partial g_m}{\partial x_1} & \dfrac{\partial^2 L}{\partial x_1^2} & \cdots & \dfrac{\partial^2 L}{\partial x_1 \partial x_n} \\
\vdots & & \vdots & \vdots & & \vdots \\
\dfrac{\partial g_1}{\partial x_n} & \cdots & \dfrac{\partial g_m}{\partial x_n} & \dfrac{\partial^2 L}{\partial x_n \partial x_1} & \cdots & \dfrac{\partial^2 L}{\partial x_n^2}
\end{pmatrix}
$$

6.11 Exercises

1.* Find all stationary points of the function $f: \mathbb{R}^2 \to \mathbb{R}$ defined by $f(x, y) = (x - 2)^2 + (y - 3)^2$.

    (i) Determine
$$\min f(x, y)$$
    subject to the constraint
$$2x + y \le 2$$

    (ii) Determine
$$\min f(x, y)$$
    subject to the constraint
$$3x + 2y \ge 6$$

2. Find all stationary points of the function $f: \mathbb{R}^2 \to \mathbb{R}$ defined by $f(x, y) = x^3 + y^3 - 3xy^2$.

    Let D be the region in $\mathbb{R}^2$ consisting of the points inside and on the boundary of the triangle with vertices $(0, 0)^\mathsf{T}$, $(0, 1)^\mathsf{T}$ and $(1, 0)^\mathsf{T}$. Find the maximum and minimum values of $f(x, y)$ subject to the constraint that $(x, y)^\mathsf{T}$ lie in D.

3.* Find the maximum and minimum values of $f = x^2 + y^2$ subject to the constraint
$$5x^2 + 6xy + 5y^2 = 8$$

4. Find the maximum and minimum values of $f = xy$ subject to the constraint
$$2x^2 + y^2 \le 1$$

5.* Find the rectangle of largest area which has given perimeter $p > 0$.

6. Find the maximum and minimum values of

$$f = 2x + y + 3z$$

subject to the constraint

$$x^2 + 4y^2 + 9z^2 = 1$$

7.* Find the minimum distance between the curves $xy = 1$ and $x + 2y = 1$ – i.e. minimise

$$(x - X)^2 + (y - Y)^2$$

subject to the constraints

$$\left. \begin{array}{r} xy = 1 \\ X + 2Y = 1 \end{array} \right\}$$

8. Find the minimum distance from the point $(0, 0, 0)^\mathsf{T}$ to the line of intersection of the planes

$$\left. \begin{array}{r} x + 2y - 2z = 3 \\ 2x + y + 2z = 6 \end{array} \right\}$$

9.* Find the constrained stationary points for xyz when the constraints are

$$\left. \begin{array}{r} x + 2y + 3z = 1 \\ 3x + 2y + z = 1 \end{array} \right\}$$

At which of these points is xyz largest? Is this a point at which xyz is maximised subject to the constraints?

10.* A consumer has the utility function

$$U(x, y) = 10x^{\frac{1}{2}}y^{\frac{1}{3}}$$

where x and y are two goods. If the prices of the goods are $p_x = 4$ and $p_y = 5$ find the amounts x and y which will maximise utility under an income constraint of 40.

11. Two consumers A and B have the following utility functions:

$$U_A(x, y) = 5x^{\frac{1}{2}}y^{\frac{1}{2}}$$
$$U_B(x, y) = 7x^{\frac{3}{4}}y^{\frac{1}{4}}$$

where x and y are two goods. If the prices of the goods are $p_x = 3$ and $p_y = 4$, find the amounts x and y which will maximise utility for each consumer under an income constraint of 50.

12. A consumer has the utility function

$$U(x, y) = 10x^{\frac{1}{2}}y^{\frac{2}{3}}$$

where x and y are two goods with prices $p_x = 2$ and $p_y = 3$ respectively. Find how a rise in income from 60 to 70 will affect the quantities x and y of the goods that she should purchase in order to maximise utility.

13. A firm's production function is

$$P(x, y) = 100 \left(\frac{3}{4}x^{-\frac{1}{4}} + \frac{1}{4}y^{-\frac{1}{4}} \right)^{-4}$$

where x and y are factor inputs which the producer purchases at prices $p_x = 20$ and $p_y = 15$. If the producer has a budget of 400, find the quantities x and y which maximise output under this constraint.

14. At a certain factory, the output produced per day is

$$P(x, y) = 50x^{\frac{1}{3}}y^{\frac{2}{3}}$$

where x denotes the total number of working hours per day of the work force in units of 100 and y denotes the capital investment in units of $1000. If the hourly wage of a worker is $20 find the maximum daily output with a budget of $80\,000.

15.*♣ Find a system of linear equations satisfied by the stationary points of

$$y = (A\mathbf{x} + \mathbf{a})^{\mathrm{T}}(B\mathbf{x} + \mathbf{b})$$

where A and B are $m \times n$ matrices and $\mathbf{a}$ and $\mathbf{b}$ are $m \times 1$ column vectors.

16.♣ Let A be a symmetric $n \times n$ matrix. Show that, if $\boldsymbol{\xi}$ is a stationary point of

$$y = \mathbf{x}^{\mathrm{T}}A\mathbf{x}$$

subject to the constraint $\mathbf{x}^{\mathrm{T}}\mathbf{x} = 1$, then $\boldsymbol{\xi}$ is an eigenvector of A, i.e.

$$A\boldsymbol{\xi} = \lambda\boldsymbol{\xi}$$

for some scalar λ. Show also that λ is the value of y when $\mathbf{x} = \boldsymbol{\xi}$.

17.*♣ Let B be an $m \times n$ matrix and let $\mathbf{c}$ be an $m \times 1$ column vector. Show that there is a unique stationary point for

$$y = \mathbf{x}^{\mathrm{T}}\mathbf{x}$$

subject to the constraint
$$B\mathbf{x} = \mathbf{c}$$

provided that the matrix BB^{T} is invertible. Find this stationary point.

18.♣ Let A be a symmetric, invertible $n \times n$ matrix. Let B be an $m \times n$ matrix and let $\mathbf{c}$ be an $m \times 1$ column vector. Show that the unique stationary point for

$$y = \mathbf{x}^{\mathrm{T}} A \mathbf{x}$$

subject to the constraint

$$B\mathbf{x} = \mathbf{c}$$

is given by

$$\mathbf{x} = A^{-1} B^{\mathrm{T}} (B A^{-1} B^{\mathrm{T}})^{-1} \mathbf{c}$$

provided that $B A^{-1} B^{\mathrm{T}}$ is invertible.

6.12 Applications (optional)

6.12.1 The Nash bargaining problem

Suppose that D is a convex region in $\mathbb{R}^2$ and that s is a specified point in D. Two individuals negotiate over which point in D should be adopted. If they agree on $\mathbf{x} = (x, y)^{\mathrm{T}}$, then the first individual receives a utility of x units and the second a utility of y units. On which point of D should they agree? If they *fail* to agree, it is to be understood that the result will be the 'status quo' point s.

The **Nash bargaining solution** for this problem is the point ξ at which

$$\max_{\mathbf{x}}(x - s_1)(y - s_2)$$

is achieved subject to the constraints that $\mathbf{x}$ lie in the region D and $x \geq s_1, y \geq s_2$. Note that ξ is always a point on the boundary of D.

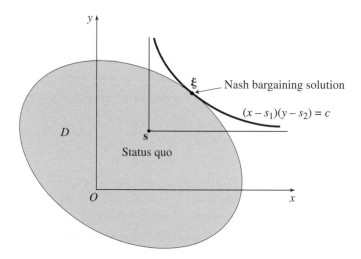

We shall not discuss the reasons which led Nash to propose ξ as the solution of the problem. Instead we shall consider some examples.

(i) Consider, to begin with, the case when D is the set of points inside or on the ellipse

$$x^2 + 4y^2 = 1$$

and $s = (0,0)^T$. Example 14 then applies and we obtain that the Nash bargaining solution is $\xi = (\frac{1}{\sqrt{2}}, \frac{1}{2\sqrt{2}})^T$. Thus the first individual receives $\frac{1}{\sqrt{2}}$ and the second $\frac{1}{2\sqrt{2}}$.

(ii) Suppose that D is the quadrilateral with vertices $(0,0)^T$, $(0,2)^T$, $(4,1)^T$, $(5,0)^T$ and $s = (1,0)^T$.

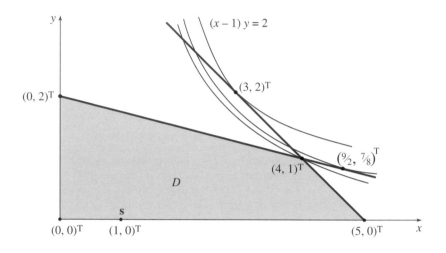

We begin by finding the point at which $(x-1)y$ is maximised subject to the constraint $x + 4y = 8$. For this purpose, we evaluate the stationary points of

$$\phi(y) = (8 - 4y - 1)y = 7y - 4y^2$$
$$\phi'(y) = 7 - 8y$$

and hence a unique stationary point occurs when $y = \frac{7}{8}$. But the point $(\frac{9}{2}, \frac{7}{8})^T$ does not lie in the region D.

Next we find the point at which $(x-1)y$ is maximised subject to $x + y = 5$.

$$\phi(y) = (5 - y - 1)y = (4 - y)y = 4y - y^2$$
$$\phi'(y) = 4 - 2y$$

Hence a unique stationary point occurs when $y = 2$. But the point $(3,2)^T$ does not lie in the region D.

We conclude that the Nash bargaining solution is $\xi = (4,1)^T$.

6.12.2 Inventory control

(i) At regular intervals, a firm orders a quantity x of a commodity which is placed in stock. This stock is depleted at a constant rate until none remains whereupon the firm immediately restocks with quantity x. The graph below shows how the amount in stock varies with time.

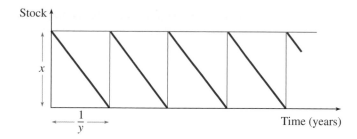

The firm requires X units of the commodity each year and, on average, the firm orders the commodity with a frequency of y times in each year. If the requirement is to be met, it is therefore necessary that

$$xy = X$$

The cost of holding one unit of the commodity in stock for a year is d. Since the average amount of the commodity held in stock is $\frac{1}{2}x$, it follows that the yearly holding cost is $\frac{1}{2}xd$. The cost of reordering the commodity is e. The yearly reordering cost is therefore ey.

The problem that firm faces is to minimise cost

$$C(x, y) = \frac{1}{2}xd + ey$$

subject to the constraint

$$xy = X$$

The Lagrangian is

$$L = \frac{1}{2}xd + ey + \lambda(xy - X)$$

$$\left. \begin{aligned} \frac{\partial L}{\partial x} &= \frac{1}{2}d + \lambda y = 0 \qquad &(1) \\[2mm] \frac{\partial L}{\partial y} &= e + \lambda x = 0 \qquad &(2) \\[2mm] \frac{\partial L}{\partial \lambda} &= xy - X = 0 \qquad &(3) \end{aligned} \right\}$$

From (1) and (2) we obtain

$$\frac{1}{2}ed = \lambda^2 xy = \lambda^2 X$$

Hence

$$\frac{1}{\lambda} = \pm\sqrt{\frac{2X}{ed}}$$

and the required solutions are

$$\left. \begin{aligned} x &= -\frac{e}{\lambda} = \sqrt{\frac{2eX}{d}} \\[3mm] y &= -\frac{d}{2\lambda} = \sqrt{\frac{dX}{2e}} \end{aligned} \right\}$$

The firm should therefore order $(2eXd^{-1})^{1/2}$ of the commodity $(dX(2e)^{-1})^{1/2}$ times a year. The yearly cost will then be

$$\sqrt{2deX}$$

(ii) Consider the previous problem but with the assumption that the firm does *not* immediately restock when its inventory is reduced to zero but waits until the demand for the commodity reaches a level of z before restocking with quantity $x + z$. Of this, z is processed immediately and the remaining x is depleted at a constant rate as before. Observe that $(x + z)y = X$.

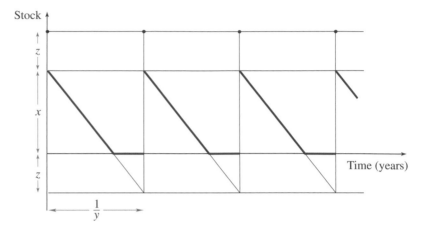

Since stock is only held for

$$\left(\frac{x}{x + z}\right)$$

of the year (*under the new arrangements*), the yearly holding cost becomes

$$\frac{1}{2}xd\left(\frac{x}{x + z}\right)$$

The yearly reordering cost remains ey but we must now also include a cost representing the penalty (*e.g. overtime or customer dissatisfaction*) resulting from the fact that a backlog is periodically built up which then has to be eliminated in a sudden flurry of activity. If it costs f to instantaneously clear one unit of backlog, then the yearly clearing cost will be

$$fyz$$

The new problem is therefore to minimise

$$C(x, y, z) = \frac{1}{2}xd\left(\frac{x}{x + z}\right) + ey + fyz \tag{4}$$

subject to the constraint $(x + z)y = X$. Make the substitutions $x + z = Xy^{-1}$ and $z = Xy^{-1} - x$. We obtain

$$D(x, y) = \frac{1}{2}dX^{-1}x^2y + ey + fX - fyx \tag{5}$$

The stationary points of this function are found by solving

$$\frac{\partial D}{\partial x} = dX^{-1}xy - fy = 0 \tag{6}$$

$$\frac{\partial D}{\partial y} = \frac{1}{2}dX^{-1}x^2 + e - fx = 0 \tag{7}$$

From (6) we obtain $y = 0$ *or* $x = fd^{-1}X$. The first possibility is incompatible with the constraint $(x + z)y = X$. Substituting the second possibility in (7), we obtain

$$\frac{1}{2}dX^{-1}(fd^{-1}X)^2 + e - f(fd^{-1}X) = 0 \qquad (8)$$

That is

$$e = \frac{1}{2}f^2d^{-1}X$$

If this equation happens not to hold, it then follows that (4) has *no* constrained stationary points. Note, however, that we have neglected to take account so far of the fact that $x \geq 0, y \geq 0$ and $z \geq 0$. It may therefore be that the minimum we are seeking occurs when $x = 0$ or $z = 0$ ($y = 0$ *does not satisfy* $(x + z)y = X$).

If $z = 0$, we are back with problem (i). If $x = 0$, we have to consider

$$C(0, y, Xy^{-1}) = ey + fX$$

Since $y = 0$ is not admissible, this quantity has no minimum but we can make the quantity as close to fX as we choose by taking y sufficiently close to 0.

As a result of this analysis, we can distinguish two cases:

(a) $\sqrt{2deX} \leq fX$. In this case a minimum is obtained by taking $z = 0$ and choosing x and y as in problem (i).

(b) $\sqrt{2deX} > fX$. In this case *no* minimum exists. However, the firm can make its cost as close to fX as it chooses by taking $x = 0$ and making y very small (*but positive*). The corresponding value of z (i.e. Xy^{-1}) will then be very large. In this case, the penalty costs are so small compared with the other costs that the firm should hold nothing in stock and delay production for as long as possible.

Note incidentally that, when equation (8) holds, case (a) applies. Substituting $e = \frac{1}{2}f^2d^{-1}X$ and $x = fd^{-1}X$ in (5), we obtain that

$$\begin{aligned} D(x, y) &= \frac{1}{2}dX^{-1}(fd^{-1}X)^2y + \frac{1}{2}f^2d^{-1}Xy + fX - fy(fd^{-1}X) \\ &= fX \end{aligned}$$

6.12.3 Least squares analysis

Suppose that a process has a single input and results in a single output. The quantity b of output will then be a function of the quantity a of input – i.e. $b = f(a)$. Suppose that we have a theory that f is an affine function – i.e.

$$b = f(a) = ma + c$$

and we would like to know the values of the two constants m and c. The obvious course of action is to run an experiment and obtain a table of corresponding values of b and a. Suppose that the table obtained is that given below.

The most notable feature about this data is that it is not consistent with the hypothesis that $b = ma + c$ – i.e. the system of equations

$$\left. \begin{aligned} -0.4 &= m1 + c \\ 0.8 &= m2 + c \\ 2.6 &= m3 + c \\ 3.2 &= m4 + c \\ 5.4 &= m5 + c \end{aligned} \right\}$$

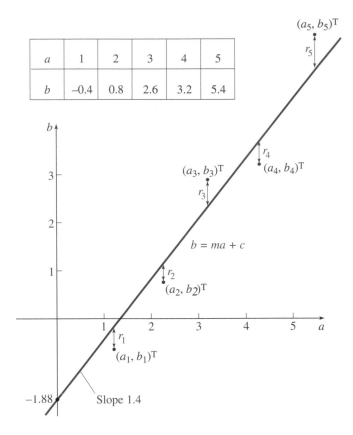

a	1	2	3	4	5
b	−0.4	0.8	2.6	3.2	5.4

is *inconsistent*. This system can be written $\mathbf{b} = A\mathbf{x}$, where

$$\mathbf{b} = \begin{pmatrix} -0.4 \\ 0.8 \\ 2.6 \\ 3.2 \\ 5.4 \end{pmatrix} \quad A = \begin{pmatrix} 1 & 1 \\ 2 & 1 \\ 3 & 1 \\ 4 & 1 \\ 5 & 1 \end{pmatrix} \quad \text{and} \quad \mathbf{x} = \begin{pmatrix} m \\ c \end{pmatrix}$$

But the fact that $\mathbf{b} = A\mathbf{x}$ is inconsistent does not necessarily imply that the theory that $b = ma + c$ is false. Perhaps the data contains errors. If the theory is correct and the errors are not too large, the data can then be used to *estimate* m and c.

The problem is then to find the straight line $b = ma + c$ which 'fits the data best'. One way of interpreting this is to seek the values of m and c which minimise

$$y = r_1^2 + r_2^2 + r_3^2 + r_4^2 + r_5^2$$

This is called the 'least squares' estimation method for obvious reasons. Observe that

$$r_k^2 = (b_k - ma_k - c)^2$$

and so

$$y = \|A\mathbf{x} - \mathbf{b}\|^2$$

The least squares method is therefore to find the value of $\mathbf{x} = (m, c)^{\mathrm{T}}$ which minimises $\|A\mathbf{x} - \mathbf{b}\|^2$.

We now look at the rationale behind this problem and find the form of the solution(s) in general.† Consider

$$y = \|A\mathbf{x} - \mathbf{b}\|^2 = (A\mathbf{x} - \mathbf{b})^{\mathrm{T}}(A\mathbf{x} - \mathbf{b}) = \mathbf{u}^{\mathrm{T}}\mathbf{u}$$

where $\mathbf{u} = A\mathbf{x} - \mathbf{b}$. Then, by the chain rule (III),‡

$$\frac{\mathrm{d}y}{\mathrm{d}\mathbf{x}} = \frac{\mathrm{d}y}{\mathrm{d}\mathbf{u}}\frac{\mathrm{d}\mathbf{u}}{\mathrm{d}\mathbf{x}} = (2\mathbf{u}^{\mathrm{T}}I)A = 2(A\mathbf{x} - \mathbf{b})^{\mathrm{T}}A$$

For the stationary points, we require that

$$\frac{\partial y}{\partial x_1} = \frac{\partial y}{\partial x_2} = \frac{\partial y}{\partial x_3} = \cdots = \frac{\partial y}{\partial x_n} = 0$$

i.e.

$$\frac{\mathrm{d}y}{\mathrm{d}\mathbf{x}} = \left(\frac{\partial y}{\partial x_1}, \frac{\partial y}{\partial x_2}, \ldots, \frac{\partial y}{\partial x_n}\right) = (0, 0, \ldots, 0) = \mathbf{0}$$

We therefore have to solve

$$2(A\mathbf{x} - \mathbf{b})^{\mathrm{T}}A = \mathbf{0}$$
$$A^{\mathrm{T}}(A\mathbf{x} - \mathbf{b}) = \mathbf{0}$$
$$A^{\mathrm{T}}A\mathbf{x} = A^{\mathrm{T}}\mathbf{b}$$

Since a minimum certainly does exist, *these* equations must be solvable. If the square matrix $A^{\mathrm{T}}A$ happens to be invertible, there is a unique solution given by

$$\mathbf{x} = (A^{\mathrm{T}}A)^{-1}A^{\mathrm{T}}\mathbf{b}$$

In the above case we may calculate the product

$$A^{\mathrm{T}}A = \begin{pmatrix} 55 & 15 \\ 15 & 5 \end{pmatrix}$$

which is invertible, so we can find a least squares solution with

$$\mathbf{x} = \begin{pmatrix} 55 & 15 \\ 15 & 5 \end{pmatrix}^{-1} \begin{pmatrix} 1 & 2 & 3 & 4 & 5 \\ 1 & 1 & 1 & 1 & 1 \end{pmatrix} \begin{pmatrix} -0.4 \\ 0.8 \\ 2.6 \\ 3.2 \\ 5.4 \end{pmatrix} = \begin{pmatrix} 1.4 \\ -1.88 \end{pmatrix}$$

6.12.4 Kuhn–Tucker conditions

Suppose that $f: \mathbb{R}^n \to \mathbb{R}$ and $\mathbf{g}: \mathbb{R}^n \to \mathbb{R}^m$ are functions. We have studied the problem of maximising or minimising

$$f(\mathbf{x})$$

subject to the constraint

$$\mathbf{g}(\mathbf{x}) = \mathbf{0}$$

One forms the Lagrangian

$$L(\mathbf{x}, \boldsymbol{\lambda}) = f(\mathbf{x}) + \boldsymbol{\lambda}^{\mathrm{T}}\mathbf{g}(\mathbf{x})$$

and examines the solutions of the *Lagrangian conditions*

$$\text{(i)} \quad \left\{ \begin{array}{l} \dfrac{\partial L}{\partial \mathbf{x}} = f'(\mathbf{x}) + \lambda^{\mathrm{T}} \mathbf{g}'(\mathbf{x}) = \mathbf{0} \\[3mm] \text{(ii)} \quad \dfrac{\partial L}{\partial \lambda} = \mathbf{g}(\mathbf{x})^{\mathrm{T}} = \mathbf{0} - \text{i.e. } \mathbf{g}(\mathbf{x}) = \mathbf{0} \end{array} \right.$$

A more general problem is to maximise or minimise

$$f(\mathbf{x})$$

subject to the constraint

$$\mathbf{g}(\mathbf{x}) \geq \mathbf{0}$$

See §1.12.4.

Recall that $\mathbf{g}(\mathbf{x}) \geq \mathbf{0}$ *means that* $g_1(\mathbf{x}) \geq 0$, $g_2(\mathbf{x}) \geq 0$, ..., *and* $g_m(\mathbf{x}) \geq 0$. *The* Lagrangian conditions then have to be replaced by the *Kuhn–Tucker conditions*. In the case of a *maximum*, these are the following:

$$\left\{ \begin{array}{ll} \text{(i)} & f'(\mathbf{x}) + \lambda^{\mathrm{T}} \mathbf{g}'(\mathbf{x}) = \mathbf{0} \\ \text{(ii)} & \mathbf{g}(\mathbf{x}) \geq \mathbf{0} \\ \text{(iii)} & \lambda^{\mathrm{T}} \mathbf{g}(\mathbf{x}) = 0 \\ \text{(iv)} & \lambda \geq \mathbf{0} \end{array} \right.$$

In the case of a *minimum*, (iv) is replaced by $\lambda \leq \mathbf{0}$.

The diagram below illustrates a possible configuration in the case $n = 2$ and $m = 3$ when $f(\mathbf{x})$ achieves a maximum (*subject to the constraint* $\mathbf{g}(\mathbf{x}) \geq \mathbf{0}$) at the point $\mathbf{x} = \tilde{\mathbf{x}}$.

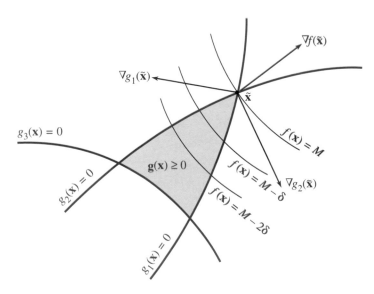

Since $g_3(\tilde{\mathbf{x}}) > 0$, *items (ii), (iii) and (iv) of the Kuhn–Tucker conditions imply that* $\lambda_3 = 0$. *Item (i) then asserts that* $\nabla f = -\lambda_1 \nabla g_1 - \lambda_2 \nabla g_2$. *Since* $\lambda_1 \geq 0$ *and* $\lambda_2 \geq 0$ *by (iv), this guarantees that* $\langle \nabla f, \mathbf{u} \rangle \leq 0$ *for each direction* $\mathbf{u}$ *which points into the region defined by* $\mathbf{g}(\mathbf{x}) \geq \mathbf{0}$.

As an example, consider the case $\mathbf{g}(\mathbf{x}) = \mathbf{x}$. Then $\mathbf{g}'(\mathbf{x}) = I$ and the Kuhn–Tucker conditions become

$$\left\{ \begin{array}{c} f'(\mathbf{x}) + \lambda^{\mathrm{T}} = \mathbf{0} \\ \lambda^{\mathrm{T}} \mathbf{x} = 0 \\ \mathbf{x} \geq \mathbf{0}, \lambda \geq \mathbf{0} \end{array} \right.$$

These imply that

$$\begin{cases} f'(\mathbf{x}) \leq \mathbf{0} & -\text{i.e.} & \nabla f(\mathbf{x}) \leq \mathbf{0} \\ \mathbf{x} \geq \mathbf{0} & & \mathbf{x} \geq \mathbf{0} \\ f'(\mathbf{x})^{\mathsf{T}}\mathbf{x} = 0 & & \langle \nabla f(\mathbf{x}), \mathbf{x} \rangle = 0 \end{cases}$$

The following diagrams illustrate the various possibilities in the case $f: \mathbb{R}^2 \to \mathbb{R}$:

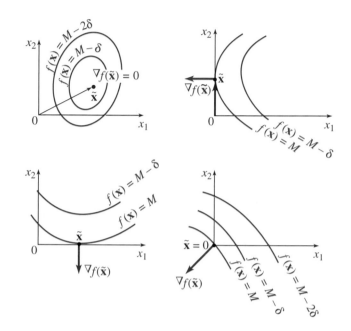

6.12.5 Linear programming

We considered linear programming in §1.12.4. Our example concerned a baker who sought to *maximise* profit

$$\mathbf{p}^{\mathsf{T}}\mathbf{x}$$

subject to the constraints

$$\left.\begin{array}{c} A\mathbf{x} \leq \mathbf{b} \\ \mathbf{x} \geq \mathbf{0} \end{array}\right\}$$

Here the vector $\mathbf{b}$ represents the baker's stock of ingredients and $A\mathbf{x}$ represents the input required to produce the output $\mathbf{x}$.

In §1.12.5, we elaborated on this theme by introducing a factory producing baked goods which sought to persuade the baker not to sell the products of his baking by quoting a price vector $\mathbf{q}$ for his ingredients. The idea of the manufacturer was to make it as attractive to the baker to sell his ingredients to the manufacturer at these prices as to bake them and sell the results of his baking. We decided that the manufacturer should choose $\mathbf{q}$ so as to *minimise*

$$\mathbf{b}^{\mathsf{T}}\mathbf{q}$$

subject to the constraints

$$\left.\begin{array}{c} A^{\mathsf{T}}\mathbf{q} \geq \mathbf{p} \\ \mathbf{q} \geq \mathbf{0} \end{array}\right\}$$

The baker's problem is a *primal* linear programming problem and the manufacturer's problem is the *dual* linear programming problem.

In this section we propose to discuss how the ideas introduced in this chapter and the previous chapter relate to these problems. Observe, to begin with, that, if the baker produces x and the manufacturer quotes price vector $\mathbf{q}$, then the baker will realise an amount

$$L(\mathbf{x}, \mathbf{q}) = \mathbf{p}^T\mathbf{x} + \mathbf{q}^T(\mathbf{b} - A\mathbf{x})$$

assuming he sells his unused ingredients (i.e. $\mathbf{b} - A\mathbf{x}$) to the manufacturer. For obvious reasons we call $L(\mathbf{x}, \mathbf{q})$ the Lagrangian of the primal problem. Note that it is also the Lagrangian of the dual problem because

$$\begin{aligned} L(\mathbf{x}, \mathbf{q}) &= L(\mathbf{x}, \mathbf{q})^T \\ &= \mathbf{x}^T\mathbf{p} + (\mathbf{b}^T - \mathbf{x}^T A^T)\mathbf{q} \\ &= \mathbf{b}^T\mathbf{q} + \mathbf{x}^T(\mathbf{p} - A^T\mathbf{q}) \end{aligned}$$

We can regard the situation as a species of 'game' (§1.12.6) in which the baker chooses x in an attempt to maximise $L(\mathbf{x}, \mathbf{q})$ and the manufacturer chooses q in an attempt to minimise $L(\mathbf{x}, \mathbf{q})$. A solution $(\tilde{\mathbf{x}}, \tilde{\mathbf{q}})$ to this game has the property that $\mathbf{x} = \tilde{\mathbf{x}}$ *maximises* $L(\mathbf{x}, \tilde{\mathbf{q}})$ subject to the constraint $\mathbf{x} \geq \mathbf{0}$ and $\mathbf{q} = \tilde{\mathbf{q}}$ *minimises* $L(\tilde{\mathbf{x}}, \mathbf{q})$ subject to the constraint $\mathbf{q} \geq \mathbf{0}$. (*We noted in §1.12.6 that such a pair $(\tilde{\mathbf{x}}, \tilde{\mathbf{q}})$ is called a Nash equilibrium for the game.*)

We considered the problem of maximising $f(\mathbf{x})$ subject to the constraint $\mathbf{x} \geq \mathbf{0}$ at the end of §6.12.4. Applying this work in the case when $f(\mathbf{x}) = L(\mathbf{x}, \tilde{\mathbf{q}})$, we obtain

(i) $\mathbf{p}^T - \tilde{\mathbf{q}}^T A \leq \mathbf{0}$ – i.e. $A^T\tilde{\mathbf{q}} \geq \mathbf{p}$

(ii) $\tilde{\mathbf{x}} \geq \mathbf{0}$

(iii) $\mathbf{p}^T\tilde{\mathbf{x}} - \tilde{\mathbf{q}}^T A\tilde{\mathbf{x}} = \mathbf{0}$ – i.e. $\mathbf{p}^T\tilde{\mathbf{x}} = \tilde{\mathbf{q}}^T A\tilde{\mathbf{x}}$

Similarly, since $\mathbf{q} = \tilde{\mathbf{q}}$ minimises $L(\tilde{\mathbf{x}}, \mathbf{q})$ subject to the constraint $\mathbf{q} \geq \mathbf{0}$

(iv) $\mathbf{b}^T - \tilde{\mathbf{x}}^T A^T \geq \mathbf{0}$ – i.e. $A\tilde{\mathbf{x}} \leq \mathbf{b}$

(v) $\tilde{\mathbf{q}} \geq \mathbf{0}$

(vi) $\mathbf{b}^T\tilde{\mathbf{q}} - \tilde{\mathbf{x}}^T A^T\tilde{\mathbf{q}} = \mathbf{0}$ – i.e. $\mathbf{b}^T\tilde{\mathbf{q}} = \tilde{\mathbf{q}}^T A\tilde{\mathbf{x}}$.

Suppose that $\mathbf{x} \geq \mathbf{0}$ and $A\mathbf{x} \leq \mathbf{b}$. Then

$$\begin{aligned} \mathbf{p}^T\mathbf{x} \leq \mathbf{p}^T\mathbf{x} + \tilde{\mathbf{q}}^T(\mathbf{b} - A\mathbf{x}) &= L(\mathbf{x}, \tilde{\mathbf{q}}) \\ &\leq \mathbf{p}^T\tilde{\mathbf{x}} + \tilde{\mathbf{q}}^T(\mathbf{b} - A\tilde{\mathbf{x}}) \\ &= \mathbf{p}^T\tilde{\mathbf{x}} \qquad \text{(by vi)} \end{aligned}$$

Since $A\tilde{\mathbf{x}} \leq \mathbf{b}$ (by iv), it follows that $\mathbf{x} = \tilde{\mathbf{x}}$ solves our primal linear programming problem. A similar argument shows that $\mathbf{q} = \tilde{\mathbf{q}}$ solves the dual linear programming problem.

The most significant feature of this analysis is that provided by (iii) and (vi) which show that

$$\mathbf{p}^T\tilde{\mathbf{x}} = \mathbf{b}^T\tilde{\mathbf{q}}$$

This result is called the **duality theorem** of linear programming. Stated in full, it asserts that if the primal and the dual problems both have solutions, then the maximum value attained in the primal problem is equal to the minimum value attained in the dual problem.

6.12.6 Saddle points

It is easy to see why maxima and minima are important but applications for saddle points are perhaps not so evident.

We therefore give an application from game theory. In a two person zero sum game, one is given a payoff function

$$L(\mathbf{p}, \mathbf{q})$$

which the first player seeks to maximise by choosing an appropriate $\mathbf{p}$ from his set P of feasible strategies, while the second player simultaneously seeks to minimise $L(\mathbf{p}, \mathbf{q})$ by choosing an appropriate $\mathbf{q}$ from his set Q of feasible strategies.

In §1.12.6, we considered the case in which

$$L(\mathbf{p}, \mathbf{q}) = \mathbf{p}^{\mathrm{T}} A \mathbf{q}$$

where A is an $m \times n$ matrix. The $m \times 1$ vector

$$\mathbf{p} = (p_1, p_2, \ldots, p_m)^{\mathrm{T}}$$

in this case represents a 'mixed' strategy in which p_k is the probability with which the kth 'pure' strategy is to be played. Thus P is the set of all vectors $\mathbf{p}$ for which $p_1 \geq 0, p_2 \geq 0, \ldots, p_m \geq 0$ and $p_1 + p_2 + \cdots + p_m = 1$. Similarly, Q is the set of all vectors $\mathbf{q}$ for which $q_1 \geq 0, q_2 \geq 0, \ldots, q_n \geq 0$ and $q_1 + q_2 + \cdots + q_n = 1$.

In §6.12.5, we considered the zero sum game whose payoff function

$$L(\mathbf{x}, \mathbf{q}) = \mathbf{p}^{\mathrm{T}}\mathbf{x} + \mathbf{q}^{\mathrm{T}}(\mathbf{b} - A\mathbf{x})$$

is the Lagrangian of the linear programming problem

$$\max \mathbf{p}^{\mathrm{T}}\mathbf{x}$$

subject to

$$\left. \begin{array}{c} A\mathbf{x} \leq \mathbf{b} \\ \mathbf{x} \geq \mathbf{0} \end{array} \right\}$$

(*Here* $\mathbf{p}$ *is a constant vector representing the prices at which the commodity bundle* $\mathbf{x}$ *can be sold.*) The baker sought to choose $\mathbf{x} \geq \mathbf{0}$ to maximise $L(\mathbf{x}, \mathbf{q})$ and the manufacturer to choose $\mathbf{q} \geq 0$ so as to minimise $L(\mathbf{x}, \mathbf{q})$.

A pair $(\tilde{\mathbf{p}}, \tilde{\mathbf{q}})$ of strategies for a zero sum game with payoff function $L(\mathbf{q}, \mathbf{p})$ is a Nash equilibrium if and only if

$$L(\mathbf{p}, \tilde{\mathbf{q}}) \leq L(\tilde{\mathbf{p}}, \tilde{\mathbf{q}}) \leq L(\tilde{\mathbf{p}}, \mathbf{q})$$

for all $\mathbf{p}$ in the set P and for all $\mathbf{q}$ in the set Q. If $(\tilde{\mathbf{p}}, \tilde{\mathbf{q}})$ is a Nash equilibrium, then the choice $\mathbf{p} = \tilde{\mathbf{p}}$ is an optimal response by the first player to the choice of $\mathbf{q} = \tilde{\mathbf{q}}$ by the second player. Simultaneously, the choice $\mathbf{q} = \tilde{\mathbf{q}}$ by the second player is an optimal response by the second player to the choice $\mathbf{p} = \tilde{\mathbf{p}}$ by the first player. Any solution to a two person zero sum game must clearly satisfy this criterion.

If $(\tilde{\mathbf{p}}, \tilde{\mathbf{q}})$ is a Nash equilibrium, then the function $L(\mathbf{p}, \tilde{\mathbf{q}})$ achieves a maximum at $\mathbf{p} = \tilde{\mathbf{p}}$ (*subject to the constraint that* $\mathbf{p}$ *lies in* P). Similarly, the function $L(\tilde{\mathbf{p}}, \mathbf{q})$ achieves a minimum at $\mathbf{q} = \tilde{\mathbf{q}}$ (*subject to the constraint that* $\mathbf{q}$ *lies in* Q). If $\tilde{\mathbf{p}}$ is an interior point of P and $\tilde{\mathbf{q}}$ is an interior point of Q, then

$$\frac{\partial L}{\partial \mathbf{p}}(\tilde{\mathbf{p}}, \tilde{\mathbf{q}}) = \mathbf{0} \quad \frac{\partial L}{\partial \mathbf{q}}(\tilde{\mathbf{p}}, \tilde{\mathbf{q}}) = \mathbf{0}$$

and hence $(\tilde{\mathbf{p}}, \tilde{\mathbf{q}})$ is a *stationary point* of L.

However, $(\tilde{\mathbf{p}}, \tilde{\mathbf{q}})$ is neither a local maximum nor a local minimum for L. It is, in fact, a *saddle point* for L.

Seven _____ Inverse functions

We begin this chapter with a fairly careful account of the inverse of a function of one variable. We have already studied the inverses of basic elementary functions in Chapter 2. We now develop the idea of a 'local inverse' of a function of one variable, to address the situation when more than one value of the independent variable x is mapped by the function to the same value of the dependent variable y. This follows on from the power functions with their inverses, the root functions and we define the 'inverse trigonometric functions'.

The chapter continues with local inverse functions in the vector case and applies this work to general coordinate systems. The ideas are explained at some length but it would be inappropriate to spend a lot of time on these sections if difficulties are being encountered elsewhere. For most purposes it is enough to have some understanding of why things go wrong when the Jacobian is zero and to be able to change the variables in differential operators as described in §7.5.

7.1 Local inverses of scalar valued functions

We say that $f^{-1}: J \to I$ is a **local inverse** for $f: \mathbb{R} \to \mathbb{R}$ if it is true that

$$x = f^{-1}(y) \text{ if and only if } y = f(x)$$

provided that $x \in I$ and $y \in J$. The existence of the local inverse therefore requires that, for each $y \in J$, the equation $y = f(x)$ has a unique solution $x \in I$.

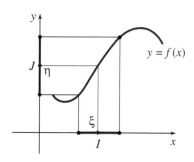
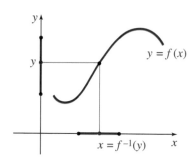

If ξ is an interior point of the interval I and $\eta = f(\xi)$ is an interior point of the interval J, we say that $f^{-1}: J \to I$ is a **local inverse for** f **at the point** $x = \xi$.

For obvious reasons, we denote a local inverse function at the point $x = \xi$ *by* f^{-1}, *but it is important to realise that there are many functions that can play this role.*

Example 1

Although the function $f : \mathbb{R} \to \mathbb{R}$ defined by

$$y = f(x) = x^2$$

has no inverse function, it does have a *local inverse function at any* $x \neq 0$.

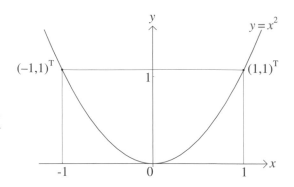

For instance it has a local inverse function at $x = 1$,

$$f_1^{-1} : [0, \infty) \to [0, \infty)$$

defined by

$$x = f_1^{-1}(y) = \sqrt{y}$$

when y is the independent variable, or

$$y = \sqrt{x}$$

when x is the independent variable.

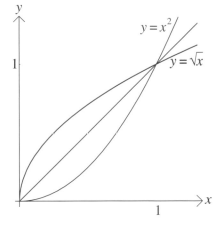

A function defined by the same formula on *any* subinterval of $[0, \infty)$ containing $x = 1$ as an interior point will also serve as a local inverse function for f at $x = 1$.

It also has a local inverse at $x = -1$, $f_2^{-1} : [0, \infty) \to (-\infty, 0]$ defined by

$$x = f_2^{-1}(y) = -\sqrt{y}$$

or

$$y = -\sqrt{x}$$

when x is the independent variable.

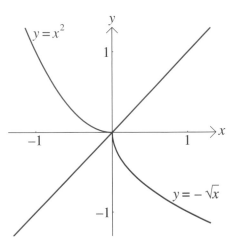

It has *no local inverse function at* $x = 0$, since $x = 0$ is not an interior point of any interval on which a local inverse may be defined.

7.1.1 Differentiability of local inverse functions

When $f'(\xi) \neq 0$, a differentiable function $f: \mathbb{R} \to \mathbb{R}$ has a local inverse at $x = \xi$ which is differentiable at $\eta = f(\xi)$ and the derivative is given by the formula

See §2.9.

$$\frac{\mathrm{d}x}{\mathrm{d}y} = \left(\frac{\mathrm{d}y}{\mathrm{d}x}\right)^{-1} \tag{1}$$

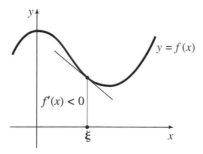

Differentiable local inverse exists at ξ

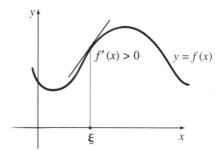

Differentiable local inverse exists at ξ

For ξ belonging to intervals where f is decreasing or increasing, a differentiable local inverse exists at ξ.

When $f'(\xi) = 0$ there are two possibilities.

Either no local inverse exists at ξ, as in the situations below. The equation $Y = f(x)$ has *no* solutions while $Z = f(x)$ has *two* solutions.

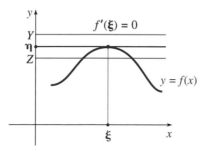

No local inverse exists at ξ

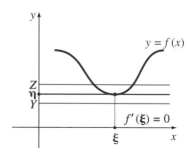

No local inverse exists at ξ

Alternatively, a local inverse may exist, as in the diagram on the left below; however its tangent at the point $\eta = f(\xi)$ is vertical and hence the local inverse is not differentiable, as in the diagram on the right.

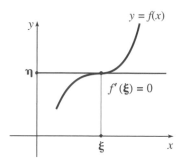

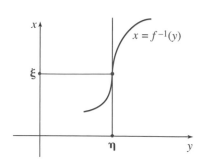

Local inverse exists at ξ Tangent to the local inverse at η is vertical

The overall result is:

When $f'(\xi) \neq 0$ a differentiable local inverse exists at ξ with derivative given by (1). When $f'(\xi) = 0$, either no local inverse exists or a local inverse exists but is not differentiable.

7.1.2 Inverse trigonometric functions

The function $f: \mathbb{R} \to \mathbb{R}$ defined by $y = f(x) = \sin x$ does *not* have an inverse function. In particular, the equation

$$2 = \sin x$$

has no solutions at all, while the equation

$$0 = \sin x$$

has an infinite number of solutions: $x = n\pi$, where $n \in \mathbb{Z}$.

However, if we restrict the range of values of x, then the 'restricted sine function' $F: [-\pi/2, \pi/2] \to [-1, 1]$ defined by $y = F(x) = \sin x$ admits an inverse function, called the **arcsine** function. It is defined by

$$x = \arcsin y \text{ if and only if } y = \sin x$$

[†] The sine function is increasing on this interval.

provided that $-\pi/2 \leq x \leq \pi/2$ and $-1 \leq y \leq 1$.[†]

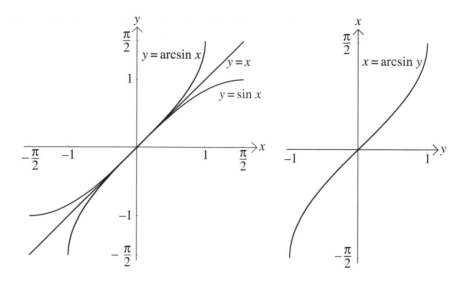

Similarly the 'restricted cosine function' defined by

$$y = \cos x$$

for $0 \le x \le \pi$ has an inverse function called the **arccosine** function, defined by

$$x = \arccos y \text{ if and only if } y = \cos x.$$

provided that $0 \le x \le \pi$ and $-1 \le y \le 1.$[†]

† The cosine function is decreasing on this interval.

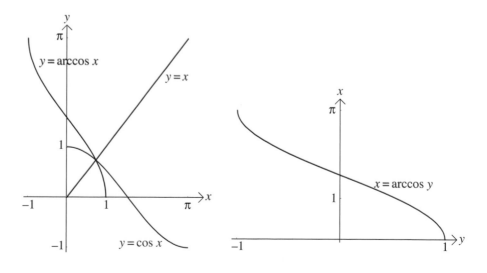

Finally, observe that the 'restricted tangent function' defined by

$$y = \tan x$$

for $-\pi/2 < x < \pi/2$ has an inverse function, the **arctangent** function defined by

$$x = \arctan y \text{ if and only if } y = \tan x$$

† The tangent function is increasing on this interval. *provided that* $-\pi/2 < x < \pi/2.$†

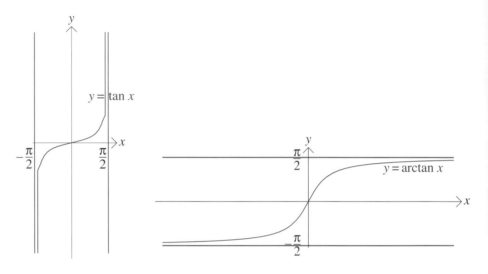

The derivatives of the arcsine, arccosine and arctangent functions are important in integration theory. We begin with the arcsine function.

If $-1 < y < 1$, we have

$$\frac{d}{dy}(\arcsin y) = \frac{dx}{dy} = \left(\frac{dy}{dx}\right)^{-1} = \frac{1}{\cos x} = \frac{1}{\sqrt{1 - \sin^2 x}} = \frac{1}{\sqrt{1 - y^2}}$$

Here we have used the fact that

$$x = \arcsin y \quad \text{if and only if} \quad y = \sin x$$

provided $-1 \le y \le 1$ *and* $-\pi/2 \le x \le \pi/2$.
  In solving $\cos^2 x + \sin^2 x = 1$ *for* $\cos x$, *we take* $\sqrt{1 - \sin^2 x}$ *rather than* $-\sqrt{1 - \sin^2 x}$ *because we know that the graph* $x = \arcsin y$ *has positive slope.*
  Similarly, if $-1 < y < 1$,

$$\frac{d}{dy}(\arccos y) = \frac{dx}{dy} = \left(\frac{dy}{dx}\right)^{-1} = \frac{1}{-\sin x} = -\frac{1}{\sqrt{1 - \cos^2 x}}$$

$$= -\frac{1}{\sqrt{1 - y^2}}$$

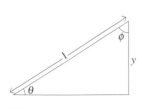

It is no surprise that the derivatives are the negative of each other, since in the triangle on the left $\theta = \pi/2 - \phi$ *where* $\theta = \arcsin y$ *and* $\phi = \arccos y$.

Finally, for all values of y,

$$\frac{d}{dy}(\arctan y) = \frac{dx}{dy} = \left(\frac{dy}{dx}\right)^{-1} = \frac{1}{\sec^2 x} = \frac{1}{1 + \tan^2 x} = \frac{1}{1 + y^2}.$$

Note that the formula $\sec^2 x = 1 + \tan^2 x$ *follows from the fact that* $\cos^2 x + \sin^2 x = 1$.

The table below summarises these results:

$$\frac{d}{dy}(\arcsin y) = \frac{1}{\sqrt{1 - y^2}} \quad (-1 < y < 1)$$

$$\frac{d}{dy}(\arccos y) = -\frac{1}{\sqrt{1 - y^2}} \quad (-1 < y < 1)$$

$$\frac{d}{dy}(\arctan y) = \frac{1}{1 + y^2}$$

7.2 Local inverses of vector valued functions

We have carried out a fairly exhaustive examination of local inverses of *real* functions $f : \mathbb{R} \to \mathbb{R}$. The situation is very similar in the case of vector functions $\mathbf{f} : \mathbb{R}^n \to \mathbb{R}^m$. Such a function can have a local inverse only if $n = m$. In this case the derivative

$$\frac{d\mathbf{y}}{d\mathbf{x}} \quad \text{or} \quad \frac{d\mathbf{f}}{d\mathbf{x}}$$

is a square matrix. If this matrix is *invertible* at $\mathbf{x} = \boldsymbol{\xi}$, then $\mathbf{f} : \mathbb{R}^n \to \mathbb{R}^n$ has a differentiable local inverse $\mathbf{f}^{-1}$ at the point $\mathbf{x} = \boldsymbol{\xi}$, with derivative at $\boldsymbol{\eta} = \mathbf{f}(\boldsymbol{\xi})$ given by the formula

$$\frac{d\mathbf{f}^{-1}}{d\mathbf{y}} = \frac{d\mathbf{x}}{d\mathbf{y}} = \left(\frac{d\mathbf{y}}{d\mathbf{x}}\right)^{-1} = \left(\frac{d\mathbf{f}}{d\mathbf{x}}\right)^{-1}$$

If the matrix

$$\frac{d\mathbf{f}}{d\mathbf{x}}$$

is not invertible at $\mathbf{x} = \boldsymbol{\xi}$, we say that $\boldsymbol{\xi}$ is a **critical point** for the function $\mathbf{f} : \mathbb{R}^n \to \mathbb{R}^n$. At a critical point, either the function has no local inverse or else the local inverse is not differentiable. Note in particular that the above formula makes no sense at a critical point.

A square matrix is not invertible if and only if its determinant is zero. It follows that $\mathbf{x} = \boldsymbol{\xi}$ is a critical point of $\mathbf{f} : \mathbb{R}^n \to \mathbb{R}^n$ if and only if

$$J = \det\left(\frac{d\mathbf{f}}{d\mathbf{x}}\right) = \begin{vmatrix} \dfrac{\partial f_1}{\partial x_1} & \dfrac{\partial f_1}{\partial x_2} & \cdots & \dfrac{\partial f_1}{\partial x_n} \\[2mm] \dfrac{\partial f_2}{\partial x_1} & \dfrac{\partial f_2}{\partial x_2} & \cdots & \dfrac{\partial f_2}{\partial x_n} \\[2mm] \vdots & \vdots & & \vdots \\[2mm] \dfrac{\partial f_n}{\partial x_1} & \dfrac{\partial f_n}{\partial x_2} & \cdots & \dfrac{\partial f_n}{\partial x_n} \end{vmatrix} = 0$$

where the partial derivatives are evaluated at $\mathbf{x} = \boldsymbol{\xi}$. The determinant, J, is called the **Jacobian** of the function $\mathbf{f}: \mathbb{R}^n \rightarrow \mathbb{R}^n$. The standard abbreviation for the Jacobian J is

$$\frac{\partial(f_1, f_2, \ldots, f_n)}{\partial(x_1, x_2, \ldots, x_n)}$$

The critical points occur where the Jacobian vanishes.

For $\mathbf{f} : \mathbb{R}^2 \rightarrow \mathbb{R}^2$ defined by

$$\mathbf{y} = \begin{pmatrix} u \\ v \end{pmatrix} = \mathbf{f}(\mathbf{x}) = \begin{pmatrix} f_1(x, y) \\ f_2(x, y) \end{pmatrix}$$

$\mathbf{x} = \boldsymbol{\xi}$ is a critical point if and only if $\begin{vmatrix} \dfrac{\partial u}{\partial x} & \dfrac{\partial u}{\partial y} \\ \dfrac{\partial v}{\partial x} & \dfrac{\partial v}{\partial y} \end{vmatrix} = 0$

We know from linear algebra that this is equivalent to the fact that the vectors $(\partial u/\partial x, \partial u/\partial y)^{\mathrm{T}}$ and $(\partial v/\partial x, \partial v/\partial y)^{\mathrm{T}}$ are linearly dependent.

This means that either ∇u and ∇v are parallel, or at least one of them is $\mathbf{0}$.

Equivalently, either $\nabla u = \lambda \nabla v$, $\lambda \neq 0$ or at least one of ∇u or ∇v is $\mathbf{0}$.

In the former case the contours of u and v touch at $\mathbf{x} = \boldsymbol{\xi}$.

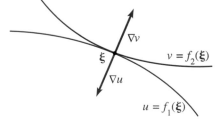

In the latter case at least one of u or v has a stationary point at $\mathbf{x} = \boldsymbol{\xi}$, which is usually (though not always) evident from a contour map.

Example 2

Consider the function $\mathbf{f}: \mathbb{R}^2 \rightarrow \mathbb{R}^2$ defined by

$$\begin{pmatrix} u \\ v \end{pmatrix} = \mathbf{f}(x, y) = \begin{pmatrix} x^2 + y^2 \\ x^2 - y^2 \end{pmatrix}$$

We have already studied this function to some extent in Example 5.1.

This function has no proper inverse. The diagram below indicates that the equations

$$\left. \begin{matrix} 5 = x^2 + y^2 \\ 3 = x^2 - y^2 \end{matrix} \right\}$$

have *four* solutions (namely $(2, 1)^{\mathrm{T}}$, $(2, -1)^{\mathrm{T}}$, $(-2, 1)^{\mathrm{T}}$ and $(-2, -1)^{\mathrm{T}}$).

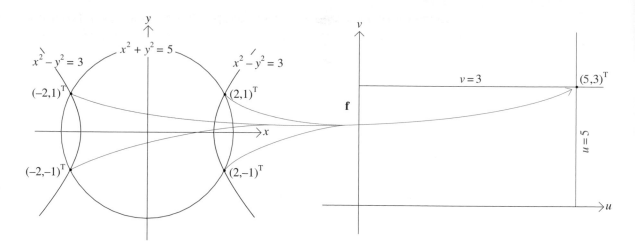

If we wish to find the points at which $\mathbf{f}$ has a *local* inverse, we must begin by calculating the derivative of $\mathbf{f}$. This is given by

$$
\left(
\begin{array}{cc}
\dfrac{\partial u}{\partial x} & \dfrac{\partial u}{\partial y} \\[2mm]
\dfrac{\partial v}{\partial x} & \dfrac{\partial v}{\partial y}
\end{array}
\right)
=
\left(
\begin{array}{cc}
2x & 2y \\
2x & -2y
\end{array}
\right)
$$

The Jacobian is therefore

$$
\left|
\begin{array}{cc}
\dfrac{\partial u}{\partial x} & \dfrac{\partial u}{\partial y} \\[2mm]
\dfrac{\partial v}{\partial x} & \dfrac{\partial v}{\partial y}
\end{array}
\right|
=
\left|
\begin{array}{cc}
2x & 2y \\
2x & -2y
\end{array}
\right|
= -8xy
$$

and so the critical points occur when $x = 0$ or $y = 0$.

At a noncritical point (i.e. where $x \neq 0$ and $y \neq 0$), a local inverse exists. We can compute the derivative of the local inverse function by observing that

$$
\left(
\begin{array}{cc}
\dfrac{\partial x}{\partial u} & \dfrac{\partial x}{\partial v} \\[2mm]
\dfrac{\partial y}{\partial u} & \dfrac{\partial y}{\partial v}
\end{array}
\right)
=
\left(
\begin{array}{cc}
\dfrac{\partial u}{\partial x} & \dfrac{\partial u}{\partial y} \\[2mm]
\dfrac{\partial v}{\partial x} & \dfrac{\partial v}{\partial y}
\end{array}
\right)^{-1}
=
\left(
\begin{array}{cc}
2x & 2y \\
2x & -2y
\end{array}
\right)^{-1}
$$

Hence

$$
\left(
\begin{array}{cc}
\dfrac{\partial x}{\partial u} & \dfrac{\partial x}{\partial v} \\[2mm]
\dfrac{\partial y}{\partial u} & \dfrac{\partial y}{\partial v}
\end{array}
\right)
= -\frac{1}{8xy}
\left(
\begin{array}{cc}
-2y & -2y \\
-2x & 2x
\end{array}
\right)
$$

In this case (although certainly not in general), it is fairly easy to find a formula for the local inverse at a particular point. To find the local inverse at $(2, -1)^T$, we begin with the equations

$$\left. \begin{array}{l} u = x^2 + y^2 \\ v = x^2 - y^2 \end{array} \right\}$$

which have to be solved to give $(x, y)^T$ in terms of $(u, v)^T$. We have

$$\left. \begin{array}{l} x^2 = \dfrac{u + v}{2} \\[2mm] y^2 = \dfrac{u - v}{2} \end{array} \right\}$$

These equations have four solutions depending on the signs we take when extracting square roots. We want x to be positive and y to be negative and so we take the solution

$$\left. \begin{array}{l} x = \sqrt{\dfrac{u + v}{2}} \\[4mm] y = -\sqrt{\dfrac{u - v}{2}} \end{array} \right\}$$

Since x and y must be real, the quantities under the square root signs must be non negative. We must therefore have

$$u + v \geq 0 \text{ and } u - v \geq 0$$

Also, from the formulas

$$\begin{pmatrix} u \\ v \end{pmatrix} = \mathbf{f}(x, y) = \begin{pmatrix} x^2 + y^2 \\ x^2 - y^2 \end{pmatrix}$$

we see that $u \geq 0$. These three inequalities specify the region in the uv plane which is the image of the $(x, y)^T$ plane under $\mathbf{f}$.

If we denote this region by Δ and the region in the $(x, y)^T$ plane specified by $x \geq 0$ and $y \leq 0$ by D, the local inverse at $(2, -1)^T$ is therefore the function $\mathbf{f}^{-1} \colon \Delta \to D$ defined by

$$\begin{pmatrix} x \\ y \end{pmatrix} = \mathbf{f}^{-1}(u, v) = \begin{pmatrix} \sqrt{\dfrac{u + v}{2}} \\[4mm] -\sqrt{\dfrac{u - v}{2}} \end{pmatrix}$$

We have already calculated the derivative of $\mathbf{f}^{-1}$. Observe in particular that

$$(\mathbf{f}^{-1})'(5, 3) = -\frac{1}{8 \cdot 2(-1)} \begin{pmatrix} -2(-1) & -2(-1) \\ -2 \cdot 2 & 2 \cdot 2 \end{pmatrix} = \begin{pmatrix} \frac{1}{8} & \frac{1}{8} \\[2mm] -\frac{1}{4} & \frac{1}{4} \end{pmatrix}$$

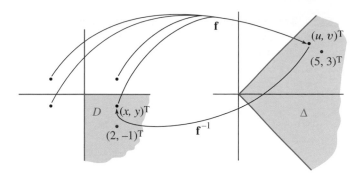

It is also instructive to see what happens at a critical point. Consider, for example, the critical point $(x, y)^T = (2, 0)^T$.

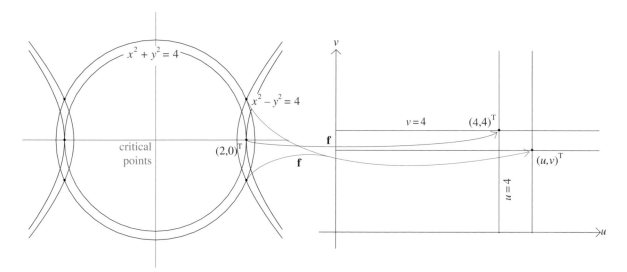

Notice that, however close $(u, v)^T$ is taken to $(4, 4)^T$ the equation $(u, v)^T = \mathbf{f}(x, y)$ has *two* solutions close to $(2, 0)^T$. *No* local inverse can therefore exist at $(2, 0)^T$.

Since the contours touch at the critical points, the critical points can be clearly seen on a contour map.

Example 3

The function

$$\begin{pmatrix} u \\ v \end{pmatrix} = \mathbf{f}(x, y) = \begin{pmatrix} x^2 \\ y \end{pmatrix}$$

has critical points given by

$$J = \begin{vmatrix} 2x & 0 \\ 0 & 1 \end{vmatrix} = 2x = 0$$

or $x = 0$ which is the y axis; *but these are not easily discernible from a contour map.*

The $(x, y)^T$ plane maps to the half plane $u \geq 0$ in which the v axis is the image of the y axis.

Consider the critical points along the y axis. However small $u > 0$, the point $(u, v)^T$ is the image of *two* points $(x, y)^T$ and $(-x, y)^T$ close to the y axis in the $(x, y)^T$ plane where $x^2 = u$ and $y = v$.

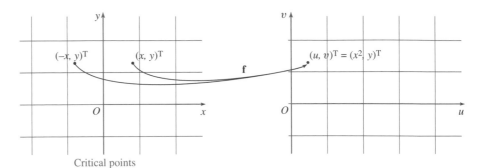

Critical points

Contours in $(x, y)^T$ plane

Half plane $u \geq 0$ with rectangular grid

Example 4

The function

$$\begin{pmatrix} u \\ v \end{pmatrix} = \mathbf{f}(x, y) = \begin{pmatrix} y - x^3 \\ y + x^3 \end{pmatrix}$$

has an inverse everywhere, but has critical points given by

$$J = \begin{vmatrix} -3x^2 & 1 \\ 3x^2 & 1 \end{vmatrix} = -6x^2 = 0$$

or $x = 0$ which is the y axis. Contours of $y - x^3$ and $y + x^3$ touch on the y axis. The inverse function

$$x = \left(\frac{v - u}{2} \right)^{\frac{1}{3}}$$

$$y = \frac{u + v}{2}$$

is not differentiable when $x = 0$ or $u = v$. Clearly, since J is 0 when $x = 0$ the derivative matrix of the inverse function does not exist.

Alternatively, entries

$$x_u = -\frac{1}{3} \left(\frac{v - u}{2} \right)^{-2/3} \quad \text{and} \quad x_v = \frac{1}{3} \left(\frac{v - u}{2} \right)^{-2/3}$$

of the derivative matrix are not defined when $u = v$.

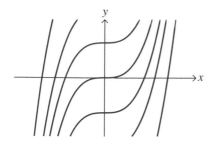

Contours of $u = y - x^3$

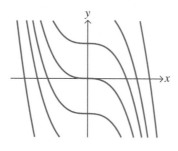

Contours of $v = y + x^3$

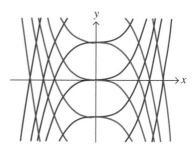

Contours touch along $x = 0$

7.3 Coordinate systems

The use of the usual system of cartesian coordinates in $\mathbb{R}^2$ amounts to locating points in the plane by reference to a rectangular grid as in the diagram below on the left.

But it is not always convenient to use a rectangular grid. If we change from the natural basis $\{(1, 0)^T, (0, 1)^T\}$ to a new basis $\{(1, 1)^T, (2, 1)^T\}$, we found in §6.1 that the new coordinates $(u, v)^T$ of a point with respect to the new basis are related to the original cartesian coordinates by the formula

$$P \begin{pmatrix} u \\ v \end{pmatrix} = \begin{pmatrix} x \\ y \end{pmatrix}$$

where

$$P = \begin{pmatrix} 1 & 2 \\ 1 & 1 \end{pmatrix}$$

So

$$\begin{pmatrix} u \\ v \end{pmatrix} = \begin{pmatrix} -1 & 2 \\ 1 & -1 \end{pmatrix} \begin{pmatrix} x \\ y \end{pmatrix} \quad - \text{i.e.} \quad \begin{cases} u = -x + 2y \\ v = x - y \end{cases}$$

To specify a point in the plane by giving its new coordinates is to locate the point with respect to the grid drawn below on the right:

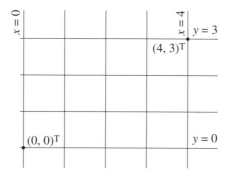

Rectangular grid

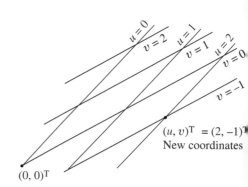

Oblique grid

There is no reason why we should restrict ourselves to straight line grids. In many cases matters can be greatly simplified by using a 'curvilinear' grid as indicated in the accompanying diagram.

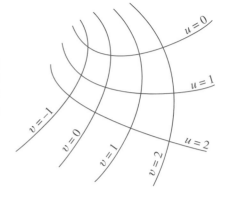

When using such 'curvilinear' coordinates, one needs of course to have a formula which expresses the new coordinates $(u, v)^T$ of a point in terms of its cartesian coordinates $(x, y)^T$.

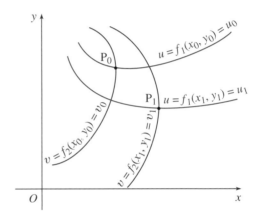

Suppose that the appropriate formula is

$$u = f_1(x, y) \\ v = f_2(x, y) \Big\}$$

Then the diagram on the right illustrates a point P_0 whose original cartesian coordinates are $(x_0, y_0)^T$ and whose new 'curvilinear' coordinates are $(u_0, v_0)^T$.

It is often helpful to introduce the function $\mathbf{f}$ defined by

$$\begin{pmatrix} u \\ v \end{pmatrix} = \mathbf{f}(x, y) = \begin{pmatrix} f_1(x, y) \\ f_2(x, y) \end{pmatrix}$$

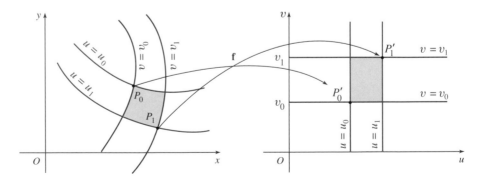

One can then not only think in terms of relabelling points P_0 and P_1 in the $(x, y)^T$ plane with new coordinates $(u, v)^T$: one can also simultaneously think in terms of transforming the points P_0 and P_1 into new points P_0' and P_1' in a new $(u, v)^\mathrm{T}$ plane. In our diagram opposite, for example, the complicated shaded region in the $(x, y)^\mathrm{T}$ plane is transformed into the simple shaded rectangular region in the $(u, v)^\mathrm{T}$ plane.

It is *not* true that *any* function $\mathbf{f} \colon \mathbb{R}^n \to \mathbb{R}^n$ can be used to define a new coordinate system in $\mathbb{R}^n$. In order that $\mathbf{f} \colon \mathbb{R}^n \to \mathbb{R}^n$ can be used to define a new coordinate system, it is essential that $\mathbf{f}$ admits at least a local *inverse* function $\mathbf{f}^{-1} \colon \Delta \to D$. The component functions of $\mathbf{f}$ can then be used to define new coordinates $(u_1, u_2, \ldots, u_n)^\mathrm{T}$ for a point $(x_1, x_2, \ldots, x_n)^\mathrm{T}$ in the set D by means of the formulas

$$\left.\begin{aligned} u_1 &= f_1(x_1, x_2, \ldots, x_n) \\ u_2 &= f_2(x_1, x_2, \ldots, x_n) \\ &\;\;\vdots \\ u_n &= f_n(x_1, x_2, \ldots, x_n) \end{aligned}\right\}$$

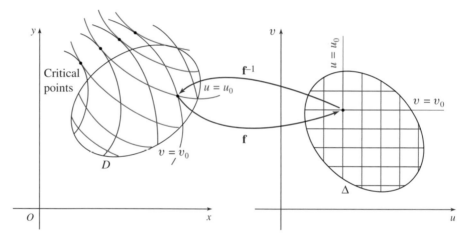

If the function $\mathbf{f}$ does *not* have an appropriate inverse, then it is useless for the purpose of introducing a new coordinate system because it will then assign the *same* new coordinates to two *different* points. It follows in particular that there is no point in trying to use a function $\mathbf{f} \colon \mathbb{R}^n \to \mathbb{R}^n$ for the purpose of defining a new coordinate system in a set D if a critical point of $\mathbf{f}$ is an interior point of D. Thus, for each interior point $\mathbf{x}$ of the set D, we require that the Jacobian

$$\det \mathbf{f}'(\mathbf{x}) \neq 0$$

Example 5

We have studied the function $\mathbf{f} \colon \mathbb{R}^2 \to \mathbb{R}^2$ defined by

$$\left.\begin{aligned} u &= x^2 + y^2 \\ v &= x^2 - y^2 \end{aligned}\right\}$$

in Example 5.1 and in Example 2. This function can be used to introduce a new coordinate system into any one of the four quadrants of the $(x, y)^\mathrm{T}$ plane. In Example 2, we focussed attention on the fourth quadrant but in the following diagram we have chosen to illustrate what happens in the first quadrant (i.e. the set D of those $(x, y)^T$ for which $x > 0$ and $y > 0$).

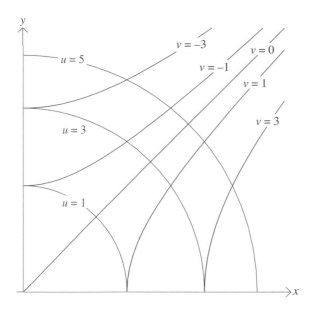

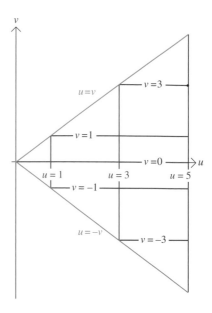

Notice that $(u, v)^T = (5, 3)^T$ corresponds to only *one* point in the first quadrant of the $(x, y)^T$ plane. In general, if $(u, v)^T$ are the new coordinates of a point in the first quadrant with old coordinates $(x, y)^T$, then

$$\left.\begin{array}{l} x = \sqrt{\dfrac{u + v}{2}} \\[4mm] y = \sqrt{\dfrac{u - v}{2}} \end{array}\right\}$$

Example 6

Consider the function $\mathbf{f} : \mathbb{R}^2 \to \mathbb{R}^2$ defined by

$$\mathbf{f}(x, y) = \begin{pmatrix} u \\ v \end{pmatrix} = \begin{pmatrix} x^2 + 4y^2 \\ 2y + 3x \end{pmatrix}$$

The critical points of $\mathbf{f}$ are given by

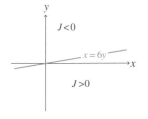

$$J = \begin{vmatrix} 2x & 8y \\ 3 & 2 \end{vmatrix} = 4x - 24y = 4(x - 6y)$$

This is equal to zero along the line $x - 6y = 0$. Critical points lie along this line which splits the $(x, y)^T$ plane into two half planes where the Jacobian is of constant sign as shown in the sketch.

The point $(x, y)^T = (1, 0)^T$ is a noncritical point, so a local inverse function of $\mathbf{f}$ exists at it. To find a formula for such a local inverse function, solve the component equations for x and y in terms of u and v.

$$u = x^2 + 4y^2$$
$$v = 2y + 3x$$

Substituting $y = \frac{1}{2}(v - 3x)$ in the first equation,

$$x^2 + (v - 3x)^2 - u = 0$$

which simplifies to

$$10x^2 - 6vx + v^2 - u = 0$$

Solving the quadratic equation

$$x = \frac{1}{10}\left(3v \pm \sqrt{9v^2 - 10(v^2 - u)}\right) = \frac{1}{10}\left(3v \pm \sqrt{10u - v^2}\right)$$

$$y = \frac{1}{2}\left(v - \frac{1}{10}\left(9v \pm 3\sqrt{10u - v^2}\right)\right) = \frac{1}{20}\left(v \mp 3\sqrt{10u - v^2}\right)$$

Since x and y are real we must have $10u > v^2$. The parabola $10u = v^2$ in the $(u, v)^{\mathrm{T}}$ plane corresponds to the line $x = 6y$ in the $(x, y)^{\mathrm{T}}$ plane.

Hence if Δ and D are the regions in the figures below, a local inverse function $\mathbf{f}^{-1}\colon \Delta \to D$ at $(x, y)^{\mathrm{T}} = (1, 0)^{\mathrm{T}}$ (for which $x > 6y$) is defined by

$$\begin{pmatrix} x \\ y \end{pmatrix} = \mathbf{f}^{-1}(u, v) = \begin{pmatrix} \frac{1}{10}\left(3v + \sqrt{10u - v^2}\right) \\ \frac{1}{20}\left(v - 3\sqrt{10u - v^2}\right) \end{pmatrix}$$

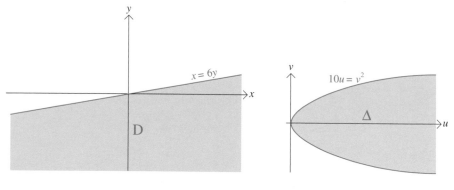

When $(x, y)^{\mathrm{T}} = (1, 0)^{\mathrm{T}}$, $(u, v)^{\mathrm{T}} = (1, 3)^{\mathrm{T}}$. The second point in the $(x, y)^{\mathrm{T}}$ plane with the same image $(u, v)^{\mathrm{T}} = (1, 3)^{\mathrm{T}}$ is given by

$$x = \frac{1}{10}\left(3 \cdot 3 - \sqrt{10 \cdot 1 - 3^2}\right) = \frac{4}{5}$$

$$y = \frac{1}{20}\left(3 + 3\sqrt{10 \cdot 1 - 3^2}\right) = \frac{3}{10}$$

We can use **f** to introduce a new coordinate system in the half plane D of the $(x, y)^T$ plane, finding images of appropriate curves in the region Δ of the $(u, v)^T$ plane.

Contours $x^2 + 4y^2 = c$ correspond to vertical lines $u = c$.

Contours $2y + 3x = c$ correspond to horizontal lines $v = c$.

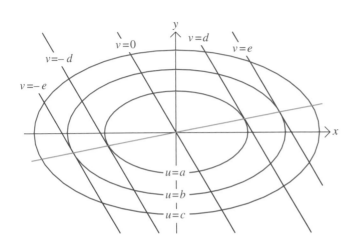

Contours in $(x, y)^T$ plane

Corresponding contours in $(u, v)^T$ plane

To find the transforms of the coordinate axes $x = 0$ and $y = 0$:

$x = 0$ implies that $u = 4y^2$, $v = 2y$

which implies $u = v^2$

$y = 0$ implies that $u = x^2$, $v = 3x$

which implies $9u = v^2$

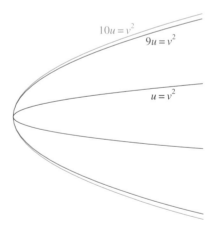

Example 7

Consider the function $\mathbf{f} : \mathbb{R}^2 \to \mathbb{R}^2$ defined by

$$\mathbf{f}(x, y) = \begin{pmatrix} u \\ v \end{pmatrix} = \begin{pmatrix} xy \\ x \end{pmatrix}$$

Its critical points are given by

$$J = \begin{vmatrix} y & x \\ 1 & 0 \end{vmatrix} = -x = 0$$

So the critical points lie along the y axis.

To find the image under $\mathbf{f}$ of the strip $1 \leq y \leq 2, x \geq 0$ we have

$$x = v, \quad y = \frac{u}{v} \quad \text{when} \quad v \neq 0$$

$$v = x = 0 \quad \text{implies} \quad u = xy = 0$$

Hence

$$1 \leq y \leq 2 \quad x \geq 0$$

if and only if

$$1 \leq \frac{u}{v} \leq 2 \quad \text{when} \quad v > 0 \quad \text{or} \quad u = v = 0$$

i.e. if and only if

$$v \leq u \leq 2v \quad \text{when} \quad v \geq 0$$

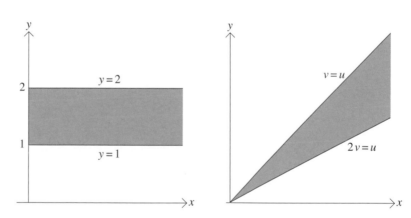

The strip is mapped to the region in the positive $(u, v)^{\mathrm{T}}$ quadrant between the lines $v = u$ and $v = u/2$.

Example 8
_____

The function $\mathbf{f} : \mathbb{R}^2 \to \mathbb{R}^2$ is defined by

$$\mathbf{f}(x, y) = \begin{pmatrix} u \\ v \end{pmatrix} = \begin{pmatrix} x^2 + 2y \\ xy \end{pmatrix}$$

The critical points of **f** are given by

$$J = \begin{vmatrix} 2x & 2 \\ y & x \end{vmatrix} = 2x^2 - 2y$$

This is equal to zero along
the parabola $y = x^2$.
When $x^2 > y, \quad J > 0$
When $x^2 < y, \quad J < 0$.

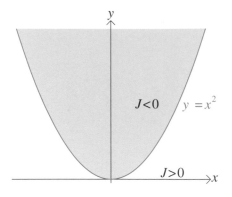

We now find ∇u and ∇v and examine the u and v contours through a critical point $(a, a^2)^{\mathrm{T}}$.

$$\nabla u = \begin{pmatrix} 2x \\ 2 \end{pmatrix} \quad \text{at } (a, a^2)^{\mathrm{T}}, \quad \nabla u = \begin{pmatrix} 2a \\ 2 \end{pmatrix} = 2 \begin{pmatrix} a \\ 1 \end{pmatrix}$$

$$\nabla v = \begin{pmatrix} y \\ x \end{pmatrix} \quad \text{at } (a, a^2)^{\mathrm{T}}, \quad \nabla v = \begin{pmatrix} a^2 \\ a \end{pmatrix} = a \begin{pmatrix} a \\ 1 \end{pmatrix}$$

So the gradients of the contours are parallel and the contours touch along the parabola $y = x^2$. The contour plots below illustrate this.

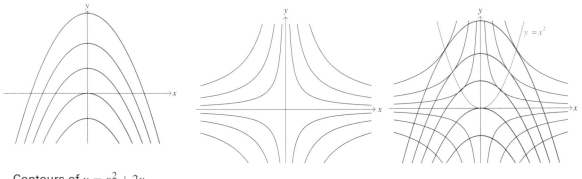

Contours of $u = x^2 + 2y$

Contours of $v = xy$

Contours touch along $y = x^2$

7.4 Polar coordinates

Often one is given the old coordinates $(x, y)^{\mathrm{T}}$ as functions of the new coordinates instead of the other way around. Consider, for example, the function $\mathbf{f} \colon \mathbb{R}^2 \to \mathbb{R}^2$ defined by

$$\left.\begin{array}{l} x = r \cos\theta \\ y = r \sin\theta \end{array}\right\}$$

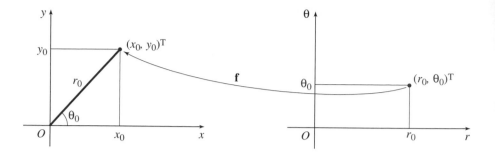

The important point to note here is the geometric interpretation of r and θ in the $(x, y)^T$ plane. This explains why r and θ are called the *polar coordinates* of the point $(x, y)^T$.

In general, the equations

$$\left.\begin{array}{l} x = r\cos\theta \\ y = r\sin\theta \end{array}\right\}$$

have many solutions for r and θ. To guarantee that each $(x, y)^T$ is assigned a *unique* pair $(r, \theta)^T$ of polar coordinates we need to restrict the values of r and θ which we are willing to admit. It is usual to require that $r > 0$ and $-\pi < \theta \le \pi$.

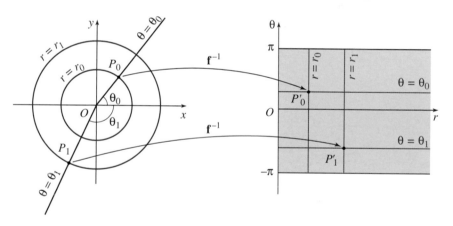

The Jacobian of $\mathbf{f}$ is given by

$$\det \mathbf{f}'(r, \theta) = \begin{vmatrix} \dfrac{\partial x}{\partial r} & \dfrac{\partial x}{\partial \theta} \\[2mm] \dfrac{\partial y}{\partial r} & \dfrac{\partial y}{\partial \theta} \end{vmatrix}$$

$$= \begin{vmatrix} \cos\theta & -r\sin\theta \\ \sin\theta & r\cos\theta \end{vmatrix} = r(\cos^2\theta + \sin^2\theta) = r$$

This is zero only when $r = 0$.

Note finally that, for $r > 0$ and $-\pi/2 < \theta < \pi/2$, the unique solution of the equations

$$\left.\begin{array}{l} x = r\cos\theta \\ y = r\sin\theta \end{array}\right\}$$

255

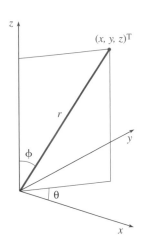

is given by

$$\left. \begin{array}{l} r = \sqrt{x^2 + y^2} \\[2mm] \theta = \arctan \dfrac{y}{x} \end{array} \right\}$$

Spherical polar coordinates are sometimes useful in $\mathbb{R}^3$. The appropriate formulas are

$$\left. \begin{array}{l} x = r \cos \theta \sin \phi \\ y = r \sin \theta \sin \phi \\ z = r \cos \phi \end{array} \right\}$$

The geometric significance of these coordinates is indicated in the margin. To guarantee a unique solution to the equations, one requires that $r > 0$,

$$-\pi < \theta \le \pi \quad \text{and} \quad 0 \le \phi < \pi$$

7.5 Differential operators♣

It is often the case that a change of variables is made to simplify a given integral or differential expression. Integrals will be left to a later chapter. In this section we shall restrict our attention to the question of how differential expressions transform when the coordinate system is changed.

The basic chain rule is that for differentiating a 'function of a function'. If $z = g(\mathbf{u})$ and $\mathbf{u} = \mathbf{f}(\mathbf{x})$ where $\mathbf{x}$ and $\mathbf{u}$ are $n \times 1$ column vectors, this rule takes the form

$$\frac{d z}{d \mathbf{x}} = \frac{d z}{d \mathbf{u}} \frac{d \mathbf{u}}{d \mathbf{x}}$$

In the case when $n = 2$, this formula reduces to

$$\left(\frac{\partial z}{\partial x_1}, \frac{\partial z}{\partial x_2} \right) = \left(\frac{\partial z}{\partial u_1}, \frac{\partial z}{\partial u_2} \right) \begin{pmatrix} \dfrac{\partial u_1}{\partial x_1} & \dfrac{\partial u_1}{\partial x_2} \\[3mm] \dfrac{\partial u_2}{\partial x_1} & \dfrac{\partial u_2}{\partial x_2} \end{pmatrix}$$

i.e.

$$\left. \begin{array}{l} \dfrac{\partial z}{\partial x_1} = \dfrac{\partial u_1}{\partial x_1} \dfrac{\partial z}{\partial u_1} + \dfrac{\partial u_2}{\partial x_1} \dfrac{\partial z}{\partial u_2} \\[4mm] \dfrac{\partial z}{\partial x_2} = \dfrac{\partial u_1}{\partial x_2} \dfrac{\partial z}{\partial u_1} + \dfrac{\partial u_2}{\partial x_2} \dfrac{\partial z}{\partial u_2} \end{array} \right\}$$

These equations are valid regardless of the function $z = g(\mathbf{u})$ to which they are applied. It is therefore common to rewrite them in **operator** notation as below:

$$\frac{\partial}{\partial x_1} = \frac{\partial u_1}{\partial x_1} \frac{\partial}{\partial u_1} + \frac{\partial u_2}{\partial x_1} \frac{\partial}{\partial u_2}$$

$$\frac{\partial}{\partial x_2} = \frac{\partial u_1}{\partial x_2} \frac{\partial}{\partial u_1} + \frac{\partial u_2}{\partial x_2} \frac{\partial}{\partial u_2}$$

The meaning of these formulas is simply that, if the left hand side is applied to z, then the result will be exactly the same as applying the right hand side to z.

Example 9

The formulas

$$u = x^2 + y^2 \atop v = x^2 - y^2 \Big\}$$

are used to change the coordinate system, in order to simplify the partial differential equation

$$\frac{1}{x}\frac{\partial z}{\partial x} + \frac{1}{y}\frac{\partial z}{\partial y} = 1$$

We have

$$\frac{\partial}{\partial x} = \frac{\partial u}{\partial x}\frac{\partial}{\partial u} + \frac{\partial v}{\partial x}\frac{\partial}{\partial v} = 2x\frac{\partial}{\partial u} + 2x\frac{\partial}{\partial v}$$

$$\frac{\partial}{\partial y} = \frac{\partial u}{\partial y}\frac{\partial}{\partial u} + \frac{\partial v}{\partial y}\frac{\partial}{\partial v} = 2y\frac{\partial}{\partial u} - 2y\frac{\partial}{\partial v}$$

It follows that

$$\frac{1}{x}\frac{\partial z}{\partial x} + \frac{1}{y}\frac{\partial z}{\partial y} = \left(2\frac{\partial z}{\partial u} + 2\frac{\partial z}{\partial v}\right) + \left(2\frac{\partial z}{\partial u} - 2\frac{\partial z}{\partial v}\right) = 4\frac{\partial z}{\partial u}$$

The given partial differential equation therefore assumes the particularly simple form

$$\frac{\partial z}{\partial u} = \frac{1}{4}$$

Matters are not quite so simple when, instead of the new coordinates being given as functions of the old coordinates, we are given the old coordinates as functions of the new coordinates – i.e. $\mathbf{x} = \mathbf{f}(\mathbf{u})$. In principle, one can of course solve the equation $\mathbf{x} = \mathbf{f}(\mathbf{u})$ to obtain $\mathbf{u} = \mathbf{f}^{-1}(\mathbf{x})$. But this is often quite difficult to do and even when it is not particularly difficult it is usually easier to use the formula

$$\frac{d\mathbf{u}}{d\mathbf{x}} = \left(\frac{d\mathbf{x}}{d\mathbf{u}}\right)^{-1}$$

instead.

Example 10

The formula

$$x = u^2 + v^2 \atop y = u^2 - v^2 \Big\}$$

is used to change the coordinate system and transform the partial differential equation

$$\frac{\partial z}{\partial x} + \frac{\partial z}{\partial y} = 1$$

We have

$$\frac{\partial}{\partial x} = \frac{\partial u}{\partial x}\frac{\partial}{\partial u} + \frac{\partial v}{\partial x}\frac{\partial}{\partial v}$$

$$\frac{\partial}{\partial y} = \frac{\partial u}{\partial y}\frac{\partial}{\partial u} + \frac{\partial v}{\partial y}\frac{\partial}{\partial v}$$

Our problem is to calculate $\partial u/\partial x$, $\partial v/\partial x$, $\partial u/\partial y$, $\partial v/\partial y$. We use the formula

$$\begin{pmatrix} \dfrac{\partial u}{\partial x} & \dfrac{\partial u}{\partial y} \\[2mm] \dfrac{\partial v}{\partial x} & \dfrac{\partial v}{\partial y} \end{pmatrix} = \begin{pmatrix} \dfrac{\partial x}{\partial u} & \dfrac{\partial x}{\partial v} \\[2mm] \dfrac{\partial y}{\partial u} & \dfrac{\partial y}{\partial v} \end{pmatrix}^{-1} = \begin{pmatrix} 2u & 2v \\ 2u & -2v \end{pmatrix}^{-1}$$

$$= -\frac{1}{8uv}\begin{pmatrix} -2v & -2v \\ -2u & 2u \end{pmatrix} = \begin{pmatrix} \dfrac{1}{4u} & \dfrac{1}{4u} \\[2mm] \dfrac{1}{4v} & -\dfrac{1}{4v} \end{pmatrix}$$

It follows that

$$\frac{\partial}{\partial x} = \frac{1}{4u}\frac{\partial}{\partial u} + \frac{1}{4v}\frac{\partial}{\partial v}$$

$$\frac{\partial}{\partial y} = \frac{1}{4u}\frac{\partial}{\partial u} - \frac{1}{4v}\frac{\partial}{\partial v}$$

Thus

$$\frac{\partial z}{\partial x} + \frac{\partial z}{\partial y} = \left(\frac{1}{4u}\frac{\partial z}{\partial u} + \frac{1}{4v}\frac{\partial z}{\partial v}\right) + \left(\frac{1}{4u}\frac{\partial z}{\partial u} - \frac{1}{4v}\frac{\partial z}{\partial v}\right) = \frac{1}{2u}\frac{\partial z}{\partial u}$$

The given partial differential equation therefore assumes the form

$$\frac{\partial z}{\partial u} = 2u$$

Example 11

If we repeat the previous example for the equation

$$\frac{\partial^2 z}{\partial x\, \partial y} = 0$$

we have

$$\frac{\partial^2 z}{\partial x\, \partial y} = \frac{\partial}{\partial x}\left(\frac{\partial z}{\partial y}\right) = \frac{\partial}{\partial x}\left(\frac{1}{4u}\frac{\partial z}{\partial u} - \frac{1}{4v}\frac{\partial z}{\partial v}\right)$$

$$= \left(\frac{1}{4u}\frac{\partial}{\partial u} + \frac{1}{4v}\frac{\partial}{\partial v}\right)\left(\frac{1}{4u}\frac{\partial z}{\partial u} - \frac{1}{4v}\frac{\partial z}{\partial v}\right)$$

$$= \frac{1}{4u}\frac{\partial}{\partial u}\left(\frac{1}{4u}\frac{\partial z}{\partial u} - \frac{1}{4v}\frac{\partial z}{\partial v}\right) + \frac{1}{4v}\frac{\partial}{\partial v}\left(\frac{1}{4u}\frac{\partial z}{\partial u} - \frac{1}{4v}\frac{\partial z}{\partial v}\right)$$

$$= \frac{1}{4u}\left\{-\frac{1}{4u^2}\frac{\partial z}{\partial u} + \frac{1}{4u}\frac{\partial^2 z}{\partial u^2} - \frac{1}{4v}\frac{\partial^2 z}{\partial u\, \partial v}\right\}$$

$$+ \frac{1}{4v}\left\{+\frac{1}{4u}\frac{\partial^2 z}{\partial v\, \partial u} + \frac{1}{4v^2}\frac{\partial z}{\partial v} - \frac{1}{4v}\frac{\partial^2 z}{\partial v^2}\right\}$$

$$= \frac{1}{16u^2}\frac{\partial^2 z}{\partial u^2} - \frac{1}{16v^2}\frac{\partial^2 z}{\partial v^2} - \frac{1}{16u^3}\frac{\partial z}{\partial u} + \frac{1}{16v^3}\frac{\partial z}{\partial v}.$$

The given equation therefore becomes

$$\frac{1}{u^2}\frac{\partial^2 z}{\partial u^2} - \frac{1}{u^3}\frac{\partial z}{\partial u} = \frac{1}{v^2}\frac{\partial^2 z}{\partial v^2} - \frac{1}{v^3}\frac{\partial z}{\partial v}$$

7.6 Exercises

1.*♣ Sketch the graphs of the functions defined by $\sinh x = \frac{1}{2}(e^x - e^{-x})$ and $\cosh x = \frac{1}{2}(e^x + e^{-x})$. Prove that

(i) $\dfrac{d}{dx}(\sinh x) = \cosh x$ (ii) $\dfrac{d}{dx}(\cosh x) = \sinh x$.

2.*♣ Explain why the equation $y = \sinh x$ always has a unique solution for x in terms of y. We write $x = \sinh^{-1} y$ if and only if $y = \sinh x$. Prove that $\cosh^2 x - \sin^2 x = 1$ and hence establish the formula

$$\frac{d}{dy}(\sinh^{-1} y) = \frac{1}{\sqrt{1 + y^2}}.$$

It is not possible to argue in a similar way from the equation $y = \cosh x$. Explain why not.

3.♣ Sketch the graph of the function defined by

$$y = \tanh x = \frac{\sinh x}{\cosh x}$$

Calculate the derivative of this function and explain how $\tanh^{-1} y$ is defined for $-1 < y < 1$. Prove that

$$\frac{d}{dy}(\tanh^{-1} y) = \frac{1}{1 - y^2} \quad (-1 < y < 1)$$

4.*♣ Differentiate $\ln(y + \sqrt{y^2 + 1})$ and discuss the relevance of your result to Exercise 2.

5.♣ Differentiate $\frac{1}{2}\ln[(1 + y)/(1 - y)]$ and discuss the relevance of your result to Exercise 3.

6.* Find a formula for a local inverse to the function $f: \mathbb{R} \to \mathbb{R}$ defined by $y = f(x) = \sin x$ at each of the points

(i) $x = \pi/6$ (ii) $x = 5\pi/6 = \pi - (\pi/6)$ (iii) $x = -7\pi/6 = -\pi - (\pi/6)$

Calculate the derivative of each local inverse directly from each formula. Evaluate these derivatives at the point $y = \frac{1}{2}$. Verify in each case that

$$\frac{dx}{dy} = \left(\frac{dy}{dx}\right)^{-1}.$$

Explain why no local inverse exists at the point $x = \pi/2$.

7. The function $\mathbf{f}: \mathbb{R}^2 \to \mathbb{R}^2$ is defined by

$$\begin{pmatrix} u \\ v \end{pmatrix} = \mathbf{f}(x, y) = \begin{pmatrix} xy \\ f(x, y) \end{pmatrix}$$

where $f(x, y) =$ (i) y (ii) $y + 2x$ and (iii) $x^2 + y^2$.

In each of the three cases find the critical points of the function $\mathbf{f}$.
Then sketch the contours of each component function. Superimpose the contours and verify that the contours touch at critical points.

8. At each of the points

(i) $(x, y)^\mathsf{T} = (2, 1)^\mathsf{T}$ (ii) $(x, y)^\mathsf{T} = (-2, 1)^\mathsf{T}$
(iii) $(x, y)^\mathsf{T} = (2, -1)^\mathsf{T}$ (iv) $(x, y)^\mathsf{T} = (-2, -1)^\mathsf{T}$

find a formula for a local inverse to the function $\mathbf{f}: \mathbb{R}^2 \to \mathbb{R}^2$ defined by

$$\begin{pmatrix} u \\ v \end{pmatrix} = \mathbf{f}(x, y) = \begin{pmatrix} x^2 + y^2 \\ x^2 - y^2 \end{pmatrix}$$

Calculate the derivative of the local inverse directly from each formula.
Evaluate each derivative at the point $(u, v)^\mathsf{T} = (5, 3)^\mathsf{T}$. Verify in each case that

$$\begin{pmatrix} \dfrac{\partial x}{\partial u} & \dfrac{\partial x}{\partial v} \\ \dfrac{\partial y}{\partial u} & \dfrac{\partial y}{\partial v} \end{pmatrix} = \begin{pmatrix} \dfrac{\partial u}{\partial x} & \dfrac{\partial u}{\partial y} \\ \dfrac{\partial v}{\partial x} & \dfrac{\partial v}{\partial y} \end{pmatrix}^{-1}$$

9.* A function $\mathbf{f}: \mathbb{R}^2 \to \mathbb{R}^2$ is defined by

$$\begin{pmatrix} u \\ v \end{pmatrix} = \mathbf{f}(x, y) = \begin{pmatrix} xe^y \\ xe^{-y} \end{pmatrix}$$

Find the derivative of this function. Determine the points at which $\mathbf{f}$ has a local inverse and find the derivative of this local inverse at these points. Evaluate

$$\left(\dfrac{\partial u}{\partial x} \right)_y \quad \text{and} \quad \left(\dfrac{\partial x}{\partial u} \right)_v$$

10.* A function $\mathbf{f}: \mathbb{R}^2 \to \mathbb{R}^2$ is defined by

$$\mathbf{f}(x, y) = \begin{pmatrix} u \\ v \end{pmatrix} = \begin{pmatrix} y^2 - x \\ x^2 - y \end{pmatrix}$$

(i) Find and sketch the set of critical points of $\mathbf{f}$, showing that they divide the $(x, y)^\mathsf{T}$ plane into three regions on each of which the Jacobian is of constant sign.

(ii) Write down the equations of the tangent flat to $\mathbf{f}$ at the point $(x, y)^\mathsf{T} = (-1, 3)^\mathsf{T}$.

(iii) Write down the equations of the u and v contours through the critical point $(x, y)^T = (\frac{1}{2}, \frac{1}{2})^T$. Find their gradient vectors and verify that the contours touch at this point. Draw these contours on your sketch from part (i).

11. A function $\mathbf{f}: \mathbb{R}^2 \to \mathbb{R}^2$ is defined by

$$\begin{pmatrix} u \\ v \end{pmatrix} = \mathbf{f}(x, y) = \begin{pmatrix} x^3 + y^2 \\ x^2 - y^3 \end{pmatrix}$$

Find the derivative of this function. Determine the points at which f has a local inverse and find the derivative of this local inverse at these points. Evaluate

$$\left(\frac{\partial v}{\partial y} \right)_x \quad \text{and} \quad \left(\frac{\partial y}{\partial v} \right)_u$$

12.* The equations

$$\left. \begin{array}{l} u = x + y \\ v = x - y \end{array} \right\}$$

are used to change the coordinate system in the $(x, y)^T$ plane. Sketch the new coordinate grid. Express

$$\frac{\partial^2 z}{\partial x^2}$$

in terms of the new coordinates.

13. A function $\mathbf{f} : \mathbb{R}^2 \to \mathbb{R}^2$ is defined by

$$\mathbf{f}(x, y) = \begin{pmatrix} u \\ v \end{pmatrix} = \begin{pmatrix} y^2 - x^2 \\ y + x^2 \end{pmatrix}$$

(i) Find the critical points of $\mathbf{f}$ and show that the $(x, y)^T$ plane splits into four regions where the Jacobian is of constant sign.

(ii) Justifying its existence, find a formula for a local inverse function of $\mathbf{f}$ at the point $(x, y)^T = (1, 1)^T$. Sketch the region in the $(x, y)^T$ plane for which this local inverse is defined.

(iii) Find, also, the derivative of the local inverse function.

14.* The equations

$$\left. \begin{array}{l} x = u + v \\ y = u - v \end{array} \right\}$$

are used to change the coordinate system in the $(x, y)^T$ plane. Prove that

$$\frac{\partial^2 z}{\partial x^2} - \frac{\partial^2 z}{\partial y^2} = \frac{\partial^2 z}{\partial u \, \partial v}$$

15. The equations

$$\left. \begin{array}{l} x = \frac{1}{2}(u^2 - v^2) \\ y = uv \end{array} \right\}$$

are used to change the coordinate system. Express

$$x\frac{\partial z}{\partial x} + y\frac{\partial z}{\partial y}$$

in terms of the new coordinates.

16.* Let $(r, \theta)^{\mathrm{T}}$ be the polar coordinates of a point $(x, y)^{\mathrm{T}}$. Prove that, for any function $f: \mathbb{R}^2 \to \mathbb{R}$,

$$\left. \begin{aligned} \frac{\partial f}{\partial r} &= \cos\theta \, \frac{\partial f}{\partial x} + \sin\theta \, \frac{\partial f}{\partial y} \\ \frac{1}{r}\frac{\partial f}{\partial \theta} &= -\sin\theta \, \frac{\partial f}{\partial x} + \cos\theta \, \frac{\partial f}{\partial y} \end{aligned} \right\}$$

What is the geometric significance of these quantities?

Put

$$\mathbf{e}_1 = (\cos\theta, \sin\theta)^{\mathrm{T}}, \ \mathbf{e}_2 = (-\sin\theta, \cos\theta)^{\mathrm{T}}$$

and consider $\langle \nabla f, \mathbf{e}_1 \rangle, \langle \nabla f, \mathbf{e}_2 \rangle$.

17. A function $\mathbf{f}: \mathbb{R}^3 \to \mathbb{R}^3$ is defined by the formulas

(i) $\left. \begin{aligned} x &= r\sin\theta \cos\phi \\ y &= r\sin\theta \sin\phi \\ z &= r\cos\theta \end{aligned} \right\}$

where $(x, y, z)^{\mathrm{T}} = \mathbf{f}(r, \theta, \phi)$. Find the Jacobian of this function and determine its critical points.

Proceed in a similar way with the formulas

(ii) $\left. \begin{aligned} x &= r\cos\theta \\ y &= r\sin\theta \\ z &= Z \end{aligned} \right\}$ (iii) $\left. \begin{aligned} x &= u^2 v^{-1} \\ y &= v^2 w^{-1} \\ z &= w^2 u^{-1} \end{aligned} \right\}$

18. The equations

$$\left. \begin{aligned} u &= \tfrac{1}{2}(x^2 - y^2) \\ v &= xy \end{aligned} \right\}$$

are used to change the coordinate system in the first quadrant of the $(x, y)^{\mathrm{T}}$ plane. Sketch the new coordinate grid. Express

$$\frac{\partial^2 z}{\partial x^2}$$

in terms of the new coordinates.

19.*♣ Introduce polar coordinates into the partial differential equation

$$\frac{\partial^2 z}{\partial x^2} + \frac{\partial^2 z}{\partial y^2} = 0$$

to transform it to the equation

$$\frac{\partial^2 z}{\partial r^2} + \frac{1}{r}\frac{\partial z}{\partial r} + \frac{1}{r^2}\frac{\partial^2 z}{\partial \theta^2} = 0$$

7.7   Application (optional): contract curve

We discussed the Edgeworth box and the contract curve in §3.10.3.

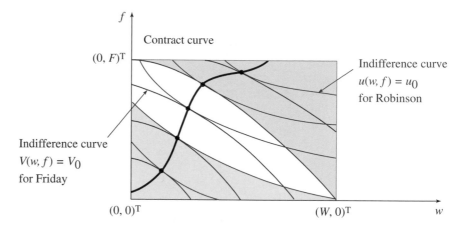

Contract curve

$(0, F)^T$

Indifference curve
$u(w, f) = u_0$
for Robinson

Indifference curve
$V(w, f) = V_0$
for Friday

Only the part outside the
shaded regions is relevant.

$(0, 0)^T$

$(W, 0)^T$

f

w

Observe that each point on the contract curve is a critical point of the function $\phi \colon \mathbb{R}^2 \to \mathbb{R}^2$ defined by

$$\begin{pmatrix} x \\ y \end{pmatrix} = \phi(w, f) = \begin{pmatrix} u(w, f) \\ V(w, f) \end{pmatrix}$$

It follows that the Jacobian is zero on the contract curve – i.e.

$$\begin{vmatrix} \dfrac{\partial x}{\partial w} & \dfrac{\partial x}{\partial f} \\[2mm] \dfrac{\partial y}{\partial w} & \dfrac{\partial y}{\partial f} \end{vmatrix} = 0$$

We can use this fact to obtain an equation for the contract curve. Consider, for example, the case

$$\left. \begin{aligned} x &= u(w, f) = wf^2 \\ y &= V(w, f) = (2 - w)^2 (1 - f) \end{aligned} \right\}$$

which we studied in §3.10.3. On the contract curve,

$$0 = \begin{vmatrix} \dfrac{\partial x}{\partial w} & \dfrac{\partial x}{\partial f} \\[2mm] \dfrac{\partial y}{\partial w} & \dfrac{\partial y}{\partial f} \end{vmatrix} = \begin{vmatrix} f^2 & 2wf \\[2mm] -2(2 - w)(1 - f) & -(2 - w)^2 \end{vmatrix}$$

$$= -(2 - w) f \{ f(2 - w) - 4w(1 - f) \}$$

Hence the contract curve has equation $3wf - 4w + 2f = 0$.

Eight	**Implicit functions**

The implicit function theorem is very easy to apply in practice provided that the material on vector derivatives given in Chapter 5 has been properly assimilated. We are therefore able to deal with it quite quickly in this chapter. Some understanding of §8.2 is necessary for the conclusion of the implicit function theorem to be meaningful but too much time should not be spent on this section; especially in respect of the theoretical discussion at the end of the section which links the proof of the implicit function theorem with the discussion of local inverses given in the previous chapter.

8.1 Implicit differentiation

Consider the equation

$$y^2 - x^3 - x^2 = 0 \tag{1}$$

Suppose that $y = g(x)$ is a solution of this equation for y in terms of x – i.e. $(g(x))^2 - x^3 - x^2 = 0$. The function $g(x)$ is said to be **implicitly defined** by equation (1).

We can then calculate $g'(x)$ using the technique of implicit differentiation, *without first finding* $g(x)$. This entails differentiating equation (1) with respect to x, treating y as a function of x, and differentiating any terms involving y using the chain rule for functions of one variable. We have

$$0 = \frac{d}{dx}(y^2 - x^3 - x^2)$$
$$= 2y\frac{dy}{dx} - 3x^2 - 2x$$

and hence

$$\frac{dy}{dx} = \frac{3x^2 + 2x}{2y}$$

i.e.

$$g'(x) = \frac{3x^2 + 2x}{2y} = \frac{3x^2 + 2x}{2g(x)}$$

The same technique can be applied in the general case using the chain rule. Consider a scalar valued function $\phi: \mathbb{R}^2 \to \mathbb{R}$ such that

$$\phi(x, y) = 0$$

has the solution $y = g(x)$ for y in terms of x. Then

$$0 = \frac{d\phi}{dx} = \frac{\partial\phi}{\partial x}\frac{dx}{dx} + \frac{\partial\phi}{\partial y}\frac{dy}{dx}$$

and so

$$\frac{dy}{dx} = -\left(\frac{\partial \phi}{\partial y}\right)^{-1} \frac{\partial \phi}{\partial x}$$

i.e.

$$g'(x) = -\frac{\phi_x(x, y)}{\phi_y(x, y)} = -\frac{\phi_x(x, g(x))}{\phi_y(x, g(x))}$$

In the case when $\phi(x, y) = y^2 - x^3 - x^2$ we have $\phi_x(x, y) = -3x^2 - 2x$ and $\phi_y(x, y) = 2y$.

8.2 Implicit functions

The account of implicit differentiation given above begs various questions. In particular, the account takes for granted that the equation $\phi(x, y) = 0$ *implicitly* defines a function $y = g(x)$ and that this function may be differentiated. But it is clear that, in general, it will *not* always be possible to solve the equation $\phi(x, y) = 0$ for y in terms of x. Moreover, where it is possible to solve the equation for y in terms of x, there may well be *several* solutions. Finally, there is no guarantee that a given solution will be differentiable.

Consider, for example, the case

$$\phi(x, y) = y^2 - x^3 - x^2$$

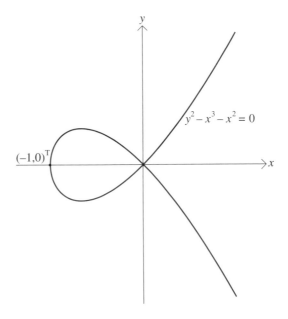

It is evident from the diagram that for values of x satisfying $x < -1$, the equation $y^2 - x^3 - x^2 = 0$ has no (real) solutions for y in terms of x. Furthermore, for all other values of x (except $x = -1$ and $x = 0$), the equation $y^2 - x^3 - x^2 = 0$ has *two* solutions for y in terms of x.

The diagrams below illustrate three different functions $g\colon [-1, \infty) \to \mathbb{R}$ which are given implicitly by $y^2 - x^3 - x^2 = 0$ – i.e. which satisfy $(g(x))^2 - x^3 - x^2 = 0$.

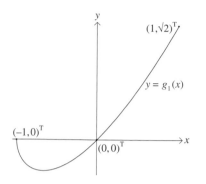
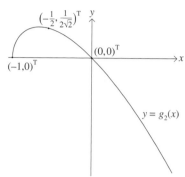
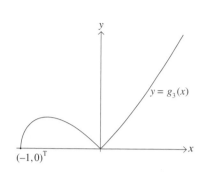

This leads us to the notion of a local solution. Suppose that $\phi\colon \mathbb{R}^2 \to \mathbb{R}$. Then we say that $g\colon I \to \mathbb{R}$ is a **local solution** of

$$\phi(x, y) = 0$$

provided that

$$\phi(x, g(x)) = 0$$

for each x in the interval I.

If ξ is an interior point of the interval I and $\eta = g(\xi)$, then we say that $g\colon I \to \mathbb{R}$ is a local solution of $\phi(x, y) = 0$ **at the point** $(\xi, \eta)^{\mathrm{T}}$.

Consider, for example, the functions g_1, g_2 and g_3 sketched above. The function g_1 is a local solution of $y^2 - x^3 - x^2 = 0$ at the point $(1, \sqrt{2})^{\mathrm{T}}$. The function g_2 is a local solution of $y^2 - x^3 - x^2 = 0$ at the point $(-\frac{1}{2}, 1/\sqrt{8})^{\mathrm{T}}$. All three functions are local solutions of $y^2 - x^3 - x^2 = 0$ at the point $(0, 0)^{\mathrm{T}}$. Note that *no* local solution of $y^2 - x^3 - x^2 = 0$ exists at the point $(-1, 0)^{\mathrm{T}}$. This is because $(-1, 0)^{\mathrm{T}}$ is not an interior point of $[-1, \infty)$ which is the largest interval on which a local inverse can be defined.

Observe that the 'bad points' on $y^2 - x^3 - x^2 = 0$ are $(-1, 0)^{\mathrm{T}}$ and $(0, 0)^{\mathrm{T}}$. At these points there are either no local solutions or too many local solutions. At all other points $(\xi, \eta)^{\mathrm{T}}$ on $y^2 - x^3 - x^2 = 0$ there is an essentially unique differentiable local solution.

A differentiable local solution is 'unique' in the sense that any two local solutions at $(\xi, \eta)^{\mathrm{T}}$ are the same for values of x sufficiently close to ξ although the intervals I on which they are defined may not be the same.

Notice that, at the 'bad points' $(-1, 0)^{\mathrm{T}}$ and $(0, 0)^{\mathrm{T}}$, we have

$$\frac{\partial \phi}{\partial y} = 2y = 0$$

and so the formula

$$\frac{\mathrm{d}y}{\mathrm{d}x} = -\left(\frac{\partial \phi}{\partial y}\right)^{-1} \left(\frac{\partial \phi}{\partial x}\right)$$

which we obtained in §8.1 ceases to make sense. It is therefore not surprising that we should find something peculiar going on at these points.

The above discussion would seem to indicate that the appropriate condition for the existence of a local solution to $\phi(x, y) = 0$ at a point $(\xi, \eta)^{\mathrm{T}}$ satisfying $\phi(\xi, \eta) = 0$ is that

$$\frac{\partial \phi}{\partial y} \neq 0$$

when $(x, y)^T = (\xi, \eta)^T$. It is quite easy to check that this is correct, as we shall now show, using the results on the existence of local inverses given in the previous chapter.

Consider the vector valued function $\mathbf{F} \colon \mathbb{R}^2 \to \mathbb{R}^2$ given by

$$\mathbf{F}(\mathbf{z}) = \mathbf{F}(x, y) = (x, \phi(x, y))^T$$

where $\mathbf{z} = (x, y)^T$. Then

$$\frac{d\mathbf{F}}{d\mathbf{z}} = \begin{pmatrix} \dfrac{\partial x}{\partial x} & \dfrac{\partial x}{\partial y} \\ \dfrac{\partial \phi}{\partial x} & \dfrac{\partial \phi}{\partial y} \end{pmatrix} = \begin{pmatrix} 1 & 0 \\ \dfrac{\partial \phi}{\partial x} & \dfrac{\partial \phi}{\partial y} \end{pmatrix}$$

Thus the Jacobian is

$$\det\left(\frac{d\mathbf{F}}{d\mathbf{z}}\right) = \frac{\partial \phi}{\partial y}$$

It follows that the critical points of $\mathbf{F}$ occur where $\phi_y = 0$. Thus, if $\phi_y(\xi, \eta) \neq 0$, then $\mathbf{F}$ admits a local inverse $\mathbf{G}$ at the point $(\xi, \eta)^T$. This means that

$$(u, v)^T = \mathbf{F}(x, y) \quad \text{if and only if} \quad (x, y)^T = \mathbf{G}(u, v) \tag{1}$$

provided $(x, y)^T$ and $(u, v)^T$ are appropriately restricted.

If we consider values of x and y which satisfy $\phi(x, y) - 0$, then

$$\mathbf{F}(x, y) = (x, \phi(x, y))^T = (x, 0)^T$$

which by (1) is equivalent to

$$(x, y)^T = \mathbf{G}(x, 0) = (G_1(x, 0), G_2(x, 0))^T$$

In particular, $y = G_2(x, 0) = g(x)$ is a unique local solution of $\phi(x, y) = 0$ at $(\xi, \eta)^T$.

8.3   Implicit function theorem

The implicit function theorem generalizes the above discussion. Suppose that $\boldsymbol{\phi} \colon \mathbb{R}^{n+m} \to \mathbb{R}^m$ has a continuous derivative and consider the equation

$$\boldsymbol{\phi}(\mathbf{x}, \mathbf{y}) = \mathbf{0}$$

in which $\mathbf{x}$ is an $n \times 1$ column vector and $\mathbf{y}$ is an $m \times 1$ column vector. Under what circumstances does there exist a unique local differentiable solution $\mathbf{y} = \mathbf{g}(\mathbf{x})$ at the point $(\boldsymbol{\xi}, \boldsymbol{\eta})^T$?

The implicit function theorem asserts that sufficient conditions are that $\boldsymbol{\phi}(\boldsymbol{\xi}, \boldsymbol{\eta}) = \mathbf{0}$ and

$$\det\left(\frac{\partial \boldsymbol{\phi}}{\partial \mathbf{y}}\right) \neq 0$$

at the point $(\boldsymbol{\xi}, \boldsymbol{\eta})^T$. The theorem may be stated as follows:

If at the point $(\xi, \eta)^{\mathsf{T}}$ for which $\boldsymbol{\phi} = \mathbf{0}$ we have $\det\left(\dfrac{\partial \boldsymbol{\phi}}{\partial \mathbf{y}}\right) \neq 0$, then there is a unique local differentiable solution $\mathbf{y} = \mathbf{g}(\mathbf{x})$ of the equation $\boldsymbol{\phi}(\mathbf{x}, \mathbf{y}) = \mathbf{0}$ at the point, with derivative given by

$$\frac{d\mathbf{y}}{d\mathbf{x}} = -\left(\frac{\partial \boldsymbol{\phi}}{\partial \mathbf{y}}\right)^{-1} \frac{\partial \boldsymbol{\phi}}{\partial \mathbf{x}}$$

Example 1

Consider the equation

$$x^2 + y^2 = 4$$

To apply the implicit function theorem, we write $\phi(x, y) = x^2 + y^2 - 4$. Then

$$\frac{\partial \phi}{\partial y} = 2y$$

It follows that a unique local differentiable solution $y = g(x)$ of $x^2 + y^2 = 4$ exists at each point $(\xi, \eta)^{\mathsf{T}}$ satisfying $\xi^2 + \eta^2 = 4$ *except* $(-2, 0)^{\mathsf{T}}$ and $(2, 0)^{\mathsf{T}}$. The derivative of this function is given by

$$\frac{dy}{dx} = -\left(\frac{\partial \phi}{\partial y}\right)^{-1}\left(\frac{\partial \phi}{\partial x}\right) = \frac{-2x}{2y}$$

Thus the derivative of the local solution $y = g_1(x)$ at $(0, 2)^{\mathsf{T}}$ is

$$g_1'(0) = \frac{-2 \times 0}{2 \times 2} = 0$$

while the derivative of the local solution $y = g_2(x)$ at $(\sqrt{2}, -\sqrt{2})^{\mathsf{T}}$ is

$$g_2'(\sqrt{2}) = \frac{-2 \times \sqrt{2}}{2 \times (-\sqrt{2})} = 1$$

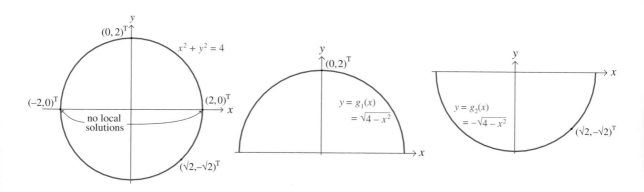

Example 2

A Cobb–Douglas production function P is defined for $x > 0$, $y > 0$ by the formula

$$P(x, y) = Ax^\alpha y^\beta$$

where x and y are the factors of production, A is a positive constant, and $0 < \alpha < 1$ and $0 < \beta < 1$.

We will use implicit differentiation to show that the contours or 'isoquants',

$$Ax^\alpha y^\beta = c$$

slope downwards. To apply the implicit function theorem we write $\phi(x, y) = Ax^\alpha y^\beta - c$. Then

$$\frac{\partial \phi}{\partial y} = A\beta x^\alpha y^{\beta-1} \neq 0 \qquad (x > 0, y > 0)$$

It follows that a unique local differentiable solution exists at each point $(x, y)^{\mathrm{T}}$ satisfying $x > 0$, $y > 0$. Differentiating the equation implicitly

$$A\alpha x^{\alpha-1} y^\beta + A\beta x^\alpha y^{\beta-1} \frac{dy}{dx} = 0$$

It follows that
$$\frac{dy}{dx} = -\frac{\alpha}{\beta} \frac{y}{x} < 0 \quad \text{since} \quad x > 0 \text{ and } y > 0$$

This implies that *an isoquant slopes downwards.*

Differentiating a second time

$$\frac{d^2 y}{dx^2} = -\frac{\alpha}{\beta} \frac{\left(x\dfrac{dy}{dx} - y\right)}{x^2} = -\frac{\alpha}{\beta} \frac{x\left(-\dfrac{\alpha}{\beta}\right)\dfrac{y}{x} - y}{x^2} = \frac{\alpha(\alpha + \beta)}{\beta^2} \frac{y}{x^2} > 0$$

We can conclude that *an isoquant is always convex.*

If we examine the derivative, we can make further observations.
When $y = x$, $\quad dy/dx = -\alpha/\beta$.

When $y > x$, $\quad dy/dx < -\alpha/\beta$. Hence $|dy/dx| > \alpha/\beta$ and the slope is steeper than when $y = x$.

When $y < x$, $\quad dy/dx > -\alpha/\beta$. Hence $|dy/dx| < \alpha/\beta$ and the slope is shallower than when $y = x$.

As $x \to \infty$, $dy/dx \to 0$.

We therefore have the characteristic shape of the Cobb–Douglas isoquant as shown in the margin.

Notice that, the larger the ratio α/β, the steeper the slope – as demonstrated below in the contour plots of Cobb–Douglas functions with varying values of this ratio.

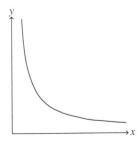

A typical isoquant

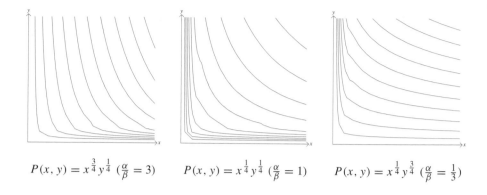

$$P(x, y) = x^{\frac{3}{4}} y^{\frac{1}{4}} \; (\tfrac{\alpha}{\beta} = 3) \qquad P(x, y) = x^{\frac{1}{4}} y^{\frac{1}{4}} \; (\tfrac{\alpha}{\beta} = 1) \qquad P(x, y) = x^{\frac{1}{4}} y^{\frac{3}{4}} \; (\tfrac{\alpha}{\beta} = \tfrac{1}{3})$$

Example 3

We repeat the exercise for a constant elasticity of substitution (CES) production function P defined for $x > 0$, $y > 0$ by

$$P(x, y) = A[\alpha x^{\beta} + (1 - \alpha) y^{\beta}]^{\frac{1}{\beta}}$$

where x and y are the factors of production and $A > 0$, $0 < \alpha < 1$ and $\beta < 1$.

An isoquant has equation

$$A[\alpha x^{\beta} + (1 - \alpha) y^{\beta}]^{\frac{1}{\beta}} = c \quad \text{where } c \text{ is constant}$$

As in the previous example, the conditions of the implicit function theorem hold. Differentiating the equation implicitly,

$$\frac{A}{\beta} [\alpha x^{\beta} + (1 - \alpha) y^{\beta}]^{\frac{(1-\beta)}{\beta}} \left[\alpha \beta x^{(\beta - 1)} + (1 - \alpha) \beta y^{(\beta - 1)} \frac{dy}{dx} \right] = 0$$

This gives

$$\frac{dy}{dx} = -\frac{\alpha}{1 - \alpha} \left(\frac{x}{y} \right)^{(\beta - 1)} < 0 \quad \text{since} \quad x > 0 \text{ and } y > 0$$

which shows that *these isoquants also slope downwards*.

Differentiating again

$$\frac{d^2 y}{dx^2} = \frac{\alpha}{1 - \alpha} (1 - \beta) \left(\frac{x}{y} \right)^{\beta - 2} \frac{\left(y - x \dfrac{dy}{dx} \right)}{y^2}$$

$$= \frac{\alpha}{1 - \alpha} (1 - \beta) \left(\frac{x}{y} \right)^{\beta - 2} \frac{\left(y + x \dfrac{\alpha}{1 - \alpha} \left(\dfrac{x}{y} \right)^{(\beta - 1)} \right)}{y^2} > 0$$

since $0 < \alpha < 1$ and $\beta < 1$.

Hence *these isoquants are also always convex.*

We now consider the effect of varying β.

The value of β cannot equal 1 for a CES function, but we can observe the effect of letting $\beta \to 1$. We are then led to the linear function

$$P(x, y) = A[\alpha x + (1 - \alpha)y]$$

As $\beta \to -\infty$

$$\frac{dy}{dx} = -\frac{\alpha}{1 - \alpha} \left(\frac{y}{x}\right)^{1-\beta} \begin{cases} \to \infty \text{ when } y > x \\ \to 0 \text{ when } y < x \end{cases}$$

In this case, an isoquant tends towards an L shaped curve, which is not differentiable when $y = x$.

The contour plots below, for a range of values of β when $\alpha = 1/4$, bear out these observations.

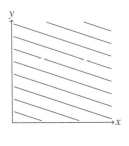

$P = \frac{1}{4}x + \frac{3}{4}y$

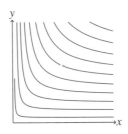

$P = [\frac{1}{4}x^{-1} + \frac{3}{4}y^{-1}]^{-1}$

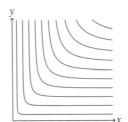

$P = [\frac{1}{4}x^{-3} + \frac{3}{4}y^{-3}]^{-\frac{1}{3}}$

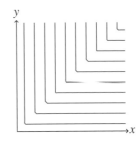

$P = [\frac{1}{4}x^{-1000} + \frac{3}{4}y^{-1000}]^{-\frac{1}{1000}}$

Economists will notice that CES production functions cover the whole spectrum between the case when the factors of production are perfect substitutes and the case when they are perfect complements.

Example 4

Consider the equation

$$xz^3 + y^2z - 2xy = 0$$

This is satisfied by $(x, y, z)^T = (1, 1, 1)^T$. Can we find a unique local differentiable solution $z = g(x, y)$ at this point?

To apply the implicit function theorem, we put $\mathbf{u} = (x, y)^T$ and write

$$\phi(\mathbf{u}, z) = \phi(x, y, z) = xz^3 + y^2z - 2xy$$

Then

$$\frac{\partial \phi}{\partial z} = 3xz^2 + y^2$$

At the point $(x, y, z)^T = (1, 1, 1)^T$, this is equal to $3 + 1 = 4$. Since this is nonzero, it follows that a unique local differentiable solution $z = g(x, y)$ exists at the point $(1, 1, 1)^T$.

The derivative of $z = g(\mathbf{u}) = g(x, y)$ is given by

$$\frac{dz}{d\mathbf{u}} = -\left(\frac{\partial \phi}{\partial z}\right)^{-1}\left(\frac{\partial \phi}{\partial \mathbf{u}}\right)$$

i.e.

$$\left(\frac{\partial z}{\partial x}, \frac{\partial z}{\partial y}\right) = -\left(\frac{\partial \phi}{\partial z}\right)^{-1}\left(\frac{\partial \phi}{\partial x}, \frac{\partial \phi}{\partial y}\right)$$

$$= -\frac{1}{3xz^2 + y^2}(z^3 - 2y, 2yz - 2x)$$

In particular,

$$g'(1, 1) = \left(\frac{1}{4}, 0\right)$$

Example 5

The equations

$$\left.\begin{array}{r} xuy + uyv + 2yvz = 0 \\ 2uyv + yvz + vzx = 0 \end{array}\right\}$$

are satisfied by $(x, y, z, u, v)^T = (1, 1, -1, 1, 1)^T$. Can we find a unique local differentiable solution

$$(u, v)^T = \mathbf{g}(x, y, z)$$

at this point?

To apply the implicit function theorem, we write $\mathbf{t} = (u, v)^T$, $\mathbf{s} = (x, y, z)^T$ and put

$$\phi(\mathbf{s}, \mathbf{t}) = \phi(x, y, z, u, v) = \left(\begin{array}{c} xuy + uyv + 2yvz \\ 2uyv + yvz + vzx \end{array}\right)$$

Then

$$\frac{\partial \phi}{\partial \mathbf{t}} = \left(\begin{array}{cc} \dfrac{\partial \phi_1}{\partial u} & \dfrac{\partial \phi_1}{\partial v} \\[2mm] \dfrac{\partial \phi_2}{\partial u} & \dfrac{\partial \phi_2}{\partial v} \end{array}\right)$$

$$= \left(\begin{array}{cc} xy + yv & uy + 2yz \\ 2yv & 2uy + yz + zx \end{array}\right)$$

and so at the point $(x, y, z, u, v)^T = (1, 1, -1, 1, 1)^T$

$$\det\left(\frac{\partial \phi}{\partial \mathbf{t}}\right) = \left|\begin{array}{cc} 2 & -1 \\ 2 & 0 \end{array}\right| = 2 \neq 0$$

273

Thus a unique local differentiable solution exists at this point. Its derivative is given by

$$\frac{d\mathbf{t}}{d\mathbf{s}} = \begin{pmatrix} \dfrac{\partial u}{\partial x} & \dfrac{\partial u}{\partial y} & \dfrac{\partial u}{\partial z} \\ \dfrac{\partial v}{\partial x} & \dfrac{\partial v}{\partial y} & \dfrac{\partial v}{\partial z} \end{pmatrix}$$

$$= -\left(\frac{\partial \phi}{\partial \mathbf{t}}\right)^{-1}\left(\frac{\partial \phi}{\partial \mathbf{s}}\right)$$

At the point $(x, y, z, u, v)^{\mathrm{T}} = (1, 1, -1, 1, 1)^{\mathrm{T}}$,

$$\left(\frac{\partial \phi}{\partial \mathbf{t}}\right)^{-1} = \begin{pmatrix} \dfrac{\partial \phi_1}{\partial u} & \dfrac{\partial \phi_1}{\partial v} \\ \dfrac{\partial \phi_2}{\partial u} & \dfrac{\partial \phi_2}{\partial v} \end{pmatrix}^{-1} = \begin{pmatrix} 2 & -1 \\ 2 & 0 \end{pmatrix}^{-1} = \frac{1}{2}\begin{pmatrix} 0 & 1 \\ -2 & 2 \end{pmatrix}$$

and

$$\frac{\partial \phi}{\partial \mathbf{s}} = \begin{pmatrix} \dfrac{\partial \phi_1}{\partial x} & \dfrac{\partial \phi_1}{\partial y} & \dfrac{\partial \phi_1}{\partial z} \\ \dfrac{\partial \phi_2}{\partial x} & \dfrac{\partial \phi_2}{\partial y} & \dfrac{\partial \phi_2}{\partial z} \end{pmatrix} = \begin{pmatrix} uy & xu + uv + 2vz & 2yv \\ vz & 2uv + vz & yv + vx \end{pmatrix}$$

$$= \begin{pmatrix} 1 & 0 & 2 \\ -1 & 1 & 2 \end{pmatrix}$$

Thus

$$\mathbf{g}'(1, 1, -1) = -\frac{1}{2}\begin{pmatrix} 0 & 1 \\ -2 & 2 \end{pmatrix}\begin{pmatrix} 1 & 0 & 2 \\ -1 & 1 & 2 \end{pmatrix}$$

$$= \begin{pmatrix} \frac{1}{2} & -\frac{1}{2} & -1 \\ 2 & -1 & 0 \end{pmatrix}$$

In this particular case it is quite simple to find a formula for the function $\mathbf{g}$ and hence to check this result. If $y \neq 0$ and $v \neq 0$, the original equations reduce to

$$\left.\begin{array}{r} xu + uv + 2vz = 0 \\ 2uy + yz + zx = 0 \end{array}\right\}$$

From the latter equation

$$u = -\frac{z(x + y)}{2y}$$

Substitute this in the other equation. Then

$$\frac{-z(x + y)}{2y}x - \frac{z(x + y)}{2y}v + 2vz = 0.$$

If $z \neq 0$ and $x \neq 3y$,

$$v\left(2 - \frac{(x + y)}{2y}\right) = \frac{(x + y)x}{2y}$$

$$v = \frac{(x + y)x}{3y - x}$$

It follows that

$$
\begin{pmatrix}
\dfrac{\partial u}{\partial x} & \dfrac{\partial u}{\partial y} & \dfrac{\partial u}{\partial z} \\[2mm]
\dfrac{\partial v}{\partial x} & \dfrac{\partial v}{\partial y} & \dfrac{\partial v}{\partial z}
\end{pmatrix}
=
\begin{pmatrix}
\dfrac{-z}{2y} & \dfrac{zx}{2y^2} & \dfrac{-(x+y)}{2y} \\[4mm]
\dfrac{(2x+y)(3y-x)+(x+y)x}{(3y-x)^2} & \dfrac{-4x^2}{(3y-x)^2} & 0
\end{pmatrix}
$$

$$
=
\begin{pmatrix}
\tfrac{1}{2} & -\tfrac{1}{2} & -1 \\[2mm]
2 & -1 & 0
\end{pmatrix}
$$

at the point $(x, y, z)^{\mathrm{T}} = (1, 1, -1)^{\mathrm{T}}$.

8.4 Exercises

1.* Sketch the curve defined by the equation

$$y^2 = x^2(1 - x^2)$$

Find the points $(\xi, \eta)^{\mathrm{T}}$ on this curve at which $y^2 = x^2(1 - x^2)$ has a unique local differentiable solution $y = g(x)$ and evaluate $g'(\xi)$.

Find formulas for local solutions at $(1/\sqrt{2}, 1/2)^{\mathrm{T}}$ and $(-1/\sqrt{2}, -1/2)^{\mathrm{T}}$.

2. Find the points $(\xi, \eta)^{\mathrm{T}}$ on the curve

$$y^2(1 - y^2) = x^2(1 - x^2)$$

at which the equation has a unique local differentiable solution $y = g(x)$ and evaluate $g'(\xi)$.

Find formulas for local solutions at $\left(\tfrac{1}{2}, \tfrac{1}{2}\right)^{\mathrm{T}}, \left(\tfrac{1}{2}, -\tfrac{1}{2}\right)^{\mathrm{T}}, \left(\tfrac{1}{2}, \sqrt{3}/2\right)^{\mathrm{T}}$ and $\left(\tfrac{1}{2}, -\sqrt{3}/2\right)^{\mathrm{T}}$.

3.* Show that the point $(x, y)^{\mathrm{T}} = (-1, 1)^{\mathrm{T}}$ lies on the curve

$$\sin(\pi x^2 y) = 1 + xy^2$$

Prove that a unique local differentiable solution $y = g(x)$ exists at $(-1, 1)^{\mathrm{T}}$ and evaluate $g'(-1)$.

4. Show that the point $(x, y)^{\mathrm{T}} = (\sqrt{e}, 1/\sqrt{e})^{\mathrm{T}}$ lies on the curve

$$xy - \ln x + \ln y = 0 \qquad (x > 0, y > 0)$$

Prove that a unique local differentiable solution $y = g(x)$ exists at $(\sqrt{e}, 1/\sqrt{e})^{\mathrm{T}}$. Evaluate $g'(x)$ and $g''(x)$ and hence show that g has a local maximum at $x = \sqrt{e}$.

5.* Suppose that the output z of a manufacturing process is related to two input factors, capital x and labour y, by the relation

$$z^5 + 2z^2x^3 + zy^4 + xyz = 1 \qquad (x > 0, y > 0, z > 0)$$

Find the partial derivatives $\partial z/\partial x$ and $\partial z/\partial y$.

6. Investigate the shape of an indifference curve of the utility function $U: \mathbb{R}^2_+ \to \mathbb{R}$ defined by

$$U(x, y) = \ln[(1 + x)^\alpha (1 + y)^\beta] \quad \text{where } 0 < \alpha < 1, 0 < \beta < 1$$

7. Show that the indifference curves of the following 'quasilinear' utility functions slope downwards and are convex:

(i)* $U(x, y) = x + y^{\frac{1}{2}}$ $(x > 0, y > 0)$

(ii) $U(x, y) = x + \ln y$ $(x > 0, y > 0)$

Show that the same is true for a general 'quasilinear' utility function defined by

$$U(x, y) = x + w(y) \qquad (x > 0, y > 0)$$

where the function $w(y)$ is increasing and concave.

8.* Find a point $(\xi, \eta, \zeta)^T$ at which the equation

$$\sin(yz) + \sin(zx) + \sin(xy) = 0$$

has a unique local differentiable solution $z = g(x, y)$. Find $g'(\xi, \eta)$.

9. Show that the point $(x, y, z)^T = (1, 1, -1)^T$ lies on the curve

$$xy^2 + yz^2 + zx^2 = 1$$

Prove that a unique local differentiable solution $z = g(x, y)$ exists at the point $(1, 1, -1)^T$ and find $g'(1, 1)$.

Also obtain a formula for $g(x, y)$.

10.* Show that the equations

$$\left. \begin{array}{l} xy^2 + y^2z^3 + z^4x^5 = 1 \\ zx^2 + x^2y^3 + y^4z^5 = -1 \end{array} \right\}$$

admit a unique local differentiable solution

$$(x, y)^T = \mathbf{g}(z)$$

at $(x, y, z)^T = (1, 1, -1)^T$. Evaluate $\mathbf{g}'(-1)$.

11. Show that the equations

$$x^4 + (x + z)y^3 - 3 = 0$$
$$x^4 + (2x + 3z)y^3 - 6 = 0$$

admit a unique local differentiable solution

$$(y, z)^{\mathrm{T}} = \mathbf{g}(x)$$

at any point $(\xi, \eta, \zeta)^{\mathrm{T}}$ which satisfies the equations. Find a formula for $\mathbf{g}'(x)$.

12.* Show that there exists a unique local differentiable solution

$$(u, v, w)^{\mathrm{T}} = \mathbf{g}(x, y, z)$$

to the equations

$$\left. \begin{array}{l} x^2 + u + e^v = 0 \\ y^2 + v + e^w = 0 \\ z^2 + w + e^u = 0 \end{array} \right\}$$

at any point which satisfies the equations. Find a formula for $\mathbf{g}'(x, y, z)$.

13. Show that there exists a unique local differentiable solution

$$(u, v)^{\mathrm{T}} = \mathbf{g}(x, y)$$

to the equations

$$\left. \begin{array}{l} x = e^u \cos v \\ v = e^y \sin x \end{array} \right\}$$

at any point which satisfies the equations. Find a formula for $\mathbf{g}'(x, y)$.

8.5 Application (optional): shadow prices

We have seen that in constrained optimisation problems of the type, find

$$\max f(\mathbf{x})$$

subject to the constraint

$$\mathbf{g}(\mathbf{x}) = \mathbf{0}$$

it is often a good idea to form the Lagrangian

$$L = f(\mathbf{x}) + \boldsymbol{\lambda}^{\mathrm{T}} \mathbf{g}(\mathbf{x})$$

See §6.9.

The stationary points of L are then found by solving the simultaneous equations

$$\begin{cases} \dfrac{\partial L}{\partial \mathbf{x}} = 0 \\[2mm] \dfrac{\partial L}{\partial \boldsymbol{\lambda}} = \mathbf{0} - \text{i.e. } \mathbf{g}(\mathbf{x}) = \mathbf{0} \end{cases}$$

If $\mathbf{x} = \tilde{\mathbf{x}}$ and $\lambda = \tilde{\lambda}$ satisfy the equations, then $\tilde{\mathbf{x}}$ is a constrained stationary point for the original problem. An added advantage to the use of the Lagrangian is that the vector $\lambda = \tilde{\lambda}$ is also an interesting quantity. The reasons for this are discussed below.

Suppose that x represents a bundle of outputs from a production process and that $f(x)$ represents the profit derived from the sale of x. We shall suppose that the input to the production process is a given commodity bundle b of raw materials. The various outputs x which can be produced using the given input b are specified by the equation $G(x) = b$. The production manager therefore seeks to maximize $f(x)$ subject to the constraint

$$g(x) = b - G(x) = 0$$

This scenario, of course, is very similar to that described in §1.12.4 and continued in §1.12.5 and §6.12.5. The optimum output $\tilde{x}$ and the corresponding $\tilde{\lambda}$ satisfy the Lagrangian equations

$$\left. \begin{aligned} \phi_1(x, \lambda, b) &= \frac{df}{dx} - \lambda^T \frac{dG}{dx} = 0 \\ \phi_2(x, \lambda, b) &= b - G(x) = 0 \end{aligned} \right\} \tag{1}$$

If we wish to know how $\tilde{x}$ and $\tilde{\lambda}$ depend on the given input b, it is necessary to solve these equations to obtain $\tilde{x}$ and $\tilde{\lambda}$ as functions of b. The implicit function theorem requires that

$$\det \begin{pmatrix} \dfrac{\partial \phi_1}{\partial x} & \dfrac{\partial \phi_1}{\partial \lambda} \\[2ex] \dfrac{\partial \phi_2}{\partial x} & \dfrac{\partial \phi_2}{\partial \lambda} \end{pmatrix} = \det \begin{pmatrix} \dfrac{\partial^2 L}{\partial x^2} & \left(\dfrac{\partial g}{\partial x}\right)^T \\[2ex] \dfrac{\partial g}{\partial x} & 0 \end{pmatrix} \neq 0$$

If the second order conditions for $\tilde{x}$ to be a local maximum for $f(x)$ subject to the constraint $g(x) = 0$ are satisfied,[†] then this determinant is indeed nonzero. A unique local differentiable solution

† See §6.10.

$$\left. \begin{aligned} \tilde{x} &= h(b) \\ \tilde{\lambda} &= k(b) \end{aligned} \right\}$$

of the equations (1) then exists and

$$\begin{pmatrix} \dfrac{d\tilde{x}}{db} \\[2ex] \dfrac{d\tilde{\lambda}}{db} \end{pmatrix} = - \begin{pmatrix} \dfrac{\partial \phi_1}{\partial x} & \dfrac{\partial \phi_1}{\partial \lambda} \\[2ex] \dfrac{\partial \phi_2}{\partial x} & \dfrac{\partial \phi_2}{\partial \lambda} \end{pmatrix}^{-1} \begin{pmatrix} \dfrac{\partial \phi_1}{\partial b} \\[2ex] \dfrac{\partial \phi_2}{\partial b} \end{pmatrix}$$

which we rewrite as

$$\begin{pmatrix} \dfrac{\partial \phi_1}{\partial x} & \dfrac{\partial \phi_1}{\partial \lambda} \\[2ex] \dfrac{\partial \phi_2}{\partial x} & \dfrac{\partial \phi_2}{\partial \lambda} \end{pmatrix} \begin{pmatrix} \dfrac{d\tilde{x}}{db} \\[2ex] \dfrac{d\tilde{\lambda}}{db} \end{pmatrix} = - \begin{pmatrix} \dfrac{\partial \phi_1}{\partial b} \\[2ex] \dfrac{\partial \phi_2}{\partial b} \end{pmatrix}$$

This yields

$$\frac{\partial \phi_2}{\partial x} \frac{d\tilde{x}}{db} + \frac{\partial \phi_2}{\partial \lambda} \frac{d\tilde{\lambda}}{db} = - \frac{\partial \phi_2}{\partial b}$$

$$\frac{dG}{dx} \frac{d\tilde{x}}{db} = I$$

This last result is more easily obtained by differentiating the equation $G(\tilde{x}(b)) = b$ with respect to b using the chain rule.

If we are interested in how a change in the input vector b will affect our profit given that we produce the optimal output $\tilde{x}$, then we need to calculate

$$\frac{d}{db}\{f(\tilde{x}(b))\} = \frac{df}{dx}\frac{d\tilde{x}}{db}$$

$$= \tilde{\lambda}^T\frac{dG}{dx}\frac{d\tilde{x}}{db} \quad \text{(using equations (1))}$$

$$= \tilde{\lambda}^T I = \tilde{\lambda}^T$$

Using the first two terms of the Taylor series expression, we have

$$f(\tilde{x}(b + \delta b)) = f(\tilde{x}(b)) + \tilde{\lambda}^T\delta b + \cdots$$

It follows that if one could purchase the commodity bundle δb of inputs, given the price vector $\tilde{\lambda}$, then the cost of so doing would be cancelled out by the extra profit which would be obtained. We call $\tilde{\lambda}$ the vector of **shadow prices** for the input commodities. One would be just as ready to start selling off one's stock b at the shadow prices $\tilde{\lambda}$ as one would be to begin a programme of optimal production.

Nine

Differentials

This is a comparatively late stage to introduce the subject of differentials but there are good reasons for the delay. Readers will probably already have met the notation dx and dy in applied subjects and have been told that these represent 'very small changes in x and y'. There are even those who tell their students that dx and dy stand for 'infinitesimally small changes in x and y'. Such statements do not help very much in explaining why manipulations with differentials give correct answers. Indeed, in some contexts, such statements can be a positive hindrance. In this chapter we have tried to provide a more accurate account of the nature of differentials without attempting anything in the way of a systematic theoretical discussion. In studying this account, readers may find it necessary to put aside some of the preconceptions they perhaps have about differentials. Those who find this hard to do may take comfort in the fact that everything which is done in this book using differentials may also be done by means of techniques described in other chapters. For instance, three alternative methods of solving a problem are given in Example 1 before the method of differentials is used. However, this is not a sound reason for neglecting differentials. Their use in both applied and theoretical work is too widespread for this to be sensible.

9.1 Matrix algebra and linear systems$^\diamond$

Matrix algebra may be regarded as a neat way of summarising the properties of linear functions. These in turn characterize the properties of systems of linear equations.

For example, to say that an $n \times n$ matrix A is invertible with inverse A^{-1} means that the linear function $L: \mathbb{R}^n \to \mathbb{R}^n$ defined by

$$\mathbf{y} = L(\mathbf{x}) = A\mathbf{x}$$

has an inverse function $M: \mathbb{R}^n \to \mathbb{R}^n$ and that this is given by

$$\mathbf{x} = M(\mathbf{y}) = A^{-1}\mathbf{y}$$

This in turn is equivalent to the assertion that the system of linear equations

$$\left.\begin{array}{l} y_1 = a_{11}x_1 + a_{12}x_2 + \cdots + a_{1n}x_n \\ y_2 = a_{21}x_1 + a_{22}x_2 + \cdots + a_{2n}x_n \\ \quad \cdots \\ y_n = a_{n1}x_1 + a_{n2}x_2 + \cdots + a_{nn}x_n \end{array}\right\}$$

has a unique solution **x** for every **y** and this solution is given by

$$
\left.
\begin{aligned}
x_1 &= b_{11}y_1 + b_{12}y_2 + \cdots + b_{1n}y_n \\
x_2 &= b_{21}y_1 + b_{22}y_2 + \cdots + b_{2n}y_n \\
&\cdots \\
x_n &= b_{n1}y_1 + b_{n2}y_2 + \cdots + b_{nn}y_n
\end{aligned}
\right\}
$$

where $B = A^{-1}$.

Although it is often easiest to translate problems involving systems of linear equations into matrix language, this is by no means invariably true. Consider, for example, the system

$$
\left.
\begin{aligned}
y_1 &= x_1 + x_2 \\
y_2 &= x_1
\end{aligned}
\right\}
$$

This can be shown to have the unique solution

$$
\left.
\begin{aligned}
x_1 &= y_2 \\
x_2 &= y_1 - y_2
\end{aligned}
\right\}
$$

simply by substituting $x_1 = y_2$ in the equation $y_1 = x_1 + x_2$. There is therefore no need to consider whether the matrix

$$
A = \begin{pmatrix} 1 & 1 \\ 1 & 0 \end{pmatrix}
$$

is invertible or to compute its inverse. The fact that the equations have a unique solution guarantees that A is invertible and the form of the solution shows that

$$
A^{-1} = \begin{pmatrix} 0 & 1 \\ 1 & -1 \end{pmatrix}
$$

Similar situations occur when dealing with the derivative of a function $\mathbf{f} \colon \mathbb{R}^n \to \mathbb{R}^m$. It is often easier not to work directly with the $m \times n$ matrix $\mathbf{f}'(\mathbf{x})$ but to use instead the system of linear equations which this matrix defines. The subject of this chapter is the special notation which is used for this purpose.

9.2 Differentials

Suppose that $\mathbf{f} \colon \mathbb{R}^n \to \mathbb{R}^m$ is differentiable at $\mathbf{x} \in \mathbb{R}^n$. Its *derivative* $\mathbf{f}'(\mathbf{x})$ is then an $m \times n$ matrix and hence defines a linear function $\mathbf{L} \colon \mathbb{R}^n \to \mathbb{R}^m$. This linear function is called the **differential** of **f** at **x**. We use the notation

$$
d\mathbf{y} = \mathbf{L}(d\mathbf{x}) = \mathbf{f}'(\mathbf{x})\, d\mathbf{x} \tag{1}
$$

In this expression the variable $d\mathbf{x}$ in $\mathbb{R}^n$ is called the **differential of x** and the variable $d\mathbf{y}$ in $\mathbb{R}^m$ is called the **differential of y**. Thus $d\mathbf{x}$ and $d\mathbf{y}$ are *variables*. We choose to denote these variables by $d\mathbf{x}$ and $d\mathbf{y}$ because (1) can then be rewritten in the form

$$
d\mathbf{y} = \left(\frac{d\mathbf{y}}{d\mathbf{x}} \right) d\mathbf{x} \tag{2}
$$

It is, of course, not true that (2) holds because the $d\mathbf{x}$ terms 'cancel out'. On the contrary, in so far as the $d\mathbf{x}$ terms can be said to 'cancel out', it is because the *notation has been rigged to give this appearance. The term* $d\mathbf{y}/d\mathbf{x}$ *represents a constant matrix, whereas the differential* $d\mathbf{x}$ *is a variable.*

In the general case of a vector function of a vector variable $\mathbf{f} \colon \mathbb{R}^n \to \mathbb{R}^m$, (2) may be written as a system of linear equations,

$$\left.\begin{aligned}
dy_1 &= \frac{\partial y_1}{\partial x_1}\, dx_1 + \frac{\partial y_1}{\partial x_2}\, dx_2 + \cdots + \frac{\partial y_1}{\partial x_n}\, dx_n \\[2mm]
dy_2 &= \frac{\partial y_2}{\partial x_1}\, dx_1 + \frac{\partial y_2}{\partial x_2}\, dx_2 + \cdots + \frac{\partial y_2}{\partial x_n}\, dx_n \\[1mm]
&\cdots \\[1mm]
dy_m &= \frac{\partial y_m}{\partial x_1}\, dx_1 + \frac{\partial y_m}{\partial x_2}\, dx_2 + \cdots + \frac{\partial y_m}{\partial x_n}\, dx_n
\end{aligned}\right\} \qquad (3)$$

The diagram below illustrates a scalar valued function of one variable $f \colon \mathbb{R} \to \mathbb{R}$ which is differentiable at x.

If dx and dy axes are introduced as illustrated, then the equation of the tangent line takes the form

$$dy = f'(x)\, dx$$

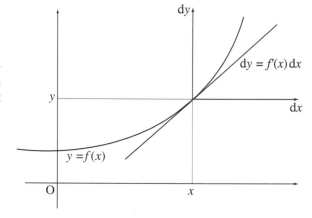

A small increase in x is often denoted by δx. It is useful to consider what happens when $dx = \delta x$, because the corresponding value of dy then provides an approximation to the change in y

$$\delta y = f(x + \delta x) - f(x)$$

caused by changing x to $x + \delta x$.

The diagram shows why dy should be expected to be a good approximation to δy when f is differentiable at x and δx is small. We call dy the linear approximation to δy.

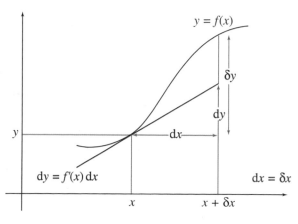

Recall that $f'(x)$ is the marginal rate of increase of f at x. Economists often talk about marginal quantities in the differentiable case as though they were dealing with discrete quantities. They then refer to $f'(x)$ as the increase or decrease in f caused by an increase or decrease of one unit in x.

This idea can be rationalised by replacing $y = f(x)$ with its linear approximation

$$dy = f'(x)\,dx$$

because $dy = f'(x)$ when $dx = 1$.

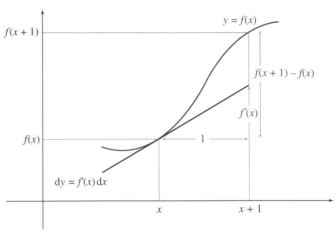

Example 1

Express

$$\left(\frac{\partial z}{\partial x}\right)_y$$

in terms of partial derivatives of z with respect to u and v when

$$\left. \begin{array}{c} x = u^2 + v^2 \\ y = u^2 - v^2 \end{array} \right\} \tag{4}$$

This problem was considered in Example 7.10. We have

$$\frac{\partial}{\partial x} = \frac{\partial u}{\partial x}\frac{\partial}{\partial u} + \frac{\partial v}{\partial x}\frac{\partial}{\partial v}$$

and

$$\begin{pmatrix} \dfrac{\partial u}{\partial x} & \dfrac{\partial u}{\partial y} \\[2mm] \dfrac{\partial v}{\partial x} & \dfrac{\partial v}{\partial y} \end{pmatrix} = \begin{pmatrix} \dfrac{\partial x}{\partial u} & \dfrac{\partial x}{\partial v} \\[2mm] \dfrac{\partial y}{\partial u} & \dfrac{\partial y}{\partial v} \end{pmatrix}^{-1}$$

$$= \begin{pmatrix} 2u & 2v \\ 2u & -2v \end{pmatrix}^{-1} = \begin{pmatrix} \dfrac{1}{4u} & \dfrac{1}{4u} \\[2mm] \dfrac{1}{4v} & -\dfrac{1}{4v} \end{pmatrix} \tag{5}$$

Thus

$$\frac{\partial z}{\partial x} = \frac{1}{4u}\frac{\partial z}{\partial u} + \frac{1}{4v}\frac{\partial z}{\partial v}$$

or, more precisely,

$$\left(\frac{\partial z}{\partial x}\right)_y = \frac{1}{4u}\left(\frac{\partial z}{\partial u}\right)_v + \frac{1}{4v}\left(\frac{\partial z}{\partial v}\right)_u$$

But this is not the only way in which one can proceed. Another possible approach is to begin by solving equations (4).[†] If we are interested only in positive u and v, the required solutions are

$$\left. \begin{array}{l} u = \sqrt{\dfrac{x+y}{2}} \\[2ex] v = \sqrt{\dfrac{x-y}{2}} \end{array} \right\}$$

Hence

$$\frac{\partial u}{\partial x} = \frac{1}{2}\left(\frac{x+y}{2}\right)^{-1/2}\frac{1}{2} = \frac{1}{4u}$$

$$\frac{\partial v}{\partial x} = \frac{1}{2}\left(\frac{x-y}{2}\right)^{-1/2}\frac{1}{2} = \frac{1}{4v}$$

and thus

$$\frac{\partial z}{\partial x} = \frac{\partial u}{\partial x}\frac{\partial z}{\partial u} + \frac{\partial v}{\partial x}\frac{\partial z}{\partial v}$$

$$= \frac{1}{4u}\frac{\partial z}{\partial u} + \frac{1}{4v}\frac{\partial z}{\partial v}$$

It is also possible to proceed using the method of implicit differentiation. Differentiating equations (4) as they stand partially with respect to x and y, we obtain

$$1 = 2u\frac{\partial u}{\partial x} + 2v\frac{\partial v}{\partial x}$$

$$0 = 2u\frac{\partial u}{\partial x} - 2v\frac{\partial v}{\partial x}$$

Thus, first adding these equations and then subtracting them,

$$1 = 4u\frac{\partial u}{\partial x} \qquad 1 = 4v\frac{\partial v}{\partial x}$$

and so

$$\frac{\partial z}{\partial x} = \frac{\partial u}{\partial x}\frac{\partial z}{\partial u} + \frac{\partial v}{\partial x}\frac{\partial z}{\partial v}$$

$$= \frac{1}{4u}\frac{\partial z}{\partial u} + \frac{1}{4v}\frac{\partial z}{\partial v}$$

Finally, we may use differentials. In this problem, equations (3) take the form

$$dx = \frac{\partial x}{\partial u}\,du + \frac{\partial x}{\partial v}\,dv$$

$$dy = \frac{\partial y}{\partial u}\,du + \frac{\partial y}{\partial v}\,dv$$

i.e.

$$\left.\begin{aligned}
\mathrm{d}x &= 2u\,\mathrm{d}u + 2v\,\mathrm{d}v \\
\mathrm{d}y &= 2u\,\mathrm{d}u - 2v\,\mathrm{d}v
\end{aligned}\right\} \tag{6}$$

These equations have the solution

$$\left.\begin{aligned}
\mathrm{d}u &= \frac{1}{4u}\,\mathrm{d}x + \frac{1}{4u}\,\mathrm{d}y \\
\mathrm{d}v &= \frac{1}{4v}\,\mathrm{d}x - \frac{1}{4v}\,\mathrm{d}y
\end{aligned}\right\}$$

But we also have

$$\left.\begin{aligned}
\mathrm{d}u &= \frac{\partial u}{\partial x}\,\mathrm{d}x + \frac{\partial u}{\partial y}\,\mathrm{d}y \\
\mathrm{d}v &= \frac{\partial v}{\partial x}\,\mathrm{d}x + \frac{\partial v}{\partial y}\,\mathrm{d}y
\end{aligned}\right\}$$

Since the final two systems of equations must hold for all $\mathrm{d}x$ and $\mathrm{d}y$, it follows that

$$\frac{\partial u}{\partial x} = \frac{1}{4u} \quad \text{and} \quad \frac{\partial v}{\partial x} = \frac{1}{4v}$$

and thus

$$\frac{\partial z}{\partial x} = \frac{1}{4u}\frac{\partial z}{\partial u} + \frac{1}{4v}\frac{\partial z}{\partial v}$$

Which of these various methods is the most appropriate to use depends on the problem in hand. In particular, one might as well use the first method rather than the fourth (or the third) if one proposes to solve the system (6) of linear equations by inverting the matrix as in (5).

Example 2

A physics application

The energy u, pressure p, volume v and temperature T of a gas are so related that each may be expressed as a function of any two of the others. Prove that

$$\left.\begin{aligned}
\left(\frac{\partial u}{\partial v}\right)_T \left(\frac{\partial v}{\partial p}\right)_T &= \left(\frac{\partial u}{\partial p}\right)_T \\
\left(\frac{\partial u}{\partial v}\right)_T \left(\frac{\partial v}{\partial T}\right)_p &= \left(\frac{\partial u}{\partial T}\right)_p - \left(\frac{\partial u}{\partial T}\right)_v
\end{aligned}\right\}$$

We are given that functions $f \colon \mathbb{R}^2 \to \mathbb{R}$, $g \colon \mathbb{R}^2 \to \mathbb{R}$ and $h \colon \mathbb{R}^2 \to \mathbb{R}$ exist such that

$$u = f(v, T) = g(p, T)$$
$$v = h(p, T)$$

Introducing differentials, we obtain

$$\mathrm{d}u = \left(\frac{\partial u}{\partial v}\right)_T \mathrm{d}v + \left(\frac{\partial u}{\partial T}\right)_v \mathrm{d}T$$

$$\mathrm{d}u = \left(\frac{\partial u}{\partial p}\right)_T \mathrm{d}p + \left(\frac{\partial u}{\partial T}\right)_p \mathrm{d}T$$

$$\mathrm{d}v = \left(\frac{\partial v}{\partial p}\right)_T \mathrm{d}p + \left(\frac{\partial v}{\partial T}\right)_p \mathrm{d}T \tag{7}$$

Eliminate $\mathrm{d}u$ from the first two equations. Then

$$\left(\frac{\partial u}{\partial v}\right)_T \mathrm{d}v = \left(\frac{\partial u}{\partial p}\right)_T \mathrm{d}p + \left\{\left(\frac{\partial u}{\partial T}\right)_p - \left(\frac{\partial u}{\partial T}\right)_v\right\} \mathrm{d}T \tag{8}$$

Equations (7) and (8) hold for all $\mathrm{d}p$ and $\mathrm{d}T$ and thus using (7) to substitute for $\mathrm{d}v$ in (8) and equating coefficients of $\mathrm{d}p$ and $\mathrm{d}T$ we obtain the required equations.

9.3 Stationary points

If $f: \mathbb{R}^n \to \mathbb{R}$ is differentiable at $\mathbf{x} = (x_1, x_2, \ldots, x_n)^{\mathrm{T}}$, then its tangent hyperplane has equation

$$\mathrm{d}y = f'(\mathbf{x})\,\mathrm{d}\mathbf{x}$$
$$= \frac{\partial f}{\partial x_1}\,\mathrm{d}x_1 + \frac{\partial f}{\partial x_2}\,\mathrm{d}x_2 + \cdots + \frac{\partial f}{\partial x_n}\,\mathrm{d}x_n$$

provided that the $\mathrm{d}x_1, \mathrm{d}x_2, \ldots, \mathrm{d}x_n$ and $\mathrm{d}y$ axes are introduced as indicated for the case $n = 2$ in the diagrams below.

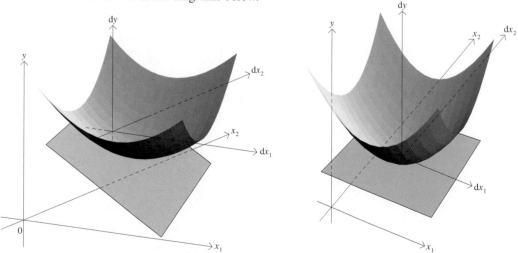

At a stationary point, the tangent hyperplane is horizontal, as in the diagram on the right – i.e. it reduces to the form

$$dy = 0$$

Thus, at a stationary point

$$\frac{\partial f}{\partial x_1} dx_1 + \frac{\partial f}{\partial x_2} dx_2 + \cdots + \frac{\partial f}{\partial x_n} dx_n = 0$$

This must hold for all $dx_1, dx_2, \ldots, dx_n$ and so we obtain the familiar condition

$$\frac{\partial f}{\partial x_1} = \frac{\partial f}{\partial x_2} = \cdots = \frac{\partial f}{\partial x_n} = 0$$

i.e.

$$f'(\mathbf{x}) = \mathbf{0}$$

See §6.9.

Consider next the case of constrained optimisation. Let $f: \mathbb{R}^n \to \mathbb{R}$ and let $\mathbf{g}: \mathbb{R}^n \to \mathbb{R}^m$ and suppose that we are interested in the constrained stationary points of f when the constraint is $\mathbf{g}(\mathbf{x}) = \mathbf{0}$. As before we require that

$$0 = dy = \frac{\partial f}{\partial x_1} dx_1 + \frac{\partial f}{\partial x_2} dx_2 + \cdots + \frac{\partial f}{\partial x_n} dx_n \tag{1}$$

but no longer is it true that this must hold for *all* $dx_1, dx_2, \ldots, dx_n$. In the presence of the constraint, we require only that $dy = 0$ for values of $dx_1, dx_2, \ldots, dx_n$ satisfying

$$\left.\begin{aligned}
0 &= \frac{\partial g_1}{\partial x_1} dx_1 + \frac{\partial g_1}{\partial x_2} dx_2 + \cdots + \frac{\partial g_1}{\partial x_n} dx_n \\[1mm]
0 &= \frac{\partial g_2}{\partial x_1} dx_1 + \frac{\partial g_2}{\partial x_2} dx_2 + \cdots + \frac{\partial g_2}{\partial x_n} dx_n \\
&\quad \cdots \\
0 &= \frac{\partial g_m}{\partial x_1} dx_1 + \frac{\partial g_m}{\partial x_2} dx_2 + \cdots + \frac{\partial g_m}{\partial x_n} dx_n
\end{aligned}\right\} \tag{2}$$

Multiplying these equations by $\lambda_1, \lambda_2, \ldots, \lambda_m$ and adding, we obtain

$$0 = \left(\frac{\partial f}{\partial x_1} + \lambda_1 \frac{\partial g_1}{\partial x_1} + \cdots + \lambda_m \frac{\partial g_m}{\partial x_n} \right) dx_1$$

$$+ \cdots + \left(\frac{\partial f}{\partial x_n} + \lambda_1 \frac{\partial g_1}{\partial x_n} + \cdots + \lambda_m \frac{\partial g_m}{\partial x_n} \right) dx_n$$

Choose $\lambda_1, \lambda_2, \ldots, \lambda_m$ to make the last m coefficients in this expression zero. The first $n - m$ coefficients must then also be zero because we can choose $dx_1, dx_2, \ldots, dx_{n-m}$ freely without violating equations (2). We therefore obtain the Lagrangian conditions

$$\left.\begin{aligned}
\frac{\partial f}{\partial x_1} + \lambda_1 \frac{\partial g_1}{\partial x_1} + \cdots + \lambda_m \frac{\partial g_m}{\partial x_1} &= 0 \\[1mm]
\frac{\partial f}{\partial x_2} + \lambda_1 \frac{\partial g_1}{\partial x_2} + \cdots + \lambda_m \frac{\partial g_m}{\partial x_2} &= 0 \\
&\quad \cdots \\
\frac{\partial f}{\partial x_n} + \lambda_1 \frac{\partial g_1}{\partial x_n} + \cdots + \lambda_m \frac{\partial g_m}{\partial x_n} &= 0
\end{aligned}\right\}$$

i.e.

$$f'(\mathbf{x}) + \boldsymbol{\lambda}^{\mathrm{T}} \mathbf{g}'(\mathbf{x}) = \mathbf{0}.$$

9.4 Small changes

It is *not* true that

$$d\mathbf{y} = f'(\mathbf{x})\,d\mathbf{x}$$

represents the 'small change' in $\mathbf{y}$ caused by **a** 'small change' $d\mathbf{x}$ in $\mathbf{x}$. What is meant when an assertion of this type is made is that, if $\delta\mathbf{x}$ is small (i.e. each of $\delta x_1, \delta x_2, \dots, \delta x_n$ is small) and $\delta\mathbf{y} = f(\mathbf{x} + \delta\mathbf{x}) - f(\mathbf{x})$ is the 'small change' in $\mathbf{y}$ induced by the 'small change' $\delta\mathbf{x}$ in $\mathbf{x}$, then

$$\delta\mathbf{y} \approx f'(\mathbf{x})\,\delta\mathbf{x}$$

i.e. $\delta\mathbf{y}$ is *approximately* equal to $f'(\mathbf{x})\,\delta\mathbf{x}$. This follows from Taylor's theorem:

$$f(\mathbf{x} + \delta\mathbf{x}) = f(\mathbf{x}) + \frac{1}{1!}f'(\mathbf{x})\delta\mathbf{x} + \cdots$$

A better approximation, of course, may be obtained by taking the second order terms in Taylor's series as well as the first order terms.

Example 3

An economics application

The number of units of output per day at a factory is

$$P(x, y) = 150\left[\frac{1}{10}x^{-2} + \frac{9}{10}y^{-2}\right]^{-\frac{1}{2}}$$

where x denotes capital investment (in units of \$1000) and y denotes the total number of hours (in units of 10) the work force is employed per day. Suppose that currently, capital investment is \$50 000 and the total number of working hours per day is 500. In order to estimate the change in output if capital investment is decreased by \$5000 and the number of working hours is increased by 10 per day, we use the differential dP, where

$$dP = \frac{\partial P}{\partial x}\,dx + \frac{\partial P}{\partial y}\,dy$$

$$= \frac{(-2) \times 150}{(-2) \times 10}x^{-3}\left[\frac{1}{10}x^{-2} + \frac{9}{10}y^{-2}\right]^{-\frac{3}{2}}\,dx$$

$$+ \frac{150 \times 9}{10}y^{-3}\left[\frac{1}{10}x^{-2} + \frac{9}{10}y^{-2}\right]^{-\frac{3}{2}}\,dy$$

In this case $x = 50 = y$, $dx = -5$ and $dy = 1$. Therefore

$$dP = \frac{15}{50^3}\left[\frac{1}{10}50^{-2} + \frac{9}{10}50^{-2}\right]^{-\frac{3}{2}}(-5) + \frac{135}{50^3}\left[\frac{1}{10}50^{-2} + \frac{9}{10}50^{-2}\right]^{-\frac{3}{2}} = 60$$

$$\tag{1}$$

Hence an estimate of the increase in production is 60 units.

Example 4

The cosine rule for a triangle asserts that

$$a^2 = b^2 + c^2 - 2bc \cos A$$

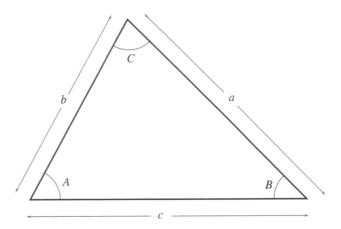

Small changes δA, δb and δc are made in A, b and c respectively. Find an approximation for the consequent change δa in a. We have

$$2a\,\mathrm{d}a = 2b\,\mathrm{d}b + 2c\,\mathrm{d}c - 2b\cos A\,\mathrm{d}c - 2c\cos A\,\mathrm{d}b + 2bc\sin A\,\mathrm{d}A$$

Hence

$$a\,\mathrm{d}a = (b - c\cos A)\,\mathrm{d}b + (c - b\cos A)\,\mathrm{d}c + bc\sin A\,\mathrm{d}A$$

We conclude that

$$\delta a \approx \left(\frac{b - c\cos A}{a}\right)\delta b + \left(\frac{c - b\cos A}{a}\right)\delta c + \frac{bc\sin A}{a}\delta A$$

9.5 Exercises

1. The number of units of output produced per day at a certain factory is

$$P(x, y) = 100\left(\frac{1}{3}x^{-\frac{1}{5}} + \frac{2}{3}y^{-\frac{1}{5}}\right)^{-5}$$

where x denotes the capital investment in units of $1000 and y denotes the total number of hours in units of 10, for which the work force is employed. If capital investment is currently $20\,000 and the number of working hours is 800, estimate the change in output if capital investment is decreased by $5000 and the number of working hours is increased by 100.

Now estimate the change in output if, instead, capital investment is increased by $5,000 and the number of working hours is decreased by 100.

2.* A company's monthly sales of a product depend on price (p), the amount spent on development (x) and advertising (y) (in units of $1000) and the total number of working hours of the sales representatives (z) (in units of 100), according to the model

$$S(p, x, y, z) = (200 - 20p)x^{\frac{1}{5}} y^{\frac{1}{4}} z^{\frac{1}{2}}$$

If price is decreased by $1, expenditure on development and advertising is each increased by $1000 and the number of working hours of the sales representatives is increased by 100, estimate the increase in sales if currently, price is $4, expenditure on development and advertising is $40 000 and $50 000 respectively and the sales representatives are employed for a total of 1200 hours.

3.* The equations

$$\left. \begin{array}{l} f(x, y, z) = 0 \\ g(x, y, z) = 0 \end{array} \right\}$$

can be solved simultaneously to give y as a function of x–i.e. $y = h(x)$. Obtain a formula for $h'(x)$ in terms of the partial derivatives of f and g.

4. The equations

$$\left. \begin{array}{l} f(x, y, z, t) = 0 \\ g(x, y, z, t) = 0 \end{array} \right\}$$

can be solved simultaneously to give z and t as functions of x and y. Find

$$\left(\frac{\partial z}{\partial y} \right)_x$$

first by using the implicit function theorem and then by using differentials.

5.* Given the equations

$$\left. \begin{array}{l} x = \frac{1}{2}(u^2 - v^2) \\ y = uv \end{array} \right\}$$

find

$$\left(\frac{\partial u}{\partial x} \right)_y \quad \text{and} \quad \left(\frac{\partial u}{\partial y} \right)_x$$

using differentials.

6. Given the equations

$$\left. \begin{array}{l} x = u^2 - v \\ y = uv^2 \end{array} \right\}$$

find

$$\left(\frac{\partial u}{\partial x} \right)_y \quad \text{and} \quad \left(\frac{\partial v}{\partial x} \right)_y$$

by at least three different methods including the use of differentials.

7.* With the assumptions of Example 2, prove that

$$
\begin{vmatrix}
\left(\dfrac{\partial T}{\partial p}\right)_u & -1 & \left(\dfrac{\partial T}{\partial u}\right)_p & 0 \\[2ex]
\left(\dfrac{\partial T}{\partial p}\right)_v & -1 & 0 & \left(\dfrac{\partial T}{\partial v}\right)_p \\[2ex]
-1 & \left(\dfrac{\partial p}{\partial T}\right)_v & 0 & \left(\dfrac{\partial p}{\partial v}\right)_T \\[2ex]
-1 & \left(\dfrac{\partial p}{\partial T}\right)_u & \left(\dfrac{\partial p}{\partial u}\right)_T & 0
\end{vmatrix} = 0
$$

8. If $f(x, y, z) = 0$, prove that

$$
\left(\frac{\partial x}{\partial y}\right)_z \left(\frac{\partial y}{\partial z}\right)_x \left(\frac{\partial z}{\partial x}\right)_y = -1
$$

9.* Given the equations

$$
\left.
\begin{array}{l}
x = \frac{1}{2}(u^2 - v^2) \\[1ex]
y = uv
\end{array}
\right\}
$$

find an approximation for the small change δv in v caused by a small change δx in x and a small change δu in u.

See Example 4.

10. The sine rule for a triangle asserts that

$$
\frac{\sin A}{a} = \frac{\sin B}{b} = \frac{\sin C}{c}
$$

Find an approximation for the small change δb in b caused by small changes $\delta a, \delta A$ and δB in a, A and B respectively.

9.6 Application (optional): Slutsky equations

An individual has a quantity I of money to spend (his *income*) and demands a commodity bundle $\mathbf{x}$ which maximises his utility $u = \phi(\mathbf{x})$ subject to the budget constraint $\mathbf{p}^T\mathbf{x} = I$ where $\mathbf{p}$ is the *price* vector.

Suppose that the bundle which maximises utility subject to the budget constraint is $\mathbf{x}^*$. This will depend on $\mathbf{p}$ and I and so we may write

$$
\mathbf{x}^* = \mathbf{f}(\mathbf{p}, I) \tag{1}
$$

The resulting utility is

$$
u^* = \phi(\mathbf{x}^*) = \phi(\mathbf{f}(\mathbf{p}, I)) \tag{2}
$$

Eliminating I between these equations yields

$$
\mathbf{x}^* = g(\mathbf{p}, u^*) \tag{3}
$$

From equation (1) we obtain

$$
d\mathbf{x}^* = \left(\frac{\partial \mathbf{x}^*}{\partial \mathbf{p}}\right)_I d\mathbf{p} + \left(\frac{\partial \mathbf{x}^*}{\partial I}\right)_{\mathbf{p}} dI \tag{4}
$$

and from equation (3) we obtain

$$dx^* = \left(\frac{\partial x^*}{\partial p}\right)_{u^*} dp + \left(\frac{\partial x^*}{\partial u^*}\right)_{p} du^* \tag{5}$$

Next we recall that x^* (and the associated shadow price λ^*) must satisfy the Lagrangian conditions

$$\left.\begin{array}{r}\phi'(x^*) - \lambda^* p^T = 0 \\ p^T x^* = I\end{array}\right\}$$

From the final equation, we obtain $p^T dx^* + x^{*T} dp = dI$ and therefore, by equation (2),

$$du^* = \phi'(x^*) dx^*$$
$$= \lambda^* p^T dx^*$$
$$= \lambda^*(dI - x^{*T} dp)$$

We substitute this result in (5) and obtain

$$dx^* = \left\{\left(\frac{\partial x^*}{\partial p}\right)_{u^*} - \lambda^* \left(\frac{\partial x^*}{\partial u^*}\right)_{p} x^{*T}\right\} dp + \lambda^* \left(\frac{\partial x^*}{\partial u^*}\right)_{p} dI$$

Compare this equation with equation (4). Since both equations hold for all dp and dI, we have

$$\left(\frac{\partial x^*}{\partial p}\right)_{I} = \left(\frac{\partial x^*}{\partial p}\right)_{u^*} - \lambda^* \left(\frac{\partial x^*}{\partial u^*}\right)_{p} x^{*T}$$

$$\left(\frac{\partial x^*}{\partial I}\right)_{p} = \lambda^* \left(\frac{\partial x^*}{\partial u^*}\right)_{p}$$

Substituting the final equation in the previous equation, we obtain

$$\left(\frac{\partial x^*}{\partial p}\right)_{I} = \left(\frac{\partial x^*}{\partial p}\right)_{u^*} - \left(\frac{\partial x^*}{\partial I}\right)_{p} x^{*T}$$

This is a matrix equation and so summarises a block of equations of the form

$$\left(\frac{\partial x_i^*}{\partial p_j}\right)_{I} = \left(\frac{\partial x_i^*}{\partial p_j}\right)_{u^*} - \left(\frac{\partial x_i^*}{\partial I}\right)_{p} x_j^*$$

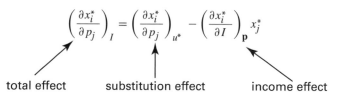

total effect   substitution effect   income effect

These are the **Slutsky equations**. If the price vector p is changed, this will lead to a change in the demand x^*. This change will depend on the tastes of the individual and on his income. The Slutsky equation separates these two effects and gives the rate at which demand for the ith commodity x_i increases with respect to the price p_j of the jth commodity in terms of a 'substitution effect' and an 'income effect'. Note that the former describes how demand changes with price, *given* that income is adjusted to keep utility constant.

Ten

Sums and integrals

This chapter is chiefly about integrating real valued functions of one real variable. Because an integral is a generalisation of the idea of a sum, it is also convenient to include some discussion of summation in this chapter. It is assumed that readers have some previous knowledge of these subjects. In particular, §10.2–§10.9 inclusive consist of an accelerated account of elementary integration theory which a newcomer to the topic would find difficult to assimilate adequately unless they had an unusual aptitude for mathematics. However, experience shows that students are often very rusty on the techniques of integration and it is strongly advised that Exercises 10.10 be used as a check on how well the reader recalls the relevant material before moving on to more advanced work. It is also suggested that all readers study §10.4 with some care. The notation for indefinite integrals can be very confusing if imperfectly understood.

The material given in §10.11–§10.12 about infinite series and integrals over infinite ranges will probably be new to most readers of this book. It is not essential for most of what follows and is best omitted if found at all difficult to understand. The same applies to §10.14 on power series. This is not to say that these are unimportant subjects: only that it may be advisable to leave their study until a later date. There remains §10.13 on differentiating integrals. Although this will be new, the technique can be very useful in some contexts.

10.1 Sums$^\diamond$

We use the notation

$$\sum_{k=1}^{n} a_k = a_1 + a_2 + a_3 + \cdots + a_{n-1} + a_n$$

Thus, for example,

$$\sum_{k=1}^{5} k^2 = 1^2 + 2^2 + 3^2 + 4^2 + 5^2$$
$$= 1 + 4 + 9 + 16 + 25 = 55$$

Some familiar but very useful summation formulas are now given:

$$\frac{1}{2}n(n+1) = \sum_{k=1}^{n} k = 1 + 2 + 3 + \cdots + n \qquad (1)$$

$$\frac{1-x^{n+1}}{1-x} = \sum_{k=0}^{n} x^k = 1 + x + x^2 + \cdots + x^n \quad (x \neq 1) \qquad (2)$$

$$(1+x)^n = \sum_{k=0}^{n} \binom{n}{k} x^k = 1 + nx + \frac{n(n-1)}{2!}x^2 + \cdots + x^n \qquad (3)$$

Formula (1) gives the sum of an 'arithmetic progression', which can be verified in various ways, of which the diagrams below indicate one:

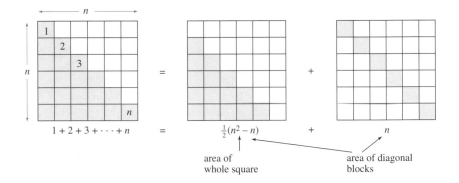

Formula (2) is that for the sum of a 'geometric progression'. This is most easily verified algebraically as below.

$$(1-x)(1+x+x^2+\cdots+x^n) = 1(1+x+x^2+\cdots+x^n)$$
$$-x(1+x+x^2+\cdots+x^n)$$
$$= 1+x+x^2+\cdots+x^n - x - x^2 - x^3$$
$$-\cdots-x^{n+1}$$
$$= 1-x^{n+1}$$

Formula (3) is the **binomial theorem**. The number

$$\binom{n}{k} = \frac{n!}{k!(n-k)!}$$

is called a **binomial coefficient**. These are often most easily calculated using 'Pascal's triangle'.

$$
\begin{array}{ccccccc}
 & & & 1 & & & \\
 & & 1 & & 1 & & \\
 & 1 & & 2 & & 1 & \\
1 & & 3 & & 3 & & 1 \\
\end{array}
$$

1
    1   1
   1   2   1
  1   3   3   1
 1   4   6   4   1
1   5 10 10   5   1
1 6 15 20 15   6   1

$$\binom{0}{0}$$
$$\binom{1}{0} \quad \binom{1}{1}$$
$$\binom{2}{0} \quad \binom{2}{1} \quad \binom{2}{2}$$
$$\binom{3}{0} \quad \binom{3}{1} \quad \binom{3}{2} \quad \binom{3}{3}$$
$$\binom{4}{0} \quad \binom{4}{1} \quad \binom{4}{2} \quad \binom{4}{3} \quad \binom{4}{4}$$
$$\binom{5}{0} \quad \binom{5}{1} \quad \binom{5}{2} \quad \binom{5}{3} \quad \binom{5}{4} \quad \binom{5}{5}$$
$$\binom{6}{0} \quad \binom{6}{1} \quad \binom{6}{2} \quad \binom{6}{3} \quad \binom{6}{4} \quad \binom{6}{5} \quad \binom{6}{6}$$

The numbers in the left hand triangle are the values of the binomial coefficients listed in the right hand triangle. Note that each number in the Pascal triangle (except for the ones) is the sum of the two numbers above it. This explains why the binomial theorem is true. We have for example that

$$
\begin{aligned}
(1+x)^5 &= (1+x)(1+x)^4 = (1+x)(1+4x+6x^2+4x^3+x^4) \\
&= 1 + (1+4)x + (4+6)x^2 + (6+4)x^3 + (4+1)x^4 + x^5 \\
&= 1 + 5x + 10x^2 + 10x^3 + 5x^4 + x^5
\end{aligned}
$$

10.2 Integrals⋄

If $b \geq a$, the notation

$$\int_a^b f(x)\,dx$$

may be interpreted as meaning the area under the graph of $y = f(x)$ between $x = a$ and $x = b$. It must be understood, however, that area beneath the x axis counts as *negative*. Thus, in the diagram below,

$$\int_a^b f(x)\,dx = A - B + C$$

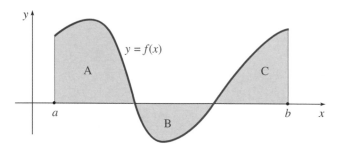

If $b < a$, we define

$$\int_a^b f(x)\,dx = -\int_b^a f(x)\,dx$$

The integral sign is an elongated 'S' from the word 'sum'. It is not hard to see what the connection is between integrals and sums. In the following diagram, the interval $[a, b]$ has been split into a large number of small subintervals. The kth interval is of length δx_k and contains the point x_k. If the function is continuous, it follows that the shaded region below will then be approximately a rectangle of area $f(x_k)\delta x_k$ and hence

$$\sum_k f(x_k)\delta x_k$$

will be approximately equal to the area under the graph between $x = a$ and $x = b$.

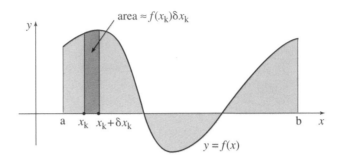

As more and more subintervals, each of smaller and smaller length, are taken, the corresponding approximating sums get closer and closer to the value of the integral

$$\int_a^b f(x)\,\mathrm{d}x$$

Thus an integral is a generalisation of the notion of a sum.

10.3 Fundamental theorem of calculus$^\diamond$

An integral is a generalisation of a sum. Similarly, a derivative is a generalisation of the notion of a difference. It is therefore not too surprising that integrals and derivatives are closely related. The theorem which links these two ideas is called the fundamental theorem of calculus.

A **primitive** (or an **antiderivative**) of f is a function F such that $F'(x) = f(x)$. The first half of the fundamental theorem of calculus asserts that the integral

$$I(x) = \int_a^x f(t)\,\mathrm{d}t$$

is a primitive for f – i.e.

$$I'(x) = \frac{\mathrm{d}}{\mathrm{d}x}\int_a^x f(t)\,\mathrm{d}t = f(x)$$

To justify this result, we observe that the shaded area $I(x + h) - I(x)$ in the right hand diagram below is nearly a rectangle of area $f(x)h$ (provided that f is continuous).

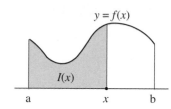 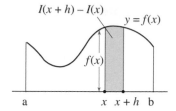

It follows that

$$I(x+h) - I(x) \approx f(x)h$$

$$\frac{I(x+h) - I(x)}{h} \approx f(x)$$

It can be shown that this approximation improves as h gets smaller (provided that f is continuous) and, in the limit,

$$I'(x) = \lim_{h \to 0} \left(\frac{I(x+h) - I(x)}{h} \right) = f(x)$$

It is important to note that I is not the only primitive for f. For example, given any constant ξ, the function J defined by

$$J(x) = \int_{\xi}^{x} f(t)\,dt$$

is also a primitive for f. Observe that

$$J(x) = \int_{a}^{x} f(t)\,dt + \int_{\xi}^{a} f(t)\,dt$$

and thus $J(x) = I(x) + c$ where c is a *constant*. This result holds in general. Any primitive F for the function f has the property that

$$F(x) = \int_{a}^{x} f(t)\,dt + c$$

for some constant c. In particular

$$F(b) - F(a) = \left\{ \int_{a}^{b} f(t)\,dt + c \right\} - \left\{ \int_{a}^{a} f(t)\,dt + c \right\}$$

This leads to the second half of the fundamental theorem of calculus – i.e.

If F is any primitive for f, then

$$\int_{a}^{b} f(t)\,dt = [F(x)]_{a}^{b} = F(b) - F(a)$$

This last result provides a very convenient way of calculating an integral providing that one can guess a primitive for the function f.

Example 1

Calculate

$$\int_1^2 x^2 \, dx.$$

We begin by guessing a primitive. The function $F(x) = \frac{1}{3}x^3$ comes to mind. To verify that this is a primitive, we differentiate and obtain

$$F'(x) = \frac{1}{3} \cdot 3x^2 = x^2 = f(x)$$

It follows from the fundamental theorem of calculus that

$$\int_1^2 x^2 \, dx = \left[\frac{1}{3} \cdot x^3 \right]_1^2 = \frac{1}{3} \cdot 2^3 - \frac{1}{3} 1^3 = \frac{7}{3}$$

Although the method given above is very useful it is not adequate for calculating *all* integrals. Consider, for example, the integral

$$\int_1^2 e^{-x^2/2} \, dx$$

It is true that there *is* a function F such that

$$F'(x) = e^{-x^2/2}$$

but this function F is *not* expressible in terms of elementary functions like the exponential, logarithm or trigonometric functions. However hard one may try to guess F, there is therefore no prospect whatever of any success if attention is confined to these simple functions.

10.4 Notation⋄

The notation

$$\int f(x) \, dx$$

is widely used to denote a primitive of the function $f(x)$. For this reason a primitive is often called an **indefinite integral**.

Although it has advantages, this notation can sometimes lead to confusion. One cause of confusion is that x is placed as though it were a 'dummy' variable of integration whereas it actually corresponds to an upper limit of an integral – i.e. for some constant ξ,

$$\int f(x) \, dx = \int_\xi^x f(t) \, dt$$

The symbol t on the right hand side is the 'dummy' variable of integration. Any other symbol (*except*, paradoxically, x) would do equally well for this 'dummy' variable.

The first half of the fundamental theorem of calculus can be written in the form

$$\frac{d}{dx} \left\{ \int f(x) \, dx \right\} = f(x)$$

but one should not suppose that this involves 'differentiating with respect to a dummy variable of integration'. It is simply a shorthand for

$$\frac{\mathrm{d}}{\mathrm{d}x}\left\{\int_{\xi}^{x} f(t)\,\mathrm{d}t\right\} = f(x)$$

To write

$$\frac{\mathrm{d}}{\mathrm{d}x}\left\{\int_{a}^{b} f(x)\,\mathrm{d}x\right\} = f(x)$$

would of course be nonsensical. The integral

$$\int_{a}^{b} f(x)\,\mathrm{d}x$$

is a *constant* which could equally well be written as

$$\int_{a}^{b} f(y)\,\mathrm{d}y$$

Example 2

(i) $\quad \dfrac{\mathrm{d}}{\mathrm{d}x}\displaystyle\int_{0}^{x}\dfrac{\mathrm{d}t}{1+t^{100}} = \dfrac{1}{1+x^{100}}$

(ii) $\quad \dfrac{\mathrm{d}}{\mathrm{d}x}\displaystyle\int_{99}^{x}\dfrac{\mathrm{d}t}{1+t^{100}} = \dfrac{1}{1+x^{100}}$

These results are instances of the first half of the fundamental theorem of calculus which may be expressed in the form

$$\frac{\mathrm{d}}{\mathrm{d}x}\int\frac{\mathrm{d}x}{1+x^{100}} = \frac{1}{1+x^{100}}$$

but it would be *nonsense* to write

$$\frac{\mathrm{d}}{\mathrm{d}t}\int_{0}^{x}\frac{\mathrm{d}t}{1+t^{100}} = \frac{1}{1+t^{100}}$$

Note also that

(iii) $\quad \dfrac{\mathrm{d}}{\mathrm{d}x}\displaystyle\int_{x}^{1}\dfrac{\mathrm{d}t}{1+t^{100}} = \dfrac{\mathrm{d}}{\mathrm{d}x}\left\{-\displaystyle\int_{1}^{x}\dfrac{\mathrm{d}t}{1+t^{100}}\right\} = -\dfrac{1}{1+x^{100}}$

A further problem with the 'indefinite integral' notation is that it tempts one to forget the *constant of integration*. The notation

$$\int f(x)\,\mathrm{d}x$$

means a primitive for $f(x)$ – i.e. a function whose derivative is $f(x)$. But $f(x)$ has *many* primitives. The general form for a primitive is

$$\int f(x)\,dx + c$$

where c is the constant of integration.

Example 3 _____

From §7.1.2 we know that

$$\frac{d}{dx}(\arcsin x) = \frac{1}{\sqrt{1-x^2}} \qquad (-1 < x < 1)$$

and so it follows that we may write

$$\int \frac{dx}{\sqrt{1-x^2}} = \arcsin x$$

But it is equally true that

$$\frac{d}{dx}(-\arccos x) = \frac{1}{\sqrt{1-x^2}} \qquad (-1 < x < 1)$$

and so we may also write

$$\int \frac{dx}{\sqrt{1-x^2}} = -\arccos x$$

But we should *not* fall into the trap of forgetting the constant of integration and deducing that $\arcsin x = -\arccos x$. In fact , as we have seen in §7.1.2,

$$\arcsin x = -\arccos x + \frac{\pi}{2}$$

_____

10.5 Standard integrals$^{\diamond}$

From the formulas for derivatives given in §2.10 and §7.1.2 we obtain the following list of 'standard integrals' – i.e. commonly occurring primitives. It is useful to know these by heart. More extensive lists will be found in the backs of books of tables but the standard integrals given here, together with the integration techniques given later, will be found adequate for most purposes:

$$\int x^\alpha \, dx = \frac{1}{\alpha + 1} x^{\alpha+1} \qquad (\alpha \neq -1)$$

$$\int \frac{dx}{ax + b} = \frac{1}{a} \ln |ax + b| \qquad (a \neq 0)$$

$$\int e^{ax} \, dx = \frac{1}{a} e^{ax} \qquad (a \neq 0)$$

$$\int \cos(ax) \, dx = \frac{1}{a} \sin(ax) \qquad (a \neq 0)$$

$$\int \sin(ax) \, dx = -\frac{1}{a} \cos(ax) \qquad (a \neq 0)$$

$$\int \frac{dx}{a^2 + x^2} = \frac{1}{a} \arctan \left(\frac{x}{a} \right) \qquad (a \neq 0)$$

$$\int \frac{dx}{\sqrt{a^2 - x^2}} = \arcsin \left(\frac{x}{a} \right) \qquad (a \neq 0)$$

Each of these results should be checked by differentiating the right hand side to confirm that the integrand is obtained. For example,

$$\frac{d}{dx} \left[\frac{1}{a} \arctan \left(\frac{x}{a} \right) \right] = \frac{1}{a} \cdot \frac{1}{1 + (x/a)^2} \cdot \frac{1}{a} = \frac{1}{a^2 + x^2}$$

Note in particular, that since the logarithmic function is only defined for positive numbers, when $ax + b < 0$ $(a \neq 0)$,

$$\int \frac{dx}{ax + b} = -\int \frac{dx}{-(ax + b)} = -\frac{1}{(-a)} \ln(-(ax + b)) = \frac{1}{a} \ln |ax + b|$$

When $ax + b > 0$, it is clear that

$$\int \frac{dx}{ax + b} = \frac{1}{a} \ln |ax + b|$$

Remember that the functions on the right hand side are only *one* of the primitives for the integrands on the left hand side. The general form of the primitive is obtained by adding a constant of integration.

Example 4

An economics example

Notation: $f_d(x)$ and $f_s(x)$ are the inverses of the demand and supply functions – see Example 12.4.

A demand relation $p = f_d(x)$ represents the price per unit consumers are willing to pay for quantities x of a commodity. Economists use consumer surplus as a rough and ready measure of the benefit consumers derive from buying x_0 at a fixed price of p_0 per unit; rather than paying the amount that they would have been willing to

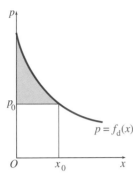

consumer's surplus

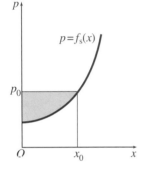

producer's surplus

pay for x_0, if the seller had been able to extract it from them. Formally, **consumer's surplus** is defined as

$$\int_0^{x_0} f_d(x)\, dx - p_0 x_0$$

(which is shaded in the diagram).

A supply relation $p = f_s(x)$ represents the unit prices at which a producer is willing to supply quantities x of the commodity. The total benefit to producers who would be willing to supply at lower than the equilibrium price (which is shaded in the diagram) is

$$\text{the \textbf{producer's surplus}} = p_0 x_0 - \int_0^{x_0} f_s(x)\, dx$$

If the demand and supply functions under pure competition are respectively

$$p = 120 - x^2 \quad \text{and} \quad p = x^2 + 2x + 8$$

we equate the supply and demand to find the equilibrium price:

$$x^2 + 2x + 8 = 120 - x^2 \quad \text{and so} \quad 2(x^2 + x - 56) = 0$$

Hence $(x + 8)(x - 7) = 0$ giving $x = 7$ for which $p = 71$.
At the equilibrium,

$$\text{consumer's surplus} = \int_0^7 (120 - x^2)\, dx - (71)(7)$$

$$= \left[120x - x^3/3\right]_0^7 - 497 = 228\frac{2}{3}$$

$$\text{producer's surplus} = 497 - \int_0^7 (x^2 + 2x + 8)\, dx$$

$$= 497 - \left[x^3/3 + x^2 + 8x\right]_0^7 = 277\frac{2}{3}$$

10.6 Partial fractions◇

If a rational function $R(x) = \dfrac{P(x)}{Q(x)}$, where P and Q are polynomials, *has the degree of P less than the degree of Q*, then R admits a **partial fraction** expansion.

If the degree of P is *not* less than the degree of Q, then P can be divided by Q to give

$$\frac{P(x)}{Q(x)} = S(x) + \frac{r(x)}{Q(x)}$$

where S and r are polynomials and the degree of r is less than the degree of Q.

Example 5

We have

$$\frac{x+1}{x+2} = \frac{x+2-1}{x+2} = 1 - \frac{1}{x+2}$$

Hence

$$\int \frac{x+1}{x+2}\,dx = \int \left(1 - \frac{1}{x+2}\right)dx$$

$$= x - \ln|x+2|$$

Now suppose that the degree of P is less than the degree of Q. There are three cases.

Case (i)

If the roots $\xi_1, \xi_2, \ldots, \xi_n$ of Q are all real and distinct, then R admits a partial fraction expansion of the form

$$\frac{P(x)}{Q(x)} = \frac{A_1}{x - \xi_1} + \frac{A_2}{x - \xi_2} + \cdots + \frac{A_n}{x - \xi_n}$$

The number A_k may be calculated from the formula

$$A_k = \lim_{x \to \xi_k} (x - \xi_k)\frac{P(x)}{Q(x)}.$$

The partial fraction expansion may be used to evaluate $\int R(x)\,dx$. Thus,

$$\int \frac{P(x)}{Q(x)}\,dx = A_1 \int \frac{dx}{x - \xi_1} + A_2 \int \frac{dx}{x - \xi_2} + \cdots + A_n \int \frac{dx}{x - \xi_n}$$

$$= A_1 \ln|x - \xi_1| + A_2 \ln|x - \xi_2| + \cdots + A_n \ln|x - \xi_n|$$

Example 6

We have

$$\frac{x-1}{x^2 - 5x + 6} = \frac{(x-1)}{(x-2)(x-3)} = \frac{A}{x-2} + \frac{B}{x-3}$$

where

$$A = \lim_{x \to 2} (x - 2)\frac{(x-1)}{(x-2)(x-3)} = \lim_{x \to 2} \frac{(x-1)}{(x-3)} = -1$$

$$B = \lim_{x \to 3} (x - 3)\frac{(x-1)}{(x-2)(x-3)} = \lim_{x \to 3} \frac{(x-1)}{(x-2)} = 2$$

305

Thus

$$\int \frac{x-1}{x^2-5x+6}\,dx = \int \left(\frac{-1}{x-2}+\frac{2}{x-3}\right)dx$$

$$= -\ln|x-2| + 2\ln|x-3|$$

$$= \ln|x-2|^{-1} + \ln(x-3)^2$$

$$= \ln \frac{(x-3)^2}{|x-2|}$$

In particular,

$$\int_4^5 \frac{x-1}{x^2-5x+6}\,dx = \left[\ln \frac{(x-3)^2}{|x-2|}\right]_4^5$$

$$= \ln \frac{4}{3} - \ln \frac{1}{2}$$

$$= \ln \left(\frac{4}{3}\cdot 2\right) = \ln \frac{8}{3}$$

Case (ii)

If the roots of Q are real but *not* all distinct, then the partial fraction expansion given above does *not* apply. If ξ_k is a root of Q with multiplicity m (i.e. ξ_k is repeated m times), then the term

$$\frac{A_k}{x-\xi_k}$$

in the partial fraction expansion must be replaced by a term of the form

$$\frac{B_1}{x-\xi_k} + \frac{B_2}{(x-\xi_k)^2} + \cdots + \frac{B_m}{(x-\xi_k)^m}$$

Fortunately, in practice m is seldom greater than 2.

Example 7

We have

$$\frac{x+1}{(x-2)(x-3)^2} = \frac{A}{x-2} + \frac{B}{(x-3)} + \frac{C}{(x-3)^2}.$$

The coefficient A may be calculated from the formula

$$A = \lim_{x\to 2}(x-2)\frac{(x+1)}{(x-2)(x-3)^2} = \lim_{x\to 2}\frac{(x+1)}{(x-3)^2} = 3$$

Similarly

$$C = \lim_{x\to 3}(x-3)^2\frac{(x+1)}{(x-2)(x-3)^2} = \lim_{x\to 3}\frac{(x+1)}{(x-2)} = 4$$

Then

$$\frac{B}{x-3} = \frac{x+1}{(x-2)(x-3)^2} - \frac{3}{x-2} - \frac{4}{(x-3)^2}$$

Taking $x = 4$ we obtain $B = -3$ (or, more easily, multiply through by $(x-3)$ and let $x \to \infty$).

It follows that

$$\int \frac{x+1}{(x-2)(x-3)^2}\, dx = \int \left(\frac{3}{x-2} - \frac{3}{(x-3)} + \frac{4}{(x-3)^2} \right) dx$$

$$= 3\ln|x-2| - 3\ln|x-3| - \frac{4}{(x-3)}$$

$$= \ln \left| \frac{x-2}{x-3} \right|^3 - \frac{4}{x-3}$$

In particular,

$$\int_4^5 \frac{x+1}{(x-2)(x-3)^2}\, dx = \left[\ln \left| \frac{x-2}{x-3} \right|^3 - \frac{4}{x-3} \right]_4^5$$

$$= \left(\ln \left(\frac{3}{2} \right)^3 - \frac{4}{2} \right) - \left(\ln \left(\frac{2}{1} \right)^3 - \frac{4}{1} \right)$$

$$= \ln \left(\frac{27}{64} \right) + 2$$

Case (iii)

See §13.6.

If Q has complex roots so that it has irreducible quadratic factors like $(x^2 + bx + c)$, the corresponding term in the partial fraction expansion of $\dfrac{P(x)}{Q(x)}$ will be $\dfrac{Ax + B}{x^2 + bx + c}$.

If Q has repeated factors like $(x^2 + bx + c)^m$, then the term

$$\frac{Ax + B}{x^2 + bx + c}$$

in the partial fraction expansion must be replaced by a term of the form

$$\frac{A_1 x + B_1}{x^2 + bx + c} + \frac{A_2 x + B_2}{(x^2 + bx + c)^2} + \cdots + \frac{A_m x + B_m}{(x^2 + bx + c)^m}$$

Example 8

On factorising the denominator we have

$$\frac{1}{x^3 - x^2 + x - 1} = \frac{A}{x-1} + \frac{Bx + C}{x^2 + 1}$$

307

The coefficient A may be calculated from the formula

$$A = \lim_{x \to 1} (x - 1) \frac{1}{x^3 - x^2 + x - 1} = \lim_{x \to 1} \frac{1}{x^2 + 1} = \frac{1}{2}$$

Then

$$\frac{Bx + C}{x^2 + 1} = \frac{1}{x^3 - x^2 + x - 1} - \frac{1}{2(x - 1)}$$

Letting $x = 0$ we obtain

$$C = -1 + \frac{1}{2} = -\frac{1}{2}$$

Letting $x = -1$ we obtain

$$\frac{-B - \frac{1}{2}}{2} = \frac{1}{(-4)} - \frac{1}{2(-2)}$$

so $B = -\frac{1}{2}$. Hence

$$\frac{1}{x^3 - x^2 + x - 1} = \frac{1}{2(x - 1)} - \frac{1}{2} \frac{(x + 1)}{(x^2 + 1)}$$

which enables the function on the left to be integrated.

10.7 Completing the square$^{\diamond}$

If $a \neq 0$, we have

$$ax^2 + bx + c = a \left(x^2 + \frac{b}{a} x + \frac{c}{a} \right)$$

$$= a \left\{ \left(x + \frac{b}{2a} \right)^2 - \frac{b^2}{4a^2} + \frac{c}{a} \right\}$$

This algebraic manipulation is called '**completing the square**'. Its most familiar use is in finding the roots of quadratic equations. If

$$ax^2 + bx + c = 0$$

then

$$\left(x + \frac{b}{2a} \right)^2 - \frac{b^2}{4a^2} + \frac{c}{a} = 0$$

Hence

$$x + \frac{b}{2a} = \pm \frac{\sqrt{b^2 - 4ac}}{2a}$$

$$x = \frac{-b \pm \sqrt{b^2 - 4ac}}{2a}$$

The next example gives another way in which it is used.

Example 9

One can use partial fractions to calculate

$$\int \frac{dx}{x^2 + x + 1}$$

but then complex numbers intrude. An alternative is to complete the square in the denominator. Then

$$\int \frac{dx}{x^2 + x + 1} = \int \frac{dx}{(x + \frac{1}{2})^2 + \frac{3}{4}}$$

$$= \left(\frac{3}{4}\right)^{-1/2} \arctan\left(\left(\frac{3}{4}\right)^{-1/2}\left(x + \frac{1}{2}\right)\right)$$

In particular,

$$\int_0^1 \frac{dx}{x^2 + x + 1} = \left[\left(\frac{3}{4}\right)^{-1/2} \arctan\left\{\left(\frac{3}{4}\right)^{-1/2}\left(x + \frac{1}{2}\right)\right\}\right]_0^1$$

$$= \frac{2}{\sqrt{3}} \arctan\left(\frac{2}{\sqrt{3}} \cdot \frac{3}{2}\right) - \frac{2}{\sqrt{3}} \arctan\left(\frac{2}{\sqrt{3}} \cdot \frac{1}{2}\right)$$

$$= \frac{2}{\sqrt{3}}\left(\arctan\sqrt{3} - \arctan\frac{1}{\sqrt{3}}\right)$$

$$= \frac{2}{\sqrt{3}}\left(\frac{\pi}{3} - \frac{\pi}{6}\right) = \frac{\pi}{3\sqrt{3}}$$

because $\tan(\pi/6) = 1/\sqrt{3}$ and $\tan(\pi/3) = \sqrt{3}$.

10.8 Change of variable◇

Suppose that $x = g(y)$. Then

$$dx = g'(y)\,dy$$

and so the formula

$$\int_a^b f(x)\,dx = \int_A^B f(g(y))g'(y)\,dy$$

where $B = g^{-1}(b)$ and $A = g^{-1}(a)$

for changing the variable in an integral is very natural. This clearly requires that g be the inverse of a function defined on the interval $[a, b]$.

To justify the formula we observe that, if F is a primitive for f, then the left hand side is equal to

$$\int_a^b f(x)\,dx = [F(x)]_a^b = F(b) - F(a) = [F(g(x))]_A^B$$

See §2.10.2. But the chain rule (IV) for differentiating a 'function of a function' implies that

$$\frac{d}{dy}\{F(g(y))\} = F'(g(y))g'(y) = f(g(y))g'(y)$$

and so we also have

$$\int_a^b f(g(y))g'(y)\,dy = [F(g(y))]_A^B$$

Example 10

Consider the integral

$$\int_1^4 \frac{\cos\sqrt{t}}{\sqrt{t}}\,dt$$

A natural change of variable is $u = \sqrt{t}$. Then

$$du = \frac{1}{2}t^{-1/2}\,dt$$

Also $u = 2$ when $t = 4$ and $u = 1$ when $t = 1$. Thus

$$\int_1^4 \frac{\cos\sqrt{t}}{\sqrt{t}}\,dt = \int_1^2 \frac{\cos\sqrt{t}}{\sqrt{t}} 2\sqrt{t}\,du$$

$$= \int_1^2 2\cos u\,du$$

$$= [2\sin u]_1^2 = 2(\sin 2 - \sin 1)$$

Example 11

Integrals involving $\sqrt{1 - t^2}$ can often be reduced to something manageable by the change of variable $t = \sin\theta$. For example,

$$\int_0^1 \sqrt{1 - t^2}\,dt = \int_0^{\pi/2} \sqrt{(1 - \sin^2\theta)}\cos\theta\,d\theta$$

$$= \int_0^{\pi/2} \cos^2\theta\,d\theta$$

$$= \int_0^{\pi/2} \frac{\cos 2\theta + 1}{2}\,d\theta$$

$$= \left[\frac{\sin 2\theta}{4} + \frac{1}{2}\theta\right]_0^{\pi/2}$$

$$= \frac{\pi}{4}$$

Similarly the change of variable $t = \tan\theta$ *is sometimes useful with integrals involving* $\sqrt{1 + t^2}$ *because* $1 + \tan^2\theta = \sec^2\theta$.

Example 12

Often integrals occur in which the integrand looks like the right hand side of the formula for changing the variable of an integral. Such integrals can be evaluated immediately provided one knows a primitive F for the function f because

$$\int_a^b f(g(y))g'(y)\,dy = [F(g(y))]_a^b$$

A particularly common application is the case with $f(x) = 1/x$. We then have

$$\int_a^b \frac{g'(y)}{g(y)}\,dy = [\ln|g(y)|]_a^b = \ln\left|\frac{g(b)}{g(a)}\right|$$

An example of this is the integral

$$\int_0^1 \frac{2x + 3}{x^2 + 3x + 2}\,dx$$

which can be evaluated using partial fractions. A better method is to observe that, if $g(x) = x^2 + 3x + 2$, then $g'(x) = 2x + 3$. Hence

$$\int_0^1 \frac{2x + 3}{x^2 + 3x + 2}\,dx = [\ln|x^2 + 3x + 2|]_0^1$$

$$= \ln\frac{6}{2} = \ln 3$$

The method is essentially the same as introducing the change of variable $y = x^2 + 3x + 2$. Then $dy = (2x + 3)\,dx$ and so

$$\int_0^1 \frac{2x + 3}{x^2 + 3x + 2}\,dx = \int_2^6 \frac{dy}{y} = [\ln y]_2^6 = \ln\frac{6}{2}$$

10.9 Integration by parts$^\diamond$

See §2.10.2.

If U is any primitive for u, the product rule (II) for differentiating a product gives

$$\frac{d}{dx}\{U(x)v(x)\} = U'(x)v(x) + U(x)v'(x)$$

$$= u(x)v(x) + U(x)v'(x)$$

and hence

$$\int_a^b u(x)v(x)\,dx = [U(x)v(x)]_a^b - \int_a^b U(x)v'(x)\,dx$$

which is the formula for **'integrating by parts'**.

Example 13

We have

$$\int_1^2 x\cos x\,dx = [x\sin x] - \int_1^2 1\sin x\,dx$$

$$= [x\sin x]_1^2 - [-\cos x]_1^2$$

$$= (2\sin 2 + \cos 2) - (1\sin 1 + \cos 1)$$

Example 14

We have

$$\int \ln x\,dx = \int 1 \cdot \ln x\,dx$$

$$= [x\ln x] - \int x \cdot \frac{1}{x}\,dx$$

$$= [x\ln x - x]$$

Therefore $x\ln x - x$ is a primitive for $\ln x$.

Example 15

Let

$$I_{mn} = \int_0^1 x^m(1-x)^n\,dx$$

Then

$$I_{mn} = \left[\frac{x^{m+1}}{(m+1)}(1-x)^n\right]_0^1 - \int_0^1 \frac{x^{m+1}}{m+1}\{-n(1-x)^{n-1}\}\,dx$$

Thus, provided $n \geq 1$,

$$I_{mn} = \frac{n}{m+1}I_{m+1,n-1}$$

$$= \frac{n(n-1)}{(m+1)(m+2)}I_{m+2,n-2}$$

$$\cdots$$

$$= \frac{n(n-1)\ldots 1}{(m+1)(m+2)\ldots(m+n)}I_{m+n,0}$$

$$= \frac{m!n!}{(m+n)!}\int_0^1 x^{m+n}\,dx = \frac{m!n!}{(m+n+1)!}$$

10.10 Exercises

1.* Use Pascal's triangle to obtain an expansion for $(1 + x)^8$.

2. Prove that, for $k = 1, 2, 3, \ldots, n + 1$,

$$\frac{(n + 1)!}{k!(n + 1 - k)!} = \frac{n!}{k!(n - k)!} + \frac{n!}{(k - 1)!(n - k + 1)!}$$

and explain what this has to do with Pascal's triangle.

3.*♣ Explain why

$$\int_0^1 f(x) \, dx = \lim_{n \to \infty} \sum_{k=0}^{n-1} f\left(\frac{k}{n}\right) \frac{1}{n}$$

provided that f is continuous. Hence evaluate

$$\int_0^1 e^x \, dx$$

without using the fundamental theorem of calculus.

4.♣ Justify the formula

$$\int_0^1 (ax + b) \, dx = \lim_{n \to \infty} \sum_{k=0}^{n-1} \left(\frac{ak}{n} + b\right) \frac{1}{n}$$

and hence evaluate the integral without using the fundamental theorem of calculus.

5.* Evaluate

(i) $\displaystyle\int_{-1/2}^{1/2} \frac{dy}{\sqrt{1 + y^2}}$

(ii) $\displaystyle\int_{-1/2}^{1/2} \frac{dy}{1 - y^2}$

using Exercises 7.6.2 and 7.6.3.

6. Show that

(i) $\dfrac{d}{dx}\left(\ln\left(\tan\frac{1}{2}x\right)\right) = \operatorname{cosec} x$ (ii) $\dfrac{d}{dx}(e^{-x^2/2}) = -xe^{-x^2/2}$

and hence evaluate

(i) $\displaystyle\int_{\pi/4}^{3\pi/4} \operatorname{cosec} x \, dx$ (ii) $\displaystyle\int_0^x xe^{-x^2/2} \, dx$

7.* Since the integrand is nonnegative, the integral

$$\int_{\pi/4}^{3\pi/4} \sec^2 x \, dx$$

should also be nonnegative. But

$$\int_{\pi/4}^{3\pi/4} \sec^2 x \, dx = [\tan x]_{\pi/4}^{3\pi/4} = -1 - 1 = -2$$

Draw graphs of $y = \sec^2 x$ and $y = \tan x$ in the relevant range and hence find the error in this reasoning.

8. Evaluate

   (i) $\dfrac{d}{dx} \displaystyle\int_0^x e^{-t^2} dt$ (ii) $\dfrac{d}{dx} \displaystyle\int_1^x e^{-t^2} dt$ (iii) $\dfrac{d}{dx} \displaystyle\int_x^2 e^{-t^2} dt$

9. A monopolist faces the demand function

$$p_d = 180 - x^2$$

His cost function is

$$C(x) = \frac{3}{2}x^2 + 92x$$

He maximises profit when

$$\text{marginal revenue} = \text{marginal cost}$$

in which case find the consumer's surplus.

If he maximises revenue instead of profits, find the change in consumer's surplus.

10. Given the demand relation

$$p_d = 116 - 4x - x^2$$

and the supply relation

$$p_s = x^2 + 2x + 8$$

find the consumer's surplus and producer's surplus at the equilibrium price under pure competition.

11. Investment $I(t)$ is the rate of change in capital K, hence $\dfrac{dK}{dt} = I$. Suppose that $I(t) = 10^3 t^{\frac{3}{2}}$, find the change in capital in five years.

12. Find the corresponding total functions from the given marginal functions:

   (i) The marginal cost function, $C'(x) = 24x^2 - 4x + 7$

   (ii) The marginal revenue function, $R'(x) = 10x^2 + 5x + 3$

   (iii) The marginal propensity to consume, $C'(Y) = \dfrac{dC}{dY} = 0.9$ where $C(Y)$ is the consumption function.

   Note that the constant of integration gives the value of the total function when the independent variable is zero.

13.* Use partial fractions to evaluate the integral

$$\int_2^3 \frac{x^3}{x^2 + x - 2} \, dx$$

14. Use partial fractions to evaluate the integrals

(i) $\displaystyle\int_2^3 \frac{dx}{x^2 + x - 2}$ (ii) $\displaystyle\int_2^3 \frac{dx}{(1 - x^2)^2}$

(iii) $\displaystyle\int_2^3 \frac{x \, dx}{x^2 + x - 2}$ (iv) $\displaystyle\int_2^3 \frac{x^2 \, dx}{x^2 + x - 2}$

15.* Complete the square in the denominator and hence evaluate

$$\int_0^1 \frac{dx}{\sqrt{7 - 2x - x^2}}$$

16. Complete the square in the denominator and hence evaluate

$$\int_0^1 \frac{dx}{x^2 + 2x + 7}$$

17.* Evaluate the integral

$$\int_0^{\pi/3} \cos^3 \theta \, d\theta$$

using the formula $\cos 3\theta = 4 \cos^3 \theta - 3 \cos \theta$.

18. Evaluate the integrals

(i) $\displaystyle\int_0^{\pi/2} \cos^2 \theta \, d\theta$ (ii) $\displaystyle\int_0^{\pi/2} \sin^2 \theta \, d\theta$

using the formulas $\cos 2\theta = 2 \cos^2 \theta - 1 = 1 - 2 \sin^2 \theta$.

19.* Make an appropriate change of variable in the integral

$$\int_1^2 \frac{\sqrt{1 + \sqrt{y}}}{\sqrt{y}} \, dy$$

and hence evaluate the integral.

20. Make an appropriate change of variable in the integrals

(i) $\displaystyle\int_{-1}^{\sqrt{2}} \frac{x \, dx}{(x^2 + 1)^3}$ (ii) $\displaystyle\int_1^2 \frac{dt}{\sqrt{3 + 5t}}$ (iii) $\displaystyle\int_0^1 x\sqrt{1 - x^2} \, dx$

For (iii), see Example 11.   and hence evaluate them.

21.* Express $\sin \theta$ and $\cos \theta$ in terms of $t = \tan \frac{1}{2}\theta$. Evaluate the integrals below by means of this change of variable.

(i) $\displaystyle\int_{\pi/4}^{\pi/2} \frac{d\theta}{\sin \theta}$ (ii) $\displaystyle\int_0^{\pi/4} \frac{d\theta}{\cos \theta}$ (iii) $\displaystyle\int_{\pi/6}^{\pi/3} \frac{d\theta}{\sin \theta + \cos \theta}$

22. Evaluate the integral

$$\int_{-1/2}^{1/2} \frac{dt}{\sqrt{1+t^2}}$$

using the change of variable $t = \tan\theta$. Also evaluate

$$\int_{-1/2}^{1/2} \frac{t\,dt}{\sqrt{1+t^2}}$$

23.* Evaluate the integral

$$\int_0^1 \frac{3x^2 + 2x + 1}{x^3 + x^2 + x + 1}\,dx$$

24. Evaluate the integral

$$\int_2^3 \frac{x\,dx}{(1-x^2)^2}$$

(i) by using partial fractions (ii) by writing $g(x) = 1 - x^2$ and observing that $x = -\frac{1}{2}g'(x)$.

25.* Integrate by parts and hence evaluate

(i) $\displaystyle\int_0^\pi x^3 \cos x\,dx$ (ii) $\displaystyle\int_0^X x^2 e^{-x}\,dx$

26. Integrate

$$I_m = \int_0^\pi \sin^m x\,dx = \int_0^\pi (\sin x)(\sin x)^{m-1}\,dx$$

by parts and hence show that

$$I_m = \left(\frac{m-1}{m}\right) I_{m-2} \qquad (m > 2)$$

Evaluate I_4 and I_5.

10.11 Infinite sums and integrals♣

In this section we discuss very briefly the idea of an 'infinite sum' or **series**

$$\sum_{n=1}^\infty a_n$$

The symbol ∞ does not, of course, stand for a *number*.

Similar remarks apply to the idea of an 'infinite integral'

$$\int_\xi^\infty f(x)\,dx$$

The series is said to **converge** if and only if there is a number s such that

$$\sum_{n=1}^N a_n \to s \quad \text{as} \quad N \to \infty$$

in which case we write

$$\sum_{n=1}^{\infty} a_n = s$$

If no such number s exists, then the series is said to **diverge**.

Similarly, the 'infinite integral' is said to **converge** if and only if there is a number I such that

$$\int_{\xi}^{X} f(x)\,dx \to I \quad \text{as} \quad X \to \infty$$

in which case we write

$$\int_{\xi}^{\infty} f(x)\,dx = I$$

If no such number I exists, the 'infinite integral' is said to **diverge**.

Example 16

(i) Suppose that $|x| < 1$. Then $x^N \to 0$ as $N \to \infty$ and so

$$\sum_{n=0}^{N} x^n = \frac{1 - x^{N+1}}{1 - x} \to \frac{1}{1 - x} \quad \text{as} \quad N \to \infty$$

Hence

$$\sum_{n=0}^{\infty} x^n = \frac{1}{1 - x} \qquad (|x| < 1)$$

(ii) Suppose that $|x| > 1$. Then $x^N \to \infty$ as $N \to \infty$ if $x > 1$ and x^N 'oscillates' if $x < -1$. In neither case does there exist a number s such that

$$\frac{1 - x^{N+1}}{1 - x} \to s \quad \text{as} \quad N \to \infty$$

It follows that, when $|x| > 1$, the series

$$\sum_{n=0}^{\infty} x^n$$

diverges. The series also diverges when $x = 1$ or $x = -1$. (Why?)

Example 17

(i) Suppose that $\alpha > 1$. Then $X^{1-\alpha} \to 0$ as $X \to \infty$ and so

$$\int_{1}^{X} \frac{dx}{x^\alpha} = \left[\frac{1}{1 - \alpha} x^{1-\alpha} \right]_{1}^{X}$$

$$= \frac{1}{\alpha - 1}(1 - X^{1-\alpha})$$

$$\to \frac{1}{\alpha - 1} \quad \text{as} \quad X \to \infty$$

Hence

$$\int_{1}^{\infty} \frac{dx}{x^\alpha} = \frac{1}{\alpha - 1} \qquad (\alpha > 1)$$

(ii)   Suppose $\alpha < 1$. Then $X^{1-\alpha} \to \infty$ as $X \to \infty$ and so the 'infinite integral'

$$\int_1^\infty \frac{\mathrm{d}x}{x^\alpha}$$

diverges when $\alpha < 1$.

(iii)   Suppose $\alpha = 1$. We then have to consider

$$\int_1^X \frac{\mathrm{d}x}{x} = [\ln x]_1^X = \ln X \to \infty \quad \text{as} \quad X \to \infty$$

Again the 'infinite integral' *diverges*.

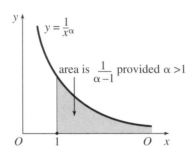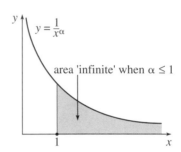

Example 18 _____

The series

$$\sum_{n=1}^{\infty} \frac{1}{n^\alpha}$$

is represented by
the shaded area
in the diagram.

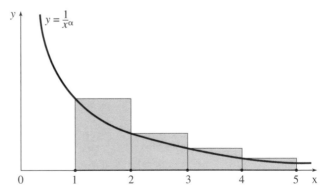

It is therefore not hard to see why the series is convergent if and only if the 'infinite integral'

$$\int_1^\infty \frac{\mathrm{d}x}{x^\alpha}$$

is convergent. Referring to the previous example, it follows that

$$\sum_{n=1}^{\infty} \frac{1}{n^\alpha}$$

converges when $\alpha > 1$ and diverges when $\alpha \leq 1$. In particular, the series

$$\sum_{n=1}^{\infty} \frac{1}{n}$$

[†] The sum of the series is $\pi^2/6$ but this is not easily proved.

diverges while the series[†]

$$\sum_{n=1}^{\infty} \frac{1}{n^2}$$

converges.

One sometimes has to deal with 'doubly infinite series' of the type

$$\sum_{n=-\infty}^{\infty} a_n$$

or with 'doubly infinite integrals' of the form

$$\int_{-\infty}^{\infty} f(x)\,dx$$

The 'doubly infinite integral' is said to converge if and only if there exists a number I such that[‡]

[‡] Note that it is important that X and Y be allowed to recede to ∞ independently.

$$\int_{-X}^{Y} f(x)\,dx \to I \quad \text{as} \quad X \to \infty \quad \text{and} \quad Y \to \infty$$

in which case we write

$$I = \int_{-\infty}^{\infty} f(x)\,dx$$

A similar definition applies in the case of 'doubly infinite series'.

Example 19

We have

$$\int_{-\infty}^{\infty} \frac{dx}{1+x^2} = \pi$$

[§] See the graph of $x = \arctan y$ given in §7.1.2.

because[§]

$$\int_{-X}^{Y} \frac{dx}{1+x^2} = [\arctan x]_{-X}^{Y}$$

$$= \arctan Y - \arctan(-X)$$

$$\to \frac{\pi}{2} - \left(-\frac{\pi}{2}\right) \quad \text{as } X \to \infty \text{ and } Y \to \infty$$

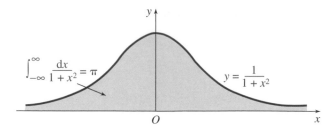

Note that

$$\int_{-\infty}^{\infty} \frac{x}{1+x^2}\, dx$$

diverges. We have

$$\int_{-X}^{Y} \frac{x}{1+x^2}\, dx \;=\; \left[\frac{1}{2}\ln(1+x^2)\right]_{-X}^{Y}$$

$$=\; \frac{1}{2}\{\ln(1+Y^2) - \ln(1+X^2)\}$$

But this tends to ∞ as $Y \to \infty$ and tends to $-\infty$ as $X \to \infty$.

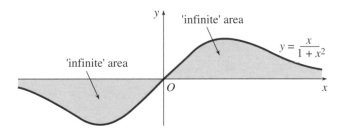

It is true that

$$\lim_{Z\to\infty} \int_{-Z}^{Z} \frac{x}{1+x^2}\, dx = 0$$

But this is not a significant fact since the upper and lower limits of integration need to recede to ∞ and $-\infty$ *independently* for the 'doubly infinite integral' to be defined.

10.12 Dominated convergence♣

We begin with the **comparison test**. In the case of integrals this takes the following form: suppose that $g(x) \geq 0$ for $x \geq \xi$ and that

$$\int_{\xi}^{\infty} g(x)\, dx$$

converges. Suppose that for all $x \geq \xi$

$$|f(x)| \leq g(x)$$

Then the 'infinite integral'

$$\int_{\xi}^{\infty} f(x)\,dx$$

Similar results hold for 'doubly infinite integrals' and for series.
converges also.

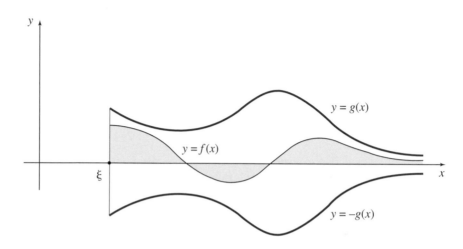

Example 20

We have

$$\int_{1}^{X} e^{-x}\,dx = [-e^{-x}]_{1}^{X} = e^{-1} - e^{-X}$$

$$\rightarrow e^{-1} \quad \text{as} \quad X \rightarrow \infty$$

It follows that the integral

$$\int_{1}^{\infty} e^{-x}\,dx$$

converges. But $x^2 \geq x$ for $x \geq 1$. Hence $-x^2 \leq -x$ for $x \geq 1$ and therefore

$$e^{-x^2} \leq e^{-x} \qquad (x \geq 1)$$

From the comparison test, it follows that the integral

$$\int_{1}^{\infty} e^{-x^2}\,dx$$

converges.

The convergence of

$$\int_{-\infty}^{\infty} e^{-x^2}\,dx$$

can be established in a similar way by splitting the range of integration into the three parts $[-\infty, -1]$, $[-1, 1]$ and $[1, \infty]$ and examining the integral over each of these intervals separately.

Suppose again that $g(x) \geq 0$ for $x \geq \xi$ and that

$$\int_{\xi}^{\infty} g(x)\, dx$$

converges. Suppose also that

$$|f(x, y)| \leq g(x) \tag{1}$$

for all $x \geq \xi$ and all y in some interval I for which η is an interior point. It follows from the comparison test that the integral

$$\int_{\xi}^{\infty} f(x, y)\, dx \tag{2}$$

converges for each y in the interval I. In view of the inequality (1), we say that the convergence is **dominated**.

It is by no means guaranteed that

$$\lim_{y \to \eta} \int_{\xi}^{\infty} f(x, y)\, dx = \int_{\xi}^{\infty} \lim_{y \to \eta} f(x, y)\, dx$$

even for those cases for which

$$\lim_{y \to \eta} f(x, y)$$

† See Exercise 10.15.5.

exists for each $x \geq \xi$.[†] However, if the convergence of (2) is *dominated*, then the reversal of the limit and integration signs *is* legitimate.

A similar problem which is more directly relevant to our concerns in this book is the question of the reversal of differentiation and integration signs. Suppose we wish to find under what circumstances it is true that

$$\frac{d}{dy} \int_{\xi}^{\infty} f(x, y)\, dx = \int_{\xi}^{\infty} \frac{\partial f}{\partial y}(x, y)\, dx$$

Here the relevant criterion is that the convergence of

$$\int_{\xi}^{\infty} \frac{\partial f}{\partial y}(x, y)\, dx \tag{3}$$

Similar results also hold for series.

be dominated.

Note that similar results hold for integrals over any range of integration (although when the limits of integration are finite, proving domination in (3) is usually too trivial to bother with).

Example 21

Consider

$$I(y) = \int_{0}^{\infty} \cos(xy)\, e^{-x^2/2}\, dx$$

We have

$$I'(y) = \frac{d}{dy} \int_0^\infty \cos(xy)\, e^{-x^2/2}\, dx$$

$$= \int_0^\infty \frac{\partial}{\partial y} \cos(xy)\, e^{-x^2/2}\, dx \qquad (4)$$

$$= \int_0^\infty -x \sin(xy)\, e^{-x^2/2}\, dx$$

Integrating by parts, we obtain

$$I'(y) = [e^{-x^2/2} \sin(xy)]_0^\infty - \int_0^\infty e^{-x^2/2}(y \cos xy)\, dx$$

$$= -yI(y)$$

Thus

$$\frac{I'(y)}{I(y)} = -y$$

and so

$$\int_0^Y \frac{I'(y)}{I(y)}\, dy = -\frac{1}{2}Y^2$$

$$\ln I(Y) - \ln I(0) = -\frac{1}{2}Y^2$$

$$I(Y) = I(0)\, e^{-Y^2/2}$$

In the next chapter we shall see that $I(0) = \sqrt{\pi/2}$ and so

$$I(Y) = \sqrt{\frac{\pi}{2}} e^{-Y^2/2}$$

Strictly speaking, we should justify the step leading to (4) although few applied mathematicians would give much attention to this point. One needs to show that the convergence of

$$\int_0^\infty -x \sin(xy)\, e^{-x^2/2}\, dx$$

is dominated for all y. This follows from the fact that

$$|-x \sin(xy)\, e^{-x^2/2}| \le xe^{-x^2/2} \qquad (x \ge 0)$$

10.13 Differentiating integrals♣

Suppose that

$$I(y) = \int_a^y f(x)\, dx$$

From the fundamental theorem of calculus,

$$\frac{dI}{dy} = \frac{d}{dy} \int_a^y f(x)\, dx = f(y)$$

If $y = g(z)$, we have

$$\frac{dI}{dz} = \frac{dI}{dy}\frac{dy}{dz} = f(y)g'(z) = f(g(z))g'(z)$$

Hence

$$\frac{d}{dz} \int_a^{g(z)} f(x)\, dx = f(g(z))g'(z)$$

On a somewhat different tack, we have from §10.12 that, under fairly general conditions,

$$\frac{d}{dy} \int_a^b f(x, y)\, dx = \int_a^b \frac{\partial f}{\partial y}(x, y)\, dx$$

These results may be combined to obtain the general formula

$$\frac{d}{dy} \int_a^{g(y)} f(x, y)\, dx = f(g(y), y)g'(y) + \int_a^{g(y)} \frac{\partial f}{\partial y}(x, y)\, dx$$

To prove this we write

$$I(u, v) = \int_a^u f(x, v)\, dx$$

See §5.4.

If $u = g(y)$ and $v = y$, it follows from the chain rule (III) that

$$\frac{dI}{dy} = \frac{\partial I}{\partial u}\frac{du}{dy} + \frac{\partial I}{\partial v}\frac{dv}{dy}$$

$$= f(u, v)g'(y) + \int_a^u \frac{\partial f}{\partial v}(x, v)\, dx$$

$$= f(g(y), y)g'(y) + \int_a^{g(y)} \frac{\partial f}{\partial y}(x, y)\, dx$$

Example 22

$$\frac{d}{dy} \int_0^{y^2} (x + y)^2\, dx = (y^2 + y)^2 2y + \int_0^{y^2} \frac{\partial}{\partial y}(x + y)^2\, dx$$

$$= (y^2 + y)^2 2y + \int_0^{y^2} 2(x + y)\, dx$$

$$= (y^4 + 2y^3 + y^2)2y + [x^2 + 2yx]_{x=0}^{y^2}$$

$$= 2y^5 + 4y^4 + 2y^3 + y^4 + 2y^3$$

$$= 2y^5 + 5y^5 + 4y^3$$

$$= y^3(2y^2 + 5y + 4)$$

10.14 Power series♣

A **power series** about the point ξ is an expression of the form

$$\sum_{n=0}^{\infty} a_n(x - \xi)^n$$

We met examples of power series with $\xi = 0$ in §2.13 when considering the Taylor series expansions of some elementary functions.

The set of values of x for which a power series converges is always an interval with midpoint ξ. This interval I is called the **interval of convergence** of the power series. This interval may be the set $\mathbb{R}$ of *all* real numbers or it may consist of *just* the single point ξ. Between these two extremes is the case when the end points are of the form $\xi - R$ and $\xi + R$ where $R > 0$. Note that the power series may *or may not* converge at these end points.

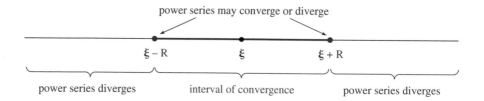

The number R is called the **radius of convergence** of the power series and can sometimes be found using the formulas,

$$\frac{1}{R} = \lim_{n \to \infty} \left| \frac{a_{n+1}}{a_n} \right| \qquad \frac{1}{R} = \lim_{n \to \infty} |a_n|^{1/n}$$

If I consists just of ξ, then $R = 0$. If I consists of all real numbers, then we say that R is 'infinite'. *The formulas given above for R remain valid provided one is willing to work with the conventions $1/0 = \infty$ and $1/\infty = 0$ but do not expect these conventions to work elsewhere.*

Differentiating and integrating power series is particularly easy. If a and b are any points in the interval of convergence (i.e. the power series converges for $x = a$ and $x = b$), then

$$\int_a^b \sum_{n=0}^{\infty} a_n(x - \xi)^n \, dx = \sum_{n=0}^{\infty} \int_a^b a_n(x - \xi)^n \, dx$$

and, if x is any point in the interior of the interval of convergence (*not* an end point)

$$\frac{d}{dx} \sum_{n=0}^{\infty} a_n(x - \xi)^n = \sum_{n=0}^{\infty} \frac{d}{dx} a_n(x - \xi)^n$$

These results may be proved using the idea of dominated convergence introduced in §10.12.

Example 23

From Example 16 we know that the power series

$$\sum_{n=0}^{\infty} x^n$$

converges to $(1-x)^{-1}$ for $|x| < 1$ and diverges for $|x| \geq 1$. Thus its interval of convergence is the set of all x satisfying $-1 < x < 1$ and its radius of convergence is $R = 1$.

If $-1 < y < 1$, it follows that

$$\int_0^y \frac{dx}{1-x} = \int_0^y \left(\sum_{n=0}^{\infty} x^n \right) dx = \sum_{n=0}^{\infty} \int_0^y x^n \, dx$$

Thus

$$[-\ln(1-x)]_0^y = \sum_{n=0}^{\infty} \frac{y^{n+1}}{(n+1)}$$

i.e.

$$\ln(1-y) = -y - \frac{y^2}{2} - \frac{y^3}{3} - \cdots \quad \text{provided } -1 < y < 1$$

Also, if $-1 < x < 1$,

$$\frac{d}{dx} \left(\frac{1}{1-x} \right) = \frac{d}{dx} \left(\sum_{n=0}^{\infty} x^n \right) = \sum_{n=0}^{\infty} \frac{d}{dx} x^n = \sum_{n=0}^{\infty} n x^{n-1}$$

i.e.

$$\frac{1}{(1-x)^2} = \sum_{n=0}^{\infty} n x^{n-1}$$

and so

$$\frac{x}{(1-x)^2} = x + 2x^2 + 3x^3 + 4x^4 + \cdots \quad \text{provided } -1 < x < 1$$

10.15 Exercises

1.*♣ The **Gamma function** is defined for $y > 0$ by

$$\Gamma(y) = \int_0^{\infty} x^{y-1} e^{-x} \, dx$$

Use Exercise 2.15.6.

Prove that the integral for $\Gamma(n)$ converges for each natural number $n = 1, 2, 3, \ldots$. Integrate by parts to show that

$$\Gamma(n+1) = n\Gamma(n) \qquad (n = 1, 2, 3, \ldots)$$

and hence prove that $\Gamma(n+1) = n!$.

2.♣ Explain why the integrals

$$\int_0^\infty (\cos bx)\, e^{-ax}\, dx \qquad \int_0^\infty (\sin bx)\, e^{-ax}\, dx$$

converge for all $a > 0$ and all b. Integrate twice by parts to prove that

(i) $\displaystyle \int_0^\infty (\cos bx)\, e^{-ax}\, dx = \frac{a}{a^2 + b^2}$

(ii) $\displaystyle \int_0^\infty (\sin bx)\, e^{-ax}\, dx = \frac{b}{a^2 + b^2}$

3.*♣ By considering

$$\frac{d^n}{dy^n} \int_0^1 x^y\, dx$$

prove that

$$\int_0^1 (\ln x)^n\, dx = (-1)^n n!$$

4.♣ By considering

$$\frac{d^n}{dy^n} \int_0^\infty e^{-yx}\, dx$$

See Exercise 10.15.1.

give an alternative proof that $\Gamma(n + 1) = n!$

5.*♣ Show that

$$\lim_{n\to\infty} \int_0^1 (n + 1)x^n\, dx \neq \int_0^1 \lim_{n\to\infty} (n + 1)x^n\, dx$$

6. Calculate

(i) $\displaystyle \frac{d}{dy} \int_0^{\ln y} \sin(xy^2)\, dx$ (ii) $\displaystyle \frac{d}{dy} \int_0^{y^3} \ln(x + y^2)\, dx$

using the formula of §10.13.

7.* Find a formula for

$$\frac{d}{dy} \int_{a(y)}^{b(y)} f(x, y)\, dx$$

8. A function f satisfies the integral equation

$$f(y) = \int_{-\infty}^y e^{x-y} f(x)\, dx$$

Prove that $f'(y) = 0$ and deduce that f is constant.

9.*♣ By integrating the equation

$$\frac{1}{1 + x^2} = 1 - x^2 + x^4 - x^6 + \cdots$$

show that

$$\arctan y = y - \frac{y^3}{3} + \frac{y^5}{5} - \frac{y^7}{7} + \cdots$$

provided $-1 < y < 1$.

10.♣ By differentiating appropriate power series, prove that

$$\frac{x(1+x)}{(1-x)^3} = 1^2 x + 2^2 x^2 + 3^2 x^3 + 4^2 x^4 + \cdots$$

provided $-1 < x < 1$.

11.*♣ Find the radii of convergence of the following power series:

(i) $\displaystyle\sum_{n=0}^{\infty} n^2 x^n$ (ii) $\displaystyle\sum_{n=0}^{\infty} 2^n x^n$ (iii) $\displaystyle\sum_{n=0}^{\infty} n^n x^n$

12.♣ Find the radii of convergence of the following power series:

(i) $\displaystyle\sum_{n=0}^{\infty} n^{-2} x^n$ (ii) $\displaystyle\sum_{n=0}^{\infty} 2^{-n} x^n$ (iii) $\displaystyle\sum_{n=0}^{\infty} n^{-n} x^n$

10.16 Applications (optional)

10.16.1 Probability

Suppose that it is not known for certain which of a number of possible events will occur. In certain circumstances it is sensible to quantify the uncertainty involved by attaching a **probability** to each possible event. Such a probability is a real number p satisfying $0 \le p \le 1$. To say that an event has probability .013, for example, means that, if the circumstances generating the event were to be repeated many times, then the event in question would be observed thirteen times in every thousand, on average.

Note that an impossible event has probability zero. But if there are an infinite number of events to be considered, it may well be that an event has zero probability without being impossible. For example, it is not impossible that a fair coin will *always* come down heads when tossed: but this event has *zero* probability.

The probability of an event E is denoted by $P(E)$.

There are various rules for manipulating probabilities. The most important of these are the two which follow.

(1) If E and F are two events which cannot *both* happen, then

$$P(E \text{ or } F) = P(E) + P(F).$$

(2) If E and F are independent events, then

$$P(E \text{ and } F) = P(E) \times P(F)$$

Consider, for example, two horses 'Punter's Folly' and 'Gambler's Ruin'. The probability that 'Punter's Folly' will win its race is $\frac{1}{2}$. The probability that 'Gambler's Ruin' will win its race is $\frac{1}{4}$. If the horses are running in the same race (and the possibility of a tie is ignored), then the probability that at least one of the two horses will win is

$$\frac{1}{2} + \frac{1}{4} = \frac{3}{4}$$

If the horses are running in different (and independent) races, then the probability that *both* will win is

$$\frac{1}{2} \times \frac{1}{4} = \frac{1}{8}$$

10.16.2 Probability density functions

A real number X whose value depends on which event out of a set of possible events actually happens is called a **random variable**. A good example of a random variable is the amount of money a gambler brings home from a casino after having used a prearranged betting system.

We shall first consider random variables X which can only take integer values (i.e. $0, \pm 1, \pm 2, \ldots$). Such random variables are called **discrete**. The function f defined by

$$f(x) = P(X = x)$$

is called the **probability density function** for the *discrete* random variable X. In order for f to qualify as a probability density function for X, it is necessary that

(i) $f(x) \geq 0 \qquad (x = 0, \pm 1, \pm 2, \ldots)$

(ii) $\displaystyle\sum_{x=-\infty}^{\infty} f(x) = 1$

The second condition expresses the requirement that it is certain that X will take one of the available values of x. Observe also that

$$P(a \leq X \leq b) = \sum_{x=a}^{b} f(x)$$

This latter observation leads us to the consideration of random variables which may take any real value. Such random variables are called **continuous**. If it is true that, for each $a \leq b$,

$$P(a \leq X \leq b) = \int_{a}^{b} f(x)\, dx$$

then the function f is called the **probability density function** for the *continuous* random variable X. In order that f qualifies as a probability density function for a continuous random variable X, it is necessary that

(i) $f(x) \geq 0 \qquad$ (all x)

(ii) $\displaystyle\int_{-\infty}^{\infty} f(x)\, dx = 1.$

The second condition expresses the requirement that it is certain that X will take some real value.

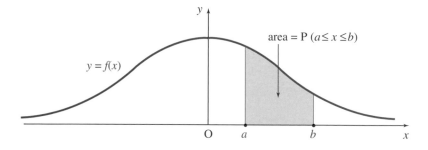

In both the discrete and the continuous case, the **probability distribution function** F for X is defined by $F(y) = P(X \leq y)$. In the discrete case

$$F(y) = \sum_{x=-\infty}^{y} f(x)$$

329

and in the continuous case,

$$F(y) = \int_{-\infty}^{y} f(x)\,dx$$

10.16.3 Binomial distribution

A weighted coin comes down heads with probability p and tails with probability $q = 1 - p$. The toss of such a coin is called a **Bernoulli** trial. A discrete random variable X is obtained by tossing the coin n times and observing the total number of heads which results. What is the probability density function for X?

We need to calculate $f(x) = P(X = x)$ for $x = 0, 1, 2, \ldots, n$. (If x is not one of these values, then $f(x) = 0$.) The tables below enumerate the possibilities in the case $n = 4$:

1. $X = 4$

   Result of 4 trials

HHHH

   Probability

pppp

2. $X = 3$

   Result of 4 trials

HHHT	HHTH	HTHH	THHH

   Probability

pppq	ppqp	pqpp	qppp

3. $X = 2$

   Result of 4 trials

HHTT	HTHT	HTTH	THHT	THTH	TTHH

   Probability

ppqq	pqpq	pqqp	qppq	qpqp	qqpp

4. $X = 1$

   Result of 4 trials

HTTT	THTT	TTHT	TTTH

   Probability

pqqq	qpqq	qqpq	qqqp

5. $X = 0$

   Result of 4 trials

TTTT

   Probability

qqqq

The probabilities attached to each possible event are calculated on the assumption that each trial is *independent* of the others. Since the events listed are mutually exclusive, we obtain for the case $n = 4$ that

$$P(X = 4) = pppp$$

$$P(X = 3) = pppq + ppqp + pqpp + qppp$$

$$P(X = 2) = ppqq + pqpq + pqqp + qppq + qpqp + qqpp$$

$$P(X = 1) = pqqq + qpqq + qqpq + qqqp$$

$$P(X = 0) = qqqq$$

Now consider the following proof of the binomial theorem in the case $n = 4$. We have

$$(p + q)^4 = (p + q)^2 \Big((p + q)(p + q) \Big)$$

$$= (p + q)(p + q)(pp + pq + qp + qq)$$

$$= (p + q)(ppp + ppq + pqp + pqq + qpp + qpq + qqp + qqq)$$

$$= (pppp) + (pppq + ppqp + pqpp + qppp)$$

$$+ (ppqq + pqpq + pqqp + qppq + qpqp + qqpp)$$

$$+ (pqqq + qpqq + qqpq + qqqp) + (qqqq)$$

$$= \binom{4}{0} p^4 + \binom{4}{1} p^3 q + \binom{4}{2} p^2 q^2 + \binom{4}{3} pq^3 + \binom{4}{4} q^4$$

Comparing the final two lines, we see that, in the case $n = 4$,

$$f(x) = P(X = x) = \binom{n}{x} p^x q^{n-x}$$

The same proof works for *all* values of n. (*Another way of obtaining the same result is by observing that*

$$\binom{n}{x}$$

is the total number of ways that x Hs and $n - x$ Ts can be arranged in a row.)

For obvious reasons, the random variable X (equal to the number of heads which appear in n trials) is said to have the **binomial distribution**. Note that $f(x) \geq 0$ $(x = 0, 1, 2, \ldots, n)$ and, since $p + q = 1$,

$$1 = (p + q)^n = \sum_{x=0}^{n} \binom{n}{x} p^x q^{n-x} = \sum_{x=0}^{n} f(x)$$

10.16.4 Poisson distribution

It often happens that it is necessary to consider a binomially distributed random variable in the case when n is very large. It is then often easier to work with a simple approximation to the binomial distribution rather than with the binomial distribution itself. Consider, for example, the following problem.

During the Second World War, the fall of rockets on London was to a large extent random. Suppose for simplicity that any two sites of equal area in London were equally likely to be hit. If n rockets fell on London and London is divided into n regions of equal area (*so that, on average, each region will be hit by one rocket*), what is the probability that a given region was actually hit exactly ten times?

The probability p of the given region being hit by any particular rocket is $p = 1/n$. The required probability is therefore

$$\binom{n}{10} \left(\frac{1}{n} \right)^{10} \left(\frac{n-1}{n} \right)^{n-10}$$

For large values of n this is clearly an awkward quantity. We therefore seek a simple approximation.

Stirling's approximation for $m!$ is

$$m! \sim \sqrt{2\pi} m^{m+1/2} e^{-m}$$

(*Here $\sim$ means that the ratio of the two sides is nearly one when m is large.*) We shall use this to obtain an approximation for

$$\binom{n}{x} p^x q^{n-x}$$

in the case when $p = \frac{\lambda}{n}$ (where λ is a constant). We have

$$\binom{n}{x} p^x q^{n-x} = \frac{n!}{x!(n-x)!} \left(\frac{\lambda}{n}\right)^x \left(1 - \frac{\lambda}{n}\right)^{n-x}$$

$$\sim \frac{\sqrt{(2\pi)} n^{n-1/2} e^{-n}}{x! \sqrt{(2\pi)} (n-x)^{n-x-1/2} e^{-(n-x)}} \left(\frac{\lambda}{n}\right)^x \left(1 - \frac{\lambda}{n}\right)^{n-x}$$

$$= \frac{\lambda^x}{x!} \frac{\left(1 - \dfrac{x}{n}\right)^{x+1/2} \left(1 - \dfrac{\lambda}{n}\right)^n}{\left(1 - \dfrac{x}{n}\right)^n e^x \left(1 - \dfrac{\lambda}{n}\right)^x}$$

$$\rightarrow \frac{\lambda^x}{x!} e^{-\lambda} \quad \text{as} \quad n \to \infty$$

For the last step, one needs to know that

$$\left(1 + \frac{y}{n}\right)^n \to e^y \quad \text{as} \quad n \to \infty$$

If λ is constant and n is large, then this argument shows that, if X is the number of 'successes' in n independent trials with the probability of 'success' in each individual trial equal to $p = \lambda/n$, then

$$P(X = x) \approx \frac{\lambda^x}{x!} e^{-\lambda}$$

In our particular problem, $\lambda = 1$ and $x = 10$ and so the required probability is approximately $1/(10! e)$

Observe that, if $\lambda > 0$, then the function

$$f(x) = \frac{\lambda^x e^{-\lambda}}{x!} \qquad (x = 0, 1, 2, \ldots)$$

qualifies as the probability density function for a discrete random variable because $f(x) \geq 0$ $(x = 0, 1, \ldots)$ and

$$\sum_{x=0}^{\infty} f(x) = \sum_{x=0}^{\infty} \frac{\lambda^x e^{-\lambda}}{x!}$$

$$= e^{-\lambda} \left(1 + \frac{\lambda}{1!} + \frac{\lambda^2}{2!} + \frac{\lambda^3}{3!} + \cdots\right)$$

$$= e^{-\lambda} e^{\lambda} = 1$$

A random variable Y with this probability density function is said to have a **Poisson distribution** with parameter λ. The random variable X of our example is therefore *approximately* Poisson.

10.16.5 Mean

The mean μ of a discrete random variable X with probability density function f is given by

$$\mu = \sum_{x=-\infty}^{\infty} x f(x).$$

Suppose that $X_1, X_2, X_3, \ldots$ are independent random variables, all of which have probability density function f. Such a sequence may be generated by repeating the trial which leads to X over and over again. We shall think of X_k as the amount a gambler wins at the kth trial. His *average* winnings over n trials will then be

$$A_n = \frac{X_1 + X_2 + \cdots + X_n}{n}$$

The law of large numbers asserts that

$$A_n \to \mu \quad \text{as} \quad n \to \infty$$

with probability one. (*This does not mean that $A_n \to \mu$ as $n \to \infty$ always: only that $A_n \to \mu$ as $n \to \infty$ happens too seldom to deserve a positive probability.*) Thus, in the long run, the gambler's average winnings will almost certainly be approximately μ. This leads us to call μ the **expectation** (or expected value) of the random variable X and we write

$$\mathcal{E}(X) = \mu = \sum_{x=-\infty}^{\infty} x f(x)$$

For a binomially distributed random variable, for example, we have

$$\mu = \sum_{x=0}^{n} x \binom{n}{x} p^x q^{n-x}$$

The value of μ can be calculated as follows. We have

$$(p+q)^n = \sum_{x=0}^{n} \binom{n}{x} p^x q^{n-x}$$

Hence

$$n(p+q)^{n-1} = \frac{\partial}{\partial p}(p+q)^n = \frac{\partial}{\partial p} \sum_{x=0}^{n} \binom{n}{x} p^x q^{n-x}$$

$$= \sum_{x=0}^{n} x \binom{n}{x} p^{x-1} q^{n-x}$$

and therefore

$$\mu = \sum_{x=0}^{n} x \binom{n}{x} p^x q^{n-x} = np(p+q)^{n-1} = np$$

The mean for the Poisson distribution is more easily calculated. We have

$$\mu = \sum_{x=0}^{\infty} x \frac{\lambda^x}{x!} e^{-\lambda} = e^{-\lambda} \sum_{x=1}^{\infty} \frac{\lambda^x}{(x-1)!}$$

$$= \lambda e^{-\lambda}\left(1 + \frac{\lambda}{1!} + \frac{\lambda^2}{2!} + \cdots\right) = \lambda e^{-\lambda} e^{\lambda} = \lambda$$

In the case of a *continuous* random variable X with probability density function f, we define the mean μ (or expectation $\mathcal{E}(X)$) by

$$\mathcal{E}(X) = \mu = \int_{-\infty}^{\infty} x f(x)\, dx$$

The interpretation is identical with that for the discrete case.

10.16.6 Variance

The **variance** var (X) of a random variable X is defined by

$$
\begin{aligned}
\mathrm{var}(X) &= \mathcal{E}(X - \mu)^2 \\
&= \mathcal{E}(X^2 - 2\mu X + \mu^2) \\
&= \mathcal{E}(X^2) - 2\mu\mathcal{E}(X) + \mu^2\mathcal{E}(1) \\
&= \mathcal{E}(X^2) - 2\mu^2 + \mu^2 \\
&= \mathcal{E}(X^2) - \mu^2
\end{aligned}
$$

See §11.7.3.

The variance is a measure of the degree of dispersion of the random variable X. It indicates the extent to which the probability density function clusters around the mean. For this reason, it is usual to write

$$\mathrm{var}(X) = \sigma^2$$

where the number σ is called the **standard deviation** of X. One would be surprised to find a random variable with small standard deviation taking values far from its mean and not at all surprised in the case of a random variable with large standard deviation.

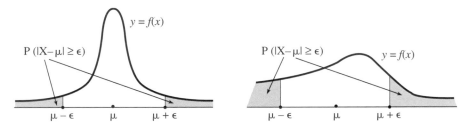

For a discrete random variable, the variance is given by

$$\sigma^2 = \sum_{x=-\infty}^{\infty} (x - \mu)^2 f(x)$$

$$= \left(\sum_{x=-\infty}^{\infty} x^2 f(x) \right) - \mu^2$$

In the case of a binomially distributed random variable we therefore have to calculate

$$\sum_{x=0}^{n} x^2 \binom{n}{x} p^x q^{n-x}$$

From §10.16.5,

$$np(p+q)^{n-1} = \sum_{x=0}^{n} x \binom{n}{x} p^x q^{n-x}$$

and so

$$n(n-1)p(p+q)^{n-2} = \frac{\partial}{\partial q} \sum_{x=0}^{n} x \binom{n}{x} p^x q^{n-x}$$

$$= \sum_{x=0}^{n} x(n-x) \binom{n}{x} p^x q^{n-x-1}$$

Thus

$$\sum_{x=0}^{n} x^2 \binom{n}{x} p^x q^{n-x} = n \sum_{x=0}^{n} x \binom{n}{x} p^x q^{n-x} - n(n-1)pq(p+q)^{n-2}$$

$$= n\mu - n(n-1)pq$$

It follows that

$$\sigma^2 = (n\mu - n(n-1)pq) - \mu^2$$

$$= n^2 p - n^2 pq + npq - n^2 p^2$$

$$= n^2 p(1-p) - n^2 pq + npq$$

$$= npq$$

The Poisson distribution is again more easily dealt with. We have

$$\sum_{x=0}^{\infty} x^2 \frac{\lambda^x}{x!} e^{-\lambda} = \sum_{x=0}^{\infty} x(x-1) \frac{\lambda^x}{x!} e^{-\lambda} + \sum_{x=0}^{\infty} x \frac{\lambda^x}{x!} e^{-\lambda}$$

$$= \lambda^2 e^{-\lambda} \sum_{x=2}^{\infty} \frac{\lambda^{x-2}}{(x-2)!} + \mu$$

$$= \lambda^2 + \lambda$$

Hence

$$\sigma^2 = (\lambda^2 + \lambda) - \mu^2 = \lambda$$

10.16.7 Standardised random variables

If a random variable X has mean μ and variance σ^2 and

$$Y = \frac{X - \mu}{\sigma}$$

then

$$\mathcal{E}(Y) = \mathcal{E}\left(\frac{X-\mu}{\sigma}\right) = \frac{1}{\sigma}\{\mathcal{E}(X) - \mu\} = 0$$

$$\text{var}(Y) = \mathcal{E}((Y-\sigma)^2) = \mathcal{E}\left(\left(\frac{X-\mu}{\sigma}\right)^2\right)$$

$$= \frac{1}{\sigma^2}\mathcal{E}((X-\mu)^2) = \frac{\sigma^2}{\sigma^2} = 1$$

It follows that the random variable Y has mean 0 and variance 1. Such a random variable is said to be **standardised**.

10.16.8 Normal distribution

In §10.16.4 we saw that the Poisson distribution provides a simple approximation to the binomial distribution in the case when n is large and $p \approx \lambda/n$ where λ is constant. A more pressing problem is to find a simple approximation for the case when n is large and p is constant.

Suppose that the probability density function for the discrete random variable X_n is binomial with parameters n and p. Consider the associated standardised random variable

$$Y_n = \frac{X_n - \mu_n}{\sigma_n} = \frac{X_n - pn}{\sqrt{npq}}$$

We shall illustrate the probability density function for Y_n by drawing a histogram. The histogram below is for Y_4 in the case $p = \frac{1}{4}$. The *area* of each rectangle represents the probability that Y_n is equal to the midpoint of its base. The mean and standard deviation for Y_4 in the case $p = \frac{1}{4}$ are 1 and $\sqrt{3}/2 \approx .87$ respectively. Thus, for example, since $(4 - 1)/\sqrt{3}/2) \approx 3.46$,

$$P(Y_4 = 3.46) = P(X_4 = 4) = \left(\frac{1}{4}\right)^4 \approx 0.0039$$

and thus the area of the rectangle whose base has midpoint 3.46 is approximately equal to 0.0039.

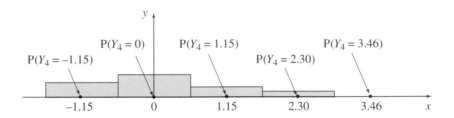

The histograms below (drawn to a different scale) show the probability density functions for Y_{10}, Y_{30} and Y_{90} in the cases $p = \frac{1}{5}$ and $p = \frac{1}{2}$.

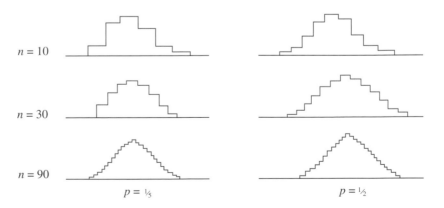

In both cases the histograms approach a smooth bell shaped curve as n becomes large:

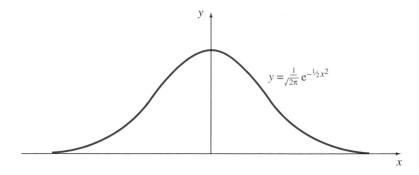

In fact, for *all* values of p with $0 < p < 1$, the appropriate histograms approach the *same* curve and this curve has equation

$$y = \frac{1}{\sqrt{2\pi}} e^{-x^2/2}$$

This can be proved with an argument similar to that used in §10.16.4.

The area shaded in the histogram for Y_n drawn below is equal to $P(Y_n \leq z)$.

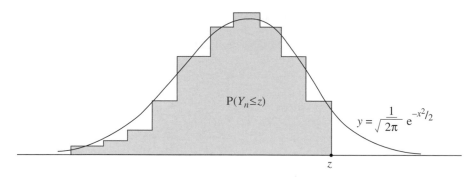

From §10.2, it follows that $P(Y_n \leq z)$ is

$$P(Y_n \leq z) \approx \frac{1}{\sqrt{2\pi}} \int_{-\infty}^{z} e^{-x^2/2} \, dx$$

The latter quantity, of course, is that illustrated below:

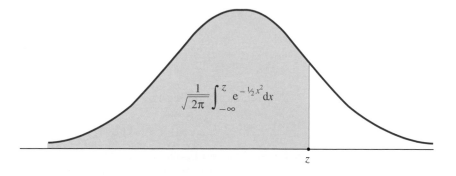

If $\sigma > 0$, a continuous random variable X with probability density function

$$f(x) = \frac{1}{\sigma\sqrt{2\pi}} e^{-\frac{1}{2}\left(\frac{x-\mu}{\sigma}\right)^2}$$

is said to have the **normal distribution**. The mean of this random variable is

$$\mathcal{E}(X) = \frac{1}{\sigma\sqrt{2\pi}} \int_{-\infty}^{\infty} x e^{-\frac{1}{2}(\frac{x-\mu}{\sigma})^2} \, dx$$

$$= \frac{1}{\sqrt{2\pi}} \int_{-\infty}^{\infty} (\sigma t + \mu) e^{-t^2/2} \, dt$$

$$= \frac{\sigma}{\sqrt{2\pi}} \int_{-\infty}^{\infty} t e^{-t^2/2} \, dt + \frac{\mu}{\sqrt{2\pi}} \int_{-\infty}^{\infty} e^{-t^2/2} \, dt$$

where we have introduced the change of variable $x = \sigma t + \mu$ which makes $dx = \sigma \, dt$. But

$$\int_{-T_1}^{T_2} t e^{-t^2/2} \, dt = [-e^{-t^2/2}]_{-T_1}^{T_2} = e^{-T_1^2/2} - e^{-T_2^2/2}$$

$$\to 0 \quad \text{as} \quad T_1 \to \infty \quad \text{and} \quad T_2 \to \infty$$

and it is shown in the next chapter that

$$\int_{-\infty}^{\infty} e^{-t^2/2} \, dt = \sqrt{2\pi} \tag{1}$$

It follows that $\mathcal{E}(X) = \mu$ (as one might reasonably expect from the use of the symbol μ). Similarly $\text{var}(X) = \sigma^2$. To prove this, we begin by making the change of variable $t = yu$ in (1). Then

$$\int_{-\infty}^{\infty} e^{-y^2 u^2/2} \, du = \frac{\sqrt{2\pi}}{y}$$

Hence

$$\frac{-\sqrt{2\pi}}{y^2} = \frac{d}{dy} \int_{-\infty}^{\infty} e^{-y^2 u^2/2} \, du = \int_{-\infty}^{\infty} \frac{\partial}{\partial y} e^{-y^2 u^2/2} \, du$$

$$= \int_{-\infty}^{\infty} -yu^2 e^{-y^2 u^2/2} \, du$$

Taking $y = 1$, we obtain

$$\int_{-\infty}^{\infty} u^2 e^{-u^2/2} \, du = \sqrt{2\pi}$$

Thus

$$\text{var}(X) = \mathcal{E}((X - \mu)^2)$$

$$= \frac{1}{\sigma\sqrt{2\pi}} \int_{-\infty}^{\infty} (x - \mu)^2 e^{-\frac{1}{2}(\frac{x-\mu}{\sigma})^2} \, dx$$

$$= \frac{1}{\sqrt{2\pi}} \int_{-\infty}^{\infty} \sigma^2 t^2 e^{-t^2/2} \, dt = \sigma^2$$

The function n defined by

$$n(x) = \frac{1}{\sqrt{2\pi}} e^{-x^2/2}$$

is therefore the probability density function of a 'standardised' normal distribution. Returning to the problem of approximating binomial random variables, we see that the 'standardised' binomial random variable Y_n is approximately the 'standardised' normal random variable Y provided n is large – i.e.

$$P(Y_n \leq z) \approx P(Y \leq z)$$

Suppose that one has to calculate $P(W \le w)$ where W is a binomial random variable with parameters p and n. One should then note that

$$P(W \le w) = P\left(\frac{W - np}{\sqrt{npq}} \le \frac{w - np}{\sqrt{npq}}\right)$$

$$= P\left(Y \le \frac{w - np}{\sqrt{npq}}\right) = \frac{1}{\sqrt{2\pi}} \int_{-\infty}^{\frac{w-np}{\sqrt{(npq)}}} e^{-x^2/2} \, dx$$

Since tables for

$$N(z) = \frac{1}{\sqrt{2\pi}} \int_{-\infty}^{z} e^{-x^2/2} \, dx$$

are readily available, this means that it is easy to obtain an approximation to $P(W \le w)$. How good is this approximation? Our explanation only indicates that it is good for large values of n but the further p is from $\frac{1}{2}$, the larger n has to be. A useful rule of thumb is that the approximation will be adequate for practical purposes when $\sigma_n^2 = np(1 - p) > 10$. (If $p = \frac{1}{2}$, this means that $n > 40$.)

10.16.9 Sums of random variables

If X and Y are two random variables it is *always* the case that

$$\mathcal{E}(X + Y) = \mathcal{E}(X) + \mathcal{E}(Y)$$

If X and Y are *independent* random variables, then it is also true that

$$\mathcal{E}(XY) = \mathcal{E}(X)\mathcal{E}(Y)$$

See §11.7.3 and §11.7.4.

Now suppose that X_1, X_2, $X_3, \ldots$ is a sequence of independent random variables each of which has the same probability distribution with mean μ and variance σ^2. We can then easily calculate the mean and variance of the random variable

$$S_n = X_1 + X_2 + \cdots + X_n$$

We have

$$\mathcal{E}(S_n) = \mathcal{E}(X_1 + X_2 + \cdots + X_n)$$

$$= \mathcal{E}(X_1) + \mathcal{E}(X_2) + \cdots + \mathcal{E}(X_n)$$

$$= n\mu$$

It is not difficult to prove that

$$\text{var}(S_n) = \text{var}(X_1) + \text{var}(X_2) + \cdots + \text{var}(X_n) = n\sigma^2$$

using the case $n = 2$. The proof is left as an exercise. Note that

$$\mathcal{E}((X_1 + X_2)^2) = \mathcal{E}(X_1^2 + X_2^2 + 2X_1X_2)$$

$$= \mathcal{E}(X_1^2) + \mathcal{E}(X_2^2) + 2\mathcal{E}(X_1)\mathcal{E}(X_2)$$

$$= \text{var}(X_1) + (\mathcal{E}(X_1))^2 + \text{var}(X_2) + (\mathcal{E}(X_2))^2 + 2\mathcal{E}(X_1)\mathcal{E}(X_2)$$

$$= \text{var}(X_1) + \text{var}(X_2) + (\mathcal{E}(X_1))^2 + 2\mathcal{E}(X_1)\mathcal{E}(X_2) + (\mathcal{E}(X_2))^2$$

$$= \text{var}(X_1) + \text{var}(X_2) + (\mathcal{E}(X_1 + X_2))^2$$

Thus
$$\text{var}(X_1 + X_2) = \text{var}(X_1) + \text{var}(X_2)$$

Suppose, for example, that a weighted coin is tossed n times with probability p on each occasion that it comes down heads and $1 - p = q$ that it comes down tails. Let X_k be 1 if the kth trial produces a head and 0 if it produces a tail. Then

$$
\begin{aligned}
\mu &= \mathcal{E}(X_k) = 1p + 0(1 - p) = p \\
\sigma^2 &= \mathcal{E}(X_k^2) - \mu^2 \\
&= \{1^2 p + 0^2(1 - p)\} - \mu^2 = p - p^2 = pq
\end{aligned}
$$

From the previous discussion it follows that the random variable

$$S_n = X_1 + X_2 + \cdots + X_n$$

has mean and variance

$$
\left.
\begin{aligned}
\mathcal{E}(S_n) &= n\mu = np \\
\text{var}(S_n) &= n\sigma^2 = npq
\end{aligned}
\right\}
$$

But S is of course a binomial random variable with parameters p and n. These calculations therefore provide an alternative method of computing the mean and variance of a binomial random variable.

Returning to our arbitrary sequence $X_1, X_2, \ldots$ of independent random variables, all of which have the same probability distribution, we introduce the 'standardised' random variable

$$T_n = \frac{S_n - n\mu}{\sigma\sqrt{n}}$$

It is a remarkable fact that under quite mild conditions it is *always* true that

$$P(T_n \le t) = \frac{1}{\sqrt{2\pi}} \int_{-\infty}^{t} e^{-x^2/2}\,dx$$

– i.e. the random variable T is approximately equal to the 'standardised' normal random variable Y when n is large. This result is called the **central limit theorem**. It remains valid even when $X_1, X_2, \ldots$ do not have quite the same probability distribution. Since random variables observed in practice often arise from the superposition of large numbers of random errors, the central limit theorem makes it clear why many random variables turn out to be normally distributed. The use of the word 'normal' indicates just how pervasive these random variables are.

Going back to our example in which X_k is the random variable which is equal to one if the kth toss of a weighted coin yields heads and equal to zero if tails, we note that the discussion of §10.16.8 is concerned with a simple special case of the central limit theorem. As an application, observe that

$$
\begin{aligned}
P\left(\frac{S_n}{n} - p \le -\epsilon\right) &= P(S_n - np \le -n\epsilon) \\[2ex]
&= P\left(\frac{S_n - np}{\sqrt{npq}} \le -\epsilon\sqrt{\frac{n}{pq}}\right) \\[2ex]
&\approx \frac{1}{\sqrt{2\pi}} \int_{-\infty}^{-\epsilon\sqrt{(n/pq)}} e^{-x^2/2}\,dx \\[2ex]
&\le \frac{1}{\sqrt{2\pi}} \int_{-\infty}^{-\epsilon\sqrt{n/pq}} \frac{2}{x^2}\,dx = \frac{2}{\epsilon}\sqrt{\frac{pq}{n}}
\end{aligned}
$$

because $e^y = 1 + y + \frac{1}{2}y^2 + \cdots \geq y \; (y \geq 0)$. A similar inequality may be obtained for

$$P\left(\frac{S_n}{n} - p \geq \epsilon\right)$$

and it follows that

$$P\left(-\epsilon < \frac{S_n}{n} - p < \epsilon\right) > 1 - \frac{4}{\epsilon}\sqrt{\frac{pq}{n}}$$

provided that n is large.

Thus, for example, if $p = \frac{1}{4}$, then the probability that the average number of heads in 100 000 000 trials will be within .01 of the probability $p = \frac{1}{4}$ of obtaining heads in one trial is at least .98. This result should be considered in conjunction with the remarks concerning the law of large numbers made in §10.16.5.

10.16.10   Cauchy distribution

A continuous random variable X with the probability density function

$$f(x) = \frac{1}{\pi}\frac{1}{1+x^2}$$

is said to have the Cauchy distribution. We mention this distribution only to point out that it *does not have a mean*. As pointed out in Example 19, the integral

$$\frac{1}{\pi}\int_{-\infty}^{\infty}\frac{x}{1+x^2}\,dx$$

diverges. For random variables which share this property with X, the law of large numbers *does not hold*.

10.16.11   Auctions

One of the triumphs of game theory in recent years has been its use in designing auctions. Telecom auctions, in which governments sell licences to use particular bands of radiospectrum in their territories, have proved a spectacular success. A British telecom auction that one of us helped to design raised more than $35 billion! However, we shall confine our attention to two more traditional auctions here – the English and the Dutch.

A house is offered for sale. The house is worth nothing to the seller if left unsold. There are n potential buyers who each value the house at something between 0 and 1 (million dollars). It is common knowledge that each valuation $v_i \; (i = 1, 2, \ldots, n)$ is drawn independently from a uniform probability distribution on $[0, 1]$, so that

$$P(a < v_i \leq b) = b - a$$

provided that $0 \leq a < b \leq 1$. However, only bidder i knows the precise value of the v_i assigned to him.

The seller would like to sell the house in a way that maximises her expected revenue. She therefore considers using a traditional English auction in which the price rises until only one bidder is left. He is then sold the house at that price. The strategic problem for a buyer is then very simple. Bidder i should stay in the auction until the price reaches his valuation v_i, when he should bail out. The house will then be sold to the buyer who has drawn the highest valuation at a price equal to the *second highest* valuation drawn by a buyer.

To find the seller's expected revenue when all the buyers bid rationally, we begin by finding the probability that the house is sold at a price between x and

$x + \delta x$. There are n ways of choosing the buyer with the highest valuation. After he has been chosen, there are $n - 1$ ways of choosing the second highest bidder. The probability we are seeking is therefore approximately

$$n(n-1)P\,(x + \delta x < v_1 \le 1 \,\&\, x < v_2 \le x + \delta x \,\&\, 0 < v_3 \le x \,\&\, \cdots \,\&\, 0 < v_n \le x)$$
$$\approx n(n-1)(1-x)\delta x\, x^{n-2}$$

when δx is sufficiently small. It follows that the seller's expected revenue with an English auction is

$$n(n-1)\int_0^1 x(1-x)x^{n-2}\,\mathrm{d}x = n(n-1)\left(\int_0^1 x^{n-1}\,\mathrm{d}x - \int_0^1 x^n\,\mathrm{d}x\right)$$
$$= n(n-1)\left(\frac{1}{n} - \frac{1}{n+1}\right) = \frac{n-1}{n+1}$$

Can the seller do better with a Dutch auction, in which the price starts high and *decreases* until a buyer calls a halt? The strategic problem for a buyer in a Dutch auction is more complicated than in an English auction. He certainly won't shout stop when the price reaches his valuation. He will hold on a little longer, hoping to win the auction at a lower price.

Suppose that a rational buyer with valuation v will stop a Dutch auction when the price gets down to $b(v) < v$. If so, then the probability that buyer 1 will win if he plans to stop the auction when the price reaches p is

$$= P\Big(b(v_2) < p \quad \& \quad b(v_3) < p \quad \& \quad \cdots \quad \& \quad b(v_n) < p\Big)$$
$$= P\Big(v_2 < b^{-1}(p) \quad \& \quad v_3 < b^{-1}(p) \quad \& \quad \cdots \quad \& \quad v_n < b^{-1}(p)\Big)$$
$$= \Big(c(p)\Big)^{n-1}$$

where $c = b^{-1}$ is the inverse function to b, provided that b is a strictly *decreasing* function (so that a bidder with a lower valuation plans to stop the auction at a lower price).

If he plans to stop the auction at price p, buyer 1's expected profit is therefore

$$(v_1 - p)\Big(c(p)\Big)^{n-1}$$

If the buyers are trying to maximise expected profit, it follows that

$$\frac{\mathrm{d}}{\mathrm{d}p}\Big((v_1 - p)\big(c(p)\big)^{n-1}\Big) = 0 \tag{2}$$

when $p = b(v_1)$ because $b(v)$ is the price at which a rational bidder with valuation v plans to stop the auction. Differentiating (2) as a product and then writing $v_1 = c(p)$, we are led to the differential equation

$$(n-1)(c-p)\frac{\mathrm{d}c}{\mathrm{d}p} = c \tag{3}$$

Solving differential equations scientifically will have to wait until Chapter 12. Here we shall simply make the inspired guess that a solution takes the form $c = kp$, where k is a constant. Substituting in (3), we obtain

See Exercise 12.13.19.

$$(n-1)(kp - p)k = kp$$
$$(n-1)(k-1) = 1$$
$$k = \frac{n}{n-1}$$

Recall that $p = b(v)$ is what we get by solving the equation $v = c(p) = b^{-1}(p)$ for p in terms of v. When $v = c = kp$, this is a particularly easy task and we find that $p = b(v) = v/k$. With our assumptions, a rational buyer with valuation v therefore plans to stop the auction when the price gets down to

$$p = \left(\frac{n-1}{n}\right) v$$

Having solved the buyer's strategic problem in our Dutch auction, we can now ask what the seller's expected revenue will be. The probability that she receives between x and $x + \delta x$ is approximately

$$n\, P\left(x < \left(\frac{n-1}{n}\right) v_1 \leq x + \delta x \ \& \ \left(\frac{n-1}{n}\right) v_2 \leq x \& \cdots \& \left(\frac{n-1}{n}\right) v_n \leq x\right)$$

$$\approx n \left(\frac{n}{n-1}\right)^n \delta x\, x^{n-1}$$

The seller's expected revenue in a Dutch auction is therefore

$$n \left(\frac{n}{n-1}\right)^n \int_0^{\frac{n-1}{n}} x \cdot x^{n-1}\, dx = n \left(\frac{n}{n-1}\right)^n \frac{1}{n+1} \left(\frac{n-1}{n}\right)^{n+1}$$

$$= \frac{n-1}{n+1}$$

The seller does no better and no worse with a Dutch auction than with an English auction. This conclusion typifies a whole class of results in auction theory called revenue equivalence theorems.

Eleven	**Multiple integrals**

This chapter is chiefly concerned with three techniques – i.e. evaluating repeated integrals, changing the order of integration in repeated integrals and changing variables in multiple integrals. The important feature of each technique is the necessity of ensuring that the correct limits of integration are used. This is fairly straightforward in the two variable cases provided one is not tempted to hurry things along by such short cuts as dispensing with a diagram.

The theoretical explanation of the formula for changing variables given in §11.3 should be omitted if this is found troublesome.

Techniques similar to those which work for multiple integrals also work for multiple sums except that things are usually rather easier in the latter case. Some discussion of multiple sums and series has been given in §11.5 but this should certainly be omitted if §10.14 has not been studied.

11.1 Introduction

Suppose that $f: \mathbb{R}^2 \to \mathbb{R}$ and that D is some region in $\mathbb{R}^2$. Then the double integral

$$\iint_D f(x, y) \, dx \, dy$$

may be interpreted as the volume beneath the surface $z = f(x, y)$ and above the region D.

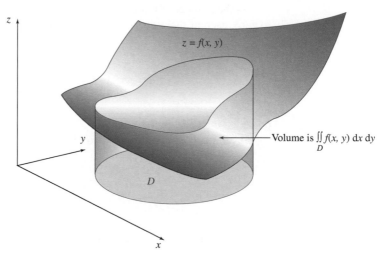

See §10.2.

As in the case of simple integrals, volume beneath the $(x, y)^{\mathrm{T}}$ plane is counted negative.

In the following diagram, the region D has been split up into a large number of small regions. The shaded region has area A_i and contains the point $(x_i, y_i)^{\mathrm{T}}$. Thus

$$f(x_i, y_i) A_i$$

is an approximation to the volume of the 'obelisk' drawn and hence

$$\sum_i f(x_i, y_i) A_i$$

is an approximation to the volume under the surface.

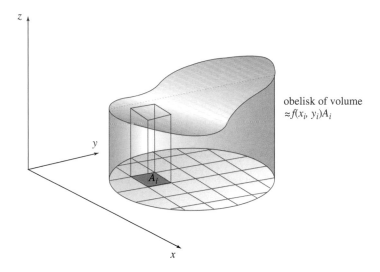

obelisk of volume
$\approx f(x_i, y_i) A_i$

As the number of regions increases and the area of each individual region shrinks to zero, each of these approximating sums approaches

$$\iint_D f(x, y) \, \mathrm{d}x \, \mathrm{d}y$$

provided that f is continuous.

Similar considerations apply to higher dimensional multiple integrals

$$\iint_D \cdots \int f(x_1, x_2, \ldots, x_n) \, \mathrm{d}x_1 \, \mathrm{d}x_2 \ldots \mathrm{d}x_n$$

except that in this case the geometry of the situation is not so easily visualised.

11.2 Repeated integrals

The most straightforward method of evaluating a double integral is to think of the region D as divided up into little rectangles as in the diagram.

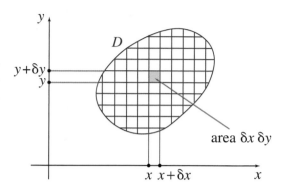

We then begin by considering each row of little rectangles separately.

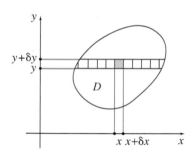

top view

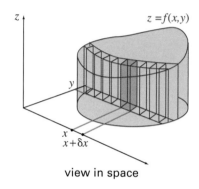

side view

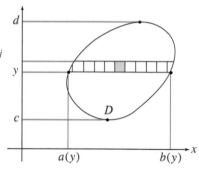

view in space

The volume above the row of the little rectangles in D between y_j and $y_j + \delta y_j$ is approximately

$$\sum_i f(x_i, y_j)\delta x_i \delta y_j = \left\{ \sum_i f(x_i, y_j)\delta x_i \right\} \delta y_j$$

This in turn is approximately

$$\left\{ \int_{a(y_j)}^{b(y_j)} f(x, y_j)\,dx \right\} \delta y_j$$

where the limits of integration are as indicated in the figure.

Having estimated the volumes obtained from each separate row, we can now estimate the total volume by adding these up. An approximation to the volume under the surface is therefore

$$\sum_j \left\{ \int_{a(y_j)}^{b(y_j)} f(x, y_j)\,dx \right\} \delta y_j$$

This in turn is approximately equal to

$$\int_c^d \left\{ \int_{a(y)}^{b(y)} f(x, y) \, dx \right\} dy$$

It is more convenient to omit the brackets and write

$$\int_c^d \int_{a(y)}^{b(y)} f(x, y) \, dx \, dy$$

All of these approximations improve as the number of little rectangles increases and the area of each individual rectangle shrinks to zero. It follows that

$$\iint_D f(x, y) \, dx \, dy = \int_c^d \int_{a(y)}^{b(y)} f(x, y) \, dx \, dy$$

Thus the double integral may be evaluated by regarding it as a repeated integral. We first calculate

$$I(y) = \int_{a(y)}^{b(y)} f(x, y) \, dx$$

for each y ($c \leq y \leq d$) and then evaluate

$$\int_c^d I(y) \, dy$$

One can equally well, of course, begin by considering contributions from each column separately (instead of each row). A similar argument then shows that

$$\iint_D f(x, y) \, dx \, dy = \int_a^b \int_{c(x)}^{d(x)} f(x, y) \, dy \, dx$$

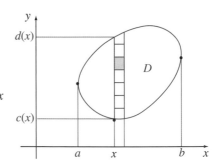

where the limits of integration are as indicated in the figure.

This means, in particular, that

$$\int_c^d \int_{a(y)}^{b(y)} f(x, y) \, dx \, dy = \int_a^b \int_{c(x)}^{d(x)} f(x, y) \, dy \, dx$$

Example 1 ───

Calculate the mass of the flat rectangular plate indicated on the right given that its density at the point $(x, y)^{\mathrm{T}}$ is $\rho(x, y) = x^2 y^3 + 3x$.

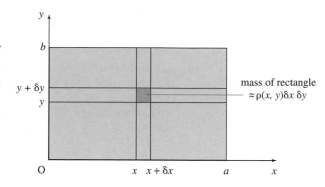

The problem reduces to evaluating

$$M = \iint_D \rho(x, y)\, dx\, dy = \iint_D (x^2 y^3 + 3x)\, dx\, dy$$

We can arrange this double integral as a repeated integral in two ways. Only one of these repeated integrals need be calculated. However, we give both calculations to check that the answers are equal.

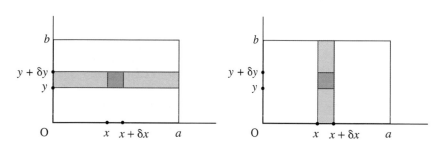

$$M = \int_0^b \int_0^a (x^2 y^3 + 3x)\, dx\, dy \qquad M = \int_0^a \int_0^b (x^2 y^3 + 3x)\, dy\, dx$$

$$= \int_0^b \left(\frac{1}{3} a^3 y^3 + \frac{3}{2} a^2 \right) dy \qquad = \int_0^a \left(\frac{1}{4} x^2 b^4 + 3xb \right) dx$$

$$= \left[\frac{1}{12} a^3 y^4 + \frac{3}{2} a^2 y \right]_0^b \qquad = \left[\frac{1}{12} x^3 b^4 + \frac{3}{2} x^2 b \right]_0^a$$

$$= \frac{1}{12} a^3 b^4 + \frac{3}{2} a^2 b \qquad = \frac{1}{12} a^3 b^4 + \frac{3}{2} a^2 b$$

349

Example 2

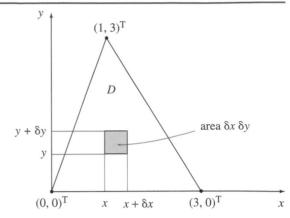

The case when D is a rectangle is particularly easy. We therefore next consider the problem of calculating the area of the triangle illustrated on the right.

To make the problem interesting we shall feign ignorance of the formula

$$\frac{1}{2}(\text{base} \times \text{height})$$

and instead evaluate the double integral

$$A = \iint\limits_{D} \mathrm{d}x \, \mathrm{d}y$$

Again we consider both repeated integrals. Note that, as in Example 1, we begin by drawing diagrams. Without the aid of such diagrams it is easy to make a mistake when determining the limits of integration in the repeated integrals.

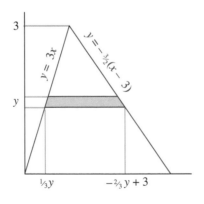

(i) By rows

(ii) By columns

In the second case it is best to proceed by splitting A into two areas A_1 and A_2 as indicated above.

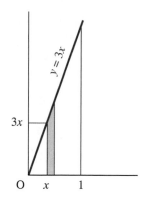

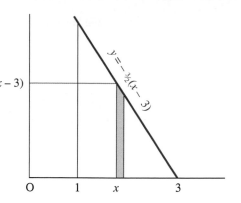

$$A = \int_0^3 \int_{y/3}^{-2y/3+3} dx\, dy \qquad\qquad A = A_1 + A_2$$

$$= \int_0^3 [x]_{y/3}^{-2y/3+3} dy \qquad\qquad = \int_0^1 \int_0^{3x} dy\, dx + \int_1^3 \int_0^{-3(x-3)/2} dy\, dx$$

$$= \int_0^3 \left(-\frac{2}{3}y + 3 - \frac{1}{3}y\right) dy \qquad = \int_0^1 [y]_0^{3x} dx + \int_1^3 [y]_0^{-3(x-3)/2} dx$$

$$= \int_0^3 (3 - y)\, dy \qquad\qquad = \int_0^1 3x\, dx + \int_1^3 -\frac{3}{2}(x - 3)\, dx$$

$$= \left[3y - \frac{1}{2}y^2\right]_0^3 \qquad\qquad = \left[\frac{3}{2}x^2\right]_0^1 + \left[-\frac{3}{2}\left(\frac{1}{2}x^2 - 3x\right)\right]_1^3$$

$$= 9 - \frac{9}{2} = 4\frac{1}{2} \qquad\qquad = \frac{3}{2}\left(1 + \frac{9}{2} - \frac{5}{2}\right) = 4\frac{1}{2}$$

Example 3

Evaluate

$$I = \iint_D xy\, dx\, dy$$

when D is the region indicated in the diagram.

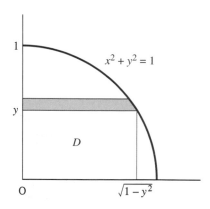

We have

$$I = \int_0^1 \int_0^{\sqrt{1-y^2}} xy \, dx \, dy$$

$$= \int_0^1 y \int_0^{\sqrt{1-y^2}} x \, dx \, dy$$

$$= \int_0^1 y \left[\frac{1}{2}x^2 \right]_0^{\sqrt{1-y^2}} dy$$

$$= \int_0^1 \frac{1}{2} y(1 - y^2) \, dy$$

$$= \left[\frac{1}{4}y^2 - \frac{1}{8}y^4 \right]_0^1 = \frac{1}{4} - \frac{1}{8} = \frac{1}{8}$$

Example 4

Change the order of integration in the repeated integral

$$J = \int_0^a \int_0^{a-\sqrt{a^2-x^2}} \frac{xe^y}{(y-a)^2} \, dy \, dx$$

where $a > 0$ and hence evaluate it.
Note that it is quite intractable in its present form.

The first and vital step is to determine the region D over which we are integrating. For a fixed value of x, y ranges between 0 and $a - \sqrt{a^2 - x^2}$. This observation leads us to the left hand diagram below. What we have to do is to rearrange the integral as indicated in the right hand diagram.

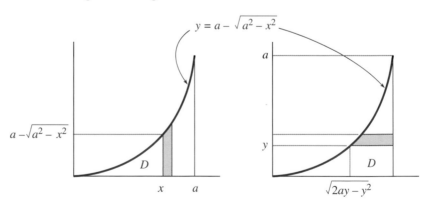

Solving the equation $y = a - \sqrt{a^2 - x^2}$ for x yields that $x = \sqrt{2ay - y^2}$. Thus when we reverse the order of integration we obtain that

$$J = \int_0^a \int_{\sqrt{2ay-y^2}}^a \frac{xe^y}{(y-a)^2} \, dx \, dy$$

$$= \int_0^a \frac{e^y}{(y-a)^2} \int_{\sqrt{2ay-y^2}}^a x \, dx \, dy$$

$$= \int_0^a \frac{e^y}{(y-a)^2} \left[\frac{1}{2}x^2\right]_{\sqrt{2ay-y^2}}^a dy$$

$$= \int_0^a \frac{e^y}{(y-a)^2} \left\{\frac{1}{2}a^2 - \frac{1}{2}(2ay - y^2)\right\} dy$$

$$= \int_0^a \frac{e^y}{(y-a)^2} \frac{1}{2}(y-a)^2 \, dy = \frac{1}{2}\int_0^a e^y \, dy = \frac{1}{2}(e^a - 1)$$

11.3 Change of variable in multiple integrals♣

In §7.3 we considered the question of changing variables in $\mathbb{R}^2$, and in §7.5 we saw the effect of it on differential operators. In this section we shall examine how such a change of variable transforms a multiple integral. Recalling the success which this technique enjoys for the evaluation of one dimensional integrals, we should anticipate that this will prove a profitable enterprise. For the one dimensional integral $\int_a^b f(x)\,dx$, if we let $x = g(u)$ so that $dx = g'(u)\,du$, we obtain

The transformed integral has a factor $g'(u)$. We shall see that a transformed multiple integral has a similar factor.

$$\int_a^b f(x)\,dx = \int_c^d f(g(u))g'(u)\,du$$

where $c = g^{-1}(a)$ and $d = g^{-1}(b)$. This clearly requires that g be the inverse of a function defined on the interval $[a, b]$ (*see* §10.8).

Considering integrals in $\mathbb{R}^2$, suppose that we wish to introduce the change of variable defined by

$$\begin{pmatrix} u \\ v \end{pmatrix} = \begin{pmatrix} h_1(x, y) \\ h_2(x, y) \end{pmatrix} \quad (= \mathbf{h}(x, y)) \tag{1}$$

into the double integral

$$\iint_D f(x, y)\,dx\,dy \tag{2}$$

Now suppose that the function $\mathbf{h}$ maps the region D onto the region Δ.

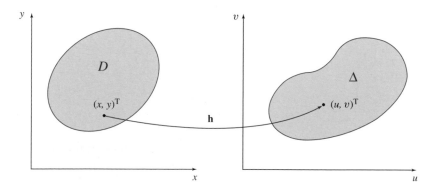

As explained in §7.3 it is necessary when changing coordinate systems in $\mathbb{R}^2$ to restrict our attention to the case when $\mathbf{h}: D \to \Delta$ has an inverse function $\mathbf{g}: \Delta \to D$:

$$\begin{pmatrix} x \\ y \end{pmatrix} = \mathbf{g}(u, v) = \begin{pmatrix} g_1(u, v) \\ g_2(u, v) \end{pmatrix}$$

We divide the region Δ into a large number of small rectangles. The lines $u = u_i$ and $v = v_i$ in the $(u, v)^{\mathrm{T}}$ plane correspond to the curves $h_1(x, y) = u_i$ and $h_2(x, y) = v_i$ in the $(x, y)^{\mathrm{T}}$ plane. Our subdivision of Δ therefore induces a corresponding subdivision of D (*see* §7.3). It is essential for our arguments that there are *no* critical points for $\mathbf{g}$ inside the region Δ.

Critical points exist where the Jacobian

$$\begin{vmatrix} \dfrac{\partial x}{\partial u} & \dfrac{\partial x}{\partial v} \\[2mm] \dfrac{\partial y}{\partial u} & \dfrac{\partial y}{\partial v} \end{vmatrix} = 0$$

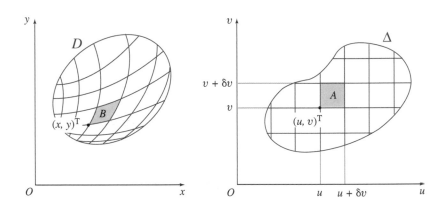

The shaded area with a corner at $(u, v)^{\mathrm{T}}$ will be denoted by A and the corresponding shaded area with a corner at $\mathbf{g}(u, v)^{\mathrm{T}} = (x, y)^{\mathrm{T}}$ will be denoted by B. The next step is to express the area B in terms of the area $A = \delta u \, \delta v$.

Recall that $\mathbf{g}'(u, v)$

$$= \begin{pmatrix} \dfrac{\partial g_1}{\partial u} & \dfrac{\partial g_1}{\partial v} \\[2mm] \dfrac{\partial g_2}{\partial u} & \dfrac{\partial g_2}{\partial v} \end{pmatrix}$$

$$= \begin{pmatrix} \dfrac{\partial x}{\partial u} & \dfrac{\partial x}{\partial v} \\[2mm] \dfrac{\partial y}{\partial u} & \dfrac{\partial y}{\partial v} \end{pmatrix}$$

(see §5.3) and

$$\det \begin{pmatrix} \dfrac{\partial x}{\partial u} & \dfrac{\partial x}{\partial v} \\[2mm] \dfrac{\partial y}{\partial u} & \dfrac{\partial y}{\partial v} \end{pmatrix}$$

$$= \dfrac{\partial(x, y)}{\partial(u, v)}$$

See §7.2.

† See §1.7.

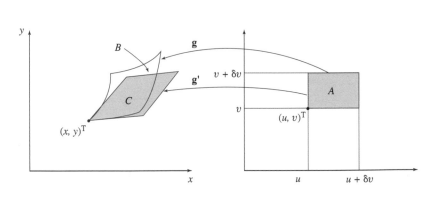

The area C of the parallelogram in the diagram, tangential to the area B, is a reasonable approximation to B at $(x, y)^{\mathrm{T}}$, provided that δu and δv are small. Since the area A is mapped to the area B by the function $\mathbf{g}$, its derivative function $\mathbf{g}'$ maps the area A to the area C, provided that δu and δv are small.†

Hence

$$
B \approx C = |\det \mathbf{g}'(u, v)| A = \left| \det \begin{pmatrix} \dfrac{\partial x}{\partial u} & \dfrac{\partial x}{\partial v} \\[2mm] \dfrac{\partial y}{\partial u} & \dfrac{\partial y}{\partial v} \end{pmatrix} \right| \delta u \, \delta v = \left| \dfrac{\partial(x, y)}{\partial(u, v)} \right| \delta u \, \delta v
$$

An approximating sum to the double integral $\iint\limits_{D} f(x, y) \, dx \, dy$ is

$$
\sum_i f(x_i, y_i) B_i = \sum_i f(g_1(u_i, v_i), g_2(u_i, v_i)) B_i
$$

$$
\approx \sum_i f\big(g_1(u_i, v_i), g_2(u_i, v_i)\big) \left| \dfrac{\partial(x, y)}{\partial(u, v)} \right| \delta u_i \, \delta v_i
$$

where $\dfrac{\partial(x, y)}{\partial(u, v)}$ is evaluated at $(u_i, v_i)^{\mathrm{T}}$. But this last sum is an approximating sum for the double integral,

$$
\iint\limits_{\Delta} f(g_1(u, v), g_2(u, v)) \left| \dfrac{\partial(x, y)}{\partial(u, v)} \right| du \, dv \tag{3}
$$

We conclude that

$$
\iint\limits_{D} f(x, y) \, dx \, dy = \iint\limits_{\Delta} f(g_1(u, v), g_2(u, v)) \left| \dfrac{\partial(x, y)}{\partial(u, v)} \right| du \, dv
$$

The general version of this in $\mathbb{R}^n$ is

$$
\iint \cdots \int\limits_{D} f(x_1 \ldots x_n) \, dx_1, \ldots, dx_n =
$$

$$
\iint \cdots \int\limits_{\Delta} f(g_1(u_1, \ldots, u_n), \ldots, g_n(u_1, \ldots, u_n)) \left| \dfrac{\partial(x_1, \ldots, x_n)}{\partial(u_1, \ldots, u_n)} \right| du_1 \ldots du_n
$$

Since by rule (IV) for vector derivatives[†]

$$
\dfrac{d\mathbf{x}}{d\mathbf{u}} = \left(\dfrac{d\mathbf{u}}{d\mathbf{x}} \right)^{-1}
$$

we obtain the really useful formula

$$
\dfrac{\partial(x_1, x_2, \ldots, x_n)}{\partial(u_1, u_2, \ldots, u_n)} = \left\{ \dfrac{\partial(u_1, u_2, \ldots, u_n)}{\partial(x_1, x_2, \ldots, x_n)} \right\}^{-1}
$$

which allows us to calculate the necessary Jacobian without first solving equations (1). This sometimes saves a lot of work.

[†] See §5.4.

Example 5

We return to Example 3 and consider the double integral

$$I = \iint_D xy \, dx \, dy$$

Since part of the boundary of D is circular, it is natural to think of using polar coordinates.

We have

$$x = r \cos \theta$$
$$y = r \sin \theta$$

from which it follows that the lines $x = 0$ and $y = 0$ correspond to $\theta = \pi/2$ and $\theta = 0$ respectively. Moreover, since $r^2 = x^2 + y^2$, the circle $x^2 + y^2 = 1$ corresponds to $r = 1$. These observations make it clear that Δ is the region indicated below:

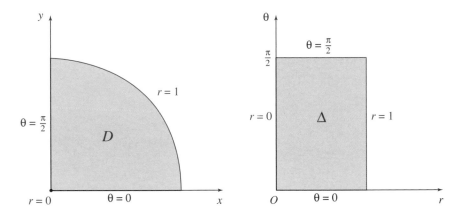

The Jacobian is given by

$$\frac{\partial(x, y)}{\partial(r, \theta)} = \begin{vmatrix} \dfrac{\partial x}{\partial r} & \dfrac{\partial x}{\partial \theta} \\ \dfrac{\partial y}{\partial r} & \dfrac{\partial y}{\partial \theta} \end{vmatrix} = \begin{vmatrix} \cos \theta & -r \sin \theta \\ \sin \theta & r \cos \theta \end{vmatrix} = r(\cos^2 \theta + \sin^2 \theta) = r$$

It follows that

$$\iint_D xy \, dx \, dy = \iint_\Delta (r \cos \theta)(r \sin \theta)|r| \, dr \, d\theta$$
$$= \iint_\Delta r^3 \cos \theta \sin \theta \, dr \, d\theta$$

Because Δ is rectangular, it is easy to evaluate this double integral. We obtain

$$\iint_D xy \, dx \, dy = \int_0^1 r^3 \int_0^{\pi/2} \cos \theta \sin \theta \, d\theta \, dr$$

$$= \int_0^1 r^3 \int_0^{\pi/2} \frac{\sin 2\theta}{2} \, d\theta \, dr$$

$$= \int_0^1 r^3 \left[-\frac{\cos 2\theta}{4} \right]_0^{\pi/2} dr$$

$$= \frac{1}{2} \int_0^1 r^3 \, dr = \frac{1}{2} \left[\frac{1}{4} r^4 \right]_0^1 = \frac{1}{8}$$

Note that there is a critical point at $(x, y)^T = (0, 0)^T$ but this is not inside the region D.

Example 6

We give a second example using polar coordinates because of the importance of the result in statistics. The problem is to evaluate

$$I = \int_{-\infty}^{\infty} e^{-x^2/2} \, dx$$

We have

$$I^2 = \left\{ \int_{-\infty}^{\infty} e^{-x^2/2} \, dx \right\} \left\{ \int_{-\infty}^{\infty} e^{-y^2/2} \, dy \right\}$$

$$= \int_{-\infty}^{\infty} \int_{-\infty}^{\infty} e^{-(x^2+y^2)/2} \, dx \, dy$$

The region Δ in the $(r, \theta)^T$ plane which corresponds to D in the case when D is the whole plane $\mathbb{R}^2$ is the set of those $(r, \theta)^T$ which satisfy $r \geq 0$ and $-\pi < \theta \leq \pi$. Hence

See §7.4.

$$I^2 = \iint_{\Delta} e^{-r^2/2} \left| \frac{\partial(x, y)}{\partial(r, \theta)} \right| dr \, d\theta$$

$$= \iint_{\Delta} r e^{-r^2/2} \, dr \, d\theta$$

$$= \int_{-\pi}^{\pi} \int_0^{\infty} r e^{-r^2/2} \, dr \, d\theta$$

$$= \int_{-\pi}^{\pi} \left[-e^{-r^2/2} \right]_0^{\infty} d\theta = \int_{-\pi}^{\pi} d\theta = 2\pi$$

It follows that

$$I = \sqrt{2\pi}$$

Example 7

Let D be the set of those $(x, y)^T$ in $\mathbb{R}_+^2$ which lie above the curve $yx^2 = a$ and between the curves $y = bx^2$ and $y = cx^2$ (where $a > 0$ and $c > b > 0$). Evaluate

$$\iint_D \frac{dx \, dy}{y^2 x^3}$$

by means of the change of variable

$$\left.\begin{array}{c} u = yx^2 \\[2mm] v = \dfrac{y}{x^2} \end{array}\right\} \qquad (4)$$

The first task is to sketch the region D and to determine the corresponding region Δ in the $(u, v)^{\mathrm{T}}$ plane.

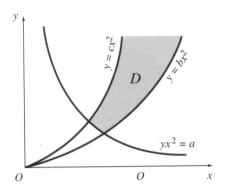 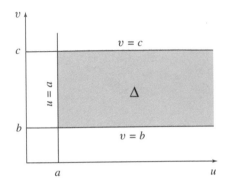

The region Δ can be identified by observing that the curves $yx^2 = a$, $y = bx^2$ and $y = cx^2$ map onto $u = a$, $v = b$ and $v = c$ respectively.

The Jacobian is

$$\frac{\partial(x, y)}{\partial(u, v)} = \left\{\frac{\partial(u, v)}{\partial(x, y)}\right\}^{-1} = \left|\begin{array}{cc} 2xy & x^2 \\[2mm] -2y & 1 \\ x^3 & x^2 \end{array}\right|^{-1}$$

$$= \left\{\frac{2y}{x} + \frac{2y}{x}\right\}^{-1} = \frac{x}{4y}$$

and so

$$\iint_{D} \frac{dx\,dy}{y^2 x^3} = \iint_{\Delta} \frac{1}{y^2 x^3} \left|\frac{\partial(x, y)}{\partial(u, v)}\right| du\,dv$$

$$= \iint_{\Delta} \frac{1}{y^2 x^3} \frac{x}{4y}\,du\,dv$$

$$= \frac{1}{4} \iint_{\Delta} \frac{1}{y^2 x^4} \frac{x^2}{y}\,du\,dv$$

$$= \frac{1}{4} \iint_{\Delta} \frac{1}{u^2} \frac{1}{v}\,du\,dv$$

$$= \frac{1}{4} \int_{a}^{\infty} \frac{1}{u^2} \int_{b}^{c} \frac{dv}{v}\,du$$

Note the delay in substituting for x and y in terms of u and v in case it should prove possible to avoid solving equations (4).

$$= \frac{1}{4} \int_a^\infty \frac{1}{u^2} [\ln v]_b^c \, du$$

$$= \frac{1}{4} \ln \frac{c}{b} \left[-\frac{1}{u} \right]_a^\infty = \frac{1}{4a} \ln \frac{c}{b}$$

Example 8

Make the change of variable

$$\left. \begin{array}{l} u = x + y \\ v = \dfrac{y}{x} \end{array} \right\}$$

in the integral

$$\int_0^1 \int_y^1 \frac{(x+y)}{x^2} e^{(x+y)} \, dx \, dy$$

and hence evaluate it.

It is first necessary, as in Example 4, to identify the region D of integration. For a fixed value of y, x ranges between y and 1 and so the region D is as indicated in the following diagram on the left. The corresponding region Δ is indicated on the right.

In determining the region Δ, one may begin by observing that the line $y = 0$ is mapped into $v = 0$ and the line $x = y$ into $v = 1$. When $x = 1$, we obtain $u = y + 1$ and $v = y$. Thus $x = 1$ becomes $u = v + 1$. Normally this information would suffice to specify Δ but in this case one has to be a little careful since, although there is no critical point *inside* D, there is a problem about what happens when $x = 0$. This problem is most easily resolved by observing that, for each α with $0 \le \alpha \le 1$, the line segment $y = \alpha x$ $(0 < x \le 1)$ is mapped to $v = \alpha$ $(0 < u \le \alpha + 1)$.

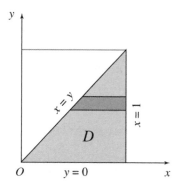
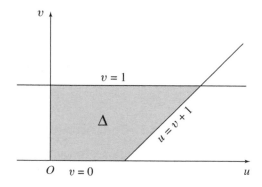

The Jacobian is

$$\frac{\partial(x, y)}{\partial(u, v)} = \left\{ \frac{\partial(u, v)}{\partial(x, y)} \right\}^{-1} = \left| \begin{array}{cc} 1 & 1 \\ -\dfrac{y}{x^2} & \dfrac{1}{x} \end{array} \right|^{-1} = \left(\frac{1}{x} + \frac{y}{x^2} \right)^{-1}$$

It follows that

$$\int_0^1 \int_y^1 \frac{(x+y)}{x^2} e^{(x+y)} \, dx \, dy = \iint_D \frac{(x+y)}{x^2} e^{(x+y)} \, dx \, dy$$

$$= \iint_\Delta \frac{(x+y)}{x^2} e^{(x+y)} \left| \frac{x^2}{x+y} \right| du \, dv$$

$$= \iint_\Delta e^u \, du \, dv$$

This may be evaluated as a repeated integral.

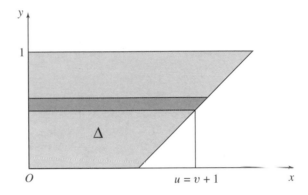

We obtain

$$\iint_\Delta e^u \, du \, dv = \int_0^1 \int_0^{v+1} e^u \, du \, dv$$

$$= \int_0^1 [e^u]_0^{v+1} \, dv = \int_0^1 (e^{v+1} - 1) \, dv$$

$$= [e^{v+1} - v]_0^1 = e^2 - e - 1$$

11.4 Unbounded regions of integration♣

In Example 7 and elsewhere we have manipulated a double integral over an unbounded region D quite freely, although our account of the theory given in §11.2 and §11.3 was restricted to the case when the integrand f is continuous and the region D is bounded. Such manipulations in the general case are legitimate, provided that the convergence of the double integral is dominated. This is guaranteed in the case of a convergent double integral of a *nonnegative* integrand. However, applied mathematicians tend to give little attention to this question since it is only in exceptional cases that things can go wrong.

See §10.12.

11.5 Multiple sums and series♣

Multiple sums and series can be manipulated in much the same way as multiple integrals. We give some examples below.

Example 9 _____

We have, for $-1 < x < 1$,

$$(1-x)^{-1} = 1 + x + x^2 + x^3 + \cdots$$
$$= \sum_{n=0}^{\infty} x^n$$

In Example 10.23, we differentiated and obtained the result

$$(1-x)^{-2} = 1 + 2x + 3x^2 + 4x^3 + \cdots$$
$$= \sum_{n=1}^{\infty} nx^{n-1}$$

provided $-1 < x < 1$. The same result can also be obtained as follows. Note that, for $-1 < x < 1$,

$$(1-x)^{-2} = (1-x)^{-1}(1-x)^{-1}$$
$$= \left\{ \sum_{m=0}^{\infty} x^m \right\} \left\{ \sum_{n=0}^{\infty} x^n \right\}$$
$$= \sum_{m=0}^{\infty} \sum_{n=0}^{\infty} x^{m+n}$$

We have

$$\sum_{n=0}^{\infty} x^{m+n} = \sum_{l=m}^{\infty} x^l$$

This is obtained by writing $l = m + n$ and observing that $l = m$ when $n = 0$. Hence

$$(1-x)^{-2} = \sum_{m=0}^{\infty} \sum_{l=m}^{\infty} x^l \tag{1}$$

We now propose to change the order of summation. As in the case of integration, a diagram is usually useful:

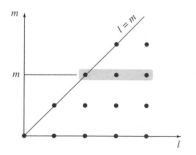
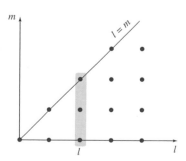

In the repeated series (1), we begin by summing over rows of points (as indicated in the left hand diagram). If we first sum over columns (as in the right hand diagram) we obtain that

$$\sum_{m=0}^{\infty}\sum_{l=m}^{\infty} x^l = \sum_{l=0}^{\infty}\sum_{m=0}^{l} x^l$$

$$= \sum_{l=0}^{\infty} x^l \sum_{m=0}^{l} 1$$

$$= \sum_{l=0}^{\infty} x^l(l+1)$$

Finally, we may write $k = l + 1$ and obtain

$$(1-x)^{-2} = \sum_{k=1}^{\infty} kx^{k-1}$$

$$= 1 + 2x + 3x^2 + \cdots$$

provided $-1 < x < 1$.

Example 10 ——————————————————————————————————————

When dealing with double series it is sometimes a good idea to begin neither with rows nor with columns but with diagonals. In the case of power series, this can be a particularly helpful technique. We have that

$$\left(\sum_{m=0}^{\infty} a_m x^m\right)\left(\sum_{n=0}^{\infty} b_n x^n\right) = \sum_{m=0}^{\infty}\sum_{n=0}^{\infty} a_m b_n x^{m+n}$$

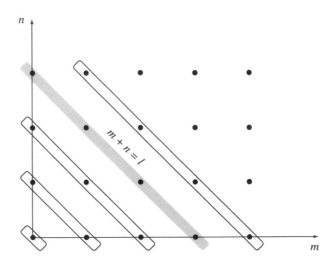

If we begin by summing along diagonals as indicated, we obtain

$$\left(\sum_{m=0}^{\infty} a_m x^m\right)\left(\sum_{n=0}^{\infty} b_n x_n\right) = \sum_{l=0}^{\infty}\sum_{m=0}^{l} a_m b_{l-m} x^l$$

$$= \sum_{l=0}^{\infty} x^l \left\{\sum_{m=0}^{l} a_m b_{l-m}\right\}$$

provided x lies inside the interval of convergence of the two power series.

Applying this result to the problem considered in the previous example, we obtain the following alternative argument:

$$(1-x)^{-2} = (1-x)^{-1}(1-x)^{-1} = \left(\sum_{m=0}^{\infty} x^m\right)\left(\sum_{n=0}^{\infty} x^n\right)$$

$$= \sum_{l=0}^{\infty} x^l \sum_{m=0}^{l} 1$$

$$= \sum_{l=0}^{\infty} (l+1)x^l$$

provided $-1 < x < 1$.

11.6 Exercises

1.* By changing the order of integration in the repeated integral, prove that

$$\int_0^{\pi/2}\int_x^{\pi/2} \frac{\sin y}{y}\, dy\, dx = 1$$

2. Change the order of integration in the repeated integral

$$\int_0^1\int_x^{2-x} \frac{x}{y}\, dy\, dx$$

and evaluate the result.

3.* If $a > 0$, prove that

$$\int_0^a y \int_{y^2/a}^y f(x, y)\, dx\, dy = \int_0^a \int_x^{\sqrt{ax}} f(x, y)\, dy\, dx$$

Deduce that

$$\int_0^a y \int_{y^2/a}^y \frac{1}{(a-x)\sqrt{ax-y^2}}\, dx\, dy = \frac{1}{2}\pi$$

4. Evaluate the double integral

$$\iint_D e^{(ax+by)}\, dx\, dy$$

where D is the triangle enclosed by the lines $x = 0$, $y = 0$ and $ax + by = 1$.

5.*♣ The identity

$$I(a, b) = \int_0^\infty (\sin bx)e^{-ax} \, dx = \frac{b}{a^2 + b^2}$$

was given in Exercise 10.15.2. By considering

$$\frac{\partial}{\partial b} \int_a^\infty I(\alpha, b) \, d\alpha$$

deduce the identity

$$J(a, b) = \int_0^\infty (\cos bx)e^{-ax} \, dx = \frac{a}{a^2 + b^2}$$

6.♣ Deduce the identity given in the previous question for $I(a, b)$ from the identity given for $J(a, b)$ by considering

$$\frac{\partial}{\partial a} \int_0^b J(a, \beta) d\beta$$

7.* The region D is the part of the positive quadrant enclosed by the curves $x^2 + 2y^2 = 1$, $x^2 + 2y^2 = 4$, $y = 2x$ and $y = 5x$. Sketch D and evaluate the double integral

$$\iint_D \frac{y}{x} \, dx \, dy$$

by introducing the change of variable

$$\left. \begin{array}{l} u = x^2 + 2y^2 \\[2mm] v = \dfrac{y}{x} \end{array} \right\}$$

8. Evaluate

$$\iint_D x^2 y^2 (y^2 - x^2) \, dx \, dy$$

where D is the region in the first quadrant enclosed by the lines $y = x + 1$, $y = x + 3$ and the curves $xy = 1$, $xy = 4$ by making the change of variable

$$\left. \begin{array}{l} u = xy \\[2mm] v = x - y \end{array} \right\}$$

9.* Consider the function $f : \mathbb{R}^2 \to \mathbb{R}^2$ defined by

$$\begin{pmatrix} u \\ v \end{pmatrix} = f(x, y) = \begin{pmatrix} xe^y \\ ye^x \end{pmatrix}$$

Sketch the region in the $(x, y)^{\mathrm{T}}$ plane which is mapped onto the square in the $(u, v)^{\mathrm{T}}$ plane enclosed by the lines $u = 1$, $u = e$, $v = 1$, $v = e$. At what points does the Jacobian of f vanish and what significance does this have?

Use the function given above to change variables in the double integral

$$\iint_D (x^2 y^3 - x^3 y^4) e^{4x+3y} \, dx \, dy$$

where D is the region enclosed by the curves $y + \ln x = 0$, $y + \ln x = 1$, $x + \ln y = 0$ and $x + \ln y = 1$. Hence evaluate the integral.

10. Sketch the region D in the first quadrant which lies above the curve $y^2 x^3 = a$ and between the curves $y^2 = bx^3$, $y^2 = cx^3$ where $a > 0$ and $0 < b < c$. By making the change of variable

$$\left. \begin{array}{l} u = y^2 x^3 \\ v = y^2 x^{-3} \end{array} \right\}$$

evaluate the double integral

$$\iint_D \frac{dx \, dy}{x^4 y^7}$$

11.* Introduce the change of variable defined by

$$\left. \begin{array}{l} x = r \cos \theta \sin \phi \\ y = r \sin \theta \sin \phi \\ z = r \cos \phi \end{array} \right\}$$

See §7.4.

into the triple integral

$$\iiint_D dx \, dy \, dz$$

where D is the inside of a sphere with radius r and centre $(0, 0, 0)^{\mathrm{T}}$. Hence calculate the volume of a sphere.

12. Evaluate

$$\iiint_D \left\{ \frac{x^2}{a^2} + \frac{y^2}{b^2} + \frac{z^2}{c^2} \right\} dx \, dy \, dz$$

where D is the inside of the ellipsoid

$$\frac{x^2}{a^2} + \frac{y^2}{b^2} + \frac{z^2}{c^2} = 1$$

Begin with the change of variable $X = x/a$, $Y = y/b$, $Z = z/c$ and then proceed as in the previous question.

13.*♣ Prove that $e^{x+y} = e^x e^y$ by considering

$$\sum_{n=0}^{\infty} \frac{(x+y)^n}{n!} = \sum_{n=0}^{\infty} \frac{1}{n!} \sum_{k=0}^{n} \binom{n}{k} x^k y^{n-k}$$

14.♣ Explain why

$$\sum_{j=1 \atop j \neq k}^{\infty} \sum_{k=1}^{\infty} \frac{1}{j^2 - k^2} \neq \sum_{k=1}^{\infty} \sum_{j=1 \atop j \neq k}^{\infty} \frac{1}{j^2 - k^2}$$

11.7 Applications (optional)

11.7.1 Joint probability distributions

Suppose that $f: \mathbb{R}^2 \to \mathbb{R}$ has the property that for all regions D in $\mathbb{R}^2$

$$\iint\limits_D f(x, y) \, dx \, dy$$

represents the probability that the pair $(X, Y)^T$ of random variables lies in the region D. We then say that f is the **joint probability density function** for the continuous random variables X and Y. Similar considerations apply in the discrete case except that the double integral is replaced by a double sum.

To qualify as a joint probability density function in the continuous case, a function $f: \mathbb{R}^2 \to \mathbb{R}$ must satisfy $f(x, y) \geq 0$ for all $(x, y)^T$. Moreover

$$\iint\limits_{\mathbb{R}^2} f(x, y) \, dx \, dy = 1$$

Similarly in the discrete case.

11.7.2 Marginal probability distributions

Suppose that $f: \mathbb{R}^2 \to \mathbb{R}$ is the joint probability density function for the continuous random variables X and Y. Then

$$P(a \leq X \leq b) = \iint\limits_D f(x, y) \, dx \, dy$$

where D is the strip in $\mathbb{R}^2$ indicated below:

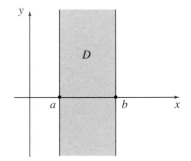

A point $(x, y)^T$ lies in this strip D provided $a \leq x \leq b$ regardless of the value of y.

Observe that

$$\iint\limits_D f(x, y) \, dx \, dy = \int_a^b \int_{-\infty}^{\infty} f(x, y) \, dy \, dx$$

and hence

$$P(a \leq X \leq b) = \int_a^b g(x) \, dx$$

where $g: \mathbb{R} \to \mathbb{R}$ is defined by

$$g(x) = \int_{-\infty}^{\infty} f(x, y) \, dy$$

Thus g is the probability density function for X. We say that g is a **marginal** density function derived from the joint density function f.

Similarly, the probability density function for Y is given by

$$h(y) = \int_{-\infty}^{\infty} f(x, y)\,dx$$

In the case of discrete random variables, the integrals must be replaced by sums.

11.7.3 Expectation, variance and covariance

An immediate consequence of the above discussion is that

$$\mathcal{E}(X + Y) = \mathcal{E}(X) + \mathcal{E}(Y)$$

which we used frequently in discussing applications of the results of the previous chapter. In the case of continuous random variables

$$
\begin{aligned}
\mathcal{E}(X + Y) &= \iint_{\mathbb{R}^2} (x + y) f(x, y)\,dx\,dy \\
&= \iint_{\mathbb{R}^2} x f(x, y)\,dx\,dy + \iint_{\mathbb{R}^2} y f(x, y)\,dx\,dy \\
&= \int_{-\infty}^{\infty} x \int_{-\infty}^{\infty} f(x, y)\,dy\,dx + \int_{-\infty}^{\infty} y \int_{-\infty}^{\infty} f(x, y)\,dx\,dy \\
&= \int_{-\infty}^{\infty} x g(x)\,dx + \int_{-\infty}^{\infty} y h(y)\,dy \\
&= \mathcal{E}(X) + \mathcal{E}(Y)
\end{aligned}
$$

Write $\mathcal{E}(X) = \mu$ and $\mathcal{E}(Y) = \nu$. Then

$$
\begin{aligned}
\text{var}(X + Y) &= \mathcal{E}\{(X + Y) - (\mu + \nu)\}^2 \\
&= \mathcal{E}(X - \mu)^2 + 2\mathcal{E}(X - \mu)(Y - \nu) + \mathcal{E}(Y - \nu)^2
\end{aligned}
$$

The quantity $\mathcal{E}(X - \mu)(Y - \nu)$ is called the **covariance** of X and Y and is denoted by

$$\text{cov}(X, Y) = \mathcal{E}(X - \mu)(Y - \nu)$$

Thus

$$\text{var}(X + Y) = \text{var}\,X + 2\text{cov}(X, Y) + \text{var}\,Y$$

The number

$$\rho(X, Y) = \frac{\text{cov}(X, Y)}{\sqrt{\text{var}\,X\,\text{var}\,Y}}$$

is called the **correlation coefficient** for X and Y. It is a measure of the tendency of the two random variables to vary together. Its largest possible value is 1 and this is achieved when $X = Y$. Its smallest possible value is zero and this is achieved when X and Y are independent. In the latter case, $\mathcal{E}(XY) = (\mathcal{E}X)(\mathcal{E}Y)$ and so $\text{cov}(X, Y) = 0$.

11.7.4 Independent random variables

Two random variables X and Y are independent if it is always the case that

$$P(\xi \le X \le x \text{ and } \eta \le Y \le y)$$
$$= P(\xi \le X \le x)P(\eta \le Y \le y)$$

In the continuous case, this means that

$$\int_\xi^x \int_\eta^y f(s, t)\, dt\, ds = \left(\int_\xi^x g(s)\, ds \right) \left(\int_\eta^y h(t)\, dt \right)$$

Differentiating both sides successively with respect to x and y, we obtain that

$$f(x, y) = g(x)h(y)$$

If the continuous random variables X and Y are independent, it follows that

$$\mathcal{E}(XY) = \iint\limits_{\mathbb{R}^2} xy f(x, y)\, dx\, dy$$

$$= \iint\limits_{\mathbb{R}^2} xy g(x)h(y)\, dx\, dy$$

$$= \left(\int_{-\infty}^{\infty} x g(x)\, dx \right) \left(\int_{-\infty}^{\infty} y h(y)\, dy \right)$$

$$= \mathcal{E}(X)\mathcal{E}(Y)$$

It is often useful to know the probability density function k for the random variable $Z = X + Y$ in the case when X and Y are independent. We have that

$$P(a \le Z \le b) = \iint\limits_{D} f(x, y)\, dx\, dy$$

where D is the region indicated in the diagram below:

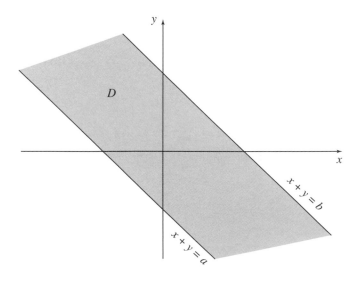

with the change of variable

$$\left.\begin{array}{c} t = x \\ z = x + y \end{array}\right\}$$

we obtain

$$\iint_D f(x, y)\, \mathrm{d}x\, \mathrm{d}y = \int_a^b \int_{-\infty}^\infty f(t, z - t)\, \mathrm{d}t\, \mathrm{d}z$$

$$= \int_a^b \int_{-\infty}^\infty g(t)h(z - t)\, \mathrm{d}t\, \mathrm{d}z$$

Hence

$$k(z) = \int_{-\infty}^\infty g(t)h(z - t)\, \mathrm{d}t$$

As an example, consider the case when X_1 and X_2 are independent standardised normal random variables. Then the probability density function for $X_1 + X_2$ is

$$k(z) = \frac{1}{2\pi} \int_{-\infty}^\infty e^{-t^2/2} e^{-(z-t)^2/2}\, \mathrm{d}t$$

$$= \frac{1}{2\pi} \int_{-\infty}^\infty e^{-(2t^2 - 2zt + z^2)/2}\, \mathrm{d}t$$

$$= \frac{1}{2\pi} e^{-z^2/4} \int_{-\infty}^\infty e^{-(t - z/2)^2}\, \mathrm{d}t$$

See §10.7.

where we have 'completed the square' in the exponent. In the final integral, make the change of variable $t - z/2 = u/\sqrt{2}$. Then

$$k(z) = \frac{1}{2\pi} e^{-z^2/4} \frac{1}{\sqrt{2}} \int_{-\infty}^\infty e^{-u^2/2}\, \mathrm{d}u$$

$$= \frac{1}{2\pi} e^{-z^2/4} \frac{1}{\sqrt{2}} \sqrt{2\pi} \qquad \text{(Example 6)}$$

$$= \frac{1}{\sqrt{4\pi}} e^{-(z/\sqrt{2})^2/2}$$

It follows that $X_1 + X_2$ is normally distributed with mean 0 and variance 2. In the case when X_1 is normally distributed with mean μ_1 and variance σ_1^2 and X_2 is independent and normally distributed with mean μ_2 and variance σ_2^2, then $X_1 + X_2$ is normally distributed with mean $\mu = \mu_1 + \mu_2$ and variance $\sigma^2 = \sigma_1^2 + \sigma_2^2$. A more complicated version of the argument given above suffices to prove this.

See also §13.9.

11.7.5 Generating functions

Suppose that X_1 is a discrete random variable with probability density function g_1 and that X_2 is an independent discrete random variable with probability density function g_2. We shall suppose that X_1 and X_2 do not take negative values so that $g_1(j) = g_2(j) = 0$ when $j = -1, -2, -3, \ldots$.

The functions G_1 and G_2 defined by the power series

$$G_1(x) = \sum_{j=0}^\infty g_1(j)x^j \qquad G_2(x) = \sum_{k=0}^\infty g_2(k)x^k$$

are called **generating functions** for g_1 and g_2. They are useful because, by Example 10,

$$G_1(x)G_2(x) = \sum_{l=0}^{\infty} h(l)x^l$$

where

$$h(l) = \sum_{j=0}^{l} g_1(j)g_2(l-j) = \sum_{j=-\infty}^{\infty} g_1(j)g_2(l-j)$$

since $g_1(j) = 0$ for $j < 0$ and $g_2(l-j) = 0$ for $j > l$. It follows from §11.7.4 that h is the probability density function of the random variable $X_1 + X_2$.

As an example, consider the case in which X_1 and X_2 are independent Poisson random variables with parameters λ_1 and λ_2 respectively. Then

$$G_1(x) = \sum_{j=0}^{\infty} \frac{\lambda_1^j e^{-\lambda_1}}{j!} x^j = e^{-\lambda_1} \sum_{j=0}^{\infty} \frac{(\lambda_1 x)^j}{j!} = e^{\lambda_1(x-1)}$$

Similarly, $G_2(x) = e^{\lambda_2(x-1)}$. Thus

$$\begin{aligned} G_1(x)G_2(x) &= e^{\lambda_1(x-1)}e^{\lambda_2(x-1)} \\ &= e^{(\lambda_1+\lambda_2)(x-1)} \\ &= e^{\lambda(x-1)} \end{aligned}$$

where $\lambda = \lambda_1 + \lambda_2$. It follows that the random variable $X_1 + X_2$ is Poisson with parameter $\lambda = \lambda_1 + \lambda_2$.

11.7.6 Multivariate normal distributions

We conclude this chapter by considering the multivariate normal distribution. This will give us the opportunity to review a number of useful techniques.

The continuous random variables $X_1, X_2, \ldots, X_n$ are said to have the **multivariate normal distribution** provided that their joint probability density function $f : \mathbb{R}^n \to \mathbb{R}$ is of the form

$$f(x_1, x_2, \ldots, x_n) = \frac{\sqrt{\det A}}{(\sqrt{2\pi})^n} e^{-(\mathbf{x}-\boldsymbol{\mu})^{\mathrm{T}} A(\mathbf{x}-\boldsymbol{\mu})/2}$$

[†] See §6.1.

where $\boldsymbol{\mu}$ is an $n \times 1$ column vector and A is a positive definite $n \times n$ matrix.[†] Since A is positive definite if and only if its eigenvalues are positive, it is legitimate to denote the eigenvalues by $s_1^2, s_2^2, \ldots, s_n^2$. Note that the determinant of a square

[‡] See §6.2.

matrix is equal to the product of the eigenvalues[‡] and so $\sqrt{\det A} = s_1 s_2 \ldots s_n$.

The first thing to check is that f qualifies as a probability density function. We require that $f(x_1, x_2, \ldots, x_n) \geq 0$ for all $(x_1, x_2, \ldots, x_n)^{\mathrm{T}}$ and that

$$\underset{\mathbb{R}^n}{\int\int \cdots \int} f(x_1, x_2, \ldots, x_n) \, dx_1 \, dx_2 \ldots dx_n = 1$$

In proving the latter condition, it is useful to recall from §6.1 that there exists an orthogonal matrix P (i.e. $P^{\mathrm{T}}P = I$) such that $P^{\mathrm{T}}AP = D$ where D is the diagonal matrix whose diagonal entries are $s_1^2, s_2^2, \ldots, s_n^2$. Write $D = S^{\mathrm{T}}S = SS$ where

$$S = \begin{pmatrix} s_1 & 0 & \ldots & 0 \\ 0 & s_2 & \ldots & 0 \\ \vdots & \vdots & & \vdots \\ 0 & 0 & \ldots & s_n \end{pmatrix}$$

Then $A = PDP^T = PS^TSP^T = (SP^T)^T(SP^T)$. This representation makes it natural to introduce the change of variable defined by

$$SP^T(\mathbf{x} - \boldsymbol{\mu}) = \mathbf{y}$$

because then

$$(\mathbf{x} - \boldsymbol{\mu})^T A(\mathbf{x} - \boldsymbol{\mu}) = \mathbf{y}^T \mathbf{y}$$

The appropriate Jacobian is

$$\det\left(\frac{d\mathbf{x}}{d\mathbf{y}}\right) = \det(PS^{-1}) = (\det \mathbf{A})^{-1/2}$$

It follows that

$$\iint_{\mathbb{R}^n} \cdots \int f(x_1, \ldots, x_n) \, dx_1 \ldots dx_n$$

$$= \frac{1}{(\sqrt{2\pi})^n} \iint_{\mathbb{R}^n} \cdots \int e^{-\mathbf{y}^T\mathbf{y}/2} \, dy_1 \ldots dy_n$$

$$= \frac{1}{(\sqrt{2\pi})^n} \iint_{\mathbb{R}^n} \cdots \int e^{-(y_1^2 + y_2^2 + \cdots + y_n^2)/2} \, dy_1 \ldots dy_n$$

$$= \left(\frac{1}{\sqrt{2\pi}} \int_{-\infty}^{\infty} e^{-y_1^2/2} \, dy_1\right) \cdots \left(\frac{1}{\sqrt{2\pi}} \int_{-\infty}^{\infty} e^{-y_n^2/2} \, dy_n\right)$$

$$= 1$$

We next show that $\mathcal{E}(\mathbf{X}) = \boldsymbol{\mu}$. This simply means that $\mathcal{E}(X_1) = \mu_1$, $\mathcal{E}(X_2) = \mu_2, \ldots, \mathcal{E}(X_n) = \mu_n$. Since $\mathbf{x} = \boldsymbol{\mu} + PS^{-1}\mathbf{y}$, the above argument gives

$$\mathcal{E}(\mathbf{X}) = \iint_{\mathbb{R}^n} \cdots \int \mathbf{x} f(x_1, \ldots, x_n) \, dx_1 \ldots dx_n$$

$$= \boldsymbol{\mu} + \frac{1}{(\sqrt{2\pi})^n} PS^{-1} \iint_{\mathbb{R}^n} \cdots \int \mathbf{y} e^{-\mathbf{y}^T\mathbf{y}/2} \, dy_1 \ldots dy_n$$

But the final term is the zero vector. This follows from the fact that, for each $j = 1, 2, \ldots, n$,

$$\iint_{\mathbb{R}^n} \cdots \int y_j e^{-\mathbf{y}^T\mathbf{y}/2} \, dy_1 \ldots dy_n$$

$$= \left(\int_{-\infty}^{\infty} e^{-y_1^2/2} \, dy_1\right) \cdots \left(\int_{-\infty}^{\infty} y_j e^{-y_j^2/2} \, dy_j\right) \cdots \left(\int_{-\infty}^{\infty} e^{-y_n^2/2} \, dy_n\right)$$

and the jth term is zero.

It remains to interpret the matrix A in terms of the random variables $X_1, X_2, \ldots, X_n$. It turns out that A is the inverse of their *covariance* matrix

$$\begin{pmatrix} \text{var}(X_1) & \text{cov}(X_1, X_2) & \cdots & \text{cov}(X_1, X_n) \\ \text{cov}(X_2, X_1) & \text{var}(X_2) & \cdots & \text{cov}(X_2, X_n) \\ \vdots & \vdots & & \vdots \\ \text{cov}(X_n, X_1) & \text{cov}(X_n, X_2) & \cdots & \text{var}(X_n) \end{pmatrix}$$

This matrix may be regarded as the expectation of the $n \times n$ matrix $(\mathbf{X} - \boldsymbol{\mu})(\mathbf{X} - \boldsymbol{\mu})^{\mathrm{T}}$. But

$$\mathcal{E}(\mathbf{X} - \boldsymbol{\mu})(\mathbf{X} - \boldsymbol{\mu})^{\mathrm{T}}$$

$$= \frac{1}{(\sqrt{2\pi})^n} \underset{\mathbb{R}^n}{\int\int} \cdots \int PS^{-1}\mathbf{y}\mathbf{y}^{\mathrm{T}}S^{-1}P^{\mathrm{T}}\mathrm{e}^{-\mathbf{y}^{\mathrm{T}}\mathbf{y}/2} \, dy_1 \ldots dy_n$$

$$= \frac{1}{(\sqrt{2\pi})^n} PS^{-1} \left(\underset{\mathbb{R}^n}{\int\int} \cdots \int \mathbf{y}\mathbf{y}^{\mathrm{T}}\mathrm{e}^{-\mathbf{y}^{\mathrm{T}}\mathbf{y}/2} \, dy_1 \ldots dy_n \right) S^{-1}P^{\mathrm{T}}$$

But

$$\underset{\mathbb{R}^n}{\int\int} \cdots \int y_i y_j \mathrm{e}^{-\mathbf{y}^{\mathrm{T}}\mathbf{y}/2} \, dy_1 \ldots dy_n = \begin{cases} (\sqrt{2\pi})^n & (i = j) \\ 0 & (i \neq j) \end{cases}$$

See §10.16.8.   and it follows that

$$\underset{\mathbb{R}^n}{\int\int} \cdots \int \mathbf{y}\mathbf{y}^{\mathrm{T}}\mathrm{e}^{-\mathbf{y}^{\mathrm{T}}\mathbf{y}/2} \, dy_1 \ldots dy_n = (\sqrt{2\pi})^n$$

Thus

$$\mathcal{E}(\mathbf{X} - \boldsymbol{\mu})(\mathbf{X} - \boldsymbol{\mu})^{\mathrm{T}} = PS^{-1}S^{-1}P^{\mathrm{T}}$$
$$= PD^{-1}P^{\mathrm{T}}$$
$$= (PDP^{\mathrm{T}})^{-1} = A^{-1}$$

Twelve

Differential equations of order one

Differential equations are of immense importance in many different areas.   In this chapter we provide only the briefest of introductions to the most elementary techniques and here we confine ourselves to differential equations of order one. None of this material is difficult and all of it is essential.

   Further material on differential equations will be found in Chapter 14.

12.1 Differential equations

A **differential equation** is an equation which contains at least one derivative of an unknown function. Some examples of differential equations are given below.

$$\frac{dy}{dx} = \cos x \tag{1}$$

$$\frac{d^2 y}{dx^2} + k^2 y = 0 \tag{2}$$

$$\left(\frac{d^2 w}{dx^2}\right)^3 - x\frac{dw}{dx} + w = 0 \tag{3}$$

$$\frac{d^2 u}{dt^2} + 7\left(\frac{du}{dt}\right)^4 - 8u = 0 \tag{4}$$

$$\frac{\partial^2 V}{\partial x^2} + \frac{\partial^2 V}{\partial y^2} = 0 \tag{5}$$

$$x\frac{\partial f}{\partial x} + y\frac{\partial f}{\partial y} = nf \tag{6}$$

   Equations like (1)–(4) which involve only one independent variable are called **ordinary differential equations**. Those that involve more than one independent variable, like (5) and (6) above, are called **partial** differential equations.

   In the examples given above, equations (1) and (6) are first order equations. The rest are second order equations. The **order** of a differential equation is the highest order of any derivative appearing in the equation.

   The **degree** of the equation is the algebraic degree with which the derivative of highest order appears in the equation. All but equation (3) are of first degree. Equation (3) is of degree 3.

   A **solution** of a differential equation is a relation between the variables involved in the equation which is free from derivatives and which is consistent with the differential equation. For example, $y = \sin x$ is a solution of equation (1) and $f = x^n + y^n$ is a solution of equation (6).

We shall concentrate almost entirely on ordinary differential equations (with only a brief look at partial differential equations in §12.10). In this chapter, the ordinary differential equations will be of order one and we shall devote no attention at all to equations of degree 2 or more. Hence our treatment is clearly very far from complete.

12.2 General solutions of ordinary equations

Any equation which connects x, y and an 'arbitrary constant' c may be differentiated. The arbitrary constant c may then be eliminated, leaving an ordinary differential equation of the first order. It follows that the original equation is a solution of the resulting ordinary differential equation for all values of the arbitrary constant c.

Consider for instance, the family of circles

$$x^2 + (y - c)^2 = c^2 \qquad c \in \mathbb{R} \tag{1}$$

This may be rewritten as

$$x^2 + y^2 - 2cy = 0 \tag{2}$$

We differentiate implicitly with respect to x and obtain

$$2x + (2y - 2c)\frac{dy}{dx} = 0$$

Hence

$$c = \left(x + y\frac{dy}{dx}\right) \Big/ \frac{dy}{dx}$$

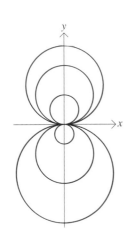

Substituting this result in equation (2) yields the differential equation

$$(x^2 + y^2)\frac{dy}{dx} = 2y\left(x + y\frac{dy}{dx}\right)$$

or

$$(x^2 - y^2)\frac{dy}{dx} - 2xy = 0 \tag{3}$$

This is a first order, ordinary differential equation. Observe that (1) is a solution of the differential equation (3) whatever the value of the constant c. Thus the differential equation does *not* have a *unique* solution. Each possible value for c generates a different solution for it.

A reversal of this reasoning leads us to the notion of a **general solution** of an ordinary differential equation of the **first order**. Such a general solution is a solution involving *one arbitrary constant*.

For example,

$$(x^2 - y^2)\frac{dy}{dx} - 2xy = 0 \tag{3}$$

has general solution

$$x^2 + (y - c)^2 = c^2 \qquad c \in \mathbb{R} \tag{1}$$

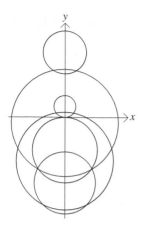

Similar reasoning leads us to define a **general solution** of an ordinary differential equation of the *n***th order** to be a solution containing *n arbitrary constants*. For example, the family of circles

$$x^2 + (y - c)^2 = r^2 \tag{4}$$

where c and r are arbitrary constants is a general solution of the second order, ordinary differential equation

$$\left(\frac{dy}{dx}\right) + \left(\frac{dy}{dx}\right)^3 - x\left(\frac{d^2 y}{dx^2}\right) = 0 \tag{5}$$

This can be checked by differentiating equation (4) twice with respect to x.

12.3 Boundary conditions

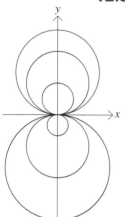

A particular circle of the family (1) is obtained by assigning a particular value to the arbitrary constant c. For this purpose, we require further information in the form of **boundary conditions** for the problem. In our case, the information may consist of a point through which the circle passes, because only one circle of the family (1) passes through any particular point $(x, y)^T \neq (0, 0)^T$. To specify $(x, y)^T$ is the same as specifying which value of c is to be substituted in (1).

The point $(x, y)^T = (0, 0)^T$ however, satisfies (1) for every c in $\mathbb{R}$*, so lies on every circle of the family. To specify that the solution of our differential equation passes through $(0, 0)^T$ will therefore not suffice as a boundary condition.*

For example, to find the particular solution to (2) passing through the point $(x, y)^T = (2, 2)^T$ *(one boundary condition)*, substitute for x and y in the general solution.

$$(2 - c)^2 + 2^2 = c^2 \quad \text{implies that} \quad c = 2$$

The particular solution is therefore

$$x^2 + (y - 2)^2 = 4$$

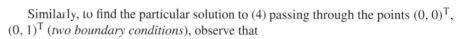

Similarly, to find the particular solution to (4) passing through the points $(0, 0)^T$, $(0, 1)^T$ *(two boundary conditions)*, observe that

$$c^2 = r^2 \quad \text{and} \quad 1 - 2c = 0 \quad \text{imply that} \quad c = r = \frac{1}{2}$$

The particular solution is

$$x^2 + \left(y - \frac{1}{2}\right)^2 = \frac{1}{4}$$

In brief, to select the particular solution of a differential equation which is relevant to the problem in hand one requires *boundary conditions*. Without appropriate boundary conditions, the problem of finding particular solutions is *indeterminate*.

12.4 Separable equations

We will now study some systematic methods for finding the general solution of certain kinds of first order, ordinary differential equations. Consider, to begin with, a differential equation that can be written in the form

$$M(x) = N(y)\frac{dy}{dx}$$

See Chapter 9.

The equation may be rewritten in the differential form

$$M(x)\,dx = N(y)\,dy$$

where the variables x and y are separated from each other. The process which leads to this form is called **separating the variables**, and equations that can be written in this form are called **separable.** On integrating both sides of the equation, we obtain the solution involving an arbitrary constant $c \in \mathbb{R}$:

Never forget the arbitrary constant!

$$\int M(x)\,dx = \int N(y)\,dy + c$$

Example 1

This equation is often written as

$$(D - \alpha)y = 0$$

(see Chapter 14).

To solve the equation

$$\frac{dy}{dx} = \alpha y \quad \alpha \in \mathbb{R}$$

we can proceed by separating the variables. When $y \neq 0$, integrating both sides of the equation, we obtain

$$\int \frac{1}{y}\,dy = \int \alpha\,dx$$

Integration yields

$$\ln y = \alpha x + \ln c$$

where the arbitrary constant of integration is written as $\ln c$ to make the final solution look neat. Then[†]

† We use the fact that $\ln x$ and e^x are inverse functions.

$$\ln y = \ln e^{\alpha x} + \ln c = \ln c\,e^{\alpha x}$$

Taking exponentials, we obtain the solution

$$y = ce^{\alpha x}$$

This is a key result for the equations of Chapter 14, and singling it out now will pay dividends later:

The general solution of $(D - \alpha)y = 0$ is $y = ce^{\alpha x}$

Two of the many instances where a differential equation of the above form arises are now illustrated:

(i) Application to compound interest

The problem is to find the amount S accumulated when $\$P$ is deposited in an account which is continuously compounded at an interest rate r.

The rate of growth is r times the amount, which leads to the differential equation

$$\frac{dS}{dt} = rS$$

with general solution $S = ce^{rt}$.

When $t = 0$, $S = P$, and so we have the boundary condition that the solution passes through the point $(t, S)^{\mathrm{T}} = (0, P)^{\mathrm{T}}$. Writing $t = 0$ in $S = ce^{rt}$ we obtain $P = c$ and so

$$S = Pe^{rt}$$

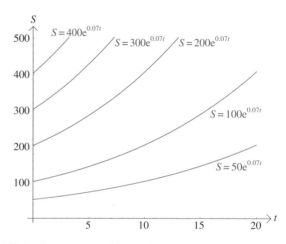

An amount invested at compound interest therefore grows exponentially. The figure on the right shows the growth in S at a rate of 7% with

$$P = 50, 100, 200, 300, 400$$

over a period of 20 years.

Suppose that we wish to find if the investment will quadruple over 20 years at a rate of 7%. To find the time necessary for an investment to quadruple, we want the smallest t for which

$$4P \le P\,e^{0.07t}$$

or

$$4 \le e^{0.07t} \quad \text{since } P > 0$$

i.e.

$$t \ge \frac{1}{0.07}\ln 4 \approx 19.8 \text{ years.}$$

Hence the sums increase by slightly more than a factor of 4 in 20 years, which is clearly seen in the graph.

(ii) Application to population models

The rate of increase with time t of a population P of a colony of bacteria is proportional to the population P at time t. So the rate of growth of P is again modelled by the differential equation

$$\frac{dP}{dt} = kP$$

where k is a particular constant. This has general solution

$$P(t) = ce^{kt}$$

where c is any arbitrary constant.

Suppose that we need to calculate the population at time $t = 100$ in the case $k = 2$. Then

$$P(100) = ce^{200}$$

If it is known that the population was 60 at time $t = 5$, the boundary condition takes the form $P(5) = 60$. It then follows that

$$60 = P(5) = ce^{2 \times 5} = ce^{10}$$

and so

$$c = 60e^{-10}$$

Thus

$$P(t) = 60e^{-10}e^{2t} = 60e^{2t-10}$$

and, in particular, $P(100) = 60e^{190}$.

Alternatively, it may be that the boundary condition takes the form $P'(1) = 3$. That is, it is known that the rate of increase of the population was 3 at time 1. Since

$$P'(t) = cke^{kt}$$

it follows that

$$3 = P'(1) = c2e^{2 \times 1} = 2ce^2$$

Thus

$$c = \frac{3}{2}e^{-2}$$

and so

$$P(t) = \frac{3}{2}e^{-2}e^{2t} = \frac{3}{2}e^{2(t-1)}$$

In particular, $P(100) = \frac{3}{2}e^{198}$.

Example 2

To find the general solution of the differential equation

$$xy\frac{dy}{dx} = 2(y + 3)$$

we begin by writing the equation in the form

$$xy\,dy = 2(y + 3)\,dx$$

Next, separate the variables. This gives

$$\frac{y\,dy}{y + 3} = \frac{2\,dx}{x}$$

or

$$\left(1 - \frac{3}{y + 3}\right)dy = \frac{2\,dx}{x}$$

and so

$$\int dy = \int \frac{3}{y+3}\, dy + 2\int \frac{dx}{x} + \ln c$$

where the arbitrary constant is written as $\ln c$, as before.

Hence the general solution is

$$y = 3\ln(y+3) + 2\ln x + \ln c = \ln(y+3)^3 + \ln x^2 + \ln c = \ln cx^2(y+3)^3$$

Taking exponentials, the general solution takes the form

$$e^y = cx^2(y+3)^3$$

We can always check that the solution is correct, by differentiating it to eliminate c, thereby recovering the differential equation.

Example 3

An application to physics

Ignoring the effect of the atmosphere, suppose we wish to find the 'escape velocity', which is the velocity with which a particle must be projected radially from the surface of the earth if it is not to fall back again.

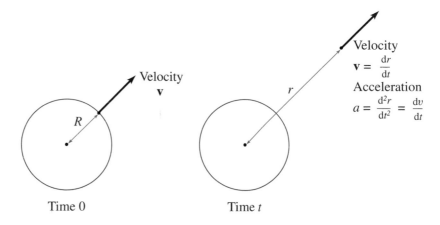

Newton's law of gravitation asserts that the (outward) acceleration a satisfies

$$a = -\frac{k}{r^2}$$

The well known constant g is approximately 981 cm s^{-2} when distances are measured in centimetres and times in seconds.

where the positive constant k is given by $k = gR^2$. Observe that

$$a = \frac{dv}{dt} = \frac{dv}{dr}\frac{dr}{dt} = v\frac{dv}{dr}$$

We therefore obtain the differential equation

$$v\frac{dv}{dr} = -\frac{k}{r^2}$$

379

We separate the variables. Then

$$v \, dv = -\frac{k}{r^2} \, dr$$

$$\int v \, dv = -k \int \frac{dr}{r^2} + c$$

$$\frac{1}{2} v^2 = \frac{k}{r} + c$$

This is the *general solution*. We require the particular solution which satisfies the boundary condition $v = V$ when $r = R$. If this is satisfied, then

$$\frac{1}{2} V^2 = \frac{k}{R} + c$$

and so

$$c = \frac{1}{2} V^2 - \frac{k}{R}$$

It follows that

$$\frac{1}{2} v^2 = k \left\{ \frac{1}{r} - \frac{1}{R} \right\} + \frac{1}{2} V^2$$

The condition that the particle never falls back to the Earth may be expressed as the requirement that $v > 0$ for all v. We therefore require that

$$k \left\{ \frac{1}{r} - \frac{1}{R} \right\} + \frac{1}{2} V^2 > 0$$

for *all* $r \geq R$. But this is true if and only if

$$\frac{1}{2} V^2 - \frac{k}{R} \geq 0$$

i.e.

$$V^2 \geq \frac{2k}{R} = 2gR$$

It follows that the escape velocity is $\sqrt{2gR}$.

Example 4

An application to economics

A **demand function** is a function of the market price p which gives the quantity demanded x_d. A **supply function** is the function of the market price which gives the quantity x_s that producers are willing to supply.

Suppose that the demand and supply functions for a commodity are modelled by the equations

$$x_d = 4 - p$$
$$x_s = 2p - 2$$

Economics textbooks argue that the price p will increase if demand exceeds supply, and decrease if supply exceeds demand. To find the equilibrium price, one should therefore set demand equal to supply. Writing $4 - p = 2p - 2$, we find that the equilibrium price is $p = 2$.

An adjustment process that corrects p to its equilibrium value is called a **tâtonnement**. Economists envisage an auctioneer who announces prices, listens to the amounts that traders would be willing to buy or sell at that price, and then moves the price up or down depending on whether the excess demand $x_d - x_s$ is positive or negative.[†]

[†] The traditional 'gold fix' operated by Rothschild's Bank in London determines the daily price of gold in just this way.

Suppose that the price p in a *tâtonnement* adjusts according to the rule

$$\frac{dp}{dt} = \frac{1}{108}(x_d - x_s)^5$$

That is

$$\frac{dp}{dt} = \frac{1}{108}(4 - p - (2p - 2))^5$$

or

$$\frac{dp}{dt} = \frac{1}{108}(6 - 3p)^5 = \frac{9}{4}(2 - p)^5$$

This separable differential equation can be integrated:

$$\int \frac{4\,dp}{9(2 - p)^5} = \int dt$$

Integrating both sides of the equation

$$-\frac{1}{(-4)}\frac{4}{9(2 - p)^4} = (t + c)$$

or

$$(2 - p)^4 = \frac{1}{9(t + c)}$$

Taking fourth roots[‡]

[‡] The other two roots are complex (*see* §13.5).

$$(2 - p) = \pm\frac{1}{\sqrt{3}(t + c)^{\frac{1}{4}}}$$

Solving for p

$$p(t) = 2 \pm \frac{1}{\sqrt{3}(t + c)^{\frac{1}{4}}} \tag{1}$$

To find c let $t = 0$:

$$p(0) = 2 \pm \frac{1}{\sqrt{3}c^{\frac{1}{4}}}$$

so

$$c^{\frac{1}{4}} = \frac{1}{\sqrt{3}(2 - p(0))}$$

and

$$c = \frac{1}{\sqrt{3}^4(2 - p(0))^4}$$

Suppose that $p(0) = 1$. Since $p(0) < 2$ for this particular solution we must choose the negative sign in (1). Then $c = \frac{1}{9}$. Hence

$$p(t) = 2 - \frac{1}{\sqrt{3}(t + \frac{1}{9})^{\frac{1}{4}}}$$

or

$$p(t) = 2 - \frac{1}{(9t+1)^{\frac{1}{4}}}$$

As $t \to \infty$, $\dfrac{1}{(9t+1)^{\frac{1}{4}}} \to 0$ so $p(t) \to 2$ from below.

Recall that $p = 2$ is the **equilibrium price** that equates supply and demand. When $p(t) = 2$ the rate of price change, $dp/dt = \frac{9}{4}(2-p)^5$, is zero, which means that price stays at this level once it has reached it.

When $p(0) = 2$, price remains constant and the time path is horizontal.

Suppose that $p(0) = 3$. Since $p(0) > 2$ we must choose the positive sign in equation (1) for this particular solution. Then $c = \frac{1}{9}$. Hence

$$p(t) = 2 + \frac{1}{(9t+1)^{\frac{1}{4}}}$$

As $t \to \infty$, $p(t) \to 2$ from above.

The graph below shows price trends with several initial prices. In every case price tends *very slowly* to the equilibrium level 2, if it does not already start at 2.

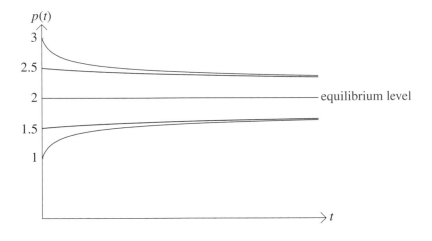

12.5 Exact equations

Consider the equation

$$f(x, y) = c$$

where c is a constant. Introducing differentials, we obtain

$$df = \frac{\partial f}{\partial x}\, dx + \frac{\partial f}{\partial y}\, dy = 0$$

Thus, if an ordinary differential equation of the first order

$$M(x, y)\, dx + N(x, y)\, dy = 0$$

has the property that

$$M = \frac{\partial f}{\partial x} \quad \text{and} \quad N = \frac{\partial f}{\partial y} \tag{1}$$

for some function $f: \mathbb{R}^2 \to \mathbb{R}$, then the general solution takes the form

$$f(x, y) = c$$

where c is an arbitrary constant. Such a differential equation is said to be **exact**.

If equations (1) are satisfied, then

$$\frac{\partial M}{\partial y} = \frac{\partial}{\partial y}\left(\frac{\partial f}{\partial x}\right) = \frac{\partial^2 f}{\partial y\, \partial x} = \frac{\partial^2 f}{\partial x\, \partial y} = \frac{\partial}{\partial x}\left(\frac{\partial f}{\partial y}\right) = \frac{\partial N}{\partial x}$$

Thus, for the equation $M\, dx + N\, dy = 0$ to be exact, we require that

$$\frac{\partial M}{\partial y} = \frac{\partial N}{\partial x}$$

Example 5 _____

Some simple exact equations together with their solutions are given below. Note that the above condition holds in each case.

$$y\, dx + x\, dy = 0 \tag{2}$$

We have $d(xy) = x\, dy + y\, dx$ and so the general solution is $xy = c$, a family of hyperbolas.

$$\frac{dx}{y} - \frac{x\, dy}{y^2} = 0 \tag{3}$$

We have

$$d\left(\frac{x}{y}\right) = \frac{y\, dx - x\, dy}{y^2}$$

and so the general solution is $x = yc$, a family of straight lines through the origin with slope $1/c$.

$$e^x \sin y\, dx + e^x \cos y\, dy = 0 \tag{4}$$

We have

$$d\left(e^x \sin y\right) = e^x \sin y\, dx + e^x \cos y\, dy$$

and so the general solution is

$$e^x \sin y = c$$

a family of curves which is periodic in y for each x.

The solution curves of these three equations are shown below:

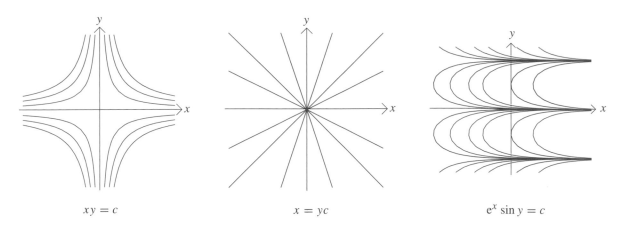

$$xy = c \qquad\qquad x = yc \qquad\qquad e^x \sin y = c$$

In general, of course, it is rather unlikely that a given equation

$$M(x, y)\,\mathrm{d}x + N(x, y)\,\mathrm{d}y = 0$$

Such a function μ is called an **integrating factor** for the equation.

will be *exact*. Sometimes, however, it is possible to find a function $\mu(x, y)$ such that

$$\mu M\,\mathrm{d}x + \mu N\,\mathrm{d}y = 0$$

is exact.

The technique of using an integrating factor is particularly important for the class of differential equations considered next.

12.6 Linear equations of order one

A **linear** differential equation of the first order has the form

$$\frac{\mathrm{d}y}{\mathrm{d}x} + P(x)y = Q(x) \tag{1}$$

To be linear, an equation must only contain terms of *degree* one in $\mathrm{d}y/\mathrm{d}x$ and y.

Treat expressions involving x alone as 'constants' and write the left hand side as a 'linear combination' of $\mathrm{d}y/\mathrm{d}x$ and y:

$$a(x)\frac{\mathrm{d}y}{\mathrm{d}x} + b(x)y = c(x)$$

Before finding the integrating factor of the equation, it is essential to divide by $a(x)$ to make the coefficient of $\mathrm{d}y/\mathrm{d}x$ equal to one, as in (1).

A suitable integrating factor for such a differential equation, with the coefficient of $\mathrm{d}y/\mathrm{d}x$ equal to one, is given by $e^{\int P(x)\,\mathrm{d}x}$.

This is easily proved directly. Note first that

$$\frac{d}{dx}(e^{\int P dx}) = Pe^{\int P dx}$$

Now multiply the differential equation through by $e^{\int P(x) dx}$. Then

$$e^{\int P(x) dx}\frac{dy}{dx} + e^{\int P(x) dx} Py = e^{\int P(x) dx} Q$$

$$\frac{d}{dx}(e^{\int P(x) dx} y) = e^{\int P(x) dx} Q$$

It follows that the general solution is

$$e^{\int P(x) dx} y = \int e^{\int P(x) dx} Q \, dx + c$$

Example 6

To find the general solution of

$$x\frac{dy}{dx} + 2y = 8x^2$$

we must first divide by x to make the coefficient of $\dfrac{dy}{dx}$ equal to one, to get

$$\frac{dy}{dx} + \frac{2y}{x} = 8x$$

† *For the last step, note that e^x and $\ln x$ are inverse functions.*

This equation is linear with $P = 2/x$ and $Q = 8x$. The integrating factor† is

$$e^{\int P dx} = e^{\int (2/x) dx} = e^{2\ln x} = e^{\ln x^2} = x^2$$

We therefore consider

$$x^2\frac{dy}{dx} + 2xy = 8x^3$$

that is,

$$\frac{d}{dx}(x^2 y) = 8x^3$$

The general solution is then

$$x^2 y = \int 8x^3 \, dx + c$$

or

$$x^2 y = 2x^4 + c$$

Example 7

To find the general solution of

$$x(x+1)\frac{dy}{dx} + y = x + 1$$

first put this into the standard form of (1), i.e.

$$\frac{dy}{dx} + \frac{1}{x(x+1)} y = \frac{1}{x}$$

This is linear with $P = \dfrac{1}{x(x+1)}$ and $Q = \dfrac{1}{x}$. Since

$$\int \frac{dx}{x(x+1)} = \int \left\{ \frac{1}{x} - \frac{1}{x+1} \right\} dx = \ln x - \ln(x+1) = \ln\left(\frac{x}{x+1}\right)$$

it follows that the appropriate integrating factor is

$$\mu = e^{\ln(x/x+1)} = \frac{x}{x+1}$$

We therefore consider

$$\frac{x}{x+1}\frac{dy}{dx} + \frac{1}{(x+1)^2} y = \frac{1}{x+1}$$

i.e.

$$\frac{d}{dx}\left(\frac{xy}{x+1}\right) = \frac{1}{x+1}$$

The general solution is then

$$\frac{xy}{x+1} = \ln(x+1) + c$$

Example 8

An economics application

Suppose the demand and supply functions of a commodity are the same linear functions as in Example 4:

$$x_d = 4 - p \quad \text{and} \quad x_s = 2p - 2$$

We now assume in this *tâtonnement* model of price adjustment that the rate of change of price, dp/dt, is twice the excess demand, $x_d - x_s$, i.e.

$$\frac{dp}{dt} = 2(x_d - x_s) = 12 - 6p$$

or

$$\frac{dp}{dt} + 6p = 12$$

a linear equation in p. Multiplying by the integrating factor $e^{\int 6\,dt} = e^{6t}$ and integrating,

$$e^{6t} p = 12 \int e^{6t}\,dt + c$$

$$= 2e^{6t} + c$$

or

$$p = c\,e^{-6t} + 2$$

When $t = 0$, $p(0) = c + 2$ and so

$$p(t) = (p(0) - 2)\,e^{-6t} + 2 \tag{2}$$

Since $e^{-6t} \to 0$ as $t \to \infty$, we have $p \to 2$ as $t \to \infty$. This is the equilibrium level, since as in Example 4, when $p = 2$, $dp/dt = 6(2 - p)$ is zero, so the price stays at this level once it has reached it.

When $p(0) = 2$, $p(t) = 2$ for all t, and so the time path is horizontal.

If $p(0) > 2$, the first term on the right hand side of equation (2) is positive and tends to 0 as $t \to \infty$. So $p(t) > 2$ and tends to 2 from above as $t \to \infty$.

If $p(0) < 2$, $p(t) < 2$ and similarly tends to 2 from below as $t \to \infty$.

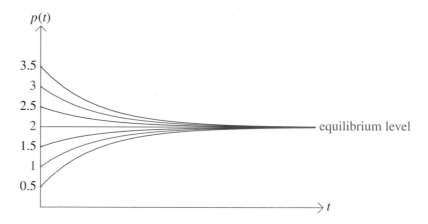

The graph above shows price trends for various initial prices. These are similar to our previous example, except that here the convergence to the equilibrium level is very rapid.

Convergence of solutions will be discussed in §14.8.

12.7 Homogeneous equations

Recall from Exercise 5.8.13 that a function $f(x, y)$ is called **homogeneous of degree n** if

$$f(\lambda x, \lambda y) = \lambda^n f(x, y)$$

It is easy to see at a glance if a function is homogeneous of degree n when the function is made up of powers of x and y. The combined degrees of x and y in each term must always be the same integer n. E.g. the combined degrees of x and y in each term of the function $x^4 - x^3 y$ are 4, and so it is homogeneous of degree 4, with

$$(\lambda x)^4 - (\lambda x)^3(\lambda y) = \lambda^4(x^4 - x^3 y)$$

Since $\lambda y/\lambda x = \lambda^0 (y/x)$, any function of y/x is always homogeneous of degree 0. The appearance of any expressions that can be written in terms of y/x therefore does not undermine any possible homogeneity in a function. Since x/y is a function of y/x, the same also goes for expressions that can be written in terms of x/y.

E.g. the function $\cos(y/x) + e^{x/y}$ is homogeneous of degree 0 since

$$\cos(\lambda y/\lambda x) + e^{\lambda x/\lambda y} = \lambda^0 \left(\cos(y/x) + e^{x/y}\right)$$

Again, the function $e^{x/y}(x^4 - x^3 y)$ is homogeneous of degree 4.

The function $\sin x \cos y$ is not homogeneous of any degree since

$$\sin(\lambda x)\cos(\lambda y) \neq \lambda^n \sin x \cos y \text{ for any } n$$

A differential equation

$$M(x, y) + N(x, y)\frac{dy}{dx} = 0$$

is called **homogeneous of degree** n if M and N are both homogeneous functions of degree n.

If a differential equation is homogeneous of any degree, the transformation $y = vx$ will 'separate' the variables v and x.

Example 9

The following differential equation is not separable, but is homogeneous of degree 3:

$$(x^3 + y^3) = 3xy^2\frac{dy}{dx} \qquad (1)$$

Let $y = vx$, so that

$$\frac{dy}{dx} = v + x\frac{dv}{dx}$$

Then

$$(x^3 + v^3 x^3) = 3x.v^2 x^2 \left(v + x\frac{dv}{dx}\right)$$

$$(1 + v^3) - 3v^3 - 3v^2 x\frac{dv}{dx} = 0$$

$$(1 - 2v^3) = 3v^2 x\frac{dv}{dx}$$

Separating variables

$$\frac{dx}{x} = \frac{3v^2\,dv}{1 - 2v^3}$$

On integrating,

$$\ln x = -\frac{1}{2}\ln(1 - 2v^3) + \ln c'$$

$$x\sqrt{1 - 2v^3} = c'$$

Remember to return to the original variables.

Substituting back for v in terms of y

$$x^2\left(1 - 2\frac{y^3}{x^3}\right) = c$$

or

$$x^3 - 2y^3 = cx \tag{2}$$

The equation does not define dy/dx for the point $(x, y) = (0, 0)$, which satisfies (2) for every c in $\mathbb{R}$. This is reflected in the fact that all curves of the family pass through this point, as shown in the following figure.

The particular solution passing through the point $(x, y)^T = (1, 0)^T$ is obtained by substituting in (2), which gives $c = 1$ and so the solution, which is shown in red in the figure is

$$x^3 - 2y^3 = x$$

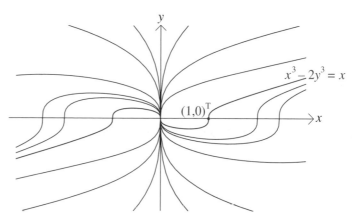

Solution curves of $(x^3 + y^3) = 3xy^2 \dfrac{dy}{dx}$

Occasionally the alternative substitution $x = vy$ results in more manageable functions. For instance, with functions of the form $f(x/y)$, like $e^{x/y}$, it gives e^v rather than $e^{1/v}$.

12.8 Change of variable

We have seen that the solution of homogeneous equations is effected by a change of variable. More generally, an unpleasant differential equation can sometimes be converted to any one of the forms we have studied by making a change of variable. Unfortunately it is not always easy to think of an appropriate change of variable. Some possible instances are now illustrated.

Example 10

We first illustrate in this example that an equation can sometimes be converted into a known form, by merely interchanging the roles of x and y. To solve

$$y \ln y + (x - \ln y) \frac{dy}{dx} = 0$$

which is not linear in y, but linear in x, interchange the roles of x and y.

$$\frac{dx}{dy} + \frac{x}{y \ln y} = \frac{1}{y}$$

This has integrating factor $e^{\int P \, dy} = e^{\int \frac{dy}{y \ln y}} = e^{\ln \ln y} = \ln y$. Hence

$$\ln y \, \frac{dx}{dy} + \frac{x}{y} = \frac{\ln y}{y}$$

or

$$\frac{d}{dy}(x \ln y) = \frac{\ln y}{y}$$

On integrating with respect to y,

$$x \ln y = \int \frac{\ln y}{y} \, dy = \frac{(\ln y)^2}{2} + c \tag{1}$$

We find the particular solution through the point $(x, y)^T = (0, 1)^T$ by substituting in (1), which yields $c = 0$. So the solution we want is

$$x \ln y = \frac{(\ln y)^2}{2}$$

which factorises to give

$$\ln y = 0 \quad \text{and} \quad x = \frac{1}{2} \ln y$$

or, as shown in red in the figure

$$y = 1 \quad \text{and} \quad y = e^{2x}$$

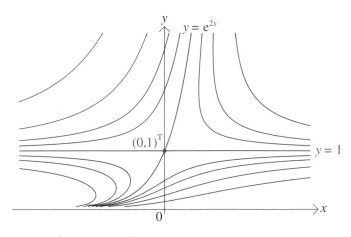

Solution curves of $y \ln y + (x - \ln y)\dfrac{dy}{dx} = 0$

Example 11

To find the general solution of

$$\frac{dy}{dx} = (9x + 4y + 1)^2$$

the change of variable $z = 9x + 4y + 1$ looks as though it might be helpful. We then have

$$\frac{dz}{dx} = 9 + 4\frac{dy}{dx}$$

and so the differential equation becomes

$$\frac{dz}{dx} = 9 + 4z^2$$

$$\frac{dz}{9 + 4z^2} = dx$$

$$\int \frac{dz}{9 + 4z^2} = \int dx + c$$

$$\frac{1}{6} \arctan\left(\frac{2z}{3}\right) = x + c$$

$$2z = 3\tan(6x + a)$$

Substituting for z, the solution is

$$2(9x + 4y + 1) = 3\tan(6x + a)$$

Example 12

To solve

$$\sin y \frac{dy}{dx} - 2\cos x \cos y + \cos x = 0$$

substitute $-\cos y = v$ so $\sin y \, dy = dv$.

The equation assumes the linear form

$$\frac{dv}{dx} + 2v \cos x = -\cos x.$$

Multiplying by the integrating factor $e^{\int 2\cos x \, dx} = e^{2\sin x}$ we get

$$e^{2\sin x} \, dv + e^{2\sin x} 2\cos x v \, dx = -e^{2\sin x} \cos x \, dx$$

or

$$d(e^{2\sin x} v) = -e^{2\sin x} \cos x \, dx$$

with solution

$$e^{2\sin x} v = -\int e^{2\sin x} \cos x \, dx = -\frac{e^{2\sin x}}{2} + c$$

In terms of the original variables the solution is

$$\cos y - \frac{1}{2} + c e^{-2 \sin x} = 0$$

12.9 Identifying the type of first order equation

It is useful to have a clear strategy for sorting equations into the different classes. To begin with, write the equation as

$$\frac{dy}{dx} = f(x, y)$$

I The equation is separable if $f(x, y)$ can be written as the product of a function of x alone and a function of y alone, so that the equation takes the form

$$\frac{dy}{dx} = g(x)h(y)$$

II The equation is homogeneous of degree n if it can be written in the form

$$\frac{dy}{dx} = \frac{g(x, y)}{h(x, y)}$$

where g and h are homogeneous of the same degree n.

III Write the equation in the differential form

$$M(x, y)\, dx + N(x, y)\, dy = 0$$

The equation is exact if

$$\frac{\partial M}{\partial y} = \frac{\partial N}{\partial x}$$

IV The equation is linear if it only contains terms of *degree* one in dy/dx and y. Treat expressions involving x alone as 'constants' and write the left hand side as a 'linear combination' of dy/dx and y:

$$a(x)\frac{dy}{dx} + b(x)y = c(x)$$

There is no reason why an equation may not belong to more than one category. The trivial example

$$\frac{dy}{dx} = -\frac{y}{x}$$

for instance, which we have met as (2) of Example 5, is of all four types. When an equation belongs to more than one class it can be a useful exercise to solve it by the different methods and compare solutions. You must realise, though, that two solutions may look very different, even though they might be equivalent. As they involve arbitrary constants, the scope for variation is endless!

12.10 Partial differential equations

We touch very briefly on this topic, and only consider the simplest type of partial differential equation, whose solution can be found directly by partial integration. The general solution of a partial differential equation of order n requires n arbitrary *functions*. We also consider equations which can be transformed to this type.

Example 13

Suppose that

$$f(x, y) = g(y)x + y \tag{1}$$

Then

$$\frac{f - y}{x} = g(y)$$

and hence, differentiating partially with respect to x,

$$\frac{\partial}{\partial x}\left\{\frac{f - y}{x}\right\} = \frac{\partial}{\partial x}\{g(y)\} = 0$$

or

$$\frac{1}{x}\frac{\partial f}{\partial x} - \frac{1}{x^2}(f - y) = 0$$

Thus equation (1) is the general solution of the partial differential equation

$$x\frac{\partial f}{\partial x} = f - y \tag{2}$$

Example 14

Suppose that we wish to find the general solution of

$$\frac{\partial f}{\partial x} = 3y^2 + x \tag{3}$$

When the partial derivative was calculated, we kept y constant. To recover f we must therefore integrate with respect to x, *keeping y constant* (*see* §11.2). The constant of integration obtained will then really be constant only if y is held constant.

When integrating partially with respect to x the 'constant of integration' is actually an arbitrary function of y.

In the case of the current problem, we obtain

$$f(x, y) = \int (3y^2 + x)\, dx + g(y)$$

$$= 3y^2 x + \frac{1}{2}x^2 + g(y)$$

where g is an arbitrary function.

This result can be checked by differentiating partially with respect to x to recover equation (3).

Example 15

To find the general solution of

$$\frac{\partial^2 f}{\partial x \, \partial y} = 3y^2 + x \tag{4}$$

Since we assume that

$$\frac{\partial^2 f}{\partial x \, \partial y} = \frac{\partial^2 f}{\partial y \, \partial x},$$

the order of integration does not strictly matter.

integrate partially with respect to x and then y, to obtain

$$\frac{\partial f}{\partial y} = \int (3y^2 + x) \, dx + g(y)$$

$$= \left(3y^2 x + \frac{1}{2}x^2 \right) + g(y)$$

$$f = \int \left(3y^2 x + \frac{1}{2}x^2 \right) dy + \int g(y) \, dy + h(x)$$

$$f(x, y) = y^3 x + \frac{1}{2}x^2 y + G(y) + h(x)$$

where G and h are arbitrary functions.

To verify the solution, differentiate partially, first with respect to x and then with respect to y. The result is equation (4).

Example 16

We now carry out the reverse process to that of Example 13. We integrate the partial differential equation (2), derived in Example 13, to obtain its solution, equation (1). This is slightly more complicated than the previous examples as besides the term $\partial f / \partial x$ the equation also involves f. The equation

$$x \frac{\partial f}{\partial x} = f - y \tag{2}$$

can be written in the form

$$\frac{\partial f}{\partial x} - \frac{1}{x} f = -\frac{y}{x}$$

which is 'linear' in the variables $\partial f / \partial x$ and f, with x the independent variable and y treated as a 'constant'. Its integrating factor is given by

See §12.6.

$$e^{-\int \frac{1}{x} dx} = e^{-\ln x} = e^{\ln x^{-1}} = \frac{1}{x}$$

On multiplication by the integrating factor the equation becomes

$$\frac{1}{x} \frac{\partial f}{\partial x} - \frac{1}{x^2} f = -\frac{y}{x^2}$$

or

$$\frac{\partial}{\partial x} \left(\frac{1}{x} f \right) = -\frac{y}{x^2}$$

On integrating partially with respect to x,

$$\frac{1}{x} f = \frac{y}{x} + g(y)$$

where g is an arbitrary function of y. Multiplying by x gives the solution in the form with which we had started in Example 13,

$$f(x, y) = g(y)x + y \tag{1}$$

12.11 Exact equations and partial differential equations

The terms $M(x, y)$ and $N(x, y)$ of an exact equation

$$M(x, y)\,\mathrm{d}x + N(x, y)\,\mathrm{d}y = 0$$

are the partial derivatives $\dfrac{\partial f}{\partial x}$ and $\dfrac{\partial f}{\partial y}$ respectively of a function f. Hence this function is a common solution of the partial differential equations

$$\left.\begin{array}{c} \dfrac{\partial f}{\partial x} = M \\[2mm] \dfrac{\partial f}{\partial y} = N \end{array}\right\}$$

The following example illustrates this.

Example 17

It is easy to check that the equation

$$(3x^2y^2 + 2yx + 4x^3)\,\mathrm{d}x + (2x^3y + x^2)\,\mathrm{d}y = 0$$

is exact.

$$\frac{\partial M}{\partial y} = 6x^2y + 2x, \qquad \frac{\partial N}{\partial x} = 6x^2y + 2x$$

and so the equation is exact. This means that a function f can be found such that

$$\left.\begin{array}{c} \dfrac{\partial f}{\partial x} = 3x^2y^2 + 2yx + 4x^3 = M \\[2mm] \dfrac{\partial f}{\partial y} = 2x^3y + x^2 = N \end{array}\right\}$$

The general solutions of these partial differential equations are

$$\left.\begin{array}{c} f = x^3y^2 + x^2y + x^4 + g(y) \\[2mm] f = x^3y^2 + x^2y + h(x) \end{array}\right\}$$

These equations hold simultaneously if $g(y) = 0$ and $h(x) = x^4$. Hence the general solution of the original equation is

$$x^3y^2 + x^2y + x^4 = c$$

Note that the arbitrary function $h(x)$ which is x^4 comes from integrating $M(x, y)$ partially with respect to x, and so can be ignored when integrating $N(x, y)$ partially with respect to y. A similar remark applies to the arbitrary function $g(y)$.

The above reasoning indicates that a quick method of finding the solution of the exact equation

$$M \, dx + N \, dy = 0$$

is the following:

Integrate M partially with respect to x and N partially with respect to y, ignoring the respective arbitrary functions of integration. The solution is obtained by equating the sum of these results to a constant, after discarding repeated terms.

Example 18

We can employ the above method to solve the following equation, after checking that it is exact:

$$(3e^{3x} y - 2x) \, dx + e^{3x} \, dy = 0$$

$$\frac{\partial M}{\partial y} = 3e^{3x} \qquad \frac{\partial N}{\partial x} = 3e^{3x}$$

so the equation is exact.

Integrating $(3e^{3x} y - 2x)$ partially with respect to x, gives $e^{3x} y - x^2$. Then integrating e^{3x} partially with respect to y, gives $e^{3x} y$. This term already appears in our previous integration, and so need not be written down again. The solution is therefore

$$ye^{3x} - x^2 = c \qquad (c \in \mathbb{R})$$

Example 19

The differential equation

$$2y^2 e^{2x} \, dx + (4ye^{2x} - y^{-2} e^{-y}) \, dy = 0$$

is not exact, but has an integrating factor of the form y^n where n is an integer.
We shall check that the equation is not exact.

$$\frac{\partial M}{\partial y} = 4ye^{2x} \qquad \frac{\partial N}{\partial x} = 8ye^{2x}$$

Therefore

$$\frac{\partial M}{\partial y} \neq \frac{\partial N}{\partial x}$$

Multiplying by the integrating factor, the equation becomes

$$2y^{n+2} e^{2x} \, dx + (4y^{n+1} e^{2x} - y^{n-2} e^{-y}) \, dy = 0$$

The equation should now be exact, and so, applying the criterion for exactness with our new M and N, we get

$$\frac{\partial M}{\partial y} = 2(n+2)y^{n+1}e^{2x} = 8y^{n+1}e^{2x} = \frac{\partial N}{\partial x}$$

Hence $n+2 = 4$ or $n = 2$; and the following equation is exact:

$$2y^4 e^{2x}\,dx + (4y^3 e^{2x} - e^{-y})\,dy = 0$$

Integrating $2y^4 e^{2x}$ partially with respect to x gives $y^4 e^{2x}$. Integrating $4y^3 e^{2x} - e^{-y}$ partially with respect to y gives $y^4 e^{2x} + e^{-y}$. Discarding the repeated term, the solution may be written as

$$y^4 e^{2x} + e^{-y} = c$$

12.12 Change of variable in partial differential equations

Example 20

It is evident that the differential equation

$$\frac{\partial f}{\partial x} = \frac{\partial f}{\partial y}$$

cannot be solved by direct partial integration. We therefore introduce the change of variable

Criteria for *choosing* an appropriate transformation lie beyond the scope of this book.

$$\left.\begin{array}{l} x = u + v \\ y = u - v \end{array}\right\}$$

which leads to

$$\left.\begin{array}{l} \dfrac{\partial}{\partial u} = \dfrac{\partial}{\partial x} + \dfrac{\partial}{\partial y} \\[2mm] \dfrac{\partial}{\partial v} = \dfrac{\partial}{\partial x} - \dfrac{\partial}{\partial y} \end{array}\right\}$$

Hence

$$\frac{\partial f}{\partial x} - \frac{\partial f}{\partial y} = \frac{\partial f}{\partial v} = 0$$

The general solution is therefore

$$f = g(u) = g\left(\frac{x+y}{2}\right) = g_1(x+y)$$

where g_1 is an arbitrary function.

We have merely shown that any function which satisfies the partial differential equation is an arbitrary function which is expressed in terms of the sum (and no other combination) of x and y. For instance, you may find it helpful to verify that e^{x+y} *and* $\sin(x + y)$ *are solutions, but* $e^x + e^y$ *and* $e^{x^2 y}$ *are not.*

Example 21

Application to physics

In a two dimensional universe, the gravitational potential $f(x, y)$ would satisfy the two dimensional equation

$$\frac{\partial^2 f}{\partial x^2} + \frac{\partial^2 f}{\partial y^2} = 0$$

If, like Newton, we are interested in the gravitational potential induced by a single point mass, it seems a good idea to introduce polar coordinates $(r, \theta)^{\mathrm{T}}$. The reason is that we know that the gravitational potential at a point will depend *only* on its distance r from the gravitating mass. Thus

$$\frac{\partial f}{\partial \theta} = 0$$

From Exercise 7.6.19 we have

$$\frac{\partial^2 f}{\partial x^2} + \frac{\partial^2 f}{\partial y^2} = \frac{\partial^2 f}{\partial r^2} + \frac{1}{r}\frac{\partial f}{\partial r} + \frac{1}{r^2}\frac{\partial^2 f}{\partial \theta^2}$$

But since f is only a function of r, the equation becomes

$$\frac{d^2 f}{dr^2} + \frac{1}{r}\frac{df}{dr} = 0$$

If we write

$$g = \frac{df}{dr}$$

this differential equation reduces to

$$\frac{dg}{dr} + \frac{g}{r} = 0$$

Separating the variables and integrating

$$\frac{dg}{g} = -\frac{dr}{r}$$
$$\ln g = -\ln r + \ln c$$
$$g = \frac{c}{r}$$

where the arbitrary constant is written as $\ln c$ for reasons mentioned before. Thus

$$\frac{df}{dr} = \frac{c}{r}$$

and so

$$f = c \ln r + b$$

where b is an arbitrary constant.

12.13 Exercises

1. Determine the order and the degree of the following ordinary differential equations:

    (i)* $\dfrac{d^3y}{dx^3} + 2xy\dfrac{dy}{dx} = \left(\dfrac{d^2y}{dx^2}\right)^4$

    (ii) $\left(\dfrac{d^2y}{dx^2}\right)^2 - 2\left(\dfrac{dy}{dx}\right)^3 + yx = 0$

2. Eliminate the arbitrary constants a and b from the following equations and hence find differential equations which they satisfy.

    (i)* $x^2y = 1 + ax$ (ii)* $y = ae^{-x} + be^{2x}$ (iii)* $y^2 = 4ax$
    (iv) $e^x \sin y = a$ (v) $y = a\cos(2x + b)$ (vi) $ax^2 + by^2 = 1$

3. Separate the variables and hence find the general solutions of the following differential equations.

    (i)* $(4 + x)\dfrac{dy}{dx} = y^3$ (ii)* $(xy + y)\dfrac{dy}{dx} = (x - xy)$

    (iii)* $x^2y\dfrac{dy}{dx} = e^y$ (iv)* $\dfrac{1}{y}\dfrac{dy}{dx} = (\ln x)(\ln y)$

    (v) $2x\dfrac{dy}{dx} = 3y$ (vi) $\dfrac{dy}{dx} = \dfrac{(y + 1)(y + 2)}{(x + 1)(x + 2)}$

    (vii) $\dfrac{dy}{dx} = xy^2$ (viii) $\dfrac{dy}{dx} = x(1 + x^2)y$

    (ix) $\tan x\dfrac{dy}{dx} = \tan y$

4. Find the general solutions of the following linear differential equations:

    (i)* $x\dfrac{dy}{dx} = e^x(x + 3) - 3y$

    (ii)* $(1 + x^2)\dfrac{dy}{dx} = 2xy + x^2 + x^4$

    (iii)* $\cos x\dfrac{dy}{dx} = \sin x\,(x\cos x + y)$

    (iv) $\dfrac{dy}{dx} = x - 2y$

    (v) $\dfrac{dy}{dx} + 2xy = 2x^3$

    (vi) $(x + 1)^2\dfrac{dy}{dx} + 3(x + 1)y = e^x$

5. Solve the following homogeneous differential equation:

    $$(x^3 + y^3) = 3xy^2\dfrac{dy}{dx}$$

    Write down, without any further calculation, the general solution of

    $$(x^3 + y^3)\dfrac{dy}{dx} = 3x^2y$$

6. Each of the following equations is separable, linear or homogeneous of degree n where $n \in \mathbb{Z}$:

(i)* $x\dfrac{dy}{dx} - y = 3xe^{\frac{2y}{x}}$

(ii) $(x+1)\dfrac{dy}{dx} = (e^x - 2y)$

(iii)* $xy^2\dfrac{dy}{dx} - x^2e^y = e^y$

(iv) $xy\cos\left(\dfrac{y}{x}\right)\dfrac{dy}{dx} = x^2 + y^2\cos\left(\dfrac{y}{x}\right)$

(v)* $\dfrac{dy}{dx} + (2x+1)e^{3x} = 3y$

(vi) $2(2x^2 + y^2)\,dx - xy\,dy = 0$

(vii) $\sin x\dfrac{dy}{dx} - 2(4\sin^2 x - y)\cos x = 0$

Classify them into these types and solve them.

7. Find which of the following equations are exact and solve those which are exact.

(i)* $3x(xy - 2)\,dx + (x^3 + 2y)\,dy = 0$

(ii)* $(x^5 + 3y)\,dx - xdy = 0$

(iii) $(1 + y^2)\,dx + (x^3y + y)\,dy = 0$

(iv) $(2x^3 - xy^2 - 2y + 3)\,dx - (x^2y + 2x)\,dy = 0$

(v) $(2xy + x^2 + x^4)\,dx - (1 + x^2)\,dy = 0$

8. Show that the equation

$$\left(xe^x + 3x(xy - 2)\right)dx + (x^3 + 2y^4)\,dy = 0$$

is exact and solve it. Without further calculation, write down the solution of

$$(y^3 + 2x^4)\,dx + \left(ye^y + 3y(xy - 2)\right)dy = 0$$

9. Classify each of the differential equations below as linear, exact or homogeneous of degree n, $n \in \mathbb{Z}$.

Find the general solutions of the equations by any appropriate method.

(i)* $(\cos y + y\cos x)\,dx + (\sin x - x\sin y)\,dy = 0$

(ii) $(6x^5y^3 + 4x^3y^5)\,dx + (3x^6y^2 + 5x^4y^4)\,dy = 0$

(iii)* $x\,dy - 4y\,dx = (x+1)\,dx$

(iv) $(y^2e^{xy^2} + 4x^3)\,dx + (2xye^{xy^2} - 3y^2)\,dy = 0$

(v)* $(1 + 2e^{x/y})\,dx + 2e^{x/y}\left(1 - \dfrac{x}{y}\right)dy = 0$

(vi) $2x\dfrac{dy}{dx} = 2\sqrt{x} + y + 2$

(vii)* $xy\dfrac{dy}{dx} - y^2 = 3x^2 e^{2\frac{y}{x}}$

10.*♣ For $n \in \mathbb{Z}$, let

$$3x\dfrac{dy}{dx} = 2x^n - 3y$$

For what value(s) of n is the equation

(i)   exact,

(ii) separable

(iii) linear and

(iv) homogeneous of a particular degree?

Find the solution to the equation in each case.

11.   If $y = C(x)$ represents the cost of producing x units in a manufacturing process, the elasticity of cost is defined to be

$$E_C(x) = \dfrac{C'(x)}{C(x)/x} = \dfrac{x}{y}\dfrac{dy}{dx}$$

Given that $E_C(x) = \dfrac{4y - 30x}{y}$ where $C(1) = 15$, find $C(x)$.

12.   Find the demand function $p = f(x)$, where x is quantity and p is price, which has a constant price elasticity of demand

$$E_d = \dfrac{p}{x}\dfrac{dx}{dp}$$

Show that the consumer's surplus for this function is a constant multiple of the revenue.

If the price elasticity of demand is always equal to -5 and the equilibrium price is $p = 2$, find the consumer's surplus.

13.* Make the change of variable

$$\left.\begin{array}{l} u = \tan x \\ v = \cos y \end{array}\right\}$$

and hence find the general solution of

$$(3\tan x - 2\cos y)\sec^2 x\,dx + \tan x \sin y\,dy = 0$$

14.   Make the changes of variable indicated and hence find the general solutions of the given differential equations.

(i) $\dfrac{d^2 y}{dx^2} + \dfrac{dy}{dx} = x$ (Let $\dfrac{dy}{dx} = v$.)

(ii) $x\,dy - y\,dx = x^3 y^6\,dx$ (Let $y^{-5} = v$.)

(iii) $(3x - xy + 2)\dfrac{dy}{dx} + y = 0$ (Interchange x and y.)

15. Solve the following differential equations with the help of suitable substitutions:

    (i) $\quad \dfrac{dy}{dx} = (y - 16x + 7)^2$

    (ii)*♣ $(1 + y^2)\,dx - (\arctan y - x)\,dy = 0$

16. Show that the differential equation

$$2xy\,dx - (4x^2 + 5y^3)\,dy = 0$$

is not exact. Given that it has an integrating factor of the form y^n, find n and hence solve the differential equation.

17.*♣ Show that the differential equation

$$(y + 2)\cos x\,dx + 3\sin x\,dy = 0$$

is not exact. Given that it has an integrating factor of the form $g(y)$, find $g(y)$ and hence solve the differential equation.

18.♣ Show that the differential equation

$$(2x\sin(x + y) + \cos(x + y))\,dx + \cos(x + y)\,dy = 0$$

is not exact. Given that it has an integrating factor of the form $g(x)$, find $g(x)$ and hence solve the differential equation.

19. Derive the solution

$$c = \frac{n}{n - 1}p$$

of the differential equation

$$(n - 1)(c - p)\frac{dc}{dp} = c$$

in either of the following ways.

    (i) Rearrange the equation to solve it as one which is linear in p.

    (ii) Write the equation in its differential form and show that it is not exact. Given that it has an integrating factor of the form c^α, find α and hence solve the equation.

20. Make the changes of variable indicated and hence find the general solutions of the given differential equations.

    (i) $\dfrac{dy}{dx} = 1 + 6xe^{x-y}$

       $x = u + y$

    (ii) $\dfrac{dy}{dx} = \dfrac{4x - 4y}{y - 4x + 3}$

$$x = X + 1, \ y = Y + 1$$

$$\text{(iii) } y \, dx + x \, dy = -\frac{x}{y^2} \, dy + \frac{dx}{y}$$

$$\left. \begin{array}{l} u = xy \\[4pt] v = \dfrac{x}{y} \end{array} \right\}$$

21.♣ Find the general solution of

$$\frac{\partial^2 f}{\partial x \, \partial y} = e^x \sin y + \frac{1}{x} \ln(xe^y)$$

22.* Make the change of variable

$$\left. \begin{array}{l} u = x^2 + y^2 \\ v = x^2 - y^2 \end{array} \right\}$$

and hence find the general solution of

$$\frac{1}{x} \frac{\partial f}{\partial x} + \frac{1}{y} \frac{\partial f}{\partial y} = 1$$

See Example 7.10.

23.*♣ Make the change of variable

$$\left. \begin{array}{l} u = y^2 - x^2 \\ v = x \end{array} \right\}$$

and hence find the general solution of

$$x \frac{\partial f}{\partial y} + y \frac{\partial f}{\partial x} = 2xyf$$

24.* Make the change of variable

$$\left. \begin{array}{l} u = xy \\ v = x \end{array} \right\}$$

and hence find the general solution of

$$x \frac{\partial f}{\partial x} - y \frac{\partial f}{\partial y} = 4y$$

25. Make the change of variable

$$\left. \begin{array}{l} x = u + v \\ y = u - v \end{array} \right\}$$

and hence find the general solution of

$$\frac{\partial^2 f}{\partial x^2} - \frac{\partial^2 f}{\partial y^2} = x$$

Thirteen

Complex numbers

The next chapter on differential and difference equations of order greater than one requires the use of complex numbers. Only a little knowledge of complex numbers is required for this purpose but it is essential that this small amount of information be properly understood. The relevant material is given in §13.1–§13.5 inclusive.

The application of complex numbers to the solution of differential equations is just one of a vast number of their uses. Some applications in statistics, for example, are given at the end of the chapter. More important, however, than all their many uses in applied work is the light an understanding of complex numbers casts on theoretical issues. For example, in §2.3 and §2.6.2 we considered the exponential and logarithmic functions and the trigonometric functions. On the face of it, these two families of functions seem to be entirely unrelated. It is not until the complex versions of these functions are studied that their close relationship becomes apparent.

13.1 Quadratic equations

The solutions to the quadratic equation

$$ax^2 + bx + c = 0 \qquad a \neq 0$$

are obtained by completing the square[†] which gives the well known formula

† See §10.7.

$$x = \frac{-b \pm \sqrt{b^2 - 4ac}}{2a}$$

If

$$\alpha = \frac{-b + \sqrt{b^2 - 4ac}}{2a} \quad \text{and} \quad \beta = \frac{-b - \sqrt{b^2 - 4ac}}{2a}$$

it is easy to check that

$$ax^2 + bx + c = a(x - \alpha)(x - \beta)$$

for all values of x. We call α and β the **roots** of the quadratic equation. If $\alpha = \beta$ the roots are said to be coincident.

When a, b and c are real and $b^2 - 4ac \geq 0$, complex numbers have no role to play in the analysis. However, if $b^2 - 4ac < 0$, we are faced with the problem of extracting the square root of a negative number. Since the squares of *real* numbers are always nonnegative, it was necessary to invent complex numbers for this purpose. They are expressed in terms of an *imaginary* number i that satisfies

$$i^2 = -1$$

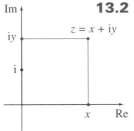

13.2 Complex numbers

Define a complex number z by

$$z = x + iy \qquad (x, y \in \mathbb{R})$$

The complex number $z = x + iy$ may then be plotted in **the complex plane**, where the horizontal and vertical axes represent real and imaginary axes respectively, as in the figure on the left.

We call x the *real part* of z and y the *imaginary part* of z, and write[†]

$$x = \operatorname{Re} z \quad \text{and} \quad y = \operatorname{Im} z.$$

[†] Note that $\operatorname{Im} z$ is not iy as one might expect, but y. The imaginary axis, though, consists of purely imaginary numbers iy.

Real numbers are simply complex numbers for which $y = 0$. A complex number with $x = 0$ is said to be **purely imaginary**.

Addition and multiplication of complex numbers are defined by the rules

$$z_1 + z_2 = x_1 + iy_1 + x_2 + iy_2 = (x_1 + x_2) + i(y_1 + y_2) \tag{1}$$

$$z_1 z_2 = (x_1 + iy_1)(x_2 + iy_2) = (x_1 x_2 - y_1 y_2) + i(x_1 y_2 + x_2 y_1) \tag{2}$$

If z_1 and z_2 are both real (i.e. $y_1 = y_2 = 0$), these formulas reduce to the usual laws of addition and multiplication for real numbers. If α is a real number and z is complex, then (2) says that

$$\alpha z = \alpha(x + iy) = \alpha x + i\alpha y$$

The rules for addition and multiplication of complex numbers are chosen so that *complex numbers satisfy all the usual laws of arithmetic*. This information, together with the definitions above, is all that is necessary for the manipulation of complex quantities.

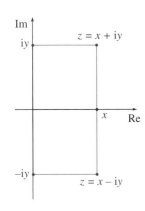

Define $\bar{z} = x - iy$ to be the **complex conjugate** of $z = x + iy$. *In the complex plane, z and its conjugate $\bar{z}$ are reflections in the real axis.*

In the case when $b^2 - 4ac < 0$, the roots α and β of the quadratic equation in §13.1 are the complex conjugate numbers

$$\alpha = \frac{-b + i\sqrt{4ac - b^2}}{2a} \quad \text{and} \quad \beta = \frac{-b - i\sqrt{4ac - b^2}}{2a}$$

Complex roots always occur in conjugate pairs in a quadratic equation with real coefficients. For example, the roots of the equation $z^2 + 1 = 0$ are the conjugate pair i and $-i$.

The following properties of the conjugate of a complex number are immediate consequences of the definition and are worth noting:

(i) $z + \bar{z} = 2x$ is real   (ii) $z - \bar{z} = i2y$ is purely imaginary

(ii) $\overline{(\bar{z})} = z$ (iv) $\overline{z_1 + z_2} = \overline{z_1} + \overline{z_2}$

(iii) $\overline{z_1 z_2} = \overline{z_1}\,\overline{z_2}$

Division of z by the complex number $w = u + iv$ is carried out by writing

$$\frac{z}{w} = \frac{z\bar{w}}{w\bar{w}}$$

i.e.

$$\frac{x + iy}{u + iv} = \frac{(x + iy)(u - iv)}{(u + iv)(u - iv)} = \frac{(xu + yv) + i(yu - xv)}{u^2 + v^2}$$

provided that u and v are not both zero. *Division by zero is never valid.*

Example 1

(i) $(1 + 2i) + (3 + 4i) = (1 + 3) + (2 + 4)i = 4 + 6i$

(ii) $(1 + 2i) - (3 + 4i) = (1 - 3) + (2 - 4)i = -2 - 2i$

(iii) $(1 + 2i)(3 + 4i) = 1(3 + 4i) + 2i(3 + 4i)$
$$= 3 + 4i + 6i + 8i^2$$
$$= 3 + 10i - 8 = -5 + 10i$$

(iv) $\dfrac{(1 + 2i)}{(3 + 4i)} = \dfrac{(1 + 2i)(3 - 4i)}{(3 + 4i)(3 - 4i)} = \dfrac{3 - 4i + 6i - 8i^2}{9 - 12i + 12i - 16i^2}$
$$= \frac{11 + 2i}{25} = \frac{11}{25} + \frac{2}{25}i$$

13.3 Modulus and argument

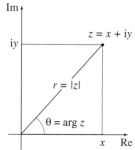

Im

$z = x + iy$

iy

$r = |z|$

$\theta = \arg z$

x Re

† *Note that $|z|^2 = z\bar{z}$.*

It is often useful to express a complex number in terms of polar coordinates, which were defined in §7.4. If $z = x + iy$, we write

$$\left. \begin{array}{l} x = r \cos \theta \\ y = r \sin \theta \end{array} \right\} \tag{1}$$

and obtain

$$z = r(\cos \theta + i \sin \theta) \tag{2}$$

The number $r \geq 0$ in this expression is called the **modulus** of z. Pythagoras' theorem implies that

$$r = |z| = \sqrt{x^2 + y^2}$$

† The number θ is called the **argument** of z, usually written

$$\theta = \arg z$$

It is clear that equations (1) do not define θ unambiguously. If θ satisfies them, then so does $\theta + 2n\pi$ for any integer $n \in \mathbb{Z}$. Usually (but not always), we select the value of θ which lies in the range $-\pi < \theta \leq \pi$, and call it the **principal argument**. Also, recall from §7.1.2 that the formula‡

$$\theta = \arctan \frac{y}{x}$$

‡ The formula cannot be used when the principal argument θ lies outside this range; that is, when $x \leq 0$.

is true only for $-\pi/2 < \theta < \pi/2$. That is, it is only valid for $x > 0$. When $x \leq 0$, θ must be calculated from (1).

In §13.7 we will establish the identity

$$\cos \theta + i \sin \theta = e^{i\theta} \tag{3}$$

which highlights the significance of the identity (2). For the moment, we shall only observe that (3) allows us to express a complex number in the form

$$r(\cos\theta + i\sin\theta) = re^{i\theta}$$

Raising each side of equation (3) to the nth power, we obtain

$$(\cos\theta + i\sin\theta)^n = (e^{i\theta})^n = e^{i(n\theta)} = \cos n\theta + i\sin n\theta.$$

This identity is known as De Moivre's theorem

$$(\cos\theta + i\sin\theta)^n = \cos n\theta + i\sin n\theta$$

Example 2

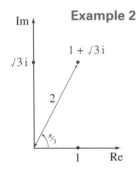

(i) $1 + \sqrt{3}i = 2e^{i\pi/3}$ We have

$$\left|1 + \sqrt{3}i\right| = \sqrt{1^2 + (\sqrt{3})^2} = 2$$

$$\arg\left(1 + \sqrt{3}i\right) = \arctan\frac{\sqrt{3}}{1} = \frac{\pi}{3}$$

(ii) $-1 + \sqrt{3}i = 2e^{i2\pi/3}$ We have

$$\left|-1 + \sqrt{3}i\right| = \sqrt{(-1)^2 + (\sqrt{3})^2} = 2$$

$$\arg\left(-1 + \sqrt{3}i\right) = \frac{2\pi}{3}$$

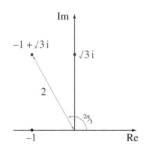

which is the solution of the equations

$$\left.\begin{array}{c} \cos\theta = -\dfrac{1}{2} \\[2mm] \sin\theta = \dfrac{\sqrt{3}}{2} \end{array}\right\}$$

Observe that $\arg\left(-1 + \sqrt{3}i\right) \neq \arctan\dfrac{\sqrt{3}}{-1} = -\dfrac{\pi}{3}$

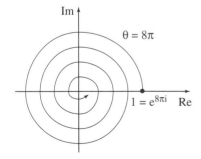

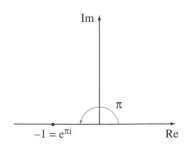

(iii) $e^{2n\pi i} = 1$ $(n \in \mathbb{Z})$ (iv) $i = e^{\pi i/2}$ (v) $e^{\pi i} = -1$

The equation $e^{\pi i} = -1$ is particularly pleasing since it incorporates all the 'mysterious' entities of elementary mathematics – that is, -1, π, e and i.

13.4 Exercises

1.* Find the real and imaginary parts of the complex numbers

    (i) $(z_1 + z_2)z_3$ (ii) $(z_1 z_2) - z_3$

    (iii) $\dfrac{z_1 - z_3}{z_2}$ (iv) $\dfrac{z_1 z_3}{z_2}$

    in the case when $z_1 = 1 - 2i$, $z_2 = -2 + 3i$, $z_3 = -3 - 4i$. Find the modulus and the argument of z_1, z_2 and z_3.

2. Repeat the previous question with $z_1 = 2-i$, $z_2 = -3+2i$, $z_3 = -4-3i$.

3.* Find the real and imaginary parts of the complex numbers

$$e^{i\frac{3\pi}{2}} \qquad e^{i\frac{2\pi}{3}} \qquad e^{i\frac{11\pi}{3}} \qquad 5e^{i\frac{7\pi}{6}} \qquad e^{-3+i5\sqrt{7}}$$

4.* Prove the following results.

    (i) $\dfrac{1}{i} = -i$ (ii) $i^4 = 1$

    (ii) $i^3 = -i$ (iv) $1 + i = \sqrt{2}e^{i\pi/4}$

5. Evaluate the following expressions by expressing the complex numbers in their exponential forms:

    (i)* $\dfrac{48(1 + i)^4}{(-1 - i)^3}$ (ii) $\dfrac{2^3(\sqrt{3} - i)^4}{3^2(-1 + i)^3}$

6. Use the exponential forms of complex numbers to establish the following results:

    (i)* The product of two complex numbers is the complex number whose modulus is the product of their moduli and whose argument is the sum of their arguments.

    (ii) The quotient of two complex numbers is the complex number whose modulus is the quotient of their moduli and whose argument is the difference between their arguments.

7. Prove the following results:

    (i) $e^{-i\theta} = \cos\theta - i\sin\theta$ (ii) $\cos\theta = \dfrac{e^{i\theta} + e^{-i\theta}}{2}$

    (iii) $\sin\theta = \dfrac{e^{i\theta} - e^{-i\theta}}{2i}$

8.*   By considering the case $n = 3$ in De Moivre's theorem, show that

$$\cos 3\theta = 4\cos^3 \theta - 3\cos \theta$$

9.   By applying the identity $\cos \theta + i \sin \theta = e^{i\theta}$ to $e^{i(A+B)}$ and $e^{iA}e^{iB}$ deduce the following basic trigonometric identities:

(i) $\sin(A + B) = \sin A \cos B + \cos A \sin B$

(ii) $\cos(A + B) = \cos A \cos B - \sin A \sin B$

10.*♣   Prove that for complex numbers z_1 and z_2

$$|z_1 + z_2| \leq |z_1| + |z_2|$$

13.5   Complex roots

The result that the equation $z^2 = -1$ has precisely two complex roots, namely i and $-i$ can be generalised. For any positive integer n and any complex number $w \neq 0$ (*which could be purely real*), the equation

$$z^n = w$$

has precisely n distinct complex solutions for z. We call these solutions the **nth roots** of w. They can be evaluated in the following way.
     Write $w = \rho e^{i\phi}$ and $z = re^{i\theta}$. Then

$$(re^{i\theta})^n = \rho e^{i\phi}$$

$$\text{i.e.} \quad r^n e^{i\theta n} = \rho e^{i\phi}$$

We deduce that $r^n = \rho$ and so $r = \rho^{1/n}$, i.e. *r is the positive real number whose nth power is* ρ.
     As each root has modulus $\rho^{\frac{1}{n}}$ all the roots lie on a circle of radius $\rho^{\frac{1}{n}}$. Also,

$$\theta n = \phi + 2k\pi \qquad (k \in \mathbb{Z})$$

Note that we cannot deduce simply that $\theta n = \phi$ for the reasons explained in §13.3. The possible values of θ are therefore

$$\theta = \frac{\phi}{n} + \frac{2k\pi}{n} \qquad (k \in \mathbb{Z})$$

Since the arguments of successive roots increase by $2\pi/n$, the n roots are equally spaced around the circle.

Observe that only n of these possibilities give rise to distinct values of $z = re^{i\theta}$. This is due to the fact that each time θ completes a revolution, subsequent values of θ are repeated together with a multiple of 2π. Since $e^{2\pi i} = 1$, these yield solutions already obtained.

Example 3

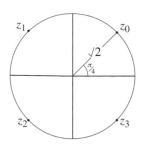

Find the complex fourth roots of -4. We write $-4 = 4e^{i\pi + i2k\pi}$ where $k \in \mathbb{Z}$. The complex fourth roots are therefore of the form $z = re^{i\theta}$ where

$$\begin{cases} r = 4^{1/4} = \sqrt{2} \\ \theta = \dfrac{\pi}{4} + \dfrac{2k\pi}{4} \quad (k \in \mathbb{Z}) \end{cases}$$

Only four of these possibilities give rise to distinct values of z and the required roots are

$$z_0 = \sqrt{2}e^{i\pi/4} \quad z_1 = \sqrt{2}e^{3i\pi/4} \quad z_2 = \sqrt{2}e^{5i\pi/4} \quad z_3 = \sqrt{2}e^{7i\pi/4}$$

Observe that

$$z_4 = \sqrt{2}e^{9i\pi/4} = \sqrt{2}e^{i\pi/4}e^{2\pi i} = \sqrt{2}e^{i\pi/4} = z_0,$$
$$z_{-1} = \sqrt{2}e^{-i\pi/4} = \sqrt{2}e^{7i\pi/4}e^{-2\pi i} = \sqrt{2}e^{7i\pi/4} = z_3$$

Example 4

To find the sixth roots of 27, write $27 = 27e^{(0\cdot i\pi + i2k\pi)}$ where $k \in \mathbb{Z}$. The sixth roots are therefore of the form $z = re^{i\theta}$ where

$$\begin{cases} r = 27^{1/6} = \sqrt{3} \\ \theta = \dfrac{k\pi}{3} \quad (k \in \mathbb{Z}) \end{cases}$$

$z_0 = \sqrt{3}$ and the arguments of successive roots increase by $\dfrac{\pi}{3}$ around the circle of radius $\sqrt{3}$, as in the figure on the left.

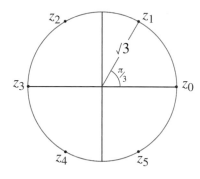

Example 5

To find the sixth roots of -27, write

$$-27 = 27e^{(i\pi + i2k\pi)} \quad \text{where} \quad k \in \mathbb{Z}$$

Here $z_0 = \sqrt{3}e^{i\frac{\pi}{6}}$ so the roots are as shown in the figure on the right.

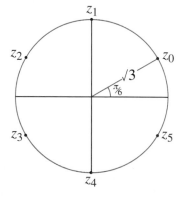

13.6 Polynomials

The fundamental theorem of algebra asserts that every polynomial

$$P(z) = a_n z^n + a_{n-1} z^{n-1} + \cdots + a_1 z + a_0$$

of degree n has n complex **roots**. That is, $P(z)$ may be expressed uniquely in the form

$$P(z) = a_n(z - \alpha_1)(z - \alpha_2) \ldots (z - \alpha_n)$$

Some of the roots $\alpha_1, \alpha_2, \ldots, \alpha_n$ may be coincident. If a root occurs k times we say that the root has **multiplicity** k.

Alternatively, when $P(z)$ has real coefficients, it may be expressed uniquely as the product of *real* linear and quadratic factors. This is a consequence of the following:

Complex roots of polynomials with real coefficients occur in conjugate pairs. It is not difficult to see why this is so, using the properties of the complex conjugate.[†]

[†] *Note that $\bar{a}_k = a_k$ because a_k is real.*

$$
\begin{aligned}
\overline{P(z)} &= \bar{a}_n \bar{z}^n + \bar{a}_{n-1} \bar{z}^{n-1} + \cdots + \bar{a}_1 \bar{z} + \bar{a}_0 \\
&= a_n \bar{z}^n + a_{n-1} \bar{z}^{n-1} + \cdots + a_1 \bar{z} + a_0 \\
&= P(\bar{z})
\end{aligned}
$$

Thus, if $P(\alpha) = 0$, then

$$P(\bar{\alpha}) = \overline{P(\alpha)} = \bar{0} = 0.$$

That is, if α is a root of $P(z)$, so also is $\bar{\alpha}$. It is obvious from the properties of the complex conjugate in §13.2 that the product

$$(z - \alpha)(z - \bar{\alpha}) = z\bar{z} - (\alpha + \bar{\alpha})z + \alpha\bar{\alpha}$$

is real. If a root α occurs more than once, $P(z)$ may be divided by the product $(z - \alpha)(z - \bar{\alpha})$ and the argument repeated with the resulting quotient. It follows that the roots which are not real fall into complex conjugate *pairs* of the form α and $\bar{\alpha}$.

Since a polynomial of *odd* degree has an *odd* number of roots, it follows immediately from the preceding result that at least one of these roots must be real.

Example 6

Let $P(z) = z^4 + 4$. From Example 3 we have

$$P(z_0) = P(z_1) = P(z_2) = P(z_3) = 0$$

and so

$$
\begin{aligned}
z^4 + 4 &= (z - \sqrt{2}e^{i\pi/4})(z - \sqrt{2}e^{3i\pi/4})(z - \sqrt{2}e^{5i\pi/4})(z - \sqrt{2}e^{7i\pi/4}) \\
&= (z - \sqrt{2}e^{i\pi/4})(z - \sqrt{2}e^{3i\pi/4})(z - \sqrt{2}e^{-3i\pi/4})(z - \sqrt{2}e^{-i\pi/4}) \\
&= [z - (1 + i)][z + (1 - i)][z + (1 + i)][z - (1 - i)] \\
&= (z^2 - 2z + 2)(z^2 + 2z + 2)
\end{aligned}
$$

It is often important to determine the roots of a given polynomial. We have just demonstrated the very special case of determining the n roots of a complex number. However, it is usually not at all easy to find the roots of a polynomial even when n is reasonably small. We shall now describe a trial and error method which is applicable only in very simple cases.

The first step is to guess a root α. Having made such a guess, it is next necessary to check that the guess is accurate by confirming that $P(\alpha) = 0$. We may then write

$$P(z) = (z - \alpha)Q(z)$$

where $Q(z)$ is a polynomial of degree one less than $P(z)$ and repeat the exercise.

Example 7

Consider the polynomial

$$P(z) = z^3 - 7z^2 + 16z - 12$$

Try $z = 1$. Then $P(1) = 1 - 7 + 16 - 12 = -2 \neq 0$ and so $z = 1$ is *not* a root. Try $z = 2$. Then $P(2) = 8 - 28 + 32 - 12 = 40 - 40 = 0$ and so $z = 2$ *is* a root. Thus

$$P(z) = (z - 2)Q(z)$$

Next we require a formula for $Q(z)$. This can be obtained by 'long division' as below.

$$
\begin{array}{r}
z^2 - 5z + 6 \\
z - 2 \overline{)\, z^3 - 7z^2 + 16z - 12} \\
\underline{z^3 - 2z^2} \\
-5z^2 + 16z - 12 \\
\underline{-5z^2 + 10z} \\
6z - 12 \\
\underline{6z - 12}
\end{array}
$$

We obtain $Q(z) = z^2 - 5z + 6 = (z - 2)(z - 3)$ and so

$$P(z) = (z - 2)(z - 2)(z - 3) = (z - 2)^2(z - 3)$$

The root $z = 2$ has multiplicity 2.

As an alternative to the use of 'long division', we may observe that

$$P(z) = (z - \alpha)Q(z) \text{ where } Q(z) = b_{n-1}z^{n-1} + \cdots + b_0 \text{ if and only if}$$

$$a_n z^n + a_{n-1}z^{n-1} + \cdots + a_1 z + a_0 = (z - \alpha)(b_{n-1}z^{n-1} + \cdots + b_1 z + b_0)$$

$$= b_{n-1}z^n + (b_{n-2} - \alpha b_{n-1})z^{n-1} + \cdots + -\alpha b_0$$

Equating coefficients of like powers of z we obtain

$$a_n = b_{n-1}, \quad a_{n-1} = b_{n-2} - \alpha b_{n-1}, \quad \ldots, \quad a_0 = -\alpha b_0$$

or, if it is understood that $b_n = b_{-1} = 0$,

$$a_k = b_{k-1} - \alpha b_k \quad (k = 0, 1, 2, \ldots, n)$$

i.e.

$$b_{k-1} = a_k + \alpha b_k \quad (k = 0, 1, 2, \ldots, n) \tag{1}$$

Example 8

It is helpful to construct a table to apply the formula. For example, this is demonstrated below with the same polynomial:

$$P(z) = z^3 - 7z^2 + 16z - 12$$

k	3	2	1	0	−1
a_k	1	−7	16	−12	0
b_k	0	1	−5	6	0

$b_{k-1} = a_k + 2b_k,$
$k = 0, 1, 2, 3$

We calculate the successive values of b_{k-1} from left to right using formula (1) above, with $\alpha = 2$, as depicted in the scheme on the left. So

$$b_2 = 1 + 2 \cdot 0 = 1 \quad b_1 = -7 + 2 \cdot 1 = -5 \quad b_0 = 16 + 2(-5) = 6$$

It is worth checking that $b_{-1} = 0$. The shaded boxes give the desired coefficients. Hence

$$Q(x) = b_2 x^2 + b_1 x + b_0 = x^2 - 5x + 6$$

13.7 Elementary functions♣

In §10.14, we considered real power series. One can equally well consider complex power series. Of particular importance are the series

$$1 + \frac{z}{1!} + \frac{z^2}{2!} + \frac{z^3}{3!} + \cdots$$

$$1 - \frac{z^2}{2!} + \frac{z^4}{4!} - \frac{z^6}{6!} + \cdots$$

$$z - \frac{z^3}{3!} + \frac{z^5}{5!} - \frac{z^7}{7!} + \cdots$$

These power series converge for all complex numbers z and we use them to define e^z, $\cos z$ and $\sin z$ respectively for *complex* values of z. By and large, these complex functions retain all of the properties of their real counterparts which can be mathematically expressed without the use of the symbols > and <. For example,

$$e^{z+w} = e^z e^w$$

because, using the binomial theorem,

$$e^{z+w} = \sum_{n=0}^{\infty} \frac{(z+w)^n}{n!} = \sum_{n=0}^{\infty} \frac{1}{n!} \sum_{k=0}^{n} \binom{n}{k} z^k w^{n-k}$$

$$= \sum_{n=0}^{\infty} \sum_{k=0}^{n} \left(\frac{z^k}{k!} \right) \left(\frac{w^{n-k}}{(n-k)!} \right)$$

$$= \sum_{k=0}^{\infty} \frac{z^k}{k!} \sum_{n=k}^{\infty} \frac{w^{n-k}}{(n-k)!}$$

$$= \sum_{k=0}^{\infty} \frac{z^k}{k!} \sum_{l=0}^{\infty} \frac{w^l}{l!} = e^z e^w$$

Similarly, we have

$$(e^z)^n = e^{nz}$$

But the complex functions also have properties which the real functions do not. In particular, the next result shows that the exponential and trigonometric functions are intimately related. If θ is real,

$$e^{i\theta} = 1 + \frac{(i\theta)}{1!} + \frac{(i\theta)^2}{2!} + \frac{(i\theta)^3}{3!} + \frac{(i\theta)^4}{4!} + \cdots$$

$$= 1 + i\theta - \frac{\theta^2}{2!} - \frac{i\theta^3}{3!} + \frac{\theta^4}{4!} - \cdots$$

$$= \left(1 - \frac{\theta^2}{2!} + \frac{\theta^4}{4!} - \cdots \right) + i \left(\theta - \frac{\theta^3}{3!} + \frac{\theta^5}{5!} - \cdots \right)$$

$$= \cos\theta + i\sin\theta$$

and we therefore obtain the formula introduced in §13.3,

$$\boxed{e^{i\theta} = \cos\theta + i\sin\theta}$$

Example 9

In Exercise 10.15.2, we considered the integrals

$$I(a,b) = \int_0^{\infty} (\sin bx) e^{-ax}\, dx \qquad J(a,b) = \int_0^{\infty} (\cos bx) e^{-ax}\, dx$$

The same integrals were considered again in Exercises 11.6.5 and 11.6.6. The integrals may be evaluated more simply by considering

$$\int_0^{\infty} e^{ibx} e^{-ax}\, dx = \left[\frac{e^{(ib-a)x}}{(ib-a)} \right]_0^{\infty}$$

$$= \frac{-1}{(ib-a)} = \frac{a+ib}{a^2+b^2}$$

Thus

$$I(a, b) = \operatorname{Im} \int_0^\infty e^{ibx} e^{-ax} \, dx = \frac{b}{a^2 + b^2}$$

$$J(a, b) = \operatorname{Re} \int_0^\infty e^{ibx} e^{-ax} \, dx = \frac{a}{a^2 + b^2}$$

The complex logarithm is less easy to deal with because the equation

$$w = e^z$$

has many solutions for z. If $w = \rho e^{i\phi}$, these solutions are of the form

$$z = \ln \rho + i\phi$$

Thus the logarithm of w is of the form

$$\ln w = \ln \rho + i\phi \tag{1}$$

but, since the value of ϕ is ambiguous, so is the value of $\ln \rho + i\phi$. To make use of formula (1), it is therefore necessary to say which value of ϕ is to be selected. Different possible selections for ϕ determine different possible 'branches' of the logarithm. We shall just observe that the branch of the logarithm for which

$$\ln(1 + w) = w - \frac{w^2}{2} + \frac{w^3}{3} - \cdots \qquad (|w| < 1)$$

is that determined by the restriction $-\pi < \arg(1 + w) \le \pi$.

These remarks are made only to indicate that the complex logarithm must be treated very carefully.

13.8   Exercises

1. Find all complex roots of the polynomials

$$\text{(i) } z^4 - 16 \qquad \text{(ii)* } z^5 + 1$$
$$\text{(iii)* } z^2 - i \qquad \text{(iv) } 125z^3 + i$$

Illustrate the roots as points in the complex plane.

2. Write the polynomials in Exercise 1 (i) and (ii) as products of linear and quadratic factors with real coefficients.

3. Illustrate the roots of the following polynomials as points of the complex plane, without calculating all the roots:

$$\text{(i) } z^6 - 729i \quad \text{and} \quad \text{(ii) } 32z^5 + 1$$

4. Factorise the polynomials

(i)* $z^3 - 3z^2 - 10z$ (ii)* $z^4 + z^3 - 3z^2 - 5z - 2$

(iii) $z^3 + 6z^2 + 11z + 6$ (iv) $z^4 + 2z^2 + 1$

(v) $z^3 - 5z^2 + 9z - 5$ (vi) $z^4 + 6z^3 + 11z^2 + 6z + 10$

5. Show that the function

$$y(x) = (1 - \sqrt{3}i)e^{(2+3i)x} + (1 + \sqrt{3}i)e^{(2-3i)x}$$

can be written in the alternative real forms

(i) $e^{2x}(a \cos 3x + b \sin 3x)$,   (ii) $re^{2x} \sin(3x + \theta)$,   (iii) $\rho e^{2x} \cos(3x - \phi)$

and find the values of the constants a, b, r, θ, ρ and ϕ.

6. Find the roots α and $\bar{\alpha}$ of the equation $z^2 - 4z + 7 = 0$.

For these values of α and $\bar{\alpha}$, find the real and imaginary parts of the following functions:

(i) $e^{\alpha x}$ (ii) $e^{\bar{\alpha} x}$ (iii) α^x (iv) $\bar{\alpha}^x$ $(x \in \mathbb{R})$

7. Show for any $\alpha \in \mathbb{C}$ and $x \in \mathbb{R}$ that the expressions $e^{\alpha x} + e^{\bar{\alpha} x}$ and $\alpha^x + \bar{\alpha}^x$ are both real.

13.9   Applications (optional)

13.9.1   Characteristic functions

In §11.7.5 we discussed generating functions for the probability density functions of discrete random variables. The notion of a **characteristic function** or **Fourier transform** is a much more powerful version of the same idea.

If X is a continuous random variable with probability density function f, we define its **characteristic function** ϕ by

$$\phi(t) = \int_{-\infty}^{\infty} e^{ixt} f(x)\, dx$$

[†] This converges for all real t because $|e^{ixt}| = 1$. See §10.13.

[†] If ψ is the characteristic function for an *independent* random variable Y with probability density function g, we have

$$\phi(t)\psi(t) = \int_{-\infty}^{\infty} e^{ixt} f(x)\, dx \int_{-\infty}^{\infty} e^{iyt} g(y)\, dy$$

$$= \int_{-\infty}^{\infty} dx \int_{-\infty}^{\infty} e^{i(x+y)t} f(x)g(y)\, dy$$

$$= \int_{-\infty}^{\infty} dx \int_{-\infty}^{\infty} e^{izt} f(x)g(z-x)\, dz$$

$$= \int_{-\infty}^{\infty} e^{izt}\, dz \int_{\infty}^{\infty} f(x)g(z-x)\, dx$$

$$= \int_{-\infty}^{\infty} e^{izt} h(z)\, dz$$

‡ See §11.7.4.

where h is the probability density function of the random variable $X + Y$.[‡] It follows that the characteristic function for $X + Y$ is the *product* of the characteristic functions for X and Y.

In the case of a discrete random variable X we define the characteristic function ϕ by

$$\phi(t) = \sum_{n=-\infty}^{\infty} e^{itn} f(n)$$

Then $\phi(t) = G(e^{it})$ where G is the corresponding generating function. The result concerning the multiplication of characteristic functions is therefore the same result that we met in §11.7.5.

As an example, we shall calculate the characteristic function ϕ of a normally distributed random variable with mean μ and variance σ^2. The appropriate formula is

$$\phi(t) = \frac{1}{\sigma\sqrt{2\pi}} \int_{-\infty}^{\infty} e^{ixt} e^{-\frac{1}{2}\left(\frac{x-\mu}{\sigma}\right)^2} dx$$

$$= \frac{1}{\sqrt{2\pi}} \int_{-\infty}^{\infty} e^{i(\mu+\sigma y)t} e^{-y^2/2} dy$$

$$= e^{i\mu t} \phi_0(\sigma t)$$

where ϕ_0 is the characteristic function of the standardised normal distribution – i.e.

$$\phi_0(t) = \frac{1}{\sqrt{2\pi}} \int_{-\infty}^{\infty} e^{ity} e^{-y^2/2} dy$$

But

$$\frac{d\phi_0}{dt} = \frac{1}{\sqrt{2\pi}} \int_{-\infty}^{\infty} iy e^{ity} e^{-y^2/2} dy$$

$$= \frac{1}{\sqrt{2\pi}} [ie^{ity}(-e^{-y^2/2})]_{-\infty}^{\infty} - \frac{1}{\sqrt{2\pi}} \int_{-\infty}^{\infty} i(ite^{ity})(-e^{-y^2/2}) dy$$

$$= \frac{i^2 t}{\sqrt{2\pi}} \int_{-\infty}^{\infty} e^{ity} e^{-y^2/2} dy = -t\phi_0(t)$$

Thus

$$\frac{d\phi_0}{dt} = -t\phi_0$$

$$\int \frac{d\phi_0}{\phi_0} = -\int t \, dt + c$$

$$\ln \phi_0 = -\frac{1}{2}t^2 + c$$

$$\phi_0(t) = ke^{-t^2/2}$$

Since $\phi_0(0) = 1$, it follows that $k = 1$. We conclude that

$$\phi(t) = e^{i\mu t} e^{-\sigma^2 t^2/2}$$

Now suppose that X_1 and X_2 are two *independent* normally distributed random variables with characteristic functions ϕ_1 and ϕ_2 respectively. Then

$$\phi_1(t)\phi_2(t) = e^{i(\mu_1+\mu_2)} e^{-(\sigma_1^2+\sigma_2^2)t^2/2}$$

which is the characteristic function of a normally distributed random variable with mean $\mu_1 + \mu_2$ and variance $\sigma_1^2 + \sigma_2^2$. It follows that $X_1 + X_2$ is such a random variable.[†]

† See §11.7.4.

13.9.2   Central limit theorem

Recall from §10.16.9 that, if $X_1, X_2, X_3, \ldots$ is a sequence of *independent*, identically distributed random variables, each with mean μ and variance σ^2, then, under very general conditions, the distribution of

$$Y = \frac{X_1 + X_2 + \cdots + X_n - n\mu}{\sigma \sqrt{n}}$$

is approximately standard normal provided that n is sufficiently large. This result is called the **central limit theorem**. We give an outline 'proof' below. This is far from being complete but it does indicate how this remarkable result comes about.

Suppose that X_k has probability density function f and characteristic function ϕ. Then

$$\phi(0) = \int_{-\infty}^{\infty} e^{i0x} f(x)\, dx = \int_{-\infty}^{\infty} f(x)\, dx = 1$$

Also

$$\left. \begin{aligned} \phi'(t) &= \int_{-\infty}^{\infty} ix e^{ixt} f(x)\, dx \\ \phi''(t) &= \int_{-\infty}^{\infty} -x^2 e^{ixt} f(x)\, dx \end{aligned} \right\}$$

and so

$$\phi'(0) = i\mu \quad \phi''(0) = -\sigma^2 - \mu^2$$

From §13.9.1 we know that the characteristic function of $S = X_1 + X_2 + \cdots + X_n$ is

$$\Phi(t) = \{\phi(t)\}^n$$

It follows that the characteristic function for $(S - a)/b$ is

$$e^{-iat/b} \Phi\left(\frac{t}{b}\right)$$

and, in particular, that Y has the characteristic function

$$\psi(t) = e^{-i\mu\sqrt{n}t/\sigma} \Phi\left(\frac{t}{\sigma\sqrt{n}}\right)$$

$$= e^{-i\mu\sqrt{n}t/\sigma} \left\{\phi\left(\frac{t}{\sigma\sqrt{n}}\right)\right\}^n$$

$$\ln \psi(t) = \frac{-i\mu\sqrt{n}t}{\sigma} + n \ln \phi\left(\frac{t}{\sigma\sqrt{n}}\right)$$

The Taylor series expansion for ϕ about the origin is given by

$$\phi(x) = \phi(0) + x\phi'(0) + \frac{x^2}{2!}\phi''(0) + \cdots$$

$$= 1 + i\mu x - \left(\frac{\sigma^2 + \mu^2}{2}\right) x^2 + \cdots$$

We also require the formula

$$\ln(1 + z) = z - \frac{z^2}{2} + \frac{z^3}{3} - \cdots \qquad (|z| < 1)$$

Systematically ignoring third and higher order terms as they arise, we obtain the approximation

$$\ln\phi\left(\frac{t}{\sigma\sqrt{n}}\right) \approx \ln\left(1 + \frac{i\mu t}{\sigma\sqrt{n}} - \left(\frac{\sigma^2+\mu^2}{2}\right)\frac{t^2}{\sigma^2 n}\right)$$

$$\approx \frac{i\mu t}{\sigma\sqrt{n}} - \left(\frac{\sigma^2+\mu^2}{2}\right)\frac{t^2}{\sigma^2 n} - \frac{1}{2}\left(\frac{i\mu t}{\sigma\sqrt{n}} - \left(\frac{\sigma^2+\mu^2}{2}\right)\frac{t^2}{\sigma^2 n^2}\right)^2$$

$$\approx \frac{i\mu t}{\sigma\sqrt{n}} - \left(\frac{\sigma^2+\mu^2}{2}\right)\frac{t^2}{\sigma^2 n} - \frac{1}{2}\left(-\frac{\mu^2 t^2}{\sigma^2 n}\right)$$

$$= \frac{i\mu t}{\sigma\sqrt{n}} - \frac{1}{2}\frac{t^2}{n}.$$

It follows that

$$\frac{-i\mu\sqrt{n}t}{\sigma} + n\ln\phi\left(\frac{t}{\sigma\sqrt{n}}\right) \approx -\frac{1}{2}t^2$$

and so we obtain the approximation

$$\psi(t) \approx e^{-t^2/2}$$

for the characteristic function of the random variable Y. But the right hand side is the characteristic function of the standard normal variable.

Fourteen   Linear differential and difference equations

Difference equations and the shift operator E are introduced in this chapter. The chapter is mainly concerned with higher order linear differential and difference equations with constant coefficients. The analogies between the two types of equation are exploited by showing that essentially the same techniques work for both differential and difference equations. For this reason, the symbol x is used to denote the variable in both types of equation, though in the case of difference equations, it only takes discrete values, which are nonnegative integers. The techniques are easy, provided some mastery of complex numbers has been garnered from the previous chapter. A brief theoretical justification for the techniques described appears in §14.3 and §14.4, which will be of particular interest to those with a knowledge of linear algebra. But readers impatient with theory will find it adequate to confine their attention in these sections to the manipulation of operators as described in the examples of §14.1 and §14.2 and to the assertions of §14.3 and §14.4 without delving into the reasons why these are correct.

It should be noted that the technique for finding particular solutions of nonhomogeneous equations described in §14.7 is only one of several. We have chosen to present this technique because it involves no essentially new ideas. But it is often quicker to find particular solutions using operators in a more adventurous way than described here.

In §14.8 there is a short discussion on the stability of the solutions of equations. This is continued in §14.9 in the context of linear systems of differential and difference equations. This material is not hard, but some readers may prefer to leave these topics until they can be studied in a more comprehensive manner at a later stage.

In the final section §14.12, which is optional, we conclude with a glance at the difference operator Δ, which provides an alternative approach to the solution of difference equations.

14.1 The operator $P(\mathrm{D})$

If $y = f(x)$, the function f is sometimes said to operate on x to produce y. This leads to functions being described as **operators** in certain contexts. In particular, this terminology is used when x and y are themselves functions.

For example, the *differential operator* D is a function which transforms a differentiable function f into its derivative f', namely, $\mathrm{D}f(x) = f'(x)$.

Now consider **linear** differential equations of the form

$$a_n \frac{d^n y}{dx^2} + a_{n-1} \frac{d^{n-1} y}{dx^{n-1}} + \cdots + a_1 \frac{dy}{dx} + a_0 y = q(x)$$

where $a_n, a_{n-1}, \ldots, a_0$ are constant. Such linear equations (*see §14.3 and §14.4*) may be written in terms of operators which are 'polynomials' in D. For example, the differential equation

$$\frac{d^2 y}{dx^2} - \frac{dy}{dx} - 2y = 0$$

can be written as $D^2 y - Dy - 2y = 0$, which we then rewrite as

$$(D^2 - D - 2I)y = 0$$

where I is the **identity operator**, which transforms everything into itself. Its use is somewhat pedantic here and we usually write 2 instead of 2I. We then obtain

$$P(D)y = 0$$

The equation

$$\left(\frac{d^3 y}{dx^3}\right)^2 = 5\frac{dy}{dx}$$

for cxample, has a derivative of degree two, so is not linear and cannot be written in this manner.

where $P(D)$ is the 'polynomial' $P(D) = D^2 - D - 2$. It is often convenient to factorise such a polynomial. For this purpose we require the roots of the **auxiliary equation**

$$z^2 - z - 2 = 0$$

We have

$$z^2 - z - 2 = (z - 2)(z + 1)$$

and thus

$$P(D) = D^2 - D - 2 = (D - 2)(D + 1)$$

Polynomials in D can be manipulated just like ordinary polynomials. We can see this by confirming that the above factors, and similarly *any* two factors, linear or quadratic in D commute:

We have

$$(D - 2)(D + 1)f = (D - 2)(f' + 1f) \qquad (1)$$
$$= (D - 2)f' + (D - 2)f$$
$$= f' - 2f' + f' - 2f$$
$$= (D^2 - D - 2)f$$

Also

$$(D + 1)(D - 2)f = (D + 1)(f' - 2f) \qquad (2)$$
$$= (D + 1)f' - (D + 1)2f$$
$$= f' + f' - 2f' - 2f$$
$$= (D^2 - D - 2)f$$

Hence

$$(D - 2)(D + 1)f = (D + 1)(D - 2)f = (D^2 - D - 2)f$$

14.2 Difference equations and the shift operator E

Difference equations involve an unknown *sequence* y_x instead of an unknown function as in differential equations. It must be stressed that the variable x is therefore to be understood to take only the discrete values

$$x = 0, 1, 2, 3, \ldots$$

Some examples of sequences are

$$1, \ 1, \ 1, \ 1, \ \ldots$$
$$1, \ 2, \ 3, \ 4, \ \ldots$$
$$1, \ -2, \ 4, \ -8, \ 16, \ \ldots$$

For the last of these sequences, we have

$$y_0 = 1 \quad y_1 = -2 \quad y_2 = 4 \quad y_3 = -8 \quad y_4 = 16 \quad \ldots$$

The three sequences can be described by the respective formulas

$$
\begin{aligned}
y_x &= 1 & (x = 0, 1, 2, \ldots) \\
y_x &= x + 1 & (x = 0, 1, 2, \ldots) \\
y_x &= (-1)^x 2^x & (x = 0, 1, 2, \ldots)
\end{aligned}
$$

The sequence obtained by omitting the first term of the sequence y_x is

$$y_{x+1} \qquad (x = 0, 1, 2, \ldots)$$

Similarly the sequence obtained by omitting the first n terms of the sequence y_x is

$$y_{x+n} \qquad (x = 0, 1, 2, \ldots)$$

A relationship between the sequences $y_x, y_{x+1}, \ldots, y_{x+n}$ is known as a **difference equation of order** n. Difference equations are also known as recurrence equations. Some typical difference equations are listed below. The first three of these equations are of order one. Equation (4) is of second order and equation (5) is of fourth order.[†] In each case $x = 0, 1, 2, \ldots$.

[†] The **order** of a difference equation is obtained by simply subtracting the smallest suffix from the largest suffix.

$$y_{x+1} - y_x = 0 \tag{1}$$
$$y_{x+1} - 2y_x = 1 \tag{2}$$
$$y_{x+1} + x^2 y_x = 2^x \tag{3}$$
$$y_{x+2} + 2y_{x+1} + y_x = 0 \tag{4}$$
$$y_{x+4}^2 + y_x y_{x+1} + x = 3 \tag{5}$$

Hence just as the solutions of differential equations involve functions, the solutions of difference equations involve sequences. As with differential equations, the general solution of a difference equation of order n contains n arbitrary constants.

Consider, for example, equation (4). We can choose y_0 and y_1 in any way we like. Thus y_0 and y_1 will serve as our arbitrary constants. But the subsequent terms of the sequence must satisfy

$$y_2 = -2y_1 - y_0$$
$$y_3 = -2y_2 - y_1 = -2(-2y_1 - y_0) - y_1 = 3y_1 + 2y_0$$
$$y_4 = -2y_3 - y_2 = -2(3y_1 + 2y_0) - (-2y_1 - y_0)$$
$$= -4y_1 - 3y_0$$

and so on. Thus the subsequent terms are all determined by the choice of y_0 and y_1.

We introduce now the shift operator E and later, in §14.12, the difference operator Δ. These operate on sequences. The shift operator is a function which transforms a sequence y_x into the sequence y_{x+1}. That is, the **shift operator** E is defined by

$$E y_x = y_{x+1}$$

It is natural to write $E^2 y_x = E(E y_x) = E y_{x+1} = y_{x+2}$ and $E^3 y_x = E(E^2 y_x) = E y_{x+2} = y_{x+3}$. In general,

$$E^n y_x = y_{x+n}$$

Difference equations can conveniently be expressed in terms of the shift operator, which leads to 'polynomials' in E. For example, the difference equation

$$y_{x+2} - 5y_{x+1} + 6y_x = 0 \quad (x = 0, 1, 2, \ldots)$$

may be rewritten in the form

$$E^2 y_x - 5E y_x + 6y_x = 0 \quad (x = 0, 1, 2, \ldots)$$

or

$$P(E) y_x = (E^2 - 5E + 6) y_x = 0$$

As in the previous example, it is usually convenient to factorise $P(E)$ and so we consider the auxiliary equation

$$z^2 - 5z + 6 = 0$$

Since

$$z^2 - 5z + 6 = (z - 3)(z - 2)$$

we obtain

$$P(E) = (E^2 - 5E + 6) = (E - 3)(E - 2)$$

Observe that just as in the case of the operator D the factors commute and hence, polynomials in E can also be manipulated like ordinary polynomials.

$$(E - 3)(E - 2) y_x = (E - 3)(y_{x+1} - 2y_x) \qquad (6)$$
$$= (E - 3) y_{x+1} - (E - 3) 2 y_x$$
$$= y_{x+2} - (3 + 2) y_{x+1} + 6 y_x$$
$$= (E^2 - 5E + 6) y_x$$

Also

$$(E - 2)(E - 3) y_x = (E - 2)(y_{x+1} - 3y_x) \qquad (7)$$
$$= (E - 2) y_{x+1} - (E - 2) 3 y_x$$
$$= y_{x+2} - (3 + 2) y_{x+1} + 6 y_x$$
$$= (E^2 - 5E + 6) y_x$$

14.3 Linear operators♣

In this section we lay down the basis for finding the general solutions of homogeneous linear equations with constant coefficients. In essence, we reason that the characteristics of matrix equations are mirrored in linear differential and difference equations.

If A is an $m \times n$ matrix, $\mathbf{x}$ and $\mathbf{y}$ are $n \times 1$ column vectors, and $c, d \in \mathbb{R}$ then

$$A(c\mathbf{x} + d\mathbf{y}) = cA\mathbf{x} + dA\mathbf{y}$$

An operator which shares this property is said to be **linear**. The differential operator D and the shift operator E are also linear, since

$$\mathrm{D}(cf + dg) = c\mathrm{D}f + d\mathrm{D}g$$
$$\mathrm{E}(cy_x + dz_x) = c\mathrm{E}y_x + d\mathrm{E}z_x$$

Not only are D and E linear, but so are any polynomials in D and E. *We have already used this fact in steps* (1) *and* (2) *in* §14.1 *and* (6) *and* (7) *in* §14.2.

The **kernel** or null space of an $m \times n$ matrix A is the set of all $n \times 1$ column vectors $\mathbf{x}$ which satisfy

$$A\mathbf{x} = \mathbf{0}$$

Since A is a linear operator, any linear combination of vectors in the kernel of A is also in the kernel. To see this let $A\mathbf{x} = A\mathbf{y} = \mathbf{0}$ and $c, d \in \mathbb{R}$ so that

$$A(c\mathbf{x} + d\mathbf{y}) = cA\mathbf{x} + dA\mathbf{y} = c\mathbf{0} + d\mathbf{0} = \mathbf{0}$$

Students with a knowledge of linear algebra will recognise that the kernel is what is known as a **vector subspace** of $\mathbb{R}^n$.

Suppose that $k \leq n$. A set of k vectors in the kernel is a **basis** for this subspace if each vector in the kernel is a unique linear combination of these vectors. Any set of k linearly independent vectors in the kernel forms a basis for it.[†]

If P is a polynomial of degree n we are interested in the kernel of an operator of the form $P(\mathrm{D})$. A function f is in the kernel of $P(\mathrm{D})$ if and only if it is a solution of the nth order differential equation

$$P(\mathrm{D})f = 0 \tag{1}$$

[†] The number k of vectors in a basis of the kernel is called the **dimension** of the kernel.

Notice that the kernel of $P(\mathrm{D})$ is a vector space of *functions (rather than vectors in* $\mathbb{R}^n$*)*.

We denote the vector subspace of functions that forms the kernel of $P(\mathrm{D})$ by ker $P(\mathrm{D})$. As with all vector spaces the dimension k of ker $P(\mathrm{D})$ is the number of elements in a basis for ker $P(\mathrm{D})$. Recall that each element in a vector space can be written as a unique linear combination of the basis elements of the space. If we can find a basis $f_1, f_2, \ldots, f_k$ of ker $P(\mathrm{D})$, any solution of (1) can therefore be written uniquely in the form

$$f = a_1 f_1 + \cdots + a_k f_k$$

Similarly, we are interested in the kernel of an operator of the form $P(\mathrm{E})$. A sequence y_x is in the kernel of $P(\mathrm{E})$ if and only if it satisfies the nth order difference equation

$$P(\mathrm{E})y_x = 0 \tag{2}$$

The operator $P(E)$ is linear and its kernel is a vector subspace of dimension k of the vector space of all sequences. Each solution of equation (2) can be written uniquely as a linear combination of sequences in a basis of the kernel of $P(E)$.

The following argument shows that $k = n$, so that the dimension of ker $P(E)$ is the same as the degree of the polynomial P. Writing all the terms of equation (2),

$$a_n y_{x+n} + a_{n-1} y_{x+n-1} + \cdots + a_1 y_{x+1} + a_0 y_x = 0 \qquad (3)$$

† We have $a_n \neq 0$ since P is a polynomial of degree n.

Since†

$$y_{x+n} = (a_n)^{-1}(a_{n-1} y_{x+n-1} + \cdots + a_1 y_{x+1} + a_0 y_x) \qquad (4)$$

once the values of $y_0, y_1, \ldots, y_{n-1}$ are known, repeated use of equation (4) allows all the remaining terms of a sequence y_x to be found satisfying $P(E)y_x = 0$. However, $y_0, y_1, \ldots, y_{n-1}$ may be chosen quite freely. Each n dimensional vector $(y_0, y_1, \ldots, y_{n-1})^{\mathrm{T}}$ therefore determines a solution of $P(E)y_x = 0$ and vice versa. That is, there is a one–one correspondence between the set of solutions and the set of n dimensional vectors. Hence the set of solutions has dimension n.

It is known that, given a point $x = x_0$, any real numbers $y_0, y_1, \ldots, y_{n-1}$ determine a unique solution f of the differential equation $P(D)f = 0$ satisfying

$$f(x_0) = y_0 \quad Df(x_0) = y_1 \quad \ldots \quad D^{n-1} f(x_0) = y_{n-1}$$

Hence, a similar argument to that for sequences establishes that the set, ker $P(D)$, of solutions of $P(D)f = 0$ also has dimension n.

We use these results in the next sections.

14.4 Homogeneous, linear, differential equations♣

A linear differential equation of order n has the form

$$a_n(x)\frac{d^n y}{dx^n} + a_{n-1}(x)\frac{d^{n-1} y}{dx^{n-1}} + \cdots + a_1(x)\frac{dy}{dx} + a_0(x)y = q(x)$$

In our study we insist that the coefficients $a_n, a_{n-1}, \ldots, a_1, a_0$ be *constant* so that the differential equation can be written in the form

$$P(D)y = q(x)$$

where P is a polynomial of degree n.

In this section we consider the **homogeneous** case when $q(x) = 0$. We know from §14.3 that the space of solutions of the homogeneous equation

$$P(D)y = 0 \qquad (1)$$

has dimension n. Our problem therefore reduces to finding a *basis* for this solution space. If $y_1, y_2, \ldots, y_n$ is such a basis, then any other solution y may be expressed uniquely as a *linear combination* of $y_1, y_2, \ldots, y_n$ – i.e.

$$y = c_1 y_1 + c_2 y_2 + \cdots + c_n y_n \qquad (2)$$

where $c_1, c_2, \ldots, c_n$ are constants. *Note that (2) involves n arbitrary constants and hence this result accords with our definition of a general solution of a differential equation of order n.*

See §12.2.

In seeking a basis for the solution space of $P(D)y = 0$, it is not enough to find n different solutions $y_1, y_2, \ldots, y_n$. We must find n *linearly independent* solutions. Fortunately, these are easy to write down provided that we can factorise the polynomial $P(D)$.

Consider the equation

$$P(D)y = (D - \alpha_1)^{m_1}(D - \alpha_2)^{m_2} \ldots (D - \alpha_k)^{m_k} y = 0$$

in which $\alpha_1, \alpha_2, \ldots, \alpha_k$ are *distinct* real or complex numbers, where the complex numbers occur in conjugate pairs. We have verified in §14.2 that the factors commute.

From §12.4 we have found that $y = e^{\alpha x}$ is a solution of $(D - \alpha)y = 0$. Now, it is easily seen that if $(D - \alpha)$ is a factor of $P(D)$, then $y = e^{\alpha x}$ is also a solution of $P(D)y = 0$. For

$$P(D)e^{\alpha x} = Q(D)(D - \alpha)e^{\alpha x} = Q(D)0 = 0$$

where $Q(D)$ is a polynomial in D of degree $n - 1$.

We next consider the equation $(D - \alpha)^2 y = 0$. We can check that $e^{\alpha x}$ and $xe^{\alpha x}$ are solutions by direct substitution. An alternative method is to write $u = (D - \alpha)y$. Then

$$(D - \alpha)^2 y = (D - \alpha)u = 0$$

and so

$$u = ce^{\alpha x}$$

i.e.

$$(D - \alpha)y = ce^{\alpha x}$$

This is a first order linear equation (§12.6) for which a suitable integrating factor is $e^{-\alpha x}$. Using this integrating factor, we obtain

$$\frac{d}{dx}(e^{-\alpha x}y) = e^{-\alpha x}\frac{dy}{dx} - \alpha e^{-\alpha x}y = c$$
$$e^{-\alpha x}y = cx + d$$
$$y = cxe^{\alpha x} + de^{\alpha x}$$

Hence, two solutions of $(D - \alpha)^2 y = 0$ are $e^{\alpha x}$ and $xe^{\alpha x}$. These solutions are linearly independent, since one is not a scalar multiple of the other.

An extension of this argument shows that each factor $(D - \alpha)^m$ contributes m linearly independent solutions to $P(D)y = 0$ – namely $e^{\alpha x}, xe^{\alpha x}, x^2 e^{\alpha x}, \ldots,$ $x^{m-1}e^{\alpha x}$. Any solution of the equation is a linear combination of these solutions.

The general solution of $(D - \alpha)^m y = 0$ is $y = (c_0 + c_1 x + \cdots + c_{m-1}x^{m-1})e^{\alpha x}$

If n is the degree of the polynomial $P(D)$, taking each factor in turn, we obtain $m_1 + m_2 + \cdots + m_k = n$ linearly independent solutions in all, which form a basis of the solution space of the equation.

E.g. consider the equation

$$P(D)y = (D - \alpha)(D - \beta)^3(D - \gamma)^2 y = 0$$

in which α, β, and γ are distinct. This has *six* linearly independent solutions. The general solution is

$$y = A_0 e^{\alpha x} + (B_0 + B_1 x + B_2 x^2) e^{\beta x} + (C_0 + C_1 x) e^{\gamma x}$$

where A_0, B_0, B_1, B_2, C_0 and C_1 are arbitrary constants.

Example 1

Find the general solution of

$$\frac{d^2 y}{dx^2} + 2 \frac{dy}{dx} - 3y = 0$$

The equation may be written as

$$P(D)y = (D^2 + 2D - 3)y = 0$$

The auxiliary equation is $P(z) = z^2 + 2z - 3 = (z + 3)(z - 1) = 0$. Thus

$$(D + 3)(D - 1)y = 0$$

so the general solution is

$$y = Ae^x + Be^{-3x}$$

where A and B are arbitrary constants.

Example 2

Consider the equation

$$\frac{d^3 y}{dx^3} + \frac{d^2 y}{dx^2} - \frac{dy}{dx} - y = 0$$

This may be written as

$$P(D)y = (D^3 + D^2 - D - 1)y = 0$$

The auxiliary equation is $P(z) = z^3 + z^2 - z - 1 = 0$. A root is $z = 1$. Thus $P(z) = (z - 1)Q(z) = (z - 1)(z + 1)^2$. Thus

$$P(D)y = (D - 1)(D + 1)^2 y = 0$$

has general solution

$$y = Ae^x + (B + Cx)e^{-x}$$

where A, B and C are arbitrary constants.

14.5 Complex roots of the auxiliary equation

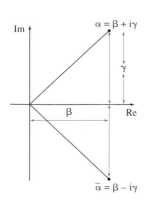

In §14.4 we explained how the general solution of a differential equation $P(\mathrm{D})f = 0$ may be expressed in terms of the roots $\alpha_1, \alpha_2, \ldots, \alpha_n$ of the auxiliary equation $P(z) = 0$. If P is a polynomial with *real* coefficients, we shall usually only be interested in *real* solutions of the differential equation. But some of the numbers $\alpha_1, \alpha_2, \ldots, \alpha_n$ may be *complex*. How can the theory be adapted in this case?

We have seen in §13.6, that if $\alpha = \beta + i\gamma$ is a root of $P(z) = 0$, then so is $\bar\alpha = \beta - i\gamma$ provided that the coefficients of P are *real*. As the nonreal roots of $P(z) = 0$ occur in *pairs* of the form $\alpha = \beta + i\gamma$ and $\bar\alpha = \beta - i\gamma$, the complex terms of the general solution of $P(\mathrm{D})y = 0$ may be paired as

$$x^k(Ae^{\alpha x} + Be^{\bar\alpha x}) \tag{1}$$

We are only interested in *real* forms of these solutions.

We need to recall that $e^{i\theta} = \cos\theta + i\sin\theta$ *(and hence* $e^{-i\theta} = \cos\theta - i\sin\theta$*).* Thus

$$e^{\alpha x} = e^{\beta x}e^{i\gamma x} = e^{\beta x}(\cos\gamma x + i\sin\gamma x)$$
$$e^{\bar\alpha x} = e^{\beta x}e^{-i\gamma x} = e^{\beta x}(\cos\gamma x - i\sin\gamma x)$$

Hence

$$x^k(Ae^{\alpha x} + Be^{\bar\alpha x}) = x^k e^{\beta x}\{(A+B)\cos\gamma x + i(A-B)\sin\gamma x\}$$
$$= x^k e^{\beta x}(a\cos\gamma x + b\sin\gamma x)$$

provided $a = A + B$ and $b = i(A - B)$. We wish only to consider *real* values of a and b. *Since we have* $A = \frac{1}{2}(a - ib)$ *and* $B = \frac{1}{2}(a + ib)$ *this will require that A and B be complex conjugates.*

See Exercise 13.8.5.

We note that

> Complex roots of the auxiliary equation yield real solutions of the form
> $$x^k e^{\beta x}(a\cos\gamma x + b\sin\gamma x)$$

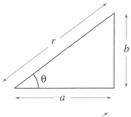

Occasionally it is useful to replace the arbitrary real constants a and b by the corresponding polar coordinates r and θ where $a = r\cos\theta$ and $b = r\sin\theta$. Making this substitution in (1), we find that it may be replaced by

$$rx^k e^{\beta x}(\cos\gamma x\cos\theta + \sin\gamma x\sin\theta) = rx^k e^{\beta x}\cos(\gamma x - \theta) \tag{2}$$

where r and θ are arbitrary real constants. Alternatively, the arbitrary constants a and b may be replaced by R and Θ where $a = R\sin\Theta$ and $b = R\cos\Theta$. We then see that (1) may be replaced by

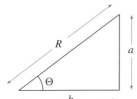

$$Rx^k e^{\beta x}(\sin\Theta\cos\gamma x + \cos\Theta\sin\gamma x) = Rx^k e^{\beta x}\sin(\gamma x + \Theta) \tag{3}$$

where R and Θ are arbitrary real constants.

Example 3

Find the general solution of

$$\frac{\mathrm{d}^2 y}{\mathrm{d}x^2} + 4y = 0$$

We write the equation in the form

$$P(\mathrm{D})y = (\mathrm{D}^2 + 4)y = 0$$

i.e.

$$(\mathrm{D} + 2\mathrm{i})(\mathrm{D} - 2\mathrm{i})y = 0$$

The general solution is therefore

$$y = A\mathrm{e}^{-2\mathrm{i}x} + B\mathrm{e}^{2\mathrm{i}x}$$

If we are interested only in real solutions, we may rewrite this in any one of the equivalent forms

$$y = a\cos 2x + b\sin 2x \quad (a, b \in \mathbb{R})$$

or

$$y = r\cos(2x - \theta) \quad (r, \theta \in \mathbb{R})$$

or

$$y = R\sin(2x + \Theta) \quad (R, \Theta \in \mathbb{R})$$

Example 4

Find the general solution of

$$\frac{\mathrm{d}^3 y}{\mathrm{d}x^3} = y$$

We write the equation in the form

$$P(\mathrm{D})y = (\mathrm{D}^3 - 1)y = 0$$

Observe that $P(z) = z^3 - 1^\dagger = (z - 1)(z^2 + z + 1)$ The roots of $z^2 + z + 1 = 0$ are given by

$$\alpha = \frac{-1 + \mathrm{i}\sqrt{3}}{2} \quad \text{and} \quad \bar{\alpha} = \frac{-1 - \mathrm{i}\sqrt{3}}{2}$$

Thus $P(z) = (z - 1)(z - \alpha)(z - \bar{\alpha})$. The general solution of $P(\mathrm{D})y = 0$ is therefore

$$y = A\mathrm{e}^x + B\mathrm{e}^{(-1+\mathrm{i}\sqrt{3})x/2} + C\mathrm{e}^{(-1-\mathrm{i}\sqrt{3})x/2}$$

which we may rewrite in the form

$$y = A\mathrm{e}^x + b\mathrm{e}^{-x/2}\cos\left(\frac{\sqrt{3}}{2}x\right) + c\mathrm{e}^{-x/2}\sin\left(\frac{\sqrt{3}}{2}x\right)$$

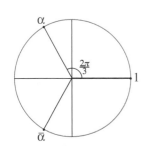

14.6 Homogeneous, linear, difference equations

Instead of considering equations of the form $P(D)y = 0$ as in §14.4, we now consider equations of the form

$$P(E)y_x = 0 \ (x = 0, 1, 2, \ldots)$$

The appropriate theory is very similar indeed to that studied in §14.4.
Suppose that

$$P(E)y_x = (E - \alpha_1)^{m_1}(E - \alpha_2)^{m_2} \ldots (E - \alpha_k)^{m_k} y_x = 0$$

where $\alpha_1, \alpha_2, \ldots, \alpha_k$ are distinct nonzero numbers. Each factor $(E - \alpha)^m$ contributes m linearly independent solutions to $P(E)y_x = 0$ – namely α^x, $x\alpha^x$, $x^2\alpha^x, \ldots, x^{m-1}\alpha^x$.

The general solution of $(E - \alpha)^m y_x = 0$ is $y_x = (c_0 + c_1 x + \cdots + c_{m-1}x^{m-1})\alpha^x$

Taking each factor in turn, we obtain $m_1 + m_2 + \cdots + m_k = n$ linearly independent solutions in all, where n is the degree of the polynomial $P(E)$.
E.g. consider the equation

$$P(E)y_x = (E - \alpha)(E - \beta)^3(E - \gamma)^2 y_x = 0$$

where α, β, and γ are distinct nonzero numbers. This has *six* linearly independent solutions. The general solution is

$$y_x = A_0\alpha^x + (B_0 + B_1 x + B_2 x^2)\beta^x + (C_0 + C_1 x)\gamma^x$$

This result can be justified along the same lines as in §14.4 using the results about difference equations obtained in §14.2.
When seeking real solutions of $P(E)y_x = 0$ in the case when the polynomial P has a pair of complex conjugate roots, the necessary argument is slightly simpler than that of §14.5. Write the roots of the auxiliary equation $P(z) = 0$ in polar coordinates; $\alpha = \rho e^{i\phi}$ and $\bar{\alpha} = \rho e^{-i\phi}$. Then as in §14.5, the expression

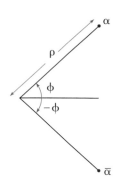

$$x^k(A\alpha^x + B\bar{\alpha}^x)$$

in the general solution of $P(E)y_x = 0$ may be replaced by the real form

$$x^k \rho^x (a \cos \phi x + b \sin \phi x)$$

where a and b are arbitrary constants. Occasionally, as for solutions of differential equations, it is useful to use one of the alternative expressions

$$r x^k \rho^x \cos(\phi x - \theta)$$

or

$$R x^k \rho^x \sin(\phi x + \Theta)$$

Example 5

[†] Alternatively, the equation could be written as
$$y_x + 3y_{x-1} + 2y_{x-2} = 0$$
$(x = 2, 3, \ldots)$.

Find the general solution of[†]

$$y_{x+2} + 3y_{x+1} + 2y_x = 0$$
$$(x = 0, 1, 2, \ldots)$$

We write this as

$$(E^2 + 3E + 2)y_x = 0$$
$$(E + 1)(E + 2)y_x = 0$$

and observe immediately that the general solution is

$$y_x = A(-1)^x + B(-2)^x \qquad (x = 0, 1, 2, \ldots)$$

Example 6

Find the general solution of

$$y_{x+3} + y_{x+1} + 10y_x = 0 \quad (x = 0, 1, 2, \ldots)$$

$$\text{or} \quad (E^3 + E + 10)y_x = 0$$

$$\text{i.e.} \quad (E + 2)(E - 1 - 2i)(E - 1 + 2i)y_x = 0$$

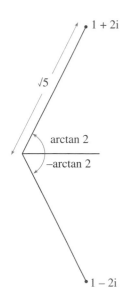

with general solution

$$y_x = A(-2)^x + B(1 + 2i)^x + C(1 - 2i)^x$$

If we are interested only in real solutions, we need to observe that

$$1 + 2i = \rho e^{i\phi} \qquad 1 - 2i = \rho e^{-i\phi}$$

where

$$|\rho| = \{1^2 + 2^2\}^{1/2} = \sqrt{5}$$

and

$$\phi = \arctan 2$$

We may then write

$$y_x = A(-2)^x + 5^{x/2}(b \cos \phi x + c \sin \phi x) \qquad (x = 0, 1, 2, \ldots)$$

where $\phi = \arctan 2$ and we restrict our attention to *real* arbitrary constants A, b and c.

Example 7

Find the general solution of

$$y_{x+3} = 2y_{x+2} \qquad (x = 0, 1, 2, \ldots)$$

We may rewrite this as

$$P(E)y = (E^3 - 2E^2)y_x = 0$$

i.e.

$$E^2(E - 2)y_x = 0$$

but this is not particularly useful since, when a root α of the auxiliary equation is zero, the solutions $y_x = \alpha^x$ given in §14.10 all reduce to zero.

It is better to put $z = x + 2$ in which case the equation becomes

$$y_{z+1} = 2y_z \quad (z = 2, 3, 4, \ldots)$$

Our equation is therefore really the first order equation

$$(E - 2)y_z = 0$$

which has the general solution

$$y_z = C2^z \quad (z = 2, 3, 4, \ldots)$$

This leaves us free to choose $y_0 = A$ and $y_1 = B$ where A and B are arbitrary.
Observe that there is a difference here with the analogous differential equation

$$\frac{d^3 y}{dx^3} = 2\frac{d^2 y}{dx^2} \quad or \quad D^2(D - 2)y = 0$$

which is genuinely of order 3, so that its general solution involves three linearly independent solutions:

$$y = A + Bx + Ce^{2x}$$

Example 8

Solve the equation

$$y_{x+2} - 4y_{x+1} + 4y_x = 0 \quad (x = 0, 1, 2, \ldots)$$

with initial conditions, $y_0 = 3$, $y_1 = 5$.
We have

$$(E - 2)^2 y_x = 0$$

with general solution

$$y_x = A2^x + Bx2^x \quad (A, B \in \mathbb{R})$$

With boundary conditions, $y_0 = 3$, $y_1 = 5$,

$$3 = A + B.0$$

and $\quad 5 = 2A + 2B \quad \Rightarrow A = 3, B = -\frac{1}{2}$

So the particular solution of the homogeneous equation satisfying the given conditions is

$$y_x = 3(2)^x - \frac{x}{2}(2)^x = 3(2)^x - x(2)^{x-1} \qquad (x = 0, 1, 2, \ldots)$$

Example 9

An application to economics

In some models of the economy, comparisons are made between consecutive time periods. So difference equations can be used to deal with the time lags between periods in analysing the changes in the system. We will denote the variable 'time' by t which is measured in discrete time periods, for instance years.

In a simple model of a closed economy without government expenditure, in year t, income Y_t is composed of consumption C_t and investment I_t:

$$Y_t = C_t + I_t \tag{1}$$

Investment in any year is proportional to income in the previous year:

$$I_t = \alpha Y_{t-1} \quad (\alpha > 0) \tag{2}$$

Consumption in any year is proportional to income in the same year:

$$C_t = \beta Y_t \quad (\beta > 0) \tag{3}$$

Equations (1), (2) and (3) give rise to a difference equation

$$Y_t = \beta Y_t + \alpha Y_{t-1} \tag{4}$$

or

$$(1 - \beta)Y_t = \alpha Y_{t-1}$$

$$Y_t = \left(\frac{\alpha}{1 - \beta}\right) Y_{t-1}$$

$$= \left(\frac{\alpha}{1 - \beta}\right)^t Y_0 \qquad (t = 1, 2, \ldots)$$

If income is to increase annually, we must have $\dfrac{\alpha}{1 - \beta} > 1$. Equations (1) and (3) imply that $\beta < 1$ so $1 - \beta > 0$. Hence the condition is equivalent to $\alpha > 1 - \beta$ or $\alpha + \beta > 1$.

14.7 Nonhomogeneous equations♣

We begin by recalling some results about systems of linear equations. Suppose that A is an $m \times n$ matrix and that $\mathbf{b}$ is an $m \times 1$ column vector. Let $\mathbf{x} = \mathbf{p}$ be any particular solution of the system

$$A\mathbf{x} = \mathbf{b}$$

Then any other $n \times 1$ column vector $\mathbf{u}$ is a solution of the system if and only if

$$\mathbf{u} = \mathbf{x}_0 + \mathbf{p}$$

where $\mathbf{x}_0$ is a solution of the *homogeneous* system

$$A\mathbf{x} = \mathbf{0}$$

(i.e. $\mathbf{x}_0$ lies in the kernel of A).

The proof is quite easy. Suppose first that $\mathbf{u} = \mathbf{x}_0 + \mathbf{p}$. Then

$$A\mathbf{u} = A(\mathbf{x}_0 + \mathbf{p}) = A\mathbf{x}_0 + A\mathbf{p} = \mathbf{0} + \mathbf{b} = \mathbf{b}$$

and so $\mathbf{u}$ is a solution of $A\mathbf{x} = \mathbf{b}$. Suppose, on the other hand, that $\mathbf{u}$ is a solution of $A\mathbf{x} = \mathbf{b}$. Then

$$A(\mathbf{u} - \mathbf{p}) = A\mathbf{u} - A\mathbf{p} = \mathbf{b} - \mathbf{b} = \mathbf{0}$$

and so $\mathbf{x}_0 = \mathbf{u} - \mathbf{p}$ is a solution of $A\mathbf{x} = \mathbf{0}$. The situation for a nonhomogeneous differential or difference equation is exactly the same.

Consider differential equations first.

14.7.1 Nonhomogeneous differential equations♣

The nonhomogeneous differential equation

$$P(\mathrm{D})y = q(x)$$

has general solution

$$y = c(x) + p(x)$$

where $c(x)$ is the general solution of the homogeneous equation $P(\mathrm{D})y = 0$ (called the **complementary function**) and $p(x)$ is a particular solution of the nonhomogeneous equation $P(\mathrm{D})y = q(x)$.

We shall provide a systematic method for locating a particular solution of a nonhomogeneous equation *only in the case when q is itself the solution of a homogeneous equation.*

Suppose that q is a solution of the homogeneous equation $Q(\mathrm{D})y = 0$ where Q has degree m; so that $Q(\mathrm{D})q = 0$. Let P have degree n. We want a particular solution p of $P(\mathrm{D})y = q$, that is

$$P(\mathrm{D})p = q \tag{1}$$

Hence, operating on both sides of equation (1) with $Q(\mathrm{D})$ we obtain

$$Q(\mathrm{D})P(\mathrm{D})p = Q(\mathrm{D})q = 0$$

Then p must be a solution of the homogeneous equation of order $(n + m)$

$$Q(\mathrm{D})P(\mathrm{D})y = 0$$

so is of the form

$$y = (c_1 y_1 + c_2 y_2 + \cdots + c_n y_n) + (c_{n+1} y_{n+1} + \cdots + c_{n+m} y_{n+m})$$

where $y_1, \ldots, y_{n+m}$ are linearly independent and $y_1, \ldots, y_n$ is a basis of ker $P(D)$.

Now the terms in the first bracket represent the complementary function $c(x)$, which vanishes under the operator $P(D)$. There is then no point in including it in our calculations, and so $p^\dagger$ may be found by substituting

$$p = c_{n+1} y_{n+1} + \cdots + c_{n+m} y_{n+m}$$

in the equation $P(D)y = q$ to obtain appropriate values of the constants $c_{n+1}, \ldots, c_{n+m}$.

The above argument indicates that $p(x)$ lies in ker $Q(D)$ when $P(D)$ and $Q(D)$ have no common factor; that is, when $c(x)$ and $q(x)$ are linearly independent. In this case, since we assume that $q(x)$ lies in ker $Q(D)$, $p(x)$ *has the same form as* $q(x)$. In §14.4 we established that the solution of the equation $P(D)y = 0$ corresponding to the factor $(D - \alpha)^m$ of $P(D)$ has the pattern $(c_0 + c_1 x + \cdots + c_{m-1} x^{m-1})e^{\alpha x}$.‡ So we can find a particular solution of $P(D)y = q$ of the same form. This suggests the rule:

> If $q(x) = (b_0 + b_1 x + \cdots + b_{m-1} x^{m-1})e^{\alpha x}$ try
> $p(x) = (d_0 + d_1 x + \cdots + d_{m-1} x^{m-1})e^{\alpha x}$

† Note that we choose a particular solution p of the nonhomogeneous equation which is linearly independent of the complementary function.

‡ If $\alpha = 0$ then $q(x)$ is just a polynomial in x.

Example 10

Find the general solution of

$$(D^2 + 9)y = xe^x$$

The complementary function is

$$c(x) = A \cos 3x + B \sin 3x$$

To find a particular solution, try a solution of the same form as xe^x:

$$p(x) = (ax + b)e^x$$
$$Dp(x) = (ax + b)e^x + ae^x = e^x(ax + a + b)$$
$$D^2 p(x) = (ax + 2a + b)e^x$$

Substituting in the differential equation

$$(ax + 2a + b + 9ax + 9b)e^x = xe^x$$

This implies

$$a + 9a = 1$$
$$2a + 10b = 0$$

yielding $a = \frac{1}{10}, b = -\frac{1}{50}$. Hence a particular solution is

$$p(x) = \left(\frac{x}{10} - \frac{1}{50}\right)e^x$$

The general solution of the nonhomogeneous equation is then

$$y = A\cos 3x + B\sin 3x + \left(\frac{x}{10} - \frac{1}{50}\right)e^x$$

When $q(x)$ and the complementary function $c(x)$ are linearly dependent, we consider a particular solution of the form of $q(x)$, multiplied by an appropriate power of x to make it linearly independent of $c(x)$.

Example 11

Solve

$$(D^2 - 1)y = e^x$$

The equation has complementary function

$$c(x) = Ae^x + Be^{-x}$$

We do not consider a particular solution of the form ce^x, since it is not independent of the complementary function. So try $p(x) = cxe^x$.
   Then $Dp(x) = c(x+1)e^x$ and $D^2p(x) = c(x+2)e^x$.
   Substituting in the differential equation

$$[c(x+2) - cx]e^x = e^x$$

so $2c = 1$ and $c = \dfrac{1}{2}$.
   The general solution of the nonhomogeneous equation is

$$y = Ae^x + Be^{-x} + \frac{x}{2}e^x \qquad (A, B \in \mathbb{R})$$

Suppose that $q(x) = e^{\beta x}\sin \gamma x$ or $e^{\beta x}\cos \gamma x$. These are solutions of the equation $((D - \beta)^2 + \gamma^2)y = 0$, so consider a general solution of this equation for $p(x)$.

If $q(x) = e^{\beta x}\sin \gamma x$ or $e^{\beta x}\cos \gamma x$ try $p(x) = e^{\beta x}(a\sin \gamma x + b\cos \gamma x)$

Example 12

Solve

$$(D^2 - 1)y = \sin 2x$$

We look for a particular solution of the form

$$p(x) = a\sin 2x + b\cos 2x$$
$$Dp(x) = a2\cos 2x - b2\sin 2x$$
$$D^2p(x) = -a2^2\sin 2x - b2^2\cos 2x$$

Substituting in the differential equation

$$-4a \sin 2x - 4b \cos 2x - a \sin 2x - b \cos 2x = \sin 2x$$

Equating coefficients of $\sin 2x$ and $\cos 2x$, since they are linearly independent functions, we have

$$-4a - a = 1$$
$$-5b = 0$$

So $a = -\frac{1}{5}$, $b = 0$. A particular solution is

$$p(x) = -\frac{1}{5} \sin 2x$$

The general solution is

$$y(x) = Ae^x + Be^{-x} - \frac{1}{5} \sin 2x$$

Consider the case when $q(x)$ is the sum of two functions so that

$$P(\mathrm{D})y = q_1(x) + q_2(x) \tag{2}$$

Suppose that $p_1(x)$ is a solution of $P(\mathrm{D})y = q_1(x)$ and that $p_2(x)$ is a solution of $P(\mathrm{D})y = q_2(x)$. Then it is evident that $p_1(x) + p_2(x)$ is a solution of equation (2) since

$$P(\mathrm{D})(p_1 + p_2) = P(\mathrm{D})p_1 + P(\mathrm{D})p_2 = q_1(x) + q_2(x)$$

Example 13

Solve

$$(\mathrm{D}^2 - 1)y = e^x + \sin 2x$$

Using the results of Examples 11 and 12, the general solution is

$$y(x) = Ae^x + Be^{-x} + \frac{x}{2}e^x - \frac{1}{5} \sin 2x$$

Example 14

An application to physics

With time t regarded as a continuous variable here, differential equations can be used to model the movement of charge in an electrical circuit, such as that shown in the figure. A changing current gives a back electromotive force (e.m.f.) $L dI/dt$ across the inductor L; in addition, there are a voltage drop IR across the resistor R and a voltage drop Q/C across the capacitor C. Setting the externally applied e.m.f. $E \cos \omega t$ equal to the total voltage drop across the circuit, we find:

$$L\frac{dI}{dt} + RI + \frac{Q}{C} = E \cos \omega t$$

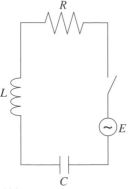

Now $I = \dfrac{dQ}{dt}$, so we obtain

$$L\frac{d^2 Q}{dt^2} + R\frac{dQ}{dt} + \frac{Q}{C} = E \cos \omega t \qquad (3)$$

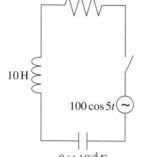

With an inductor L of 10H, a resistor R of $100\,\Omega$ and a capacitor C of 8×10^{-4}F, fed with a generator of electromotive force $100 \cos 5t$, equation (3) becomes

$$10\frac{d^2 Q}{dt^2} + 100\frac{dQ}{dt} + \frac{10^4}{8}Q = 100 \cos 5t$$

or

$$\frac{d^2 Q}{dt^2} + 10\frac{dQ}{dt} + 125Q = 10 \cos 5t \qquad (4)$$

Suppose that we wish to find the charge at time t if, initially, there are no charge and no current. We must find the particular solution of this equation which satisfies these boundary conditions.

The auxiliary equation is

$$z^2 + 10z + 125 = 0$$

with roots $z = -5 \pm 10i$. Hence the complementary solution is

$$c(t) = e^{-5t}(A \cos 10t + B \sin 10t)$$

This solution is oscillatory but is 'damped' by the negative exponential factor, so $c(t) \to 0$ as $t \to \infty$ as displayed in the graph below.

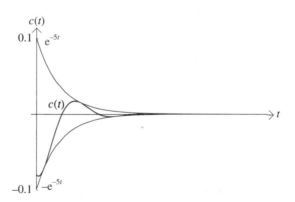

We seek a particular solution of the form

$$p(t) = a \cos 5t + b \sin 5t$$

Differentiating

$$\frac{Dp(t)}{dt} = -5a \sin 5t + 5b \cos 5t$$

and

$$\frac{D^2 p(t)}{dt^2} = -25a \cos 5t - 25b \sin 5t$$

Substituting in equation (4)

$$(-25a \cos 5t - 25b \sin 5t) + 10(-5a \sin 5t + 5b \cos 5t)$$
$$+125(a \cos 5t + b \sin 5t) = 10 \cos 5t$$

The linear independence of the functions $\cos 5t$ and $\sin 5t$ gives the equations

$$\left. \begin{array}{l} -25a + 50b + 125a = 10 \\ -25b - 50a + 125b = 0 \end{array} \right\}$$

whose solution is $a = \dfrac{2}{25}$, $b = \dfrac{1}{25}$. The particular solution

$$p(t) = \frac{1}{25}(2 \cos 5t + \sin 5t)$$

is oscillatory. Its graph shows stable oscillations.

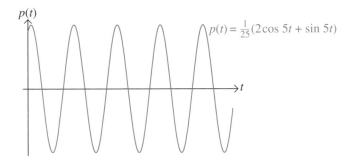

Hence the general formula for the charge in the circuit is

$$Q(t) = e^{-5t}(A \cos 10t + B \sin 10t) + \frac{1}{25}(2 \cos 5t + \sin 5t) \qquad (5)$$

As there is no initial charge

$$0 = Q(0) = A + \frac{2}{25} \quad \text{and} \quad A = -\frac{2}{25}$$

Differentiating equation (5)

$$\frac{dQ(t)}{dt} = e^{-5t}\{(-5A + 10B) \cos 10t + (-10A - 5B) \sin 10t\}$$
$$+ \frac{1}{25}(-10 \sin 5t + 5 \cos 5t)$$

With no initial current $I(0) = \dfrac{dQ(0)}{dt} = 0$ so we have

$$0 = \frac{dQ(0)}{dt} = (-5)\left(\frac{-2}{25}\right) + 10B + \frac{1}{5}, \quad \text{giving } B = -\frac{3}{50}$$

Hence, with the given initial conditions the formula for the charge is

$$Q(t) = \frac{1}{50}e^{-5t}(-4 \cos 10t - 3 \sin 10t) + \frac{1}{25}(2 \cos 5t + \sin 5t)$$

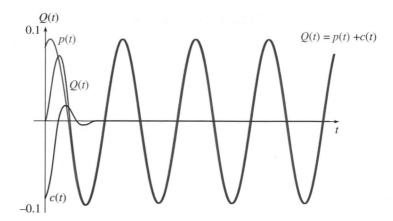

The graph of $Q(t)$ indicates that $c(t)$ initially dampens the oscillations. But as it dies away, $Q(t)$ does not converge but oscillates finitely.

14.7.2   Nonhomogeneous difference equations♣

Analogous statements are true for difference equations with E replacing D.

$$P(E)y_x = q_x \qquad (x = 0, 1, 2, \ldots)$$

has general solution

$$y_x = c_x + p_x \qquad (x = 0, 1, 2, \ldots)$$

where c_x is the complementary sequence and p_x is a particular solution of the nonhomogeneous equation. Restricting ourselves to the case when q_x is itself the solution of a homogeneous difference equation and using the pattern for a solution of a difference equation found in §14.6 we can suggest:

> If $q_x = (b_0 + \cdots + b_{m-1}x^{m-1})\alpha^x$ try
> $p_x = (d_0 + \cdots + d_{m-1}x^{m-1})\alpha^x$.

Example 15

Solve

$$(E - 1)y_x = x2^x \quad (x = 0, 1, 2, \ldots)$$

The complementary sequence

$$c_x = A(1)^x = A$$

We seek a particular solution of the form $p_x = (ax + b)2^x$:

$$Ep_x = p_{x+1} = [a(x + 1) + b]2^{x+1}$$

Substituting in the difference equation

$$[ax + a + b]2^{x+1} - [ax + b]2^x = x2^x$$

so

$$2ax + 2a + 2b - ax - b = x$$

and $a = 1$, $b = -2$.

The particular solution is $p_x = (x - 2)2^x$. The general solution of the equation is

$$y_x = A + (x - 2)2^x \qquad (x = 0, 1, 2, \ldots)$$

Example 16

Solve

$$(E - 1)y_x = 5x + 4 + x2^x \quad (x = 0, 1, 2, \ldots)$$

First consider the polynomial $5x + 4$. Since the complementary sequence $c_x = A$ we cannot try a particular solution with a constant term, so set

$$p_x = x(ax + b)$$
$$Ep_x = (x + 1)(ax + a + b) = ax^2 + (2a + b)x + a + b$$
$$ax^2 + (2a + b)x + a + b - ax^2 - bx = 5x + 4$$
$$\Rightarrow a = \frac{5}{2} \quad b = \frac{3}{2}$$
$$p_x = \frac{x}{2}(5x + 3)$$

The particular solution for the term $x2^x$ was found in Example 15 to be $(x - 2)2^x$. The general solution of the equation is then

$$y_x = A + \frac{x}{2}(5x + 3) + (x - 2)2^x \qquad (x = 0, 1, 2, \ldots)$$

Example 17

An application to economics – Samuelson multiplier–accelerator model

In this model of a closed economic system, national income Y_t at time t depends on consumption C_t, investment I_t and government expenditure G (assumed constant). The appropriate equations are

$$Y_t = C_t + I_t + G$$
$$C_{t+1} = \gamma Y_t$$
$$I_{t+1} = \alpha(C_{t+1} - C_t)$$

Here γ $(0 < \gamma < 1)$ is a **multiplier** and α $(\alpha > 0)$ is an **accelerator**. The idea is that consumption is proportional to income in the previous period and that investment is proportional to the *increase* in current consumption as compared with consumption in the previous period.

In our illustration of the model, government expenditure G is assumed to be the constant 900; the multiplier is 0.8 and the accelerator 0.375. So

$$Y_t = C_t + I_t + 900 \tag{6}$$
$$C_{t+1} = 0.8Y_t \tag{7}$$
$$I_{t+1} = 0.375(C_{t+1} - C_t) \tag{8}$$

We may eliminate I_t and C_t from the equations to obtain a difference equation of order 2 in the single variable Y_t.

Substituting in equation (6) from equations (7) and (8)

$$Y_t = 0.8Y_{t-1} + 0.375(C_t - C_{t-1}) + 900 \qquad (t = 2, 3, \ldots)$$

Substituting for C_t from equation (6) this becomes

$$Y_t = 0.8Y_{t-1} + 0.375 \times 0.8(Y_{t-1} - Y_{t-2}) + 900$$
$$= 0.8Y_{t-1} + 0.3Y_{t-1} - 0.3Y_{t-2} + 900$$

This difference equation can be rewritten as

$$(E^2 - 1.1E + 0.3)Y_t = 900 \qquad (t = 0, 1, 2, \ldots)$$

or

$$(E - 0.6)(E - 0.5)Y_t = 900$$

which has complementary sequence

$$c_t = A(0.6)^t + B(0.5)^t$$

If there is an equilibrium level of income Y^* then

$$Y_t = Y_{t-1} = Y_{t-2} = Y^*$$

which is given by

$$Y^* - 1.1Y^* + 0.3Y^* = 900$$

and so $0.2 \times Y^* = 900$ and $Y^* = 4500$. Hence the general solution is

$$Y_t = A(0.6)^t + B(0.5)^t + 4500$$

Since $|0.6| < 1$ and $|0.5| < 1$,

$$c_t = A(0.6)^t + B(0.5)^t \to 0 \text{ as } t \to \infty$$

So, as $t \to \infty$, $Y_t \to 4500$. This indicates that in the long run national income Y_t is stable and converges to its equilibrium level 4500.

From equation (7) we conclude that C_t converges to $0.8 \times 4500 = 3600$. Finally, equations (6) and (8) indicate that the investment I_t converges to 0.

In any particular situation, in order to find the values of the constants A and B it is necessary to know the initial income levels Y_0 and Y_1. Suppose that $Y_0 = 9500$ and $Y_1 = 9000$. Substitution in the formula for Y_t yields the equations

$$A + B = 5000$$

$$(0.6)A + (0.5)B = 4500$$

which have the solution $A = 20\,000, \ B = -15\,000$. Hence in this case

$$Y_t = 20\,000(0.6)^t - 15\,000(0.5)^t + 4500$$

A different approach which treats the equations as a system of two first order difference equations in two sequences appears in Example 21.

14.8 Convergence and divergence♣

The behaviour of solutions of differential and difference equations when x is very large is often of interest since it indicates what will happen 'in the long run'.

The diagrams below indicate the behaviour of the functions which arise in the solution of the differential equations considered in this chapter.

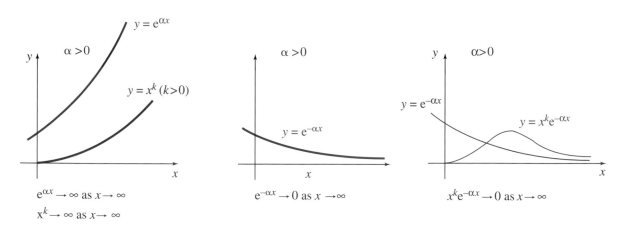

Note in particular that $x^k e^{-\alpha x} \to 0$ as $x \to \infty$ no matter how large the constant k may be and no matter how small the positive constant α may be. This is an example of the important fact that 'exponentials dominate powers'. The same phenomenon is also evident in the first diagram where the graph $y = e^{\alpha x}$ grows much faster than that of $y = x^k$.

These results are easy to confirm using the Taylor series expansion for the exponential function. We have

$$e^{\alpha x} = 1 + \alpha x + \frac{(\alpha x)^2}{2!} + \cdots + \frac{(\alpha x)^{k+1}}{(k+1)!} + \cdots$$

and so

$$e^{\alpha x} > \frac{(\alpha x)^{k+1}}{(k+1)!} \quad (x \geq 0)$$

It follows that

$$x^k e^{-\alpha x} < \frac{(k+1)!}{\alpha^{k+1}} \cdot \frac{1}{x} \to 0 \quad \text{as} \quad x \to \infty$$

When the auxiliary equation has a complex root $\alpha = \beta + i\gamma$, the solutions oscillate. When $\beta = 0$ they oscillate finitely.

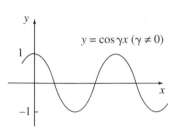

cos γx oscillates finitely as $x \to \infty$

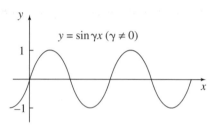

sin γx oscillates finitely as $x \to \infty$

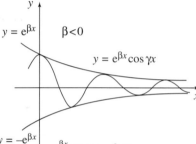

$e^{\beta x} \cos \gamma x \to 0$ as $x \to \infty$

When $\beta < 0$ the solutions converge
as $x \to \infty$.
When $\beta > 0$ they oscillate infinitely
as $x \to \infty$.

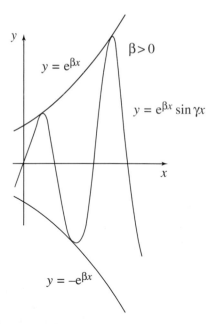

The situation for the solutions of difference equations is very similar except that terms of the form $e^{\alpha x}$ are replaced by terms of the form α^x.

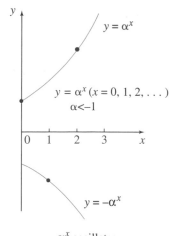

$y = \alpha^x \ (x = 0, 1, 2, \dots)$

$\alpha < -1$

$y = -\alpha^x$

$y = \alpha^x$ oscillates

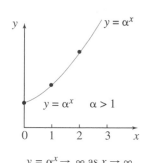

$y = \alpha^x \qquad \alpha > 1$

$y = \alpha^x \to \infty$ as $x \to \infty$

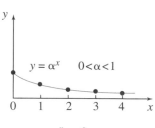

$y = \alpha^x \qquad 0 < \alpha < 1$

$y = \alpha^x \to 0$ as $x \to \infty$

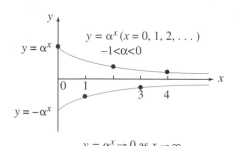

$y = \alpha^x \ (x = 0, 1, 2, \dots)$

$-1 < \alpha < 0$

$y = \alpha^x \to 0$ as $x \to \infty$

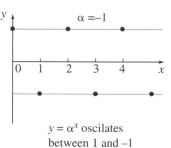

$\alpha = -1$

$y = \alpha^x$ oscilates
between 1 and –1

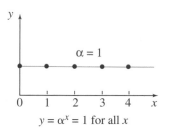

$\alpha = 1$

$y = \alpha^x = 1$ for all x

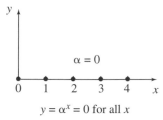

$\alpha = 0$

$y = \alpha^x = 0$ for all x

Example 18 _____

In Example 1 we found that the general solution of

$$\frac{d^2 y}{dx^2} + 2\frac{dy}{dx} - 3y = 0$$

was given by

$$y = Ae^x + Be^{-3x}$$

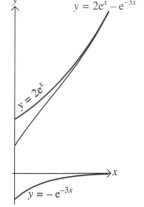

To investigate the possible behaviour of a solution, we must initially observe that if $A \neq 0$, the dominant term is the first.

> If $A > 0$, then $y \to \infty$ as $x \to \infty$.
> If $A < 0$, then $y \to -\infty$ as $x \to \infty$.
> If $A = 0$, then $y \to 0$ as $x \to \infty$

Suppose that we wish to find the particular solution which satisfies the boundary conditions $y(0) = 1$ and $y(1) = 2e - e^{-3}$. Imposing these conditions on the solution, we obtain the simultaneous equations

$$A + B = 1 \qquad \text{and} \qquad Ae + Be^{-3} = 2e - e^{-3}$$

which give $A = 2$ and $B = -1$. The solution we want is

$$y = 2e^x - e^{-3x}$$

The first term dominates and the sketch of the solution confirms the conclusion reached above, that $y \to \infty$ as $x \to \infty$ when $A > 0$.

Example 19

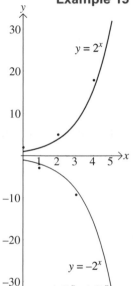

$y = 2^x$

$y = -2^x$

$y_x = (-1)^x + (-2)^x$

In Example 5 we found that the general solution of

$$y_{x+2} + 3y_{x+1} + 2y_x = 0 \qquad (x = 0, 1, 2, \ldots)$$

was given by

$$y_x = A(-1)^x + B(-2)^x \qquad (x = 0, 1, 2, \ldots)$$

If $B \neq 0$, the dominant term is the second and y *oscillates infinitely* as $x \to \infty$. If $B = 0$, y *oscillates finitely* between A and $-A$ as $x \to \infty$.

Suppose that we wish to find the particular solution which satisfies the boundary conditions $y_0 = 2$ and $y_1 = -3$. On applying these two conditions to the general solution we obtain the simultaneous equations

$$A + B = 2 \quad \text{and} \quad -A - 2B = -3$$

which yield $A = B = 1$. Hence the solution satisfying these constraints is

$$y_x = (-1)^x + (-2)^x \quad (x = 0, 1, 2, \ldots)$$

14.9 Systems of linear equations♣

Consider the system of simultaneous differential equations

$$
\begin{aligned}
y_1' &= m_{11}y_1 + m_{12}y_2 + c_1 \\
y_2' &= m_{21}y_1 + m_{22}y_2 + c_2
\end{aligned}
\tag{1}
$$

Recall that $y' = \dfrac{dy}{dx}$.

in which y_1 and y_2 are unknown functions of x and the other quantities are constant. We may rewrite this system in the matrix form

$$\mathbf{y}' = M\mathbf{y} + \mathbf{c} \tag{2}$$

If the matrix M is invertible, the system has a unique constant vector solution $\mathbf{p}$ satisfying

$$\mathbf{p}' = M\mathbf{p} + \mathbf{c} = \mathbf{0}$$

called the **equilibrium solution**. So equation (2) may be written as

$$\mathbf{y}' = M(\mathbf{y} - \mathbf{p}) \tag{3}$$

447

In general, it is not enough to know that a system has an equilibrium solution. One also needs to know something about the *stability* of the system. Ideally one would like *all* nonconstant solutions $\mathbf{y}$ of the system to converge on the equilibrium solution – i.e. $\mathbf{y} \to \mathbf{p}$ as $\mathbf{x} \to \infty$. One can then be sure that, if the system is disturbed by some external shock, the system will act to restore the lost equilibrium. We now study the conditions under which stability is guaranteed.

We begin by converting the nonhomogeneous system to a homogeneous one with the translation

$$\mathbf{z} = \mathbf{y} - \mathbf{p}$$

Then equation (2) becomes

$$\mathbf{z}' = M\mathbf{z} \tag{4}$$

The system can be solved if each component equation involves a single variable. This can be achieved by diagonalising the matrix M with a change to a basis consisting of eigenvectors of M, say $\mathbf{b}$ and $\tilde{\mathbf{b}}$. If P is the matrix whose columns are the eigenvectors, as defined in §6.1, that is,

$$P = \begin{pmatrix} | & | \\ \mathbf{b} & \tilde{\mathbf{b}} \\ | & | \end{pmatrix}$$

write

$$\mathbf{z} = P\mathbf{w}$$

Then substituting in equation (4) and noting that P is a constant matrix

$$(P\mathbf{w})' = P(\mathbf{w}') = MP\mathbf{w}$$

Premultiplying the latter equality by P^{-1}

$$\mathbf{w}' = P^{-1}MP\mathbf{w}$$

where the matrix $P^{-1}MP$ is diagonal, with entries the *eigenvalues* λ and μ of M, which are the roots of $\det(M - \lambda I) = 0$. If these are distinct, we may conclude that

$$\begin{pmatrix} w'_1 \\ w'_2 \end{pmatrix} = \begin{pmatrix} \lambda & 0 \\ 0 & \mu \end{pmatrix} \begin{pmatrix} w_1 \\ w_2 \end{pmatrix} = \begin{pmatrix} \lambda w_1 \\ \mu w_2 \end{pmatrix}$$

[†] If $\lambda = \mu$, $e^{\mu x}$ must be replaced by $xe^{\lambda x}$.

Solving the component equations by separating the variables as in §12.4, we obtain[†]

$$\begin{pmatrix} w_1 \\ w_2 \end{pmatrix} = \begin{pmatrix} ce^{\lambda x} \\ de^{\mu x} \end{pmatrix}$$

Therefore

$$\begin{pmatrix} z_1 \\ z_2 \end{pmatrix} = P \begin{pmatrix} ce^{\lambda x} \\ de^{\mu x} \end{pmatrix} = \begin{pmatrix} | & | \\ \mathbf{b} & \tilde{\mathbf{b}} \\ | & | \end{pmatrix} \begin{pmatrix} ce^{\lambda x} \\ de^{\mu x} \end{pmatrix} = c\mathbf{b}e^{\lambda x} + d\tilde{\mathbf{b}}e^{\mu x}$$

Returning to our original variables, we are finally in a position to state that the general solution of the system is[‡]

[‡] An identical argument shows that the same results hold in n dimensions.

$$\mathbf{y} = \mathbf{p} + c\mathbf{b}e^{\lambda x} + d\tilde{\mathbf{b}}e^{\mu x}$$

It is now evident that the requisite condition for stability is

$$\left.\begin{array}{l} \lambda < 0 \quad \text{and} \quad \mu < 0 \quad \text{when } \lambda \text{ and } \mu \text{ are real,} \\ \text{Re } \lambda = \text{Re } \mu < 0 \quad \text{when } \lambda \text{ and } \mu \text{ are complex.} \end{array}\right\} \quad (5)$$

Under this condition it is guaranteed that $\mathbf{y} \to \mathbf{p}$ as $x \to \infty$.

If λ and μ are real and if $\lambda > 0$ and $\mu < 0$, then a solution with $c = 0$ will also converge to $\mathbf{p}$. In general, however, one must expect that a solution that does not satisfy the condition (5) will not converge to $\mathbf{p}$, and it is frequently the case that it behaves very wildly indeed.

Example 20

A physics application

Consider the electrical circuit in Example 14 with the variable electromotive force $E = 100 \cos 5t$ replaced by a constant battery B of voltage 360 volts. The equations can be written as a system of first order differential equations in the two variables I and Q:

$$I' = -10I - 125Q + 36$$
$$Q' = I$$

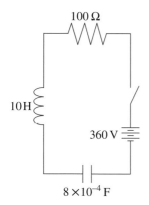

which we write in matrix form as

$$\begin{pmatrix} I' \\ Q' \end{pmatrix} = \begin{pmatrix} -10 & -125 \\ 1 & 0 \end{pmatrix} \begin{pmatrix} I \\ Q \end{pmatrix} + \begin{pmatrix} 36 \\ 0 \end{pmatrix}$$

The equilibrium solution $(I_p, Q_p)^{\mathrm{T}}$ is found by solving

$$\begin{pmatrix} I'_p \\ Q'_p \end{pmatrix} = \begin{pmatrix} -10 & -125 \\ 1 & 0 \end{pmatrix} \begin{pmatrix} I_p \\ Q_p \end{pmatrix} + \begin{pmatrix} 36 \\ 0 \end{pmatrix} = \begin{pmatrix} 0 \\ 0 \end{pmatrix}$$

Then

$$\begin{pmatrix} I_p \\ Q_p \end{pmatrix} = \begin{pmatrix} -10 & -125 \\ 1 & 0 \end{pmatrix}^{-1} \begin{pmatrix} -36 \\ 0 \end{pmatrix}$$

$$= \frac{1}{125} \begin{pmatrix} 0 & 125 \\ -1 & -10 \end{pmatrix} \begin{pmatrix} -36 \\ 0 \end{pmatrix} = \begin{pmatrix} 0 \\ \frac{36}{125} \end{pmatrix}$$

which gives $I_p = 0$ and $Q_p = \dfrac{36}{125}$. To discuss stability we require the eigenvalues of the matrix. These are found by solving

$$\begin{vmatrix} -10 - \lambda & -125 \\ 1 & -\lambda \end{vmatrix} = \lambda^2 + 10\lambda + 125 = 0$$

which gives $\lambda = -5 + 10\mathrm{i}$ and $\mu = -5 - 10\mathrm{i}$. Since Re $\lambda < 0$ and Re $\mu < 0$, the system is *stable*; that is, *any* solution $(I, Q)^{\mathrm{T}} \to (I_p, Q_p)^{\mathrm{T}}$ as $t \to \infty$.

We can obtain the general solution explicitly by computing eigenvectors $\mathbf{b}$ and $\tilde{\mathbf{b}}$ corresponding to λ and μ. To find $\mathbf{b}$, we consider

$$\begin{pmatrix} -10 & -125 \\ 1 & 0 \end{pmatrix} \begin{pmatrix} b_1 \\ b_2 \end{pmatrix} = (-5 + 10\mathrm{i}) \begin{pmatrix} b_1 \\ b_2 \end{pmatrix}$$

The use of complex numbers in matrices is shown here to indicate the flavour of more advanced applications.

449

i.e.

$$-10b_1 - 125b_2 = (-5 + 10i)b_1$$
$$b_1 = (-5 + 10i)b_2$$

It is easily checked that the first equation gives the same solution as the second and hence the eigenvectors are of the form

$$\begin{pmatrix} (-5 + 10i)b_2 \\ b_2 \end{pmatrix}$$

We choose the eigenvector $\mathbf{b}$ with $b_2 = 1$, namely $\begin{pmatrix} -5 + 10i \\ 1 \end{pmatrix}$

Similarly, the other eigenvector can be written as

$$\begin{pmatrix} -5 - 10i \\ 1 \end{pmatrix}$$

It follows that

$$\begin{pmatrix} I \\ Q \end{pmatrix} = \begin{pmatrix} 0 \\ \frac{36}{125} \end{pmatrix} + c \begin{pmatrix} -5 + 10i \\ 1 \end{pmatrix} e^{(-5+10i)t} + d \begin{pmatrix} -5 - 10i \\ 1 \end{pmatrix} e^{(-5-10i)t}$$

Hence

$$Q(t) = e^{-5t}(A \cos 10t + B \sin 10t) + \frac{36}{125}$$

where A and B are arbitrary constants. The presence of the negative exponential term ensures that $Q(t) \to \frac{36}{125}$ as $t \to \infty$. As observed earlier, the system is stable and converges to its equilibrium solution.

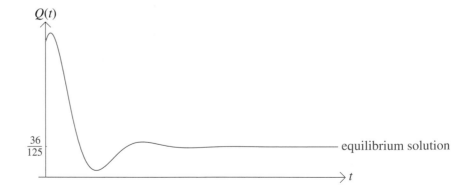

Systems of difference equations can be treated in precisely the same way. For stability in this case, however, we require that $|\lambda_1| < 1$, $|\lambda_2| < 1$, ..., $|\lambda_n| < 1$ instead of $\operatorname{Re} \lambda_1 < 0$, $\operatorname{Re} \lambda_2 < 0$, ..., $\operatorname{Re} \lambda_n < 0$.

Example 21

An economics application – Samuelson multiplier accelerator model

By eliminating Y_t from the equations of the Samuelson multiplier accelerator problem of Example 17 we obtain a system of first order difference equations in the variables C_t and I_t. This can be solved by the methods of this section. Although this involves a considerable amount of calculation compared with solving a second order difference equation as in Example 17, it is instructive and worth pursuing. To begin with we eliminate Y_t from the equations:

$$Y_t = C_t + I_t + G \tag{6}$$
$$C_t = 0.8Y_{t-1} \tag{7}$$
$$I_t = 0.375(C_t - C_{t-1}) \tag{8}$$

From equations (6) and (7)

$$C_{t+1} = 0.8(C_t + I_t + 900) = 0.8C_t + 0.8I_t + 720$$

From equation (8)

$$\begin{aligned} I_{t+1} &= 0.375(C_{t+1} - C_t) \\ &= 0.375(0.8(C_t + I_t + 900) - C_t) \\ &= -0.075C_t + 0.3I_t + 270 \end{aligned}$$

We can then write

$$\begin{pmatrix} C_{t+1} \\ I_{t+1} \end{pmatrix} = E \begin{pmatrix} C_t \\ I_t \end{pmatrix} = \begin{pmatrix} 0.8 & 0.8 \\ -0.075 & 0.3 \end{pmatrix} \begin{pmatrix} C_t \\ I_t \end{pmatrix} + \begin{pmatrix} 720 \\ 270 \end{pmatrix}$$

To discuss stability we require the eigenvalues of the matrix. These are found by solving

$$\begin{vmatrix} 0.8 - \lambda & 0.8 \\ -0.075 & 0.3 - \lambda \end{vmatrix} = \lambda^2 - 1.1\lambda + 0.3 = 0$$

which gives $\lambda = 0.6$ and 0.5. Since $|0.6| < 1$ and $|0.5| < 1$, the system is stable. The equilibrium solution $(C_p, I_p)^T$ is found by solving

$$E \begin{pmatrix} C_p \\ I_p \end{pmatrix} = \begin{pmatrix} C_p \\ I_p \end{pmatrix} = \begin{pmatrix} 0.8 & 0.8 \\ -0.075 & 0.3 \end{pmatrix} \begin{pmatrix} C_p \\ I_p \end{pmatrix} + \begin{pmatrix} 720 \\ 270 \end{pmatrix}$$

That is,

$$\begin{pmatrix} -0.2 & 0.8 \\ -0.075 & -0.7 \end{pmatrix} \begin{pmatrix} C_p \\ I_p \end{pmatrix} + \begin{pmatrix} 720 \\ 270 \end{pmatrix} = \begin{pmatrix} 0 \\ 0 \end{pmatrix}.$$

or

$$\begin{aligned} \begin{pmatrix} C_p \\ I_p \end{pmatrix} &= \begin{pmatrix} -0.2 & 0.8 \\ -0.075 & -0.7 \end{pmatrix}^{-1} \begin{pmatrix} -720 \\ -270 \end{pmatrix} \\ &= \frac{1}{0.2} \begin{pmatrix} -0.7 & -0.8 \\ 0.075 & -0.2 \end{pmatrix} \begin{pmatrix} -720 \\ -270 \end{pmatrix} = \begin{pmatrix} 3600 \\ 0 \end{pmatrix} \end{aligned}$$

Hence the consumption C_t stabilises to 3600 and the investment I_t converges to 0. From equation (7) we deduce that Y_t stabilises to $\frac{3600}{0.8} = 4500$.

If we want to obtain the general solution explicitly we must calculate eigenvectors corresponding to 0.6 and 0.5. We have

$$\begin{pmatrix} 0.8 & 0.8 \\ -0.075 & 0.3 \end{pmatrix} \begin{pmatrix} b_1 \\ b_2 \end{pmatrix} = 0.6 \begin{pmatrix} b_1 \\ b_2 \end{pmatrix}$$

and

$$\begin{pmatrix} 0.8 & 0.8 \\ -0.075 & 0.3 \end{pmatrix} \begin{pmatrix} \tilde{b}_1 \\ \tilde{b}_2 \end{pmatrix} = 0.5 \begin{pmatrix} \tilde{b}_1 \\ \tilde{b}_2 \end{pmatrix}$$

After the usual calculations, eigenvectors are found to be

$$\mathbf{b} = \begin{pmatrix} 4 \\ -1 \end{pmatrix} \qquad \tilde{\mathbf{b}} = \begin{pmatrix} 8 \\ -3 \end{pmatrix}$$

and the general solution is

$$\begin{pmatrix} C_t \\ I_t \end{pmatrix} = \begin{pmatrix} 3600 \\ 0 \end{pmatrix} + c \begin{pmatrix} 4 \\ -1 \end{pmatrix} (0.6)^t + d \begin{pmatrix} 8 \\ -3 \end{pmatrix} (0.5)^t$$

$$C_t = 3600 + 4c(0.6)^t + 8d(0.5)^t \tag{9}$$

$$I_t = 0 - c(0.6)^t - 3d(0.5)^t \tag{10}$$

where c and d are arbitrary constants. These results endorse the conclusions arrived at above. Under the initial conditions of Example 17, we obtain

$$C_1 = 0.8Y_0 = 0.8 \times 9500 = 7600$$

and

$$C_2 = 0.8Y_1 = 0.8 \times 9000 = 7200$$

Equation (9) then yields simultaneous equations with solution

$$c = \frac{20\,000}{3} \quad \text{and} \quad d = -3000$$

The particular solution in this case is then

$$C_t = 3600 + \frac{80\,000}{3}(0.6)^t - 24\,000(0.5)^t$$

$$I_t = 0 - \frac{20\,000}{3}(0.6)^t + 9000(0.5)^t$$

Using the formula for Y_t obtained in Example 17, the sequences Y_t, C_t and I_t are graphed below:

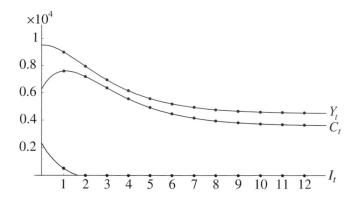

14.10 Change of variable♣

We conclude by noting that any differential or difference equation which can be converted by a change of variable to a linear form can be solved using the methods of this chapter. The following example illustrates this.

Example 22

The nonlinear difference equation

$$y_{x+1} = \frac{y_x - 1}{y_x + 3}$$

may be transformed into a linear equation by introducing the change of variable

$$y_x = \frac{1}{z_x} - 1$$

The equation reduces to

$$\frac{1 - z_{x+1}}{z_{x+1}} = \frac{1 - 2z_x}{1 + 2z_x}$$

$$1 + 2z_x - z_{x+1} - 2z_x z_{x+1} = z_{x+1} - 2z_x z_{x+1}$$

$$z_{x+1} - z_x = \frac{1}{2}$$

The complementary sequence is

$$c_x = a(1)^x = a$$

Considering the particular solution $p_x = bx$ and substituting in the difference equation, we get $b(x+1) - bx = \frac{1}{2}$ so $b = \frac{1}{2}$. The general solution of the transformed equation is

$$z_x = a + \frac{1}{2}x$$

The general solution of the original equation is therefore

$$\frac{1}{1+y_x} = a + \frac{1}{2}x$$

or

$$y_x = \frac{2}{2a+x} - 1$$

Finally, as shown in §14.9, when there is an equilibrium solution **p**, a nonhomogeneous equation or system of equations can always be made homogeneous by the translation

$$\mathbf{z} = \mathbf{y} - \mathbf{p}$$

This technique is used in the various versions of the cobweb model in the Applications. Observe for instance, that it is applicable to Example 17 with the equilibrium solution 4500, but not to Examples 14, 15 and 16, which do not have equilibrium solutions.

14.11 Exercises

1.   Find the general solutions of the following differential equations:

(i) $\dfrac{d^2 y}{dx^2} + 2\dfrac{dy}{dx} + 10y = 0$

(ii)* $3\dfrac{d^3 y}{dx^3} + 5\dfrac{d^2 y}{dx^2} - 2\dfrac{dy}{dx} = 0$

(iii)* $\dfrac{d^3 y}{dx^3} - 4\dfrac{d^2 y}{dx^2} + \dfrac{dy}{dx} + 6y = 0$

(iv)* $\dfrac{d^4 y}{dx^4} - 2\dfrac{d^3 y}{dx^3} - 3\dfrac{d^2 y}{dx^2} + 4\dfrac{dy}{dx} + 4y = 0$

(v)* $\dfrac{d^5 y}{dx^5} + 3\dfrac{d^4 y}{dx^4} + 7\dfrac{d^3 y}{dx^3} + 13\dfrac{d^2 y}{dx^2} + 12\dfrac{dy}{dx} + 4y = 0$

2.* Find and sketch the particular solution to equation (i) of Exercise 1 which satisfies the boundary conditions $y(\frac{\pi}{3}) = 0$ and $y'(\frac{\pi}{3}) = -3e^{\frac{\pi}{3}}$.

3.   Find the general solutions of the following differential equations:

(i) $\dfrac{d^2 y}{dx^2} - 2\dfrac{dy}{dx} + 10y = 0$

(ii) $\dfrac{d^2 y}{dx^2} + 2\dfrac{dy}{dx} - 3y = 0$

(iii) $\dfrac{d^4 y}{dx^4} + 2\dfrac{d^3 y}{dx^2} + \dfrac{d^2 y}{dx^2} = 0$

(iv) $\dfrac{d^2 y}{dx^2} + \dfrac{dy}{dx} + y = 0$

(v) $\dfrac{d^3y}{dx^3} - 3\dfrac{d^2y}{dx^2} + 9\dfrac{dy}{dx} + 13y = 0$

Discuss the behaviour of the solutions as $x \to \infty$.

4.   Find and sketch the solution of equation (i) of Exercise 3 which satisfies the boundary conditions $y(0) = 1$ and $y(\frac{\pi}{6}) = 0$.

5.   Solve Exercise 3 and without further calculation write down the general solutions of the following difference equations:

(i) $y_{x+2} - 2y_{x+1} + 10y_x = 0$

(ii)* $y_{x+2} + 2y_{x+1} - 3y_x = 0$

(iii)* $y_{x+4} + 2y_{x+3} + y_{x+2} = 0$

(iv)* $y_{x+2} + y_{x+1} + y_x = 0$

(v)* $y_{x+3} - 3y_{x+2} + 9y_{x+1} + 13y_x = 0$

Discuss the behaviour of the solutions as $x \to \infty$.
Illustrate with a sketch the first four terms of the solution to equation (ii), which satisfies the boundary conditions $y_0 = 1$, $y_1 = -1$.

6.   Find the general solutions of the following difference equations and deduce the general solutions of the differential equations in Exercise 1:

(i)* $y_{x+2} + 2y_{x+1} + 10y_x = 0$

(ii) $3y_{x+3} + 5y_{x+2} - 2y_{x+1} = 0$

(iii) $y_{x+3} - 4y_{x+2} + y_{x+1} + 6y_x = 0$

(iv) $y_{x+4} - 2y_{x+3} - 3y_{x+2} + 4y_{x+1} + 4y_x = 0$

(v) $y_{x+5} + 3y_{x+4} + 7y_{x+3} + 13y_{x+2} + 12y_{x+1} + 4y_x = 0$

7.* Find the general solution of the differential equation

$$3\dfrac{d^3y}{dx^3} + 5\dfrac{d^2y}{dx^2} - 2\dfrac{dy}{dx} = q(x)$$

when

(i) $q(x) = e^x$

(ii) $q(x) = \cos x$

(iii) $q(x) = x$

(iv) $q(x) = e^{-2x}$

(v) $q(x) = \cos x + e^{-2x}$

8.   Find the general solution of the differential equation

$$3\dfrac{d^3y}{dx^3} + 5\dfrac{d^2y}{dx^2} - 2\dfrac{dy}{dx} = q(x)$$

when

(i) $q(x) = e^{-x}$

(ii) $q(x) = \sin x$

(iii) $q(x) = 4e^{-x} + 2$

(iv) $q(x) = e^{-x/3}$

(v) $q(x) = \sin x + 2$

9.* Find the general solution of the difference equation

$$y_{x+3} - 5y_{x+2} + 8y_{x+1} - 4y_x = 2^x + x \qquad (x = 0, 1, 2, \ldots)$$

8. Find the general solution of the difference equation

$$y_{x+3} - 3y_{x+2} + 9y_{x+1} + 13y_x = q_x \qquad (x = 0, 1, 2, \ldots)$$

when

(i) $q_x = 1$

(ii) $q_x = x$

(iii) $q_x = (-1)^x$

(iv) $q_x = 1 + x + (-1)^x$

11.*♣ Find a solution of the differential equation of Exercise 1 (ii) which satisfies the boundary conditions

(i) $y(0) = 0$

(ii) $y(x) \to 1 \quad$ as $\quad x \to \infty$

12.♣ If y_x is a solution of

$$y_{x+2} = \frac{1}{2}(y_{x+1} + y_x) \quad (x = 0, 1, 2, \ldots)$$

which satisfies the boundary conditions

(i) $y_0 = a$

(ii) $y_1 = b$

prove that $y_x \to \frac{1}{3}(a + 2b)$ as $x \to \infty$.

13.* Write down a linear differential equation $P(D)y = 0$ whose solutions include the function e^{3x} and another such linear differential equation whose solutions include the function $x^2 e^{-x}$.

Hence write down a homogeneous linear differential equation whose solutions include both e^{3x} and $x^2 e^{-x}$. Write down the *general* solution of this equation.

14. Write down a homogeneous linear differential equation whose solutions include the function $xe^{-x}\cos(2x)$. Write down its general solution.

15. A student takes out a \$2000 loan to purchase a second hand car. He is being charged a monthly interest rate of 1% on the loan, which is to be discounted monthly as it is repaid in equal monthly installments of $b. Let y_x be the loan balance outstanding in the xth month.

   (i) Show that y_x satisfies the difference equation

$$y_{x+1} = 1.01 y_x - b \qquad (x \in \mathbb{N})$$

   (ii) Solve the equation in terms of b.

   (iii) Find b if the loan must be repaid in three years.

16.*♣ Show that the system of differential equations

$$y_1' = y_1 - y_2 - 1$$
$$y_2' = y_1 + y_2 - 1$$

is not stable. Obtain formulas for the solutions which satisfy the boundary conditions

   (i) $y_1(0) = 1$

   (ii) $y_2(0) = 1$

17.♣ Show that the system of differential equations

$$y_1' = -y_1 - y_2 + 1$$
$$y_2' = -y_1 + y_2 + 1$$

is not stable. Obtain formulas for the solutions which satisfy the boundary conditions

   (i) $y_1(0) = 1$

   (ii) $y_2(0) = 1$

18.♣ Show that the system of difference equations

$$\left. \begin{array}{l} y_{x+1} = -y_x - 3z_x + 1 \\ z_{x+1} = -\tfrac{1}{4} y_x - z_x + 1 \end{array} \right\} \qquad (x = 0, 1, 2, \ldots)$$

is stable. Obtain formulas for the solutions which satisfy the boundary conditions

   (i) $y_0 = 4$

   (ii) $z_0 = 1$

In addition, all the exercises in Exercises 14.13 to follow, from Exercise 3 onwards, can also be attempted by the methods studied so far.

14.12 The difference operator (optional)♣

In this section, we again stress the similarities between differential and difference equations, with the introduction of the **difference operator** Δ (*more properly called the forward difference operator*). This is defined by

$$\Delta y_x = y_{x+1} - y_x$$

We also define $\Delta^2 y_x = \Delta(\Delta y_x)$, $\Delta^3 y_x = \Delta(\Delta^2 y_x)$ and so on. Observe that

$$Ey_x = y_{x+1} = y_{x+1} - y_x + y_x = \Delta y_x + y_x = (\Delta + 1)y_x$$

We express this result by simply writing $E = \Delta + 1$. More useful is the generalization

$$E^n = (\Delta + 1)^n$$

In the case $n = 2$, for example, we have

$$E^2 y_x = (\Delta + 1)^2 y_x = (\Delta^2 + 2\Delta + 1)y_x = \Delta^2 y_x + 2\Delta y_x + y_x$$

These results allow us to express any difference equation in terms of the difference operator Δ. In the case of the equation $y_{x+2} + 2y_{x+1} + y_x = 0$, for example, we begin by introducing the shift operator and obtain

$$(E^2 + 2E + 1)y_x = 0$$

Using the fact that $E = (\Delta + 1)$ we then obtain the equation in the form

$$((\Delta + 1)^2 + 2(\Delta + 1) + 1)y_x = 0$$
$$(\Delta^2 + 4\Delta + 4)y_x = 0$$

The difference operator Δ has strong similarities to the differential operator D. In particular, a version of the fundamental theorem of calculus holds with Δ replacing D and Σ replacing $\int$. We have

$$\sum_{x=0}^{X-1} \Delta y_x = (\cancel{y_1} - y_0) + (\cancel{y_2} - \cancel{y_1}) + (\cancel{y_3} - \cancel{y_2})$$
$$+ \cdots + (y_X - \cancel{y_{X-1}}) = y_X - y_0$$

Thus, if F_x is a 'primitive' for f_x – i.e. $\Delta F_x = f_x$ – the difference equation

$$\Delta y_x = f_x$$

has general solution

$$y_x = F_x + c$$

where c is an arbitrary constant. To see this, it is only necessary to observe that

$$0 = \sum_{x=0}^{X-1}(\Delta y_x - f_x) = \sum_{x=0}^{X-1} \Delta(y_x - F_x)$$
$$= (y_X - F_X) - (y_0 - F_0)$$
$$= y_X - F_X - c$$

The following table lists some simple sequences f_x with their primitives F_x (i.e. $\Delta F_x = f_x$).

$f_x = 0$	$F_x = c$	
$f_x = b$	$F_x = bx + c$	
$f_x = ax$	$F_x = \frac{1}{2}ax(x-1) + c$	
$f_x = r^x$	$F_x = (r-1)^{-1}r^x + c$	$(r \neq 1)$

Example 23

The difference equation

$$y_{x+1} = y_x + 3x + 1$$

may be written in the form

$$\Delta y_x = 3x + 1$$

Taking $a = 3$ and $b = 1$ in the above table, we obtain the general solution

$$y_x = \frac{3}{2}x(x-1) + x + c$$

$$= \frac{3}{2}x^2 - \frac{1}{2}x + c$$

Example 24

The difference equation

$$y_{x+2} - 2y_{x+1} + y_x = 0$$

may be written in the form

$$(E^2 - 2E + 1)y_x = 0$$
$$(E - 1)^2 y_x = 0$$
$$\Delta^2 y_x = 0$$

From this we deduce that

$$\Delta y_x = c$$
$$y_x = cx + d$$

where c and d are arbitrary constants.

Example 25

A linear difference equation of order one may be expressed in the form

$$\Delta y_x + P_x y_x = Q_x$$

We shall consider only the special case in which the difference equation takes the form

$$y_{x+1} - ry_x = Q_x$$

Multiplying through by the 'summation factor' $\mu_x = r^{-x-1}$, we obtain

$$r^{-x-1}y_{x+1} - r^{-x}y_x = r^{-x-1}Q_x$$

i.e.

$$\Delta(r^{-x}y_x) = r^{-x-1}Q_x$$

If a 'primitive' for the right hand side is known, then the difference equation can be solved.

A significant special case arises in the case of capital accumulation under compound interest. Suppose that $1000 is invested at 10% interest per year. How much will have accumulated after 40 years? The appropriate difference equation is

$$y_{x+1} = y_x + \frac{10}{100}y_x \quad (x = 0, 1, 2, \ldots)$$

which we rewrite as

$$y_{x+1} - ry_x = 0$$

where $r = 1.1$. From the analysis given above, we know that this difference equation may be expressed in the form

$$\Delta(r^{-x}y_x) = 0$$

which has general solution $r^{-x}y_x = c$, i.e.

$$y_x = cr^x$$

where c is an arbitrary constant. To decide on an appropriate value of c for the problem in hand, we appeal to the *boundary condition* $y_0 = 1000$. This gives $y_0 = 1000 = cr^0$ and so

$$y_x = 1000(1.1)^x$$

In particular $y_{40} = \$45\,259.26$.

Suppose that a further $1000 is invested each year. What then will be the capital accumulation after 40 years? In this case the appropriate difference equation is

$$y_{x+1} = y_x + \frac{10}{100}y_x + 1000$$

which we rewrite as

$$y_{x+1} - ry_x = Q \tag{1}$$

where $r = 1.1$ and $Q = 1000$. From the analysis given above, this difference equation may be expressed in the form

$$\Delta(r^{-x}y_x) = r^{-x-1}Q.$$

A 'primitive' for r^{-x} is $(r^{-1} - 1)^{-1}r^{-x}$ and so the general solution is

$$r^{-x}y_x = r^{-1}Q(r^{-1} - 1)^{-1}r^{-x} + c$$

$$y_x = \frac{Q}{1 - r} + cr^x$$

To decide on the proper value of c, we use the boundary condition $y_0 = 1000 = Q$. Then

$$Q = y_0 = \frac{Q}{1 - r} + cr^0$$

$$c = Q\left(1 - \frac{1}{1 - r}\right) = \frac{rQ}{r - 1} = 11\,000$$

Thus

$$y_x = -10\,000 + 11\,000(1.1)^x$$

In particular, $y_{40} = \$487\,851.82$.

14.13 Exercises

1.* Determine the order of each of the following difference equations:

    (i) $y_{x+1} + y_x = 0$ (ii) $y_{x+1} - y_x = x$

    (iii) $y_{x+1}y_x = 1$ (iv) $y_{x+2}^2 y_x^3 = 1$

    (v) $y_{x+2} + y_{x+1} + y_x = 0$ (vi) $x^2 y_{x+3} + 2^x y_x^4 = x^3$

2. Prove the following results:

    (i) $\Delta(\frac{1}{2}ax(x - 1) + bx + c) = ax + b$

    (ii) $\Delta((r - 1)^{-1}r^x + c) = r^x$ $(r \neq 1)$

    (iii) $\Delta((r - 1)^{-1}xr^x - (r - 1)^{-2}r^{x+1}) = xr^x$ $(r \neq 1)$

3.* Find general solutions for the equations

    (i) $y_{x+1} - y_x = 2x$ (ii) $y_{x+1} - y_x = 2^x$

    (iii) $y_{x+1} + 2y_x = x$ (iv) $y_{x+1} - 2y_x = 3^x$

4. Find general solutions for the equations

    (i) $y_{x+1} - y_x = 1 + x$ (ii) $y_{x+1} - y_x = x2^x$

    (iii) $3y_{x+1} - y_x = 1$ (iv) $2y_{x+1} + 3y_x = x$

    In the case of (ii) and (iv) use the result of Exercise 2(iii).

5.* Make the change of variable $y_x = 2^{z_x}$ and hence find the general solution of the difference equation

$$y_{x+1}y_x = 2$$

6. Make the change of variable $z_x = y_x/x$ and hence find the general solution of the difference equation

$$x y_{x+1} + (x+1) y_x = 0$$

7.* A sum of money y_0 is invested at an interest rate of 10% per year. Translate the following questions into problems concerning difference equations with suitable boundary conditions and hence solve them.

(i) How much was originally invested if the amount which has accumulated after 100 years is $1 000 000?

(ii) Suppose that $y_0 = \$10$ and the average amount held over the first X years is $15.93. How much is held at the beginning of the $(X+1)$th year?

8. An investor has $1 000 000 invested at an interest rate of 10% per year. He transfers $1000 from this account to a second account each year for ten consecutive years. In the second account, the interest rate is 20%. What will his capital be at the beginning of the eighth year?

14.14 Applications (optional)

14.14.1 Cobweb models

Cobweb model I

We shall be concerned with the price P_x, the demand D_x and the supply S_x of a commodity at successive time periods $x = 0, 1, 2, \ldots$. These quantities are assumed to be related by the equations

$$
\left.
\begin{aligned}
D_x &= a - b P_x \\
S_x &= D_x \\
S_{x+1} &= c + d P_x
\end{aligned}
\right\} \quad (x = 0, 1, 2, \ldots)
\qquad
\begin{aligned}
&(1) \\
&(2) \\
&(3)
\end{aligned}
$$

Since $b > 0$, demand falls as price increases.

where a, b, c and d are constants with $b > 0$ and $d > 0$. The story that accompanies these equations is the following. Having produced S_x of the commodity, the supplier decides on a price P_x. This determines the demand D_x as described in equation (1). The supplier chooses the price to ensure that all of the amount S_x that he has produced is sold in period x. This explains equation (2). The supplier then decides how much to produce for the *next* time period. This will depend on his production costs and his predictions about what price he will be able to obtain for the amount he chooses to supply. It is assumed that the supplier does not understand the way that the market works but simply makes a prediction about the price P_{x+1} for the next period which is based *solely on the current price* P_x. The amount S_{x+1} supplied in the next period will then be a function of the single variable P_x. Equation (3) is the simplest form such a function can take.

Since $d > 0$, supply in the next period rises as current price increases.

The following diagram makes it clear why this model is called the **cobweb model**.

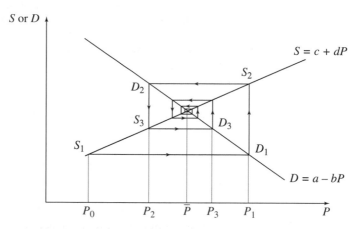

Suppose that P_0 is given. Then S_1 may be determined from the equation $S_1 = c + dP_0$. But then D_1 may be found because $D_1 = S_1$. Thus P_1 can be determined from the equation $D_1 = a - bP_1$. But then S_2 can be calculated from the equation $S_2 = c + dP_1$ and so on.

The quantity $\overline{P}$ in the diagram is the **equilibrium price**. It is chosen so that

$$a - b\overline{P} = c + d\overline{P}$$

i.e. $\overline{P} = (a - c)/(b + d)$. If $P_0 = \overline{P}$, it is clear that the price, demand and supply will all remain constant for all time periods with $P_x = \overline{P}$ and $S_x = D_x = a - b\overline{P} = c + d\overline{P}$.

The diagrammatic method is adequate for many purposes but often an analytic solution for this type of problem is necessary. From equations (1), (2) and (3), we obtain the *difference equation*

$$a - bP_{x+1} = c + dP_x$$

Now if $Q_x = P_x - \overline{P}$ the equation becomes

$$a - bQ_{x+1} - b\overline{P} = c + dQ_x + d\overline{P}$$

or

$$bQ_{x+1} + dQ_x + \overline{P}(d + b) + c - a = 0$$

that is,

$$Q_{x+1} + \frac{d}{b}Q_x = 0$$

with solution

$$Q_x = k\left(-\frac{d}{b}\right)^x$$

Hence

$$P_x = k\left(-\frac{d}{b}\right)^x + \frac{a - c}{b + d}$$

Observe that P_x is alternately larger and smaller than $\overline{P}$. Thus P_x 'oscillates' about $\overline{P}$ (and so D_x and S_x oscillate about $\overline{D}$ and $\overline{S}$). If $d < b$, the magnitude of these oscillations decreases as x grows larger and P_x approaches closer and closer to $\overline{P}$. We say that, in this case, the oscillations are **damped** and that P_x **converges** to $\overline{P}$. (*The diagram above illustrates this case.*)

If $d > b$, the oscillations become wilder and wilder and we say that P_x **oscillates infinitely**. If $d = b$, then P_x hops back and forward between $\overline{P} - k$ and $\overline{P} + k$. In this case, we say that P_x **oscillates finitely**. In neither of these cases does P_x converge. We say that P_x **diverges**.

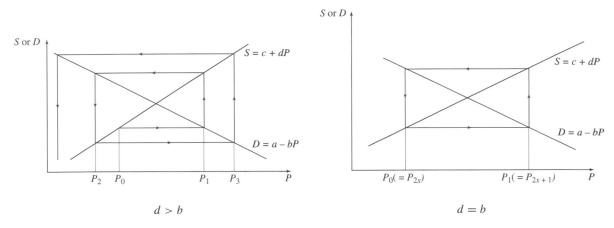

$$d > b \qquad\qquad\qquad\qquad d = b$$

The manner in which the suppliers predict future prices in the above model is not very enlightened. They never notice for example that the current price P_x is always on the wrong side of $\overline{P}$ relative to P_{x+1}. This does not matter too much in the case when $d < b$ since then we obtain convergence to the equilibrium price $\overline{P}$ in spite of the obtuseness of the suppliers. But when $d > b$, it is clear that the system will suffer a breakdown before very long.

Cobweb model II

In the first discrete version, stability occurs only for a restricted range of the parameters of the model. The equations for a more advanced discrete version are

$$
\begin{aligned}
D_x &= a - bP_x & &(4) \\
S_x &= D_x & (x = 0, 1, 2, \ldots) & \quad(5) \\
S_{x+2} &= c + d\left(\frac{P_x + P_{x+1}}{2}\right) & &(6)
\end{aligned}
$$

Here the current price is predicted as the average of the prices in the two previous periods.

These equations lead to the second order difference equation

$$2bQ_{x+2} + dQ_{x+1} + dQ_x = 0$$

in which $Q_x = P_x - \overline{P}$ where $\overline{P}$ is the equilibrium price (i.e. $a - b\overline{P} = c + d\overline{P}$). Writing the difference equation in the form

$$(2b\mathrm{E}^2 + d\mathrm{E} + d)Q = 0,$$

we see that we require the roots of the auxiliary equation $2bz^2 + dz + d = 0$. These are given by

$$z = \frac{-d \pm \sqrt{d^2 - 8bd}}{4b}$$

We leave the case $d^2 - 8bd \geq 0$ to the reader and study the case $d^2 - 8bd < 0$ (i.e. $d < 8b$) since this leads to complex roots of the auxiliary equation. If these complex roots are $\alpha = \lambda e^{i\mu}$ and $\beta = \lambda e^{-i\mu}$, then the general solution of the difference equation is

$$Q = A\lambda^x \cos \mu x + B\lambda^x \sin \mu x$$

For stability, we require that $\lambda < 1$. But

$$\lambda^2 = \frac{1}{16b^2}\{d^2 + (8bd - d^2)\} = \frac{d}{2b}$$

Stability is therefore obtained when $d < 2b$. This result should be compared with that obtained in the original cobweb model in which the condition for stability is $d < b$.

What sort of predictor for the current price will always lead to stability in the discrete case? We refine the previous model to one in which the current price is predicted as a weighted average of the prices over the previous n periods. The appropriate equations are

$$
\left.
\begin{aligned}
D_x &= a - bP_x \\
S_x &= D_x \\
S_{x+n} &= c + d\left(\frac{P_{x+n-1} + \lambda P_{x+n-2} + \cdots + \lambda^{n-1}P_x}{n}\right)
\end{aligned}
\right\}
\quad (x = 0, 1, 2, \ldots)
$$

$$(7)$$
$$(8)$$
$$(9)$$

Here the 'weights' are based on a discounting factor λ which we assume to satisfy $0 < \lambda < 1$. These equations lead to the difference equation

$$bQ_{x+n} + \frac{d}{n}(Q_{x+n-1} + \lambda Q_{x+n-2} + \cdots + \lambda^{n-1}Q_x) = 0$$

in which $Q_x = P_x - \bar{P}$ where $\bar{P}$ is the equilibrium price given by

$$\alpha - b\bar{P} = c + d\bar{P}\left(\frac{1 - \lambda^n}{1 - \lambda}\right)$$

We see that it is necessary to examine the roots of the auxiliary equation

$$bz^n + \frac{d}{n}(z^{n-1} + \lambda z^{n-2} + \cdots + \lambda^{n-1}) = 0$$

This proof that all the roots have modulus less than one is due to David Cartwright.

For stability, these roots should all have modulus less than one. Suppose that a root z of this equation lies on the circle $|z| = r$. It certainly has modulus less than one if $r \leq \lambda$ since $\lambda < 1$. So we need only consider the case when $r > \lambda$. We have

$$z^n = -\frac{d}{bn}(z^{n-1} + \lambda z^{n-2} + \cdots + \lambda^{n-1})$$

Taking moduli, this gives

$$r^n = \frac{d}{bn}|z^{n-1} + \lambda z^{n-2} + \cdots + \lambda^{n-1}|$$

$$\leq \frac{d}{bn}(r^{n-1} + \lambda r^{n-2} + \cdots + \lambda^{n-1}) = \frac{d}{bn}\frac{r^n - \lambda^n}{r - \lambda}$$

See Exercise 13.4.10.

$$< \frac{d}{bn}\frac{r^n}{(r - \lambda)}$$

It follows that $\dfrac{d}{bn(r-\lambda)} > 1$ or since $r - \lambda > 0$, $\dfrac{d}{bn} > r - \lambda$. That is

$$\frac{d}{bn} + \lambda > r$$

Since $0 < \lambda < 1$ and $d > 0$, $b > 0$ are given constants, we can choose $n > \dfrac{d}{b(1-\lambda)}$ to ensure that $\dfrac{d}{bn} + \lambda < 1$ and hence that $r < 1$. So all roots of the equation have modulus less than one.

We can conclude that so long as we consider a sufficient number of previous years' prices we can be sure that the model is stable.

Cobweb model III

Following the two discrete versions of the cobweb model, we now consider a sophisticated economy in which difference equations are replaced by differential equations. In this continuous version, stability is obtained for all values of the parameters.

The equations for this version of the cobweb model, for $x > 0$, are

$$D(x) = a - bP(x) \tag{10}$$

$$S(x) = D(x) \tag{11}$$

$$S(x) = c + d \int_{-\infty}^{x} e^{y-x} P(y)\,dy \tag{12}$$

Instead of being a discrete variable restricted to the values $0, 1, 2, \ldots x$ now represents a continuous variable which may take any positive real value. In the third equation

$$\int_{-\infty}^{x} e^{y-x} P(y)\,dy$$

may be regarded as a prediction of the price at time x calculated as a 'weighted average' over prices at all previous times. Note that the prices in the recent past are given a much larger weight than prices in remote times. The weight e^{y-x} is sometimes called a **discounting factor**.

The three equations together yield the *integral equation*

$$a - bP(x) = c + d \int_{-\infty}^{x} e^{y-x} P(y)\,dy$$

At equilibrium, the price $P(x)$ will be equal to a constant $\overline{P}$. This constant satisfies

$$a - b\overline{P} = c + d \int_{-\infty}^{x} e^{y-x} \overline{P}\,dy = c + d\overline{P}$$

and so $\overline{P} = (a - c)/(b + d)$ as before. The change of variable $Q(x) = P(x) - \overline{P}$ reduces the integral equation to the simpler form

$$-bQ(x) = d \int_{-\infty}^{x} e^{y-x} Q(y)\,dy$$

$$= de^{-x} \int_{-\infty}^{x} e^{y} Q(y)\,dy$$

† Equally, differential
equations can sometimes be
solved by converting them
into integral equations.

Integral equations can often be solved by converting them into differential equations.[†] Using the formula for differentiating a product in conjunction with the fundamental theorem of calculus (§10.3), we obtain

$$-b\frac{dQ}{dx} = -de^{-x}\int_{-\infty}^{x} e^{y}Q(y)\,dy + de^{-x}e^{x}Q(x)$$
$$= -(-bQ) + dQ$$
$$= (b+d)Q$$

Separating the variables, we obtain

$$\frac{dQ}{Q} = -\left(\frac{b+d}{b}\right)dx$$
$$\int \frac{dQ}{Q} = -\left(\frac{b+d}{b}\right)\int dx + k$$
$$\ln Q = -\left(\frac{b+d}{b}\right)x + k$$
$$P(x) - \overline{P} = Q(x) = Ke^{-x(b+d)/b}$$

This is a very much more satisfactory result than that obtained in the original version of the cobweb model. This is not too surprising since the prediction method is much less naive. As long as

$$\frac{b+d}{b} > 0$$

which will always be the case if $b > 0$ and $d > 0$, we have

$$e^{-x(b+d)/b} \to 0 \quad \text{as} \quad x \to \infty$$

and so $P(x)$ converges smoothly to $\overline{P}$.

It may be instructive to go through the above analysis *without* making the change of variable $Q(x) = P(x) - \overline{P}$. The differential equation then obtained is linear and may be solved by the method of §12.6.

Note, incidentally, that a more natural continuous cobweb model is given by

$$\left. \begin{aligned} D(x) &= a - bP(x) \\ S(x) &= D(x) \\ S(x) &= c + d\frac{1}{x}\int_{0}^{x} P(y)\,dy \end{aligned} \right\} \quad (x > 0) \qquad \begin{aligned} &(13)\\ &(14)\\ &(15) \end{aligned}$$

It may be of interest to solve this problem and to compare the result with that obtained above. Also of interest is the case in which (15) is replaced by

$$S(x) = c + d\int_{0}^{x} \frac{2y}{x^2} P(y)\,dy$$

14.14.2 Gambler's ruin

A gambler enters a casino with $\$r$ in his pocket. If this is lost, he is ruined. He decides to continue gambling until he has won $\$s$ or else is ruined. He plays the same game repeatedly. At each play he wins $\$1$ with probability p $(0 < p < 1)$ and loses $\$1$ with probability q $(p + q = 1)$. What is the probability that he will be ruined?

Let the probability that the gambler will eventually be ruined, given that his winnings so far amount to x, be y_x. We want to calculate y_0. The diagram on the left below illustrates the definition of y_x. The diagram on the right examines the situation in the case $-r < x < s$ more closely. Either the gambler will win the next play and his total winnings will become $(x + 1)$ or else he will lose and his total winnings will become $(x - 1)$. The probability of ruin via the first route is py_{x+1} and the probability of ruin via the second route is qy_{x-1}. The total probability of ruin is therefore

$$y_x = py_{x+1} + qy_{x-1}$$

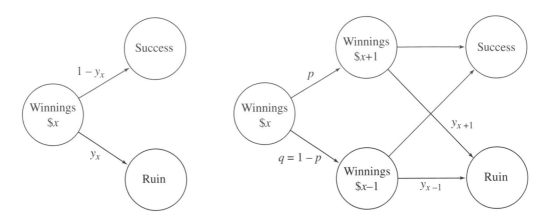

We rewrite this difference equation in the form

$$y_{x+1} = py_{x+2} + qy_x$$

i.e.

$$py_{x+2} - y_{x+1} + qy_x = 0$$
$$(pE^2 - E + q)y = 0$$

This has to be solved in accordance with the *boundary conditions*

$$y_{-r} = 1$$
$$y_s = 0$$

which are obtained from the observation that the gambler is certain of ruin when his winnings are $ - r$ and certain of success when his winnings have hit the prespecified target of s.

Since $p + q = 1$, the auxiliary equation is

$$pz^2 - z + 1 - p = 0$$
$$p(z^2 - 1) - (z - 1) = 0$$
$$p(z - 1)(z + 1) - (z - 1) = 0$$
$$(z - 1)(pz + p - 1) = 0$$
$$p(z - 1)(z - qp^{-1}) = 0$$

The roots are $\alpha = 1$ and $\beta = qp^{-1}$. Two cases need to be distinguished. The first is when the roots are unequal $(p \neq q)$ and the second is when the roots are equal $(p = q = \frac{1}{2})$.

Case (i) $p \neq q$

In this case the general solution is

$$y_x = A + B\beta^x$$

We use the boundary conditions to evaluate the constants A and B. We have that

$$1 = A + B\beta^{-r}$$
$$0 = A + B\beta^s$$

and so

$$A = -\beta^{r+s}(1 - \beta^{r+s})^{-1}, \quad B = \beta^r(1 - \beta^{r+s})^{-1}$$

It follows that

$$y_x = \frac{\beta^{r+x} - \beta^{r+s}}{1 - \beta^{r+s}}$$

and so the probability of ruin on entering the casino is

$$y_0 = \frac{\beta^r - \beta^{r+s}}{1 - \beta^{r+s}}$$

provided that $\beta \neq 1$ (i.e. $p \neq q$)

Suppose, for example, that the game chosen by the gambler is favourable to him in the sense that $p > q$ (i.e. $\beta < 1$). Then $1 - \beta^{r+s} > 1 - \beta$ and $1 - \beta^s < 1$. Thus

$$y_0 = \beta^r \frac{(1 - \beta^s)}{(1 - \beta^{r+s})} < \frac{\beta^r}{1 - \beta}$$

This means that, if r is reasonably large, the probability of ruin will be negligible no matter how large the target sum of \$$s$ is made. In particular, if $p = \frac{2}{3}, q = \frac{1}{3}$ and $r = 101$, we have that

$$y_0 < \left(\frac{1}{2}\right)^{100}$$

Case (ii) $p = q = \frac{1}{2}$

In this case the general solution is

$$y_x = A + Bx$$

Using the boundary conditions to evaluate A and B we obtain that

$$A = \frac{s}{r+s} \quad B = -\frac{1}{r+s}$$

and so

$$y_x = \frac{s - x}{s + r}$$

In particular

$$y_0 = \frac{s}{r+s}$$

This is perhaps the more interesting case since when $p = q = \frac{1}{2}$ the game is 'fair'. Suppose that the gambler risks \$9000 in such a game in an attempt to win \$1000. His probability of success is then

$$1 - \frac{s}{r+s} = \frac{r}{r+s} = \frac{9000}{9000 + 1000} = \frac{9}{10}$$

This large probability explains why the idle rich are able to boast of paying for their holidays in the casinos at Las Vegas or Monte Carlo. But they are not to be congratulated on their good business sense. All that they have done is to balance a high probability of a small win against a low probability of a large loss. Note that their expected gain is

$$1000 \times \frac{9}{10} + (-9000) \times \frac{1}{10} = 0$$

It is also instructive to observe that the probability of being successful on each occasion if they attempt the coup on ten successive occasions is only

$$\left(\frac{9}{10}\right)^{10} = \left(1 - \frac{1}{10}\right)^{10} \approx \frac{1}{e} \approx 0.37$$

One must also, of course, take into account that this is the conclusion based on chance alone, which disregards the fact that casinos incur expenses and must make a profit.

Returning to the general case, we observe that the duration of the game is also a variable of interest. If $\mathcal{E}_x$ denotes the expected duration of the play to come, given that $\$x$ has been won so far, then we obtain the difference equation

$$\mathcal{E}_x = p\mathcal{E}_{x+1} + q\mathcal{E}_{x-1} + 1$$

with boundary conditions $\mathcal{E}_s = 0$, $\mathcal{E}_{-r} = 0$. In the 'fair' case with $p = q = \frac{1}{2}$, this leads to

$$\mathcal{E}_0 = rs$$

Observe that a gambler starting with $1 hoping to win $1000 will last for 1000 plays on average!

Answers to starred exercises with some hints and solutions

Chapter 1. Vectors and matrices

Exercises 1.2

1. (i) $\quad 2D = 2\begin{pmatrix} 1 \\ 2 \\ 3 \end{pmatrix} = \begin{pmatrix} 2 \\ 4 \\ 6 \end{pmatrix}$

   (iv) $\quad C + D$ is nonsense because C is 3×3 and D is 3×1.

   (vii) $\quad B$ is not square, so $B + 3I$ is not defined.

   (x) $\quad AB = \begin{pmatrix} 6 & 1 & 0 \\ 1 & 0 & 3 \end{pmatrix}\begin{pmatrix} 3 & 1 \\ 1 & 0 \\ 0 & 2 \end{pmatrix} = \begin{pmatrix} 19 & 6 \\ 3 & 7 \end{pmatrix}$

   (xiii) $\quad CA$ is nonsense because C is 3×3 and A is 2×3.

   (xvi) $\quad CD = \begin{pmatrix} 4 & 1 & 1 \\ 0 & 1 & 0 \\ -2 & 0 & 1 \end{pmatrix}\begin{pmatrix} 1 \\ 2 \\ 3 \end{pmatrix} = \begin{pmatrix} 9 \\ 2 \\ 1 \end{pmatrix}$

   (xviii) $\quad \det(A)$ is nonsense because A is not square.

   (xix) $\quad \det(B)$ is nonsense because B is not square.

   (xxi) $\quad \det(AB) = \begin{vmatrix} 19 & 6 \\ 3 & 7 \end{vmatrix} = 19 \times 7 - 3 \times 6 = 115$

   (xxii) $\quad \det(CD)$ is nonsense because CD is not square.

   (xxiv) $\quad AA^{\mathrm{T}} = \begin{pmatrix} 37 & 6 \\ 6 & 10 \end{pmatrix}$

   (xxv) $\quad D^{\mathrm{T}}A$ is not defined since D^{T} is 1×3 and A is 2×3.

   (xxvii) $\quad B^{\mathrm{T}}C = \begin{pmatrix} 12 & 4 & 3 \\ 0 & 1 & 3 \end{pmatrix}$

   (xxviii) $\quad AC^{-1} = \begin{pmatrix} 1 & 0 & -1 \\ \frac{7}{6} & -\frac{7}{6} & \frac{11}{6} \end{pmatrix}$

   (xxx) $\quad \det(C - 3I) = 0$, so $C - 3I$ is not invertible.

3. (ii) -250 (iii) $28t - 40$ (vi) -90
   For (iii) to be zero, $t = \frac{10}{7}$.

4. $t = -2$ or 6.

8. $A^{\mathrm{T}}B = (1 \times n) \times (n \times 1) = 1 \times 1$. Similarly, $B^{\mathrm{T}}A$ is 1×1.
   $A^{\mathrm{T}}B = (B^{\mathrm{T}}A)^{\mathrm{T}}$.
   AB^{T} is $(n \times 1) \times (1 \times n) = n \times n$.
   Similarly, BA^{T} is $n \times n$. $AB^{\mathrm{T}} = (BA^{\mathrm{T}})^{\mathrm{T}}$.

9. $\det(2A) = 2^3 \det(A) = 56$
$\det(A^2) = 49$
$\det(3A^{-1}) = 27/7$
$\det((3A)^{-1}) = 1/189$

11. If $A = \begin{pmatrix} a & b \\ c & d \end{pmatrix}$, then $A^2 = 0$ implies $\begin{pmatrix} a^2 + bc & ab + bd \\ ac + cd & bc + d^2 \end{pmatrix} = 0$. Solving the four simultaneous equations, the possible forms are

$$\begin{pmatrix} a & b \\ -\dfrac{a^2}{b} & -a \end{pmatrix} \quad \text{or} \quad \begin{pmatrix} 0 & 0 \\ c & 0 \end{pmatrix}$$

14. $A(B - C) = 0$.
A invertible implies

$$B - C = A^{-1}A(B - C) = A^{-1}0 = 0$$

so $B = C$, a contradiction.
$B - C$ invertible similarly implies $A = 0$, a contradiction.

15. $\det(A) \neq 0$ implies A^{-1} exists, so

$$X = A^{-1}(AX) = A^{-1}(0) = 0$$

a contradiction.

16.

$$AX = \lambda I X$$

or

$$(A - \lambda I)X = 0 \quad (X \neq 0)$$

By Exercise 15, $\det(A - \lambda I) = 0$.

Exercises 1.4

1. (i) $(-3, 3)^T = \frac{3}{2}(-2, 2)^T$
(iii) $(-4, -6)^T$

2. (i) For $\mathbf{u}$ and $\mathbf{v}$, $\cos\theta = \frac{1}{\sqrt{5}}$, so $\theta \approx 1.107$; $\mathbf{v}$ and $\mathbf{w}$ are perpendicular.
(iii) $(168, -116)^T$, $(-19\frac{1}{2}, 19)^T$
(iv) $5\sqrt{5}$, $10 + \sqrt{5}$, $20 + 5\sqrt{5}$, $5\sqrt{13}$
(vi) 1, $(2, 1)^T = (0.8, -0.6)^T + (1.2, 1.6)^T$

3. Distance is
$$\sqrt{(v_1 - w_1)^2 + (v_2 - w_2)^2} \quad = \sqrt{\|\mathbf{v}\|^2 + \|\mathbf{w}\|^2 - 2\langle \mathbf{v}, \mathbf{w} \rangle}$$
$$= \sqrt{\|\mathbf{v}\|^2 + \|\mathbf{w}\|^2 - 2\|\mathbf{v}\|\,\|\mathbf{w}\|\cos\theta}.$$

For $0 \leq \theta \leq \pi$, $\cos\theta$ decreases as θ increases, so the distance increases.

4. (i) $\langle \mathbf{v}, \mathbf{v} \rangle = v_1^2 + v_2^2 = \|\mathbf{v}\|^2$
(ii) $\langle \mathbf{v}, \mathbf{w} \rangle = v_1 w_1 + v_2 w_2 = w_1 v_1 + w_2 v_2 = \langle \mathbf{w}, \mathbf{v} \rangle$

(iii) $\langle \alpha\mathbf{u} + \beta\mathbf{v}, \mathbf{w}\rangle = (\alpha u_1 + \beta v_1)w_1 + (\alpha u_2 + \beta v_2)w_2$
$$= \alpha(u_1 w_1 + u_2 w_2) + \beta(v_1 w_1 + v_2 w_2)$$
$$= \alpha\langle\mathbf{u}, \mathbf{w}\rangle + \beta\langle\mathbf{v}, \mathbf{w}\rangle$$

5. Use $\left\langle \begin{pmatrix} 3 \\ 5 \end{pmatrix}, \begin{pmatrix} a \\ b \end{pmatrix} \right\rangle = 0 = \left\langle \begin{pmatrix} 3 \\ b \end{pmatrix}, \begin{pmatrix} a \\ 5 \end{pmatrix} \right\rangle$ or any other rearrangement.

6. (i)

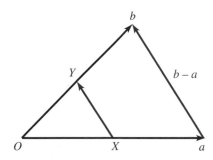

$$\overrightarrow{XY} = \overrightarrow{XO} + \overrightarrow{OY} = -\frac{\mathbf{a}}{2} + \frac{\mathbf{b}}{2} = \frac{1}{2}(\mathbf{b} - \mathbf{a})$$

7. (ii) $\dfrac{1}{m+n}(m\mathbf{b} + n\mathbf{a})$

Exercises 1.8

1. Use any point on the line and any multiple of $(2, 1)^T$.

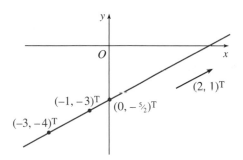

E.g if $t = 1$, then $(-1, -3)^T$ is seen to be on the line, so $\mathbf{x} = (-1, -3)^T + t(4, 2)^T$ is an equation. Or, $\mathbf{x} = (0, -\frac{5}{2})^T + t(-2, -1)^T$.

3. $\mathbf{u} = (-\frac{3}{5}, \frac{4}{5})^T$.

$(4, -2)^T = (1, 2)^T + t \cdot \mathbf{u}$

Now $t = -5$ is the consistent solution of the component equations, so the point lies on the line.

Distance between $(1, 2)^T$ and $(4, -2)^T$ is 5 which is the modulus of -5.

Distance between $(1, 2)^T$ and $(-\frac{1}{5}, \frac{18}{5})^T$ is 2.

4. We first need a vector $\mathbf{v}$ orthogonal to $(1, 2)^T$ – i.e. $(v_1, v_2)^T$ such that $1v_1 + 2v_2 = 0$. The choice $v_1 = -2$, $v_2 = 1$ suffices, though any nonzero multiple of

473

it would do. The equation of a line orthogonal to $(1, 2)^T$ is then

$$\frac{x - \xi_1}{-2} = \frac{y - \xi_2}{1}$$

Lines with $\xi = a, b$ or c would do.

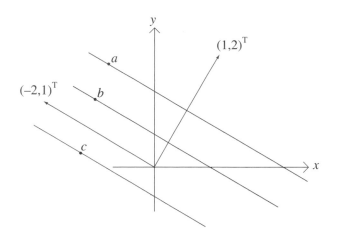

5. $(-7, 2)^T$ Parallel unit vectors are $\pm\frac{1}{\sqrt{53}}(-7, 2)^T$. If $(a, b)^T$ is orthogonal to the line, $-7a + 2b = 0$. So $\frac{a}{b} = \frac{2}{7}$. Perpendicular unit vectors are $\pm\frac{1}{\sqrt{53}}(2, 7)^T$.

8. $u^T v = (1, 0, -1) \begin{pmatrix} 2 \\ -1 \\ 3 \end{pmatrix}$ of order 1×1, a scalar. So $u^T v = (u^T v)^T = v^T u$.

$$uv^T = \begin{pmatrix} 1 \\ 0 \\ -1 \end{pmatrix} (2 \quad -1 \quad 3) = \begin{pmatrix} 2 & -1 & 3 \\ 0 & 0 & 0 \\ -2 & 1 & -3 \end{pmatrix}$$

10. $u = (1, 0, 0)^T$, $v = (2, 1, 1)^T$, $w = (1, 1, 1)^T$.

13. (i) Collinear, since $(2, 1, 4)^T - (4, 4, -1)^T = (-2, -3, 5)^T$ has the same direction as $(4, 4, -1)^T - (6, 7, -6)^T = (-2, -3, 5)^T$.

14. (i) If $(1, -1, 2)^T + t(1, 2, 3)^T = (1, 17, 6)^T + s(7, -4, -3)^T$, the component scalar equations are inconsistent.

(ii) Lines l_1 and l_2 intersect at $(2, 1, 5)^T$ at an angle ≈ 1.078.

(iii) Lines l_2 and l_3 intersect orthogonally at $(8, 13, 3)^T$.

17. The second plane is the further, at a distance of $\frac{6}{\sqrt{26}}$.

19. (i) Parallel

(ii) Neither – angle is $\pi/3$

(iii) Perpendicular

22. The required plane contains the directions of both the vectors $(1, 0, 6)^T - (-2, 3, -1)^T = (3, -3, 7)^T$ and $(5, -1, 1)^T$. Hence their vector product $4(1, 8, 3)^T$ is normal to the required plane, which has vector equation $\langle (1, 8, 3)^T, \mathbf{x} - (1, 0, 6)^T \rangle = 0$. The cartesian equation is $x + 8y + 3z = 19$.

24. $(x, y)^T$ plane, $z = 0$. $(y, z)^T$ plane, $x = 0$. $(x, z)^T$ plane, $y = 0$.

   (i) A horizontal vector: $(x, y, 0)^T$

   (ii) A vertical vector: $(0, 0, z)^T$

   (iii) A horizontal plane: $z = Z$

   (iv) A vertical plane: $ax + by = c$

28. (i)
$$\mathbf{v} \times \mathbf{w} = \left(\begin{vmatrix} v_2 & v_3 \\ w_2 & w_3 \end{vmatrix}, -\begin{vmatrix} v_1 & v_3 \\ w_1 & w_3 \end{vmatrix}, \begin{vmatrix} v_1 & v_2 \\ w_1 & w_2 \end{vmatrix} \right)^T$$
If $\mathbf{u} = (u_1, u_2, u_3)^T$, then
$$\langle \mathbf{u}, \mathbf{v} \times \mathbf{w} \rangle = u_1 \begin{vmatrix} v_2 & v_3 \\ w_2 & w_3 \end{vmatrix} - u_2 \begin{vmatrix} v_1 & v_3 \\ w_1 & w_3 \end{vmatrix} + u_3 \begin{vmatrix} v_1 & v_2 \\ w_1 & w_2 \end{vmatrix}$$
$$= \begin{vmatrix} u_1 & u_2 & u_3 \\ v_1 & v_2 & v_3 \\ w_1 & w_2 & w_3 \end{vmatrix}$$

   Now $\mathbf{u}, \mathbf{v}, \mathbf{w}$ are coplanar if and only if the normal to the plane, $\mathbf{v} \times \mathbf{w}$, is orthogonal to $\mathbf{u}$. This is true if and only if $\langle \mathbf{u}, \mathbf{v} \times \mathbf{w} \rangle = 0$. (ii) $t = -4$.

29. (i) A normal to the plane is
$$(5, -5, 4)^T \times (3, 1, 2)^T = (-14, 2, 20)^T = 2(-7, 1, 10)^T$$
   Equation of plane is $\langle (-7, 1, 10)^T, \mathbf{x} - (0, 4, -1)^T \rangle = 0$, which gives $-7x + y + 10z = -6$.

   (ii)
$$ax + by + cz + d = 0$$
$$a \cdot 0 + 4b - c + d = 0$$
$$5a - b + 3c + d = 0$$
$$2a - 2b + c + d = 0$$
$$\begin{pmatrix} x & y & z & 1 \\ 0 & 4 & -1 & 1 \\ 5 & -1 & 3 & 1 \\ 2 & -2 & 1 & 1 \end{pmatrix} \begin{pmatrix} a \\ b \\ c \\ d \end{pmatrix} = 0$$
   Since $(a, b, c, d)^T \neq 0$, by Exercise 1.2.15 we have
$$\begin{vmatrix} x & y & z & 1 \\ 0 & 4 & -1 & 1 \\ 5 & -1 & 3 & 1 \\ 2 & -2 & 1 & 1 \end{vmatrix} = 0$$

Exercises 1.11

3. $l_1:$ $\dfrac{x_1}{-2} = \dfrac{x_2 - 1}{1} = \dfrac{x_3 - 1}{2} = \dfrac{x_4 - 1}{1}$

$l_2:$ $\dfrac{x_1 - 3}{4} = \dfrac{x_2 - 1}{1} = \dfrac{x_4 - 1}{1}, x_3 = 0$

$l_3:$ $\quad x_2 = -x_4, x_1 = x_3 = 0$

A point on l_1: $(-2t, t + 1, 2t + 1, t + 1)^{\mathrm{T}}$
A point on l_2: $(4s + 3, s + 1, 0, s + 1)^{\mathrm{T}}$

We have that l_1 and l_2 intersect in the point $(1, \frac{1}{2}, 0, \frac{1}{2})^{\mathrm{T}}$. The other two pairs of lines do not intersect. None of the lines are parallel; l_1 and l_3 have orthogonal directions, as do l_2 and l_3.

10. (i) The scalar equations reduce to the equation of the plane $x + y + z = 2$.
    (ii) The scalar equations are inconsistent, so represent the empty set.

12. (i) $\left(\dfrac{5}{8}, \dfrac{3}{4}, -\dfrac{43}{8}\right)^{\mathrm{T}}$

    (v) The empty set

    (vi) The line $\mathbf{x} = t \begin{pmatrix} 1 \\ -2 \\ 9 \end{pmatrix}$

13. (i) $(1, 0, 3, -1)^{\mathrm{T}}$
    (iii) $(0, 1, 3, 1, 6)^{\mathrm{T}}$

Chapter 2. Differentiation

Exercise 2.11

2. (iii) Put $z = \sqrt{y}$ and $y = x^2 + 1$. Then

$$\frac{d}{dx}\{\sqrt{x^2 + 1}\} = \frac{dz}{dy}\frac{dy}{dx} = \frac{1}{2\sqrt{y}}2x = \frac{x}{\sqrt{x^2 + 1}}$$

(iv) Put $z = y^{3/2}$ and $y = x + 1$. Then

$$\frac{d}{dx}\{(x + 1)^{3/2}\} = \frac{dz}{dy}\frac{dy}{dx} = \frac{3}{2}y^{1/2} = \frac{3}{2}\sqrt{x + 1}$$

(vi) Put $y = \tan x$ and $z = x$. Then

$$\frac{d}{dx}\left\{\frac{\tan x}{x}\right\} = \frac{1}{z^2}\left\{z\frac{dy}{dx} - y\frac{dz}{dx}\right\} = \frac{1}{x^2}\{x \sec^2 x - \tan x\}$$

3. (i) (a) $\dfrac{1}{\sqrt{1 + x^2}}$ (c) $\dfrac{x^2 + 1}{(x^2 + x - 1)^2}$

   Tangent line: $y = 1/\sqrt{2}(x - 1)$

4. Each of these results is obtained by using the formula

$$\frac{dz}{dx} = \frac{dz}{dy}\frac{dy}{dx}$$

with $y = ax + b$. (In case (ii), $a = -1$ and $b = 0$.)

5. The required price is $f'(2) = \frac{1}{2}\{(2-1)^2 + 1\}^{-1/2}2(2-1) = 1/\sqrt{2}$.

6. (i) $\dfrac{1}{x \ln x}$ $(x > 1)$

7. (i) $-\cot x \operatorname{cosec} x$ $(x \neq n\pi)$

    (iii) $\cot x$ $(0 < x < \pi)$

8. (i) $\dfrac{x^{2/3} \sec^2((x+1)^{1/3})}{3(1+x)^{2/3}} + \dfrac{2\tan((1+x)^{1/3})}{3x^{1/3}}$

    (ii) $\exp(x\sqrt{4-x^2})\left(\sqrt{4-x^2} - \dfrac{x^2}{\sqrt{4-x^2}}\right)$

    (iii) $\dfrac{11}{2\sqrt{x-2}(5x+1)^{3/2}}$

10. The result follows since the slope $\dfrac{1}{\sqrt{1+x^2}}$ satisfies $0 < \dfrac{1}{\sqrt{1+x^2}} \leq 1$. The points with slope t, for each $t \in (0, 1]$ are given by $x = \pm\dfrac{\sqrt{1+x^2}}{t}\sqrt{1-t^2}$.

    Points with slope $2/3$ are $\left(-\dfrac{\sqrt{5}}{2}, \ln\left(\dfrac{3-\sqrt{5}}{2+2\sqrt{2}}\right)\right)$ and $\left(\dfrac{\sqrt{5}}{2}, \ln\left(\dfrac{3+\sqrt{5}}{2+2\sqrt{2}}\right)\right)$.

11. Equation of tangent when $x = x_1$ is $y = (3\cos x_1 - 1)x + 3(\sin x_1 - x_1 \cos x_1)$. Using the periodicity of $\sin x$ and $\cos x$, it can be observed that at each $x = 2n\pi \pm \frac{\pi}{2}$ where $n \in \mathbb{Z}$, the line $y = -x \pm 3$ is tangent to the curve. There are other tangents with this property. If x_1 is such that $\tan x_1 = x_1$, the tangent line is $y = (3\cos x_1 - 1)x = (3\cos(-x_1) - 1)x$, which is tangent when $x = \pm x_1$.

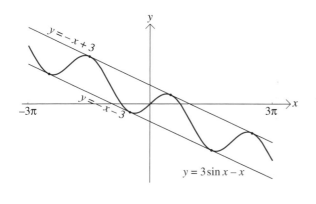

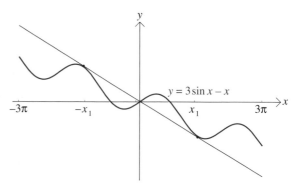

12. Solving equations of a line through $(-1, 9)^T$ gives $y = 9 + m(x + 1)$. Putting $y = \dfrac{3x}{x - 2}$ we obtain the quadratic $mx^2 + (6 - m)x - 2m - 18 = 0$. The line is tangent if the roots are equal, and the condition '$b^2 - 4ac$' gives

$$(6 - m)^2 + 4m(2m + 18) = 0$$

Solving for m, we find two tangent lines through $(-1, 9)^T$, $y = 3 - 6x$ and $y = 8\frac{1}{3} - \frac{2}{3}x$ with points of contact $(1, -3)^T$ and $(5, 5)^T$ respectively.

13. $\dfrac{\mathrm{d}}{\mathrm{d}x}(Af(x)) = \dfrac{\mathrm{d}}{\mathrm{d}x}\left(\dfrac{f(x)}{x}\right) = \dfrac{xf'(x) - f(x)}{x^2}$

$Af(x) < Mf(x) \Leftrightarrow \dfrac{f(x)}{x} < f'(x) \Leftrightarrow \dfrac{\mathrm{d}}{\mathrm{d}x}(Af(x)) > 0 \Leftrightarrow Af(x)$ is increasing.

$Af(x) > Mf(x) \Leftrightarrow Af(x)$ is decreasing.

$Af(x) = Mf(x) \Leftrightarrow Af(x)$ is stationary.

14. Revenue $R(t) = p(t) \cdot x(t)$. Annual rate of growth of revenue is

$$\dfrac{\mathrm{d}}{\mathrm{d}t}(\ln R(t)) = \dfrac{\mathrm{d}}{\mathrm{d}t}(\ln p(t)) + \dfrac{\mathrm{d}}{\mathrm{d}t}(\ln x(t))$$
$$= -0.15 + 0.25 = 0.1 \text{ or } 10\%$$

15. (i) Domain is $x \geq 0$.

$MC(x) = \dfrac{2(x^4 + 4500x^2 + 3.10^6)}{(x^2 + 2000)^2}$

$MC(100) = 2.055\,56$

$C(101) - C(100) = 2.055\,32$

Exercises 2.15

1. (ii) Near $x = 2$, $e^x \sin x \approx e^2(\sin 2 + (x - 2)(\cos 2 + \sin 2) + (x - 2)^2 \cos 2)$.

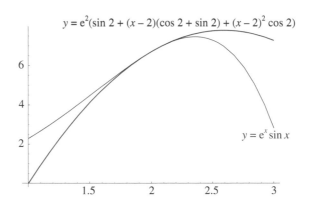

$y = e^2(\sin 2 + (x - 2)(\cos 2 + \sin 2) + (x - 2)^2 \cos 2)$

$y = e^x \sin x$

(iv) Near $x = \pi/4$, $\tan x \approx 1 + 2(x - \frac{\pi}{4}) + 2(x - \frac{\pi}{4})^2$.

Near $x = -\pi/4$, $\tan x \approx -1 + 2(x + \frac{\pi}{4}) - 2(x + \frac{\pi}{4})^2$.

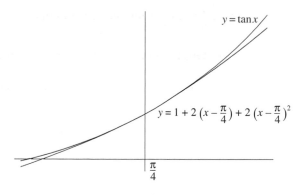

(vi) Near $x = 4$, $\sqrt{x} - \ln x \approx 2 - \ln 4 + \frac{1}{64}(x-4)^2$.

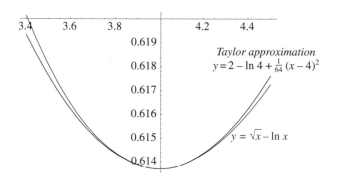

2. (i) Near $x = 1$, $e^x \approx e + e(x-1) + \dfrac{e(x-1)^2}{2!} + \cdots + \dfrac{e(x-1)^n}{n!} + \cdots$

(iii) Near $x = 1$, $\sqrt{x} = 1 + \dfrac{1}{2}(x-1) - \dfrac{1}{2}\dfrac{1}{2}\dfrac{(x-1)^2}{2!} + \dfrac{1}{2}\dfrac{1}{2}\dfrac{3}{2}\dfrac{(x-1)^3}{3!} - \cdots$

$$\cdots + (-1)^{n+1}\frac{1}{2}\frac{1}{2}\frac{3}{2}\cdots\frac{2n-3}{2}\frac{(x-1)^n}{n!} + \cdots$$

(v) Near $x = \frac{1}{2}$, $e^{-x} \approx e^{-1/2} - e^{-1/2}(x - 1/2) + e^{-1/2}\dfrac{(x-1/2)^2}{2!} - \cdots$

$$\cdots + (-1)e^{-1/2}\frac{(x-1/2)^n}{n!} + \cdots$$

3. (i)

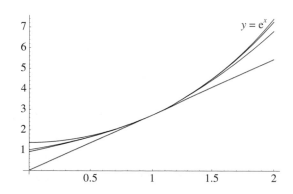

4. (i) $e^x \approx 1 + x + \dfrac{x^2}{2!} + \dfrac{x^3}{3!} + \cdots$

(iii) $(1 + x)^\alpha = 1 + \alpha x + \dfrac{\alpha(\alpha - 1)}{2} x^2 + \dfrac{\alpha(\alpha - 1)(\alpha - 2)}{3!} x^3 + \cdots$

$$\cdots + \alpha x^{\alpha - 1} + x^\alpha \quad (\alpha \in \mathbb{N})$$

(v) $\dfrac{1}{1 - x} \approx 1 + x + x^2 + x^3 + \cdots$

5. (i) $e^{-x} \approx 1 - x + \dfrac{x^2}{2!} - \dfrac{x^3}{3!} + \cdots$

(iii) $(1 + x)^{-2} \approx 1 - 2x + 3x^2 - 4x^3 + 5x^4 - 6x^5 + \cdots$

(v) $\dfrac{\sin x \cos x}{x} = \dfrac{\sin 2x}{2x} \approx \dfrac{1}{2x}\left[2x - \dfrac{(2x)^3}{3!} + \dfrac{(2x)^5}{5!} - \dfrac{(2x)^7}{7!} + \cdots\right]$

$$= 1 - \dfrac{4x^2}{3!} + \dfrac{16x^4}{5!} - \dfrac{64x^6}{7!} + \cdots$$

6. As $x > 0$, all the terms of the series are positive, so $e^x > \dfrac{x^{n+1}}{(n + 1)!}$, which means that $\dfrac{x^{n+1}}{e^x} < (n + 1)!$ as required. Hence $\dfrac{x^n}{e^x} < \dfrac{(n + 1)!}{x}$ which tends to zero as $x \to \infty$.

7.

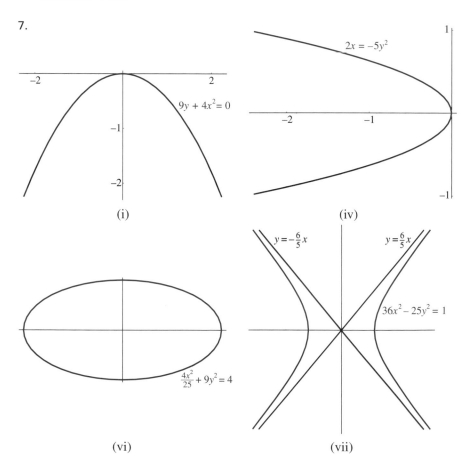

(i)

(iv)

(vi)

(vii)

8. (i) Slope of tangent, $\dfrac{dy}{dx} = \dfrac{x}{2a}$.

(ii) $\langle \overrightarrow{PT}, \overrightarrow{PQ} \rangle = x$, so $\cos \angle QPT = \dfrac{x}{\sqrt{x^2 + 4a^2}}$

$\langle \overrightarrow{PF}, \overrightarrow{PT'} \rangle = ax + xy$, so $\cos \angle FPT' = \dfrac{x(a+y)}{\sqrt{x^2 + (y-a)^2}\sqrt{x^2 + 4a^2}}$

$$= \dfrac{\dfrac{a+y}{\sqrt{4ay + (y-a)^2}} \cdot \dfrac{x}{\sqrt{x^2 + 4a^2}}}{}$$

9. (i) $2x(1 - e^2) + 2y\dfrac{dy}{dx} = 0$

(ii) $\cos \angle FPT = \dfrac{-xy + aey + xy - xye^2}{\sqrt{x^2(1 - e^2)^2 + y^2}\sqrt{y^2 + (xae)^2}}$

$\cos \angle F'PT' = \dfrac{xy + aey - xy(1 - e^2)}{\sqrt{x^2(1 - e^2)^2 + y^2}\sqrt{y^2 + (x + ae)^2}}$

10. (i) (b)   (iii) (d)   (vii) (g)

11. (i) $x = 3$, $y = 0$

(iv) $y = -x + 1$

(vii) $x = 0$, $x = -1$, $y = 0$

Chapter 3. Functions of several variables

Exercises 3.7

1. (i) (b)(i)   (ii) (a)(g)   (iii) (e)(k)   (iv) (d)(l)   (v) (c)(j)   (vi) (f)(h)

2. These results confirm that the partial derivatives give the rate of change of the function in the direction of each axis.

3. (i) $\dfrac{\partial f}{\partial x} = 2x \quad \dfrac{\partial f}{\partial y} = 6y$

(ii) $\dfrac{\partial f}{\partial x} = \dfrac{(x+y)\cdot 1 - (x-y)\cdot 1}{(x+y)^2} = \dfrac{2y}{(x+y)^2}$

$\dfrac{\partial f}{\partial y} = \dfrac{(x+y)(-1) - (x-y)\cdot 1}{(x+y)^2} = \dfrac{-2x}{(x+y)^2}$

(iii) $\dfrac{\partial f}{\partial x} = y^2 e^{xy^2} \quad \dfrac{\partial f}{\partial y} = 2xye^{xy^2}$

(iv) $\dfrac{\partial f}{\partial x} = 2xy^5 \quad \dfrac{\partial f}{\partial y} = 5x^2y^4$

4. (ii) $\dfrac{\partial f}{\partial x} = -y^2 \sin(xy^2) \quad \dfrac{\partial f}{\partial y} = -2xy \sin(xy^2)$

(iii) $\dfrac{\partial f}{\partial x} = \dfrac{1}{y} \cos\left(\dfrac{x}{y}\right) \quad \dfrac{\partial f}{\partial y} = -\dfrac{x}{y^2} \cos\left(\dfrac{x}{y}\right)$

5. The equation of the tangent plane to the surface $z = f(x, y)$ where $x = 1$ and $y = 1$ is $z - Z = f_x(1, 1)(x - 1) + f_y(1, 1)(y - 1)$, where $Z = f(1, 1)$. The required equations are therefore

(i) $z - 4 = 2(x - 1) + 6(y - 1)$ $(2, 6)^T$

(ii) $z = \frac{1}{2}(x - 1) - \frac{1}{2}(y - 1)$ $(\frac{1}{2}, -\frac{1}{2})^T$

(iii) $z - e = e(x - 1) + 2e(y - 1)$ $(e, 2e)^T$

(iv) $z - 1 = 2(x - 1) + 5(y - 1)$ $(2, 5)^T$

7. The unit vector in the direction of $(-3, 4)^T$ is $\mathbf{u} = (-\frac{3}{5}, \frac{4}{5})^T$. The rate of increase of f at $(1, 1)^T$ in the direction $(-3, 4)^T$ is then

$$\langle \nabla f, \mathbf{u} \rangle = -\frac{3}{5} f_x(1, 1) + \frac{4}{5} f_y(1, 1)$$

The direction of maximum rate of increase at $(1, 1)^T$ is ∇f, and the rate of increase in this direction is $\|\nabla f\|$.

(i) When $(x, y)^T = (1, 1)^T$, $\nabla f = (2x, 6y)^T = (2, 6)^T$. Hence $\langle \nabla f, \mathbf{u} \rangle = \frac{18}{5}$ The direction of maximum rate of increase is $\nabla f = (2, 6)^T$ and the rate is $\|\nabla f\| = \sqrt{40}$.

(ii) When $(x, y)^T = (1, 1)^T$, $\nabla f = \left(\dfrac{2y}{(x + y)^2}, -\dfrac{2x}{(x + y)^2} \right)^T = \left(\dfrac{1}{2}, -\dfrac{1}{2} \right)^T$.
Hence $\langle \nabla f, \mathbf{u} \rangle = -\dfrac{7}{10}$. The direction of maximum rate of increase is $\nabla f = \left(\dfrac{1}{2}, -\dfrac{1}{2} \right)^T$ and the rate is $\|\nabla f\| = 1/\sqrt{2}$.

(iii) When $(x, y)^T = (1, 1)^T$, $\nabla f = (y^2 e^{xy^2}, 2xy e^{xy^2})^T = (e, 2e)^T$. Hence $\langle \nabla f, \mathbf{u} \rangle = e$. The direction of maximum rate of increase is $\nabla f = (e, 2e)^T$ – which is the same direction as $(1, 2)^T$ – and the rate is $\|\nabla f\| = e\sqrt{5}$.

(iv) When $(x, y)^T = (1, 1)^T$, $\nabla f = (2xy^5, 5x^2 y^4)^T = (2, 5)^T$. Hence $\langle \nabla f, \mathbf{u} \rangle = \frac{14}{5}$. The direction of maximum rate of increase is $\nabla f = (2, 5)^T$ and the rate is $\|\nabla f\| = \sqrt{29}$.

9. (i) $\begin{pmatrix} -2x \\ 10y \end{pmatrix}$ $\begin{pmatrix} -2 \\ 10 \end{pmatrix}$ $\dfrac{6}{\sqrt{5}}$.

(ii) Find $\mathbf{u} = (u_1, u_2)^T$ where $\|\mathbf{u}\| = 1$ and $\langle \mathbf{u}, \nabla f(2, 0) \rangle = -2$. This gives $u_1 = \frac{1}{2}$ and $u_2 = \pm \frac{1}{2}\sqrt{3}$.

$-\|\nabla f(2, 0)\| = -4$, so the rate of change at $(2, 0)^T$ is never -5.

10. (ii) $\dfrac{\partial P}{\partial x} = 12x^{-1/5} y^{7/8}$ $\dfrac{\partial P}{\partial y} = \dfrac{105}{8} x^{4/5} y^{-1/8}$

$$z - 15 = 12(x - 1) + \frac{105}{8}(y - 1)$$

(iv) $\dfrac{\partial P}{\partial x} = \dfrac{20}{3} x^{-5/4} \left(\dfrac{1}{3} x^{-1/4} + \dfrac{2}{3} y^{-1/4} \right)^{-5}$

$\dfrac{\partial P}{\partial y} = \dfrac{40}{3} y^{-5/4} \left(\dfrac{1}{3} x^{-1/4} + \dfrac{2}{3} y^{-1/4} \right)^{-5}$

$$z - 20 = \frac{20}{3}(x - 1) + \frac{40}{3}(y - 1)$$

12. (i) $\nabla f = \left(\dfrac{x}{\sqrt{x^2 + y^2 + 2y + 1}}, \dfrac{y+1}{\sqrt{x^2 + y^2 + 2y + 1}}, z - \sqrt{a^2 + b^2 + 2b + 1} \right)^{\mathrm{T}}$

$= \dfrac{a}{\sqrt{a^2 + b^2 + 2b + 1}}(x - a) + \dfrac{b+1}{\sqrt{a^2 + b^2 + 2b + 1}}(y - b)$

(ii) $\left(\dfrac{a}{\sqrt{a^2 + b^2 + 2b + 1}}, \dfrac{b+1}{\sqrt{a^2 + b^2 + 2b + 1}}, -1 \right)^{\mathrm{T}}$

(iii) Tangent planes at $(0, 0)^{\mathrm{T}}$, $(1, 0)^{\mathrm{T}}$ and $(-1, 0)^{\mathrm{T}}$ respectively are

$$z - 1 = y$$
$$\sqrt{2}\,z - 2 = x - 1 + y$$
$$\sqrt{2}\,z - 2 = -x - 1 + y$$

Solve the system of equations to obtain $(0, -1, 0)^{\mathrm{T}}$. Substitute $(x, y, z)^{\mathrm{T}} = (0, -1, 0)^{\mathrm{T}}$ in (ii).

Exercises 3.9

1. (i) $f_x = 2x \quad f_y = z \quad f_z = y$

(ii) We have $f(1, 0, 2) = 1$ and so the required tangent hyperplane is $u - 1 = 2(x - 1) + 2(y - 0) + 0(z - 2)$, that is, $u = 2x + 2y - 1$.

(iii) The normal to the contour $1 = x^2 + yz$ at $(1, 0, 2)^{\mathrm{T}}$ is just the value of the gradient $\nabla f = (2x, z, y)^{\mathrm{T}}$ at the point, which is $(2, 2, 0)^{\mathrm{T}}$. The tangent plane to the contour therefore has equation $x + y = 1$.

(iv) The unit vector pointing in the same direction as $(2, -1, 2)^{\mathrm{T}}$ is $\mathbf{u} = (\frac{2}{3}, -\frac{1}{3}, \frac{2}{3})^{\mathrm{T}}$.
The rate of increase of f in the direction $(2, -1, 2)^{\mathrm{T}}$ is therefore $\langle \nabla f, \mathbf{u} \rangle = \frac{2}{3}$.

4. (i) $(2xy^2 z \cos(x^2 z), 2y \sin(x^2 z), x^2 y^2 \cos(x^2 z))^{\mathrm{T}}$

(ii) $u = -8\pi(x - 1) - 4(z - \pi)$

(iii) $-8\pi(x - 1) - 4(z - \pi) = 0$ with normal $(-8\pi, 0, -4)^{\mathrm{T}}$

(iv) Zero. Not the minimum rate which is $-\|\nabla f\| = -\sqrt{64\pi^2 + 16}$

5. (i) $\nabla f = \dfrac{1}{x^2 + y^2 + z^2}(2x, 2y, 2z)^{\mathrm{T}}$

(ii) $u = \dfrac{2}{\sqrt{3}}(x + y + z - \sqrt{3})$

(iii) $x + y + z = \sqrt{3}$ with normal $(1, 1, 1)^{\mathrm{T}}$

(iv) $\left\langle \nabla f, \dfrac{1}{\sqrt{x^2 + y^2 + z^2}}(x, y, z)^{\mathrm{T}} \right\rangle = 1$, or $x^2 + y^2 + z^2 = 4$

7. 147:186:98

9. (i) $(6x_0 + 2y_0, 2x_0 + 6y_0, -1)^T$ $(2x_0, 2y_0, 4z_0)^T$

    (ii) Their scalar product is $4(3x_0^2 + 2x_0y_0 + 3y_0^2) - 4z_0 = 0$.

    (iii) Show that the point $(1, -1, 4)^T$ satisfies the equations of both surfaces.

    (iv) $x - y + 8z = 34$

    (v) $(-1, 1, -4)^T$

10. (i) $\nabla f = (2x, 8y, 4z)^T$ $\nabla g = (2x, 2y, -4z)^T$

    (ii) $x^2 + 4y^2 + 2z^2 = 27$ is an ellipsoid.

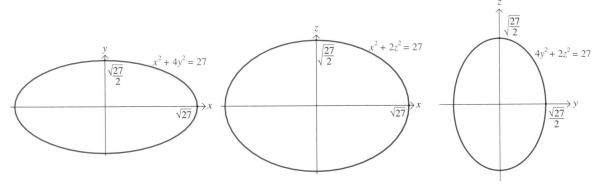

$x^2 + y^2 - 2z^2 = 11$ is a hyperboloid of one sheet.

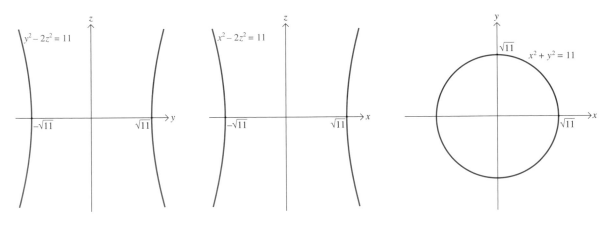

    (iii) $6(x - 3) - 16(y + 2) + 4(z - 1) = 0$ $6(x - 3) - 4(y + 2) - 4(z - 1) = 0$ The normals $(3, -8, 2)^T$ and $(3, -2, -2)^T$ differ, so the surfaces do not touch tangentially.

    (iv) The vector product of the normals gives the direction of the tangent. The equation is
    $$\frac{x - 3}{10} = \frac{y + 2}{6} = \frac{z - 1}{9}$$

    (v) Zero. This implies that the vector is along a contour of f.

11. $\left\langle \nabla \tau, \dfrac{1}{\sqrt{10^2 + 6^2 + 9^2}} (10, 6, 9)^{\mathrm{T}} \right\rangle = \dfrac{38}{\sqrt{217}}$

12. (iii) Normals $(6, 4b, 2c)^{\mathrm{T}}$ and $(-6, 2b - 6, 2c - 8)^{\mathrm{T}}$ are parallel. Hence
$-4b = 2b - 6$ and so $b = 1$
$-2c = 2c - 8$ and so $c = 2$

(iv) $\left\langle \begin{pmatrix} -6 \\ -4 \\ 4 \end{pmatrix}, \begin{pmatrix} 6 \\ -2d \\ 4e \end{pmatrix} \right\rangle = 0, \quad -3 + d + 4e = 4.$

Hence $d = 2$ and $e = \frac{5}{4}$.

(v) $(-6, -4, 4)^{\mathrm{T}}, \quad 2\sqrt{17}$

(vi) $\langle (-6, -4, 4)^{\mathrm{T}}, \quad \frac{1}{3}(1, 2, 2)^{\mathrm{T}} \rangle = -2$

Chapter 4. Stationary points

Exercises 4.5

1.   We have $f(x) = 6 - x - x^2 = (3 + x)(2 - x)$ and $f'(x) = -1 - 2x$. This is zero only when $x = -\frac{1}{2}$, so it follows that $x = -\frac{1}{2}, y = 6\frac{1}{4}$ is the single stationary point.

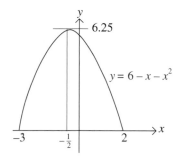

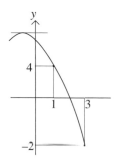

(i) $6\frac{1}{4}$ (ii) 0   (iii) 4   (iv) -6

3.   (ii) $x \leq 3$ Local maximum at $x = 2$.

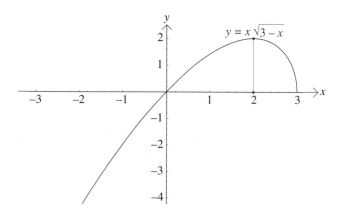

4. (i) $x = -1$ (local minimum); $x = 0$ (point of inflection); $x = 1$ (local maximum).

    (iii) $x = 2 - \sqrt{5}$ (local minimum); $x = 2 + \sqrt{5}$ (local maximum).

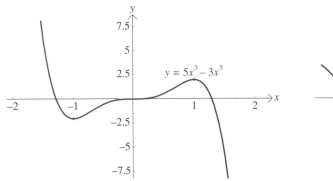

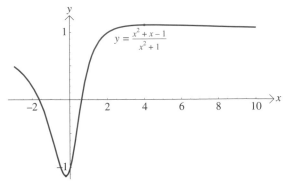

5. (i) Since the *one variable* function has only one stationary point $x = 1$, which is a local maximum, it must be the global maximum.

    (ii) Similarly, the global maximum is $x = \dfrac{1}{n^{1/n}}$.

6. The profit is $\pi(x) = -x^3 + 3x^2 + (p-3)x$ and so

    $$\pi'(x) = -3x^2 + 6x + (p-3).$$

    The stationary points are therefore found by solving

    $$-3x^2 + 6x + (p-3) = 0$$

    i.e.

    $$x^2 - 2x + \left(1 - \frac{p}{3}\right) = 0$$

    $$x = 1 \pm \sqrt{\left(1 - \left(1 - \frac{p}{3}\right)\right)} = 1 \pm \sqrt{\frac{p}{3}}$$

    The two cases to be distinguished are illustrated below.

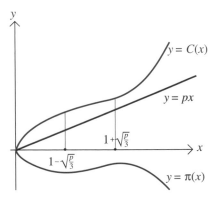

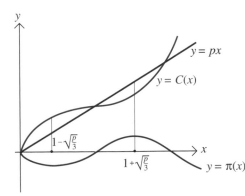

The difference in the two cases is that, in the second, there are points $x > 0$ such that $\pi(x) = 0$ – i.e.

$$\pi(x) = -x^3 + 3x^2 + (p-3)x = 0$$

has solutions $x > 0$. But then

$$x^2 - 3x + (3-p) = 0$$

has solutions and so '$b^2 - 4ac \geqslant 0$' – i.e.

$$9 - 4(3-p) \geqslant 0$$
$$4p \geqslant 3$$
$$p \geqslant \frac{3}{4}$$

Thus, if $0 \leqslant p < \frac{3}{4}$, then case (i) applies and

$$\max_{x \geqslant 0} \pi(x) = \pi(0) = 0$$

and, if $p \geqslant \frac{3}{4}$, then case (ii) applies and

$$\max_{x \geqslant 0} \pi(x) = \pi(\xi)$$

where $\xi = 1 + \sqrt{p/3}$.

8. $x = 3$ (local maximum), $\quad x = 13$ (local minimum).

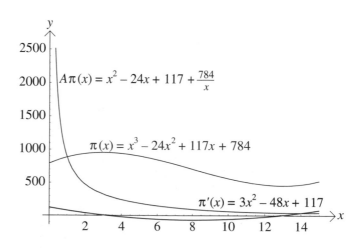

9. (i)

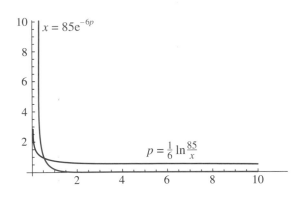

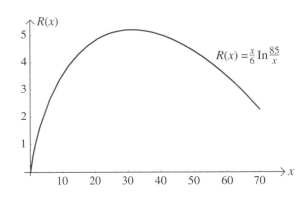

$$R(x) = \frac{x}{6} \ln \frac{85}{x} \qquad x = \frac{85}{e} \qquad p = \frac{1}{6}$$

12. Since $p = 1/\sqrt{x}$, we have $\pi(x) = px - C(x) = \sqrt{x} - 2x^2$.

Thus $\pi'(x) = \frac{1}{2}x^{-1/2} - 4x$.

The stationary points are found by solving $\frac{1}{2}x^{-1/2} - 4x = 0$, which gives $x = \frac{1}{4}$.

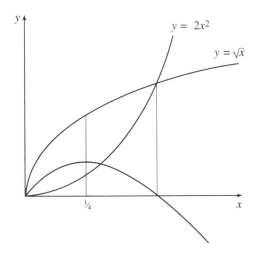

Profit is therefore maximised when $x = \frac{1}{4}$. The corresponding price is $p = 2$.

14. The stationary points are given by a polynomial equation of degree 2 which has between 0 and 2 real roots.

The possible points of inflection are given by a polynomial equation of degree one which has exactly one root.

For two stationary points, we need $b^2 > 3ac$;

for one stationary point, we need $b^2 = 3ac$;

for no stationary points, we need $b^2 < 3ac$.

(i) $b^2 < 3ac$, $a > 0$ (ii) $b^2 < 3ac$, $a < 0$

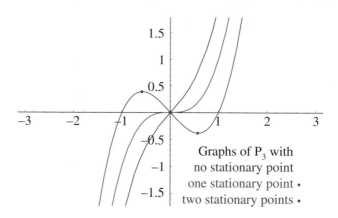

Graphs of P_3 with
no stationary point
one stationary point •
two stationary points •

16. If n is even, stationary points are given by a polynomial equation of odd degree, which has at least one real root. The points of inflection are given by a polynomial equation of even degree, which may not have any real roots. The same reasoning gives the results for n odd.

(i)

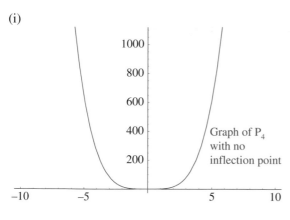

Graph of P_4 with no inflection point

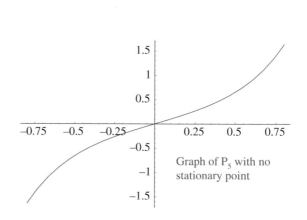

Graph of P_5 with no stationary point

(ii)

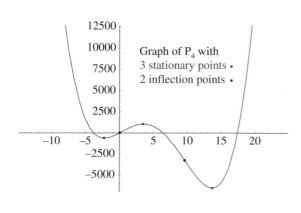

Graph of P_4 with
3 stationary points •
2 inflection points •

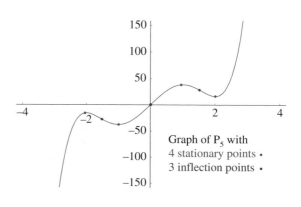

Graph of P_5 with
4 stationary points •
3 inflection points •

17. (i) Concave when $x < \frac{5}{3}$ and convex when $x > \frac{5}{3}$.

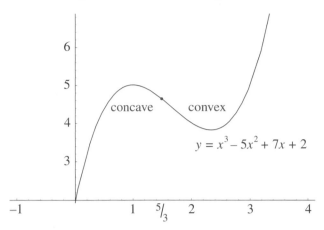

18. (i) $\dfrac{dy}{dx} = \sin x + x \cos x = 0$ at stationary points. Since if $\cos x = 0$ at a solution of the equation, then $\sin x = 0$ also, which is impossible, $\cos x \neq 0$ at a stationary point. Hence the equation may be written as $\tan x = -x$, which has an infinite number of solutions as shown by the graph.

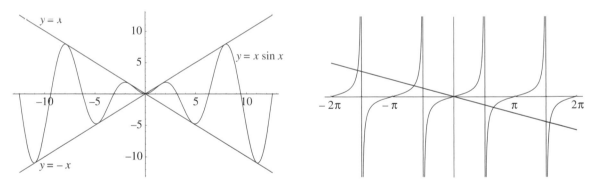

The largest negative stationary point is a local maximum at approximately $-2.028\,76$.

Exercises 4.9

1. (i) $(0, 0)^{\mathrm{T}}$ (iii) $(0, 0)^{\mathrm{T}}$ (v) $(-1, 1)^{\mathrm{T}}, (1, -1)^{\mathrm{T}}$ (vii) $(\frac{3}{5}, 2)^{\mathrm{T}}, (\frac{3}{5}, -2)^{\mathrm{T}}$
   (ix) $(\frac{1}{2}, 1)^{\mathrm{T}}, (-\frac{1}{2}, 1)^{\mathrm{T}}, (-\frac{1}{2}, -1)^{\mathrm{T}}, (\frac{1}{2}, -1)^{\mathrm{T}}, (0, 1)^{\mathrm{T}}, (0, -1)^{\mathrm{T}}$
   (xi) $(0, 0)^{\mathrm{T}}, (0, 2)^{\mathrm{T}}, (\frac{4}{9}, \frac{2}{3})^{\mathrm{T}}, (\frac{4}{3}, 0)^{\mathrm{T}}$

2. (i) No stationary points.   (ii) No stationary points.

3. (i) We have $\nabla(e^{f(\mathbf{x})}) = e^{f(\mathbf{x})} \nabla f(\mathbf{x})$. Since $e^{f(\mathbf{x})} \neq 0$, $\nabla(e^{f(\mathbf{x})}) = \mathbf{0}$ if and only if $\nabla f(\mathbf{x}) = \mathbf{0}$. Hence $f(\mathbf{x})$ and $e^{f(\mathbf{x})}$ have the same stationary points.

(ii) We have $\nabla(\ln(f(\mathbf{x}))) = \dfrac{1}{f(\mathbf{x})}\nabla f(\mathbf{x})$ and $\nabla((f(\mathbf{x}))^{\alpha}) = \alpha(f(\mathbf{x}))^{\alpha-1}\nabla f(\mathbf{x})$.

If $f(\mathbf{x}) > 0$, $\dfrac{1}{f(\mathbf{x})}$ and $f(\mathbf{x})^{\alpha-1} \neq 0$, and so $\nabla(\ln(f(\mathbf{x}))) = \mathbf{0}$ and $\nabla((f(\mathbf{x}))^{\alpha}) = \mathbf{0}$ if and only if $\nabla f(\mathbf{x}) = \mathbf{0}$.

(iii) We have $f(\mathbf{x}) = c$ if and only if $e^{f(\mathbf{x})} = e^{c}$, $\ln f(\mathbf{x}) = \ln c$, and $f(\mathbf{x})^{\alpha} = c^{\alpha}$. So, for example, a contour of value c of $f(\mathbf{x})$ is also a contour of $e^{f(\mathbf{x})}$ of value e^{c}.

(iv) We have $\nabla(\sin f(\mathbf{x})) = \cos f(\mathbf{x})\nabla f(\mathbf{x})$, so $\nabla f(\mathbf{x}) = \mathbf{0} \Rightarrow \nabla(\sin f(\mathbf{x})) = \mathbf{0}$. But $\nabla(\sin f(\mathbf{x})) = \mathbf{0}$ also when $\cos f(\mathbf{x}) = 0$ or $f(\mathbf{x}) = n\pi$, for any $n \in \mathbb{Z}$. Hence the stationary points of $f(\mathbf{x})$ are among the stationary points of $\sin f(\mathbf{x})$ (and also among the stationary points of $\cos f(\mathbf{x})$).
We have that $f(\mathbf{x}) = c$ implies $\sin f(\mathbf{x}) = \sin c$, so contours of $f(\mathbf{x})$ are contours of $\sin f(\mathbf{x})$. But if $f(\mathbf{x}) = c + 2n\pi$, then $\sin f(\mathbf{x}) = \sin c$ as well, so contours of $\sin f(\mathbf{x})$ for one value correspond to an infinite set of contours of $f(\mathbf{x})$.

4. (i) (b)(h)   (ii) (d)(k)   (iii) (a)(g)   (iv) (e)(j)   (v) (f)(i)   (vi) (c)(l)

5. (i) $(-1, \frac{1}{2}, \frac{1}{2})^{\mathrm{T}}$
   (iv) The line $x = y = z$.

6. (i) $(4, 2)^{\mathrm{T}}$ and $(4, -2)^{\mathrm{T}}$
   (iii) $(1, -2)^{\mathrm{T}}$

Chapter 5. Vector functions

Exercises 5.8

1.

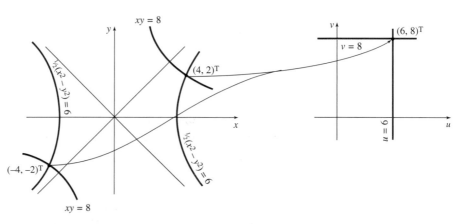

4. $\begin{pmatrix} \dfrac{\partial u}{\partial x} & \dfrac{\partial u}{\partial y} & \dfrac{\partial u}{\partial z} \\ \dfrac{\partial v}{\partial x} & \dfrac{\partial v}{\partial y} & \dfrac{\partial v}{\partial z} \end{pmatrix} = \begin{pmatrix} y^{-1} & -xy^{-2} & 0 \\ 0 & z^{-1} & -yz^{-2} \end{pmatrix}$

It follows that

$$f'(1, 2, 3) = \begin{pmatrix} \dfrac{1}{2} & -\dfrac{1}{4} & 0 \\ 0 & \dfrac{1}{3} & -\dfrac{2}{9} \end{pmatrix}$$

The tangent flat at $(1, 2, 3)^{\mathrm{T}}$ is given by

$$\begin{pmatrix} u - \dfrac{1}{2} \\ v - \dfrac{2}{3} \end{pmatrix} = \begin{pmatrix} \dfrac{1}{2} & -\dfrac{1}{4} & 0 \\ 0 & \dfrac{1}{3} & -\dfrac{2}{9} \end{pmatrix} \begin{pmatrix} x - 1 \\ y - 2 \\ z - 3 \end{pmatrix}$$

which gives the two equations

$4u = 2x - y + 2$

$9v = 3y - 2z + 6$

5. (iii) Derivative is $\begin{pmatrix} 1 & 0 \\ 0 & 1 \end{pmatrix}$

$$\begin{pmatrix} r - 1 \\ \theta \end{pmatrix} = \begin{pmatrix} 1 & 0 \\ 0 & 1 \end{pmatrix} \begin{pmatrix} x - 1 \\ y \end{pmatrix}$$

6. Put $t = x^2 + y^2$ so $z = f(t)$. If $\mathbf{u} = (x, y)^{\mathrm{T}}$, we have that

$$\left(\dfrac{\partial z}{\partial x}, \dfrac{\partial z}{\partial y} \right) = \dfrac{dz}{d\mathbf{u}} = \dfrac{dz}{dt} \dfrac{dt}{d\mathbf{u}} = f'(t) \left(\dfrac{\partial t}{\partial x}, \dfrac{\partial t}{\partial y} \right)$$

and so $\dfrac{\partial z}{\partial x} = f'(t) \dfrac{\partial t}{\partial x} = f'(x^2 + y^2) 2x$

$\dfrac{\partial z}{\partial y} = f'(t) \dfrac{\partial t}{\partial y} = f'(x^2 + y^2) 2y$

These formulas can also, of course, be obtained directly.

Eliminating $f'(x^2 + y^2)$, we obtain $\dfrac{1}{x} \cdot \dfrac{\partial z}{\partial x} = \dfrac{1}{y} \cdot \dfrac{\partial z}{\partial y}$ as required.

8. $\dfrac{dP(x(t), y(t))}{dt} = A \left(2^{1/3} \dfrac{2}{3} t^{-1/3} (t^3 + 1)^{2/3} + 2^{1/3} t^{2/3} \dfrac{2}{3} (t^3 + 1)^{-1/3} 3t^2 \right)$

By the chain rule, $\dfrac{dP}{dt} = A \dfrac{1}{3} x^{-2/3} y^{2/3} 4t + A \dfrac{2}{3} x^{1/3} y^{-1/3} 3t^2$

Substitution in terms of t and simplification show the expressions to be equal.

9. (i) $\dfrac{dz}{dt} = -8t^7 \sin(t^8)$

(ii) $\dfrac{dz}{dt} = -2 \cos t \, \sin^6 t + 5 \cos^3 t \, \sin^4 t$

10. $\begin{pmatrix} 0 & -e^{-y} \\ -e^{-y} & xe^{-y} \end{pmatrix}$

Near $(1, 0)^{\mathrm{T}}$,

$xe^{-y} \approx 1 + (1, \quad -1) \begin{pmatrix} x - 1 \\ y \end{pmatrix} + 1/2 (x - 1, \quad y) \begin{pmatrix} 0 & -1 \\ -1 & 1 \end{pmatrix} \begin{pmatrix} x - 1 \\ y \end{pmatrix}$

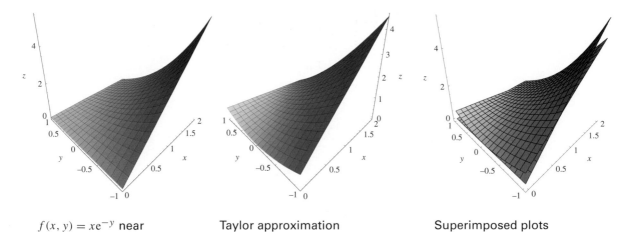

$f(x, y) = xe^{-y}$ near
$(1, 0)^{\mathrm{T}}$

Taylor approximation
near $(1, 0)^{\mathrm{T}}$

Superimposed plots

13. $f(X, Y) = \lambda^n f(x, y)$

Let $X = \lambda x, Y = \lambda y$. Then $\dfrac{\mathrm{d}f(X, Y)}{\mathrm{d}\lambda} = \dfrac{\partial f}{\partial X}\dfrac{\mathrm{d}X}{\mathrm{d}\lambda} + \dfrac{\partial f}{\partial Y}\dfrac{\mathrm{d}Y}{\mathrm{d}\lambda}.$

Hence $n\lambda^{n-1} f(x, y) = xf_X(X, Y) + yf_Y(X, Y).$

Let $\lambda = 1$ to deduce the theorem.

Chapter 6. Optimisation of scalar valued functions

Exercises 6.5

1. (i) $(2, 3)^{\mathrm{T}}$ local minimum

(iii) $(0, 0)^{\mathrm{T}}$ saddle point

(v) $(-1, 1)^{\mathrm{T}}, (1, -1)^{\mathrm{T}}$ saddle points

(vii) $(\frac{3}{5}, 2)^{\mathrm{T}}, (\frac{3}{5}, -2)^{\mathrm{T}}$ saddle points

(ix) $(\frac{1}{2}, 1)^{\mathrm{T}}, (-\frac{1}{2}, 1)^{\mathrm{T}}$ local minima, $(-\frac{1}{2}, -1)^{\mathrm{T}}, (\frac{1}{2}, 1)^{\mathrm{T}}, (0, 1)^{\mathrm{T}}$ saddle points,
$(0, -1)^{\mathrm{T}}$ local maximum

(xi) $(0, 0)^{\mathrm{T}}, (0, 2)^{\mathrm{T}}, (\frac{4}{3}, 0)^{\mathrm{T}}$ saddle points, $(\frac{4}{9}, \frac{2}{3})^{\mathrm{T}}$ local minimum

(xiii) $(0, 0)^{\mathrm{T}}$ saddle point

(xiv) We know from Example 4.4 that the stationary points are

$$(0, 0)^{\mathrm{T}} (1, 0)^{\mathrm{T}} (0, 1)^{\mathrm{T}} (0, -1)^{\mathrm{T}} \left(\frac{2}{5}, \frac{1}{\sqrt{5}}\right)^{\mathrm{T}} \left(\frac{2}{5}, -\frac{1}{\sqrt{5}}\right)^{\mathrm{T}}$$

The second derivative is

$$\begin{pmatrix} f_{xx} & f_{xy} \\ f_{yx} & f_{yy} \end{pmatrix} = \begin{pmatrix} 2y & 2x + 3y^2 - 1 \\ 2x + 3y^2 - 1 & 6xy \end{pmatrix}$$

The principal minors are

$$\Delta = \begin{vmatrix} 2y & 2x + 3y^2 - 1 \\ 2x + 3y^2 - 1 & 6xy \end{vmatrix} \quad \text{and} \quad \delta = 2y$$

(a) At $(0, 0)^T$,

$$\Delta = \begin{vmatrix} 0 & -1 \\ -1 & 0 \end{vmatrix} = -1 < 0$$

and hence $(0, 0)^T$ is a saddle point.

(b) At $(1, 0)^T$,

$$\Delta = \begin{vmatrix} 0 & 1 \\ 1 & 0 \end{vmatrix} = -1 < 0$$

and hence $(1, 0)^T$ is a saddle point.

(c) At $(0, 1)^T$,

$$\Delta = \begin{vmatrix} 2 & 2 \\ 2 & 0 \end{vmatrix} = -4 < 0$$

and hence $(0, 1)^T$ is a saddle point.

(d) At $(0, -1)^T$,

$$\Delta = \begin{vmatrix} -2 & 2 \\ 2 & 0 \end{vmatrix} = -4 < 0$$

and hence $(0, -1)^T$ is a saddle point.

(e) At $(\frac{2}{5}, 1/\sqrt{5})^T$,

$$\Delta = \begin{vmatrix} \dfrac{2}{\sqrt{5}} & \dfrac{2}{5} \\ \dfrac{2}{5} & \dfrac{12}{5\sqrt{5}} \end{vmatrix} = \frac{4}{5} > 0.$$

Also $\delta = 2/\sqrt{5} > 0$. Hence $(\frac{2}{5}, 1/\sqrt{5})^T$ is a local minimum.

(f) At $(\frac{2}{5}, -1/\sqrt{5})^T$,

$$\Delta = \begin{vmatrix} \dfrac{-2}{\sqrt{5}} & \dfrac{2}{5} \\ \dfrac{2}{5} & \dfrac{-12}{5\sqrt{5}} \end{vmatrix} = \frac{4}{5} > 0.$$

Also $\delta = -2/\sqrt{5} < 0$. Hence $(\frac{2}{5}, -1/\sqrt{5})^T$ is a local maximum.

(xvii) $(-1, 1)^T (1, 1)^T$ saddle points, $(0, 0)^T$ local maximum, $(0, 2)^T$ local minimum

2. (i) We have $|f''(x, y, z)| = \begin{vmatrix} 2y & 2x & 2z \\ 2x & 2z & 2y \\ 2z & 2y & 2x \end{vmatrix} = 0$ at the unique stationary point $(0, 0, 0)^T$.

Now $f(0, 0, 0) = 0$ and for small $h, k > 0$

$$f(0, h, k) = h^2 k > 0 \text{ and } f(0, h, -k) = -h^2 k < 0$$

Hence near $(0, 0, 0)^T$, $f(0, h, k) > f(0, 0, 0,)$ and $f(0, h, -k) < f(0, 0, 0)$, so $(0, 0, 0)^T$ is a saddle point.

(iv) $|f''(x, y, z)| = \begin{vmatrix} 2y & 2x & 0 \\ 2x & 4z & 4y \\ 0 & 4y & 0 \end{vmatrix} = -32y^3$. The signs of the principal minors show that the two stationary points $\pm\left(\frac{1}{\sqrt{2}}, \sqrt{2}, -\frac{1}{8\sqrt{2}}\right)$ are both saddle points.

(vi) We have $|f'(x, y, z)|$ is a multiple of $\begin{vmatrix} 6x & -6 & 0 \\ -6 & -12y & 0 \\ 0 & 0 & -18z \end{vmatrix} = 0$ at the unique stationary point $(-2^{2/3}, 2^{1/3}, 0)^T$.

$f(-2^{2/3}, 2^{1/3}, 0) = 1/2 \ln 8$.

For small $h > 0$, $f(-2^{2/3}, 2^{1/3}, h) = 1/2 \ln(8 - 3h^3) < f(-2^{2/3}, 2^{1/3}, 0)$ and $f(-2^{2/3}, 2^{1/3}, -h) = 1/2 \ln(8 + 3h^3) > f(-2^{2/3}, 2^{1/3}, 0)$.

Hence near $(-2^{2/3}, 2^{1/3}, 0)^T$, $f(x, y, z) > f(-2^{2/3}, 2^{1/3}, 0)$ sometimes and $f(x, y, z) < f(-2^{2/3}, 2^{1/3}, 0)$ sometimes, so the point is a saddle point.

3. (i) We have $f_{xx} = (62 - 32x + 4x^2) \exp(8x + 12y - x^2 - y^3) < 0$ when $x = 4$.

$f_{yy} = 12 \exp(8x - 12y - x^2 - y^3) > 0$ when $y = -2$, and

$f_{yy} = (-12) \exp(8x - 12y - x^2 - y^3) < 0$ when $y = 2$

$f_{xy} = (-2x + 8)(-3y^2 + 12) \exp(8x - 12y - x^2 - y^3) = 0$ at $(4, \pm 2)^T$.

Hence $\Delta > 0$, $f_{xx} < 0$ when $y = 2$ and so $(4, 2)^T$ is a local maximum.

$\Delta < 0$ when $y = -2$ and so $(4, -2)^T$ is a saddle point.

(iii) $f_x = \dfrac{2x - 4}{10 - 4x - 4y + x^2 + y^4}$ $f_y = \dfrac{4y^3 - 4}{10 - 4x - 4y + x^2 + y^4}$

so $f_{xy} = \dfrac{-(2x - 4)(4y^3 - 4)}{(10 - 4x - 4y + x^2 + y^4)^2}$ which is 0 at $(2, 1)^T$,

$f_{xx} = \dfrac{2(10 - 4x - 4y + x^2 + y^4) - (2x - 4)^2}{(10 - 4x - 4y + x^2 + y^4)^2}$ which is 2/3 at $(2, 1)^T$,

$f_{yy} = \dfrac{12y^2(10 - 4x - 4y + x^2 + y^4) - (4y^3 - 4)^2}{(10 - 4x - 4y + x^2 + y^4)^2}$ which is 4 at $(2, 1)^T$.

At $(2, 1)^T$, $\Delta = \begin{vmatrix} 2/3 & 0 \\ 0 & 4 \end{vmatrix} > 0$ and $f_{xx} = 2/3 > 0$, so the point is a local minimum.

4. $\Delta = \begin{vmatrix} \alpha(\alpha - 1)x^{\alpha-2}y^{\beta} & \alpha\beta x^{\alpha-1}y^{\beta-1} \\ \alpha\beta x^{\alpha-1}y^{\beta-1} & \beta(\beta - 1)x^{\alpha}y^{\beta-2} \end{vmatrix} = \alpha\beta(1 - \alpha - \beta)x^{2\alpha-2}y^{2\beta-2} \geq 0$ when $\alpha + \beta \leq 1$.

Also, $P_{xx} = \alpha(\alpha - 1)x^{\alpha-2}y^{\beta} \leq 0$ since $\alpha > 0$. Hence, under these conditions the graph is a concave surface.

When $\alpha + \beta > 1$, $\Delta < 0$ in which case the graph is neither convex nor concave.

5. $P_{yy} = -\alpha(1 - \alpha)(1 - \beta)x^{\beta}y^{\beta-2}[\alpha x^{\beta} + (1 - \alpha)y^{\beta}]^{\frac{1}{\beta}-2}$

$P_{xy} = \alpha(1 - \alpha)(1 - \beta)x^{\beta-1}y^{\beta-1}[\alpha x^{\beta} + (1 - \alpha)y^{\beta}]^{\frac{1}{\beta}-2}$

6. The stationary points are found by solving the simultaneous equations

$$0 = \frac{\partial f}{\partial x} = 10x - 6z \qquad \text{– i.e. } 3z = 5x$$

$$0 = \frac{\partial f}{\partial y} = 10y - 12z \qquad \text{– i.e. } 5y = 6z \qquad\qquad (1)$$

$$0 = \frac{\partial f}{\partial z} = 18z - 6x - 12y \qquad \text{– i.e. } 3z - x - 2y = 0$$

Substituting for x and y in the third equation, we obtain that

$$3z - \frac{3}{5}z - \frac{12}{5}z = 0$$

which is satisfied for *all* values of z. Thus, if we write $z = 5\alpha$, we obtain that the point

$$\alpha(3, 6, 5)^{\mathrm{T}}$$

is a stationary point for all values of α.

The second derivative is

$$\begin{pmatrix} f_{xx} & f_{xy} & f_{xz} \\ f_{yx} & f_{yy} & f_{yz} \\ f_{zx} & f_{zy} & f_{zz} \end{pmatrix} = \begin{pmatrix} 10 & 0 & -6 \\ 0 & 10 & -12 \\ -6 & -12 & 18 \end{pmatrix}$$

Since the determinant of this matrix is zero (which can be deduced, incidentally, from the fact that equations (1) have multiple solutions), the standard method does not work.

However, in this case, the given function is a quadratic form. In fact

$$f(x, y, z) = (x, y, z) \begin{pmatrix} 5 & 0 & -3 \\ 0 & 5 & -6 \\ -3 & -6 & 9 \end{pmatrix} \begin{pmatrix} x \\ y \\ z \end{pmatrix}$$

The eigenvalues of the matrix are obtained by solving

$$\begin{vmatrix} 5 - \lambda & 0 & -3 \\ 0 & 5 - \lambda & -6 \\ -3 & -6 & 9 - \lambda \end{vmatrix} = 0$$

i.e.

$$(5 - \lambda)^2(9 - \lambda) - (5 - \lambda)(9 + 36) = 0.$$

Thus $\lambda = 5$ *or*

$$(5 - \lambda)(9 - \lambda) - 45 = 0$$
$$45 - 14\lambda + \lambda^2 - 45 = 0$$
$$\lambda(\lambda - 14) = 0$$

and so

$$\lambda = 0 \quad \text{or} \quad \lambda = 14.$$

By an appropriate change of variable, we may therefore write

$$f(x, y, z) = 0X^2 + 5Y^2 + 14Z^2$$

from which it follows that f has local minima where $Y = Z = 0$ – i.e. along the line where $(x, y, z)^{\mathrm{T}} = \alpha(3, 6, 5)^{\mathrm{T}}$.

8. We have that

$$z = xy = (x, y) \begin{pmatrix} 0 & \frac{1}{2} \\ \frac{1}{2} & 0 \end{pmatrix} \begin{pmatrix} x \\ y \end{pmatrix}$$

The eigenvalues of the matrix are obtained by solving

$$0 = \begin{vmatrix} -\lambda & \frac{1}{2} \\ \frac{1}{2} & -\lambda \end{vmatrix} = \lambda^2 - \frac{1}{4}$$

We obtain that $\lambda_1 = \dfrac{1}{2}$ and $\lambda_2 = -\dfrac{1}{2}$. Thus

$$z = \frac{1}{2}X^2 - \frac{1}{2}Y^2.$$

To find the directions in which the X and Y axes point, we require eigenvectors $\mathbf{e}_1$ and $\mathbf{e}_2$ corresponding to λ_1 and λ_2. The equation for $\mathbf{e}_1$ is

$$\begin{pmatrix} 0 & \frac{1}{2} \\ \frac{1}{2} & 0 \end{pmatrix} \begin{pmatrix} e \\ f \end{pmatrix} = \frac{1}{2} \begin{pmatrix} e \\ f \end{pmatrix}$$

which yields the result $e = f$. Thus the X axis passes through $(1, 1)^T$. The equation for $\mathbf{e}_2$ is

$$\begin{pmatrix} 0 & \frac{1}{2} \\ \frac{1}{2} & 0 \end{pmatrix} \begin{pmatrix} e \\ f \end{pmatrix} = -\frac{1}{2} \begin{pmatrix} e \\ f \end{pmatrix}$$

which yields the result $e = -f$. Thus the Y axis passes through $(-1, 1)^T$.

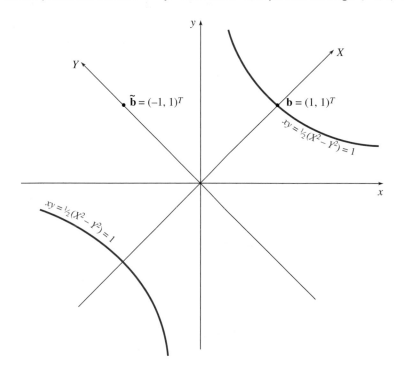

We have

$$f(x, y) = g(X, Y) = \frac{1}{2}X^2 - \frac{1}{2}Y^2.$$

Observe that $g(X, 0)$ has a maximum at $X = 0$ while $g(0, Y)$ has a minimum at $Y = 0$. Thus $(0, 0)^T$ is a *saddle point*.

10. $x = 687/61 \approx 11$ $y = 425/61 \approx 7$ $z = 3106/183 \approx 17$

12. The eigenvalues are $\lambda = A \pm B/2$ with eigenvectors $(1/\sqrt{2}, 1/\sqrt{2})^T$ and $(-1/\sqrt{2}, 1/\sqrt{2})^T$ respectively. The result follows.

13.

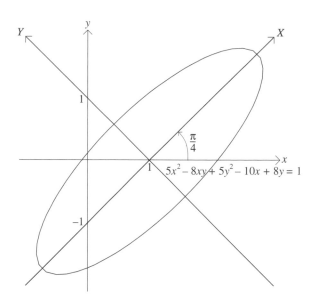

$$5x^2 - 8xy + 5y^2 - 10x + 8y = 1$$

14. (i)   A parabola

(ii) $\lambda = 5, -1$

The equation transforms to

$$(X,\ Y)\begin{pmatrix} 5 & 0 \\ 0 & -1 \end{pmatrix}\begin{pmatrix} X \\ Y \end{pmatrix} + 3\sqrt{2}(Y\sqrt{2}) = 5$$

This can be rearranged as the hyperbola $(Y-3)^2 - 5X^2 = 4$.

(iii) $\lambda = \frac{1}{2}, -\frac{1}{2}$

The equation transforms to $\frac{1}{2}X^2 - \frac{1}{2}Y^2 + \sqrt{2}Y = 1$ or $X^2 - (Y - \sqrt{2})^2 = 0$.
This is the pair of straight lines $X = Y - \sqrt{2}$ and $X = -Y + \sqrt{2}$.

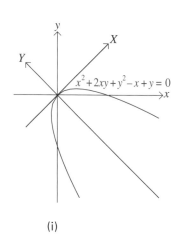

$$x^2 + 2xy + y^2 - x + y = 0$$

(i)

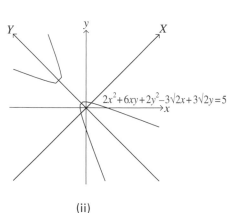

$$2x^2 + 6xy + 2y^2 - 3\sqrt{2}x + 3\sqrt{2}y = 5$$

(ii)

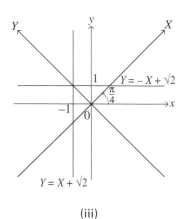

$$Y = -X + \sqrt{2}$$
$$Y = X + \sqrt{2}$$

(iii)

Exercises 6.11

1. We have $f(x, y) = (x - 2)^2 + (y - 3)^2$ and so

$$\left.\begin{array}{l} \dfrac{\partial f}{\partial x} = 2(x - 2) \\[2mm] \dfrac{\partial f}{\partial y} = 2(y - 3) \end{array}\right\}$$

The stationary points are found by solving the simultaneous equations

$$\left.\begin{array}{l} x - 2 = 0 \\ y - 3 = 0 \end{array}\right\}$$

and so there is a single stationary point, namely $(2, 3)^\mathrm{T}$.

(i) Since $2 \times 2 + 3 = 7 > 2$, the point $(2, 3)^\mathrm{T}$ does not lie in the region defined by $2x + y \leqslant 2$. If the required minimum exists, it therefore lies on the line $2x + y = 2$. We therefore seek the stationary points of $f(x, y)$ subject to the constraint $2x + y = 2$. These are obtained from the stationary points of

$$\phi(x) = f(x, 2 - 2x) = (x - 2)^2 + (2x + 1)^2.$$

But

$$\phi'(x) = 2(x - 2) + 4(2x + 1) = 10x$$

and hence ϕ has a single stationary point at $x = 0$. The corresponding value of y is $y = 2 - 2 \times 0 = 2$ and so the required minimum is attained at the point $(0, 2)^\mathrm{T}$. Thus

$$\min_{2x+y \leqslant 2} f(x, y) = (0 - 2)^2 + (2 - 3)^2 = 4 + 1 = 5.$$

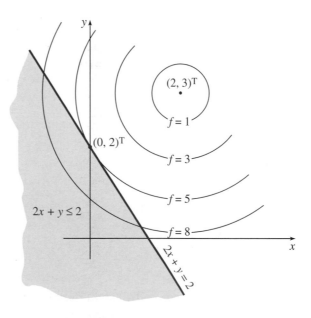

(ii) Since $3 \times 2 + 2 \times 3 = 12 > 6$, the point $(2, 3)^T$ lies in the region defined by $3x + 2y \geqslant 6$. Since $(2, 3)^T$ is a global minimum, it also minimises $f(x, y)$ subject to the constraint $3x + 2y \geqslant 6$. Thus

$$\min_{3x+2y \geqslant 6} f(x, y) = (2 - 2)^2 + (3 - 3)^2 = 0.$$

A *local* maximum will be found by evaluating the stationary points of $f(x, y)$ subject to the constraint $3x + 2y = 6$.

3. The Lagrangian is

$$L = x^2 + y^2 + \lambda(5x^2 + 6xy + 5y^2 - 8).$$

We therefore have to solve the simultaneous equations

$$\begin{cases} 0 = \dfrac{\partial L}{\partial x} = 2x + \lambda(10x + 6y) = (10\lambda + 2)x + 6\lambda y \\[2mm] 0 = \dfrac{\partial L}{\partial y} = 2y + \lambda(6x + 10y) = 6\lambda x + (10\lambda + 2)y \\[2mm] 0 = \dfrac{\partial L}{\partial \lambda} = 5x^2 + 6xy + 5y^2 - 8 \end{cases}$$

From the first two equations we obtain

$$\frac{y}{x} = -\frac{(10\lambda + 2)}{6\lambda} = \frac{x}{y}.$$

Thus $y^2 = x^2$ and so $y = \pm x$.

Case (i). $y = x$. We substitute this result in the third equation and obtain

$$5x^2 + 6x^2 + 5x^2 = 8$$
$$16x^2 = 8$$
$$x^2 = \frac{1}{2}$$
$$x = \pm\frac{1}{\sqrt{2}}$$

This case therefore yields two constrained stationary points, namely $(1/\sqrt{2}, 1/\sqrt{2})^T$ and $(-1/\sqrt{2}, -1/\sqrt{2})^T$.

Case (ii). $y = -x$. We substitute this result in the third equation and obtain

$$5x^2 - 6x^2 + 5x^2 = 8$$
$$4x^2 = 8$$
$$x = \pm\sqrt{2}$$

This case therefore yields two constrained stationary points, namely $(\sqrt{2}, -\sqrt{2})^T$ and $(-\sqrt{2}, \sqrt{2})^T$.

Now $f\left(\dfrac{1}{\sqrt{2}}, \dfrac{1}{\sqrt{2}}\right) = f\left(-\dfrac{1}{\sqrt{2}}, -\dfrac{1}{\sqrt{2}}\right) = \dfrac{1}{2} + \dfrac{1}{2} = 1$,

and $f(\sqrt{2}, -\sqrt{2}) = f(-\sqrt{2}, \sqrt{2}) = 2 + 2 = 4$.

The required maximum is therefore 4 and the minimum is 1.

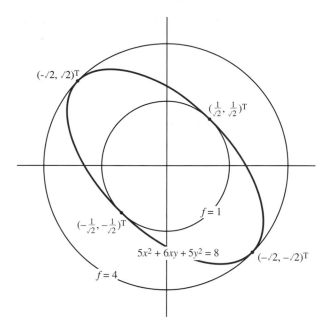

The curves in the figure are labelled $(-\sqrt{2}, \sqrt{2})^{\mathrm{T}}$, $(\tfrac{1}{\sqrt{2}}, \tfrac{1}{\sqrt{2}})^{\mathrm{T}}$, $(-\tfrac{1}{\sqrt{2}}, -\tfrac{1}{\sqrt{2}})^{\mathrm{T}}$, $f = 1$, $5x^2 + 6xy + 5y^2 = 8$, $(-\sqrt{2}, -\sqrt{2})^{\mathrm{T}}$, $f = 4$.

5. We have to maximise $A = xy$ subject to the constraint $2x + 2y = p$.

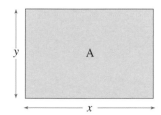

The Lagrangian is

$$L = xy + \lambda(2x + 2y - p)$$

and so we solve the simultaneous equations

$$0 = \frac{\partial L}{\partial x} = y + 2\lambda$$

$$0 = \frac{\partial L}{\partial y} = x + 2\lambda$$

$$0 = \frac{\partial L}{\partial \lambda} = 2x + 2y - p$$

From the first two equations we obtain that $x = y$. Substituting in the third equation yields

$$4x = p$$

$$x = \frac{1}{4}p$$

The rectangle of largest area with perimeter p is therefore the square of side $\frac{1}{4}p$.

7. The Lagrangian is

$$L = (x - X)^2 + (y - Y)^2 + \lambda(xy - 1) + \mu(X + 2Y - 1)$$

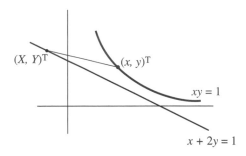

$(X, Y)^T$ $(x, y)^T$

$xy = 1$

$x + 2y = 1$

and so we solve the simultaneous equations

$$0 = \frac{\partial L}{\partial x} = 2(x - X) + \lambda y \tag{1}$$

$$0 = \frac{\partial L}{\partial y} = 2(y - Y) + \lambda x \tag{2}$$

$$0 = \frac{\partial L}{\partial X} = -2(x - X) + \mu \tag{3}$$

$$0 = \frac{\partial L}{\partial Y} = -2(y - Y) + 2\mu \tag{4}$$

$$0 = \frac{\partial L}{\partial \lambda} = xy - 1 \tag{5}$$

$$0 = \frac{\partial L}{\partial \mu} = X + 2Y - 1 \tag{6}$$

From equations (1) and (2),

$$\frac{x - X}{y - Y} = \frac{y}{x}$$

and from equations (3) and (4),

$$\frac{x - X}{y - Y} = \frac{1}{2} \tag{7}$$

Thus

$$\frac{1}{2} = \frac{y}{x} \quad \text{and} \quad x = 2y$$

Substitute this result in equation (5). Then $2y^2 = 1$, so $y = \pm\dfrac{1}{\sqrt{2}}$. We obtain

Case (i) $x = \sqrt{2}$ $y = \dfrac{1}{\sqrt{2}}$

or Case (ii) $x = -\sqrt{2}$ $y = -\dfrac{1}{\sqrt{2}}$

In case (i), we substitute in equation (7) and obtain that

$$2(\sqrt{2} - X) = \frac{1}{\sqrt{2}} - Y$$

$$2X - Y = \frac{3}{\sqrt{2}}$$

$$Y = 2X - \frac{3}{\sqrt{2}}$$

Substituting in equation (6)

$$X + 4X - 6/\sqrt{2} = 1$$

$$5X = 1 + 3\sqrt{2}$$

$$X = \frac{1}{5}(1 + 3\sqrt{2})$$

Hence $Y = \frac{1}{2}(1 - X) = \frac{1}{2}\left(1 - \frac{1}{5} - \frac{3}{5}\sqrt{2}\right) = \frac{1}{10}(4 - 3\sqrt{2})$.

Case (i) therefore gives rise to the constrained stationary point

$$(x, y, X, Y)^{\mathrm{T}} = \left(\sqrt{2}, \frac{1}{\sqrt{2}}, \frac{1}{5}(1 + 3\sqrt{2}), \frac{1}{10}(4 - 3\sqrt{2})\right)^{\mathrm{T}}$$

Case (ii) gives rise to a second constrained stationary point but we need not calculate this since case (i) is clearly the result we are seeking.

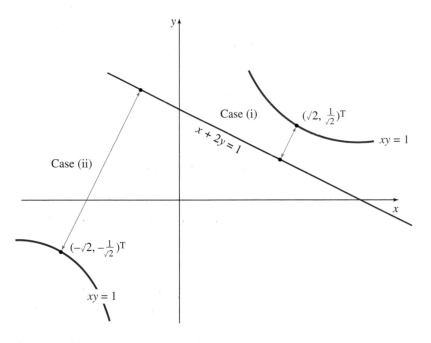

The required minimum is therefore

$$\left\{\sqrt{2} - \frac{1}{5}(1 + 3\sqrt{2})\right\}^2 + \left\{\frac{1}{\sqrt{2}} - \frac{1}{10}(4 - 3\sqrt{2})\right\}^2$$

The minimum distance is the square root of this.

9. The Lagrangian is

$$L = xyz + \lambda(x + 2y + 3z - 1) + \mu(3x + 2y + z - 1)$$

and so we have to solve the simultaneous equations

$$0 = \frac{\partial L}{\partial x} = yz + \lambda + 3\mu \tag{8}$$

$$0 = \frac{\partial L}{\partial y} = xz + 2\lambda + 2\mu \tag{9}$$

$$0 = \frac{\partial L}{\partial z} = xy + 3\lambda + \mu \tag{10}$$

$$0 = \frac{\partial L}{\partial \lambda} = x + 2y + 3z - 1 \tag{11}$$

$$0 = \frac{\partial L}{\partial \mu} = 3x + 2y + z - 1 \tag{12}$$

These equations can be solved without much difficulty but Lagrange's method is actually not the simplest way to proceed for this problem. It is easier to use the method of §6.6.

We begin by solving the constraint equations for x and z in terms of y. This yields

$$4x = 4z = 1 - 2y \tag{13}$$

Substituting for x and z in the expression xyz, we then seek the *unconstrained* stationary points for

$$\phi(y) = (1 - 2y)^2 y/16$$

These unconstrained stationary points occur where $\phi'(y) = 0$. We therefore require the values of y for which

$$2(1 - 2y)(-2)y + (1 - 2y)^2 = 0$$

Thus $y = \dfrac{1}{2}$ or $-4y + 1 - 2y = 0$, *i.e.* $y = \dfrac{1}{2}$ or $y = \dfrac{1}{6}$.

Case (i). $y = \dfrac{1}{2}$. From (13), we obtain that $\left(0, \dfrac{1}{2}, 0\right)^{\mathrm{T}}$ is a constrained stationary point for the original problem. Note that

$$f\left(0, \frac{1}{2}, 0\right) = 0.$$

Case (ii). $y = \dfrac{1}{6}$. From (13), we obtain that $\left(\dfrac{1}{6}, \dfrac{1}{6}, \dfrac{1}{6}\right)^{\mathrm{T}}$ is a constrained stationary point for the original problem. Note that

$$f\left(\frac{1}{6}, \frac{1}{6}, \frac{1}{6}\right) = \frac{1}{216}$$

Observe that xyz is larger in case (ii) but that xyz is *not* maximised at $\left(\dfrac{1}{6}, \dfrac{1}{6}, \dfrac{1}{6}\right)^{\mathrm{T}}$ subject to the constraints. To see this, note that $\phi(y)$ can be made as large as we choose by making y sufficiently large and positive.

See §5.4.

10. $x = 6 \quad y = 16/5$

15. The stationary points are found by solving the equation

$$\frac{dy}{dx} = \mathbf{0}$$

Observe that, using rules (II) and (VI),

$$\begin{aligned}
\frac{dy}{dx} &= \frac{d}{dx}(A\mathbf{x} + \mathbf{a})^{\mathrm{T}}(B\mathbf{x} + \mathbf{b}) \\
&= (B\mathbf{x} + \mathbf{b})^{\mathrm{T}} A + (A\mathbf{x} + \mathbf{a})^{\mathrm{T}} B \\
&= (\mathbf{x}^{\mathrm{T}} B^{\mathrm{T}} + \mathbf{b}^{\mathrm{T}})A + (\mathbf{x}^{\mathrm{T}} A^{\mathrm{T}} + \mathbf{a}^{\mathrm{T}})B \\
&= \mathbf{x}^{\mathrm{T}}(B^{\mathrm{T}} A + A^{\mathrm{T}} B) + \mathbf{b}^{\mathrm{T}} A + \mathbf{a}^{\mathrm{T}} B
\end{aligned}$$

The required result is therefore

$$\mathbf{x}^{\mathrm{T}}(B^{\mathrm{T}} A + A^{\mathrm{T}} B) = -(\mathbf{b}^{\mathrm{T}} A + \mathbf{a}^{\mathrm{T}} B)$$

or, transposing,

$$(A^{\mathrm{T}} B + B^{\mathrm{T}} A)\mathbf{x} = -(A^{\mathrm{T}} \mathbf{b} + B^{\mathrm{T}} \mathbf{a})$$

17. The Lagrangian is

$$L = \mathbf{x}^{\mathrm{T}}\mathbf{x} + \lambda^{\mathrm{T}}(B\mathbf{x} - \mathbf{c})$$

The constrained stationary points are found by solving

$$\left. \begin{aligned}
\frac{\partial L}{\partial \mathbf{x}} &= \mathbf{0} \\
\frac{\partial L}{\partial \lambda} &= \mathbf{0}
\end{aligned} \right\}$$

Now

$$\frac{\partial L}{\partial \mathbf{x}} = 2\mathbf{x}^{\mathrm{T}} + \lambda^{\mathrm{T}} B$$

Since L is a scalar, $L = L^{\mathrm{T}} = \mathbf{x}^{\mathrm{T}}\mathbf{x} + (B\mathbf{x} - \mathbf{c})^{\mathrm{T}}\lambda$ and thus

$$\frac{\partial L}{\partial \lambda} = (B\mathbf{x} - \mathbf{c})^{\mathrm{T}}$$

After transposing these results, we obtain

$$2\mathbf{x} + B^{\mathrm{T}}\lambda = \mathbf{0} \tag{14}$$

$$B\mathbf{x} - \mathbf{c} = \mathbf{0} \tag{15}$$

As always, the second equation is just the constraint equation.
Multiplying through (14) by B, we obtain that

$$2B\mathbf{x} + BB^{\mathrm{T}}\lambda = \mathbf{0}$$

Substituting from (15) yields that

$$2\mathbf{c} + BB^{\mathrm{T}}\lambda = \mathbf{0}$$

$$\lambda = -2(BB^{\mathrm{T}})^{-1}\mathbf{c}$$

This result may be substituted back in (14),

$$2\mathbf{x} - 2B^\mathsf{T}(BB^\mathsf{T})^{-1}\mathbf{c} = \mathbf{0}$$

and so the problem has the unique solution

$$\mathbf{x} = B^\mathsf{T}(BB^\mathsf{T})^{-1}\mathbf{c} = B^{-1}\mathbf{c}$$

Chapter 7. Inverse functions

Exercises 7.6

1.

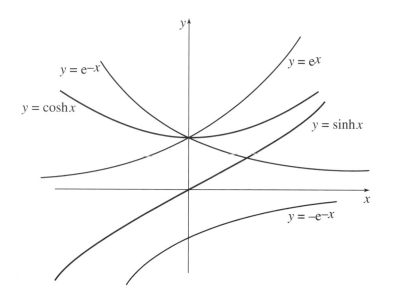

(i) $\dfrac{d}{dx}(\sinh x) = \dfrac{d}{dx}\left\{\dfrac{1}{2}(e^x - e^{-x})\right\} = \dfrac{1}{2}(e^x - (-1)e^{-x}) = \cosh x$

(ii) $\dfrac{d}{dx}(\cosh x) = \dfrac{d}{dx}\left\{\dfrac{1}{2}(e^x + e^{-x})\right\} = \dfrac{1}{2}(e^x - e^{-x}) = \sinh x$

2. The equation $y = \sinh x$ always has a unique solution for x in terms of y because $D(\sinh x)$ is always positive. *Note that $y = \cosh x$ does not always have a unique solution for x in terms of y.*

$$\cosh^2 x - \sinh^2 x = \frac{1}{4}(e^{2x} + 2 + e^{-2x}) - \frac{1}{4}(e^{2x} - 2 + e^{-2x})$$

$$= \frac{2}{4} + \frac{2}{4} = 1$$

Write $y = \sinh x$. Then $x = \sinh^{-1} y$. Hence

$$\frac{d}{dy}(\sinh^{-1} y) = \frac{dx}{dy} = \left(\frac{dy}{dx}\right)^{-1} = \frac{1}{\cosh x} = \frac{1}{\sqrt{1 + \sinh^2 x}} = \frac{1}{\sqrt{1 + y^2}}$$

(*We take the positive square root because we know that* $\sinh^{-1} y$ *has a positive derivative.*)

We have seen that the function $f: \mathbb{R} \to \mathbb{R}$ defined by $f(x) = \cosh x$ does *not* have an inverse function $g: \mathbb{R} \to \mathbb{R}$. To proceed in a similar way with $\cosh x$ it would be necessary to restrict attention to values of x and y satisfying $x > 0$ and $y > 1$.

4.
$$\frac{d}{dy} \ln\{y + \sqrt{y^2 + 1}\}$$

$$= \frac{1}{\{y + \sqrt{y^2 + 1}\}} \left\{ 1 + \frac{1}{2}(y^2 + 1)^{-1/2} 2y \right\} = \frac{1}{\sqrt{y^2 + 1}}$$

Observe that

$$\frac{d}{dy} \ln\{y + \sqrt{y^2 + 1}\} = \frac{d}{dy} \sinh^{-1} y$$

In fact,

$$\ln\{y + \sqrt{y^2 + 1}\} = \sinh^{-1} y$$

To verify this we consider

$$\sinh(\ln\{y + \sqrt{y^2 + 1}\}) = \frac{1}{2}\left(e^{\ln\left(y + \sqrt{y^2+1}\right)} - e^{-\ln\left(y + \sqrt{y^2+1}\right)} \right)$$

$$= \frac{1}{2}\left(\{y + \sqrt{y^2 + 1}\} - \frac{1}{\{y + \sqrt{y^2 + 1}\}} \right)$$

$$= \frac{y^2 + 2y\sqrt{y^2 + 1} + (y^2 + 1) - 1}{2(y + \sqrt{y^2 + 1})}$$

$$= \frac{2y(y + \sqrt{y^2 + 1})}{2(y + \sqrt{y^2 + 1})} = y$$

6.

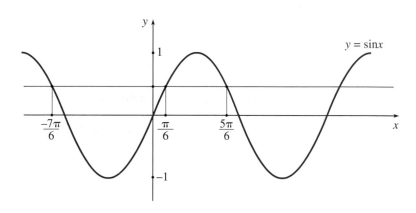

Sketched below are three appropriate local inverses

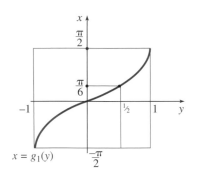

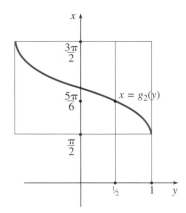

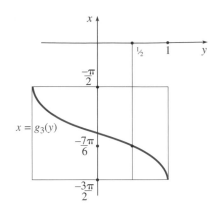

(i) $f_1^{-1} : [-1, 1] \rightarrow [-\pi/2, \pi/2]$

(ii) $f_2^{-1} : [-1, 1] \rightarrow [\pi/2, 3\pi/2]$

(iii) $f_3^{-1} : [-1, 1] \rightarrow [-\pi/2, -3\pi/2]$

The required formulas are

(i) $x = f_1^{-1}(y) = \arcsin y$

(ii) $x = f_2^{-1}(y) = \pi - \arcsin y$

(iii) $x = f_3^{-1}(y) = -\pi - \arcsin y$

We have

(i) $\dfrac{d}{dy}\left(f_1^{-1}(y)\right) = \dfrac{1}{\sqrt{1 - y^2}}$ $\quad (-1 < y < 1)$

Thus

$$\frac{d}{dy}\left(f_1^{-1}\left(\frac{1}{2}\right)\right) = \frac{2}{\sqrt{3}}$$

(ii) $\dfrac{d}{dy}\left(f_2^{-1}(y)\right) = \dfrac{d}{dy}(\pi - \arcsin y) = -\dfrac{1}{\sqrt{1 - y^2}}$ $\quad (-1 < y < 1)$

Thus

$$\frac{d}{dy}\left(f_1^{-1}\left(\frac{1}{2}\right)\right) = -\frac{2}{\sqrt{3}}$$

(iii) $\dfrac{d}{dy}\left(f_3^{-1}(y)\right) = \dfrac{d}{dy}(-\pi - \arcsin y) = -\dfrac{1}{\sqrt{1 - y^2}}$ $\quad (-1 < y < 1)$

Thus

$$\frac{d}{dy}\left(f_1^{-1}\left(\frac{1}{2}\right)\right) = -\frac{2}{\sqrt{3}}$$

If $y = \sin x$, then $dy/dx = \cos x$. Hence the formula

$$\frac{dx}{dy} = \left(\frac{dy}{dx}\right)^{-1}$$

reduces to

$$\frac{dx}{dy} = \frac{1}{\cos x}$$

Observe that

(i) $\quad \left(\cos \dfrac{\pi}{6}\right)^{-1} = \left(\dfrac{\sqrt{3}}{2}\right)^{-1} = \dfrac{d}{dy}\left(f_1^{-1}\left(\dfrac{1}{2}\right)\right)$

(ii) $\quad \left(\cos \dfrac{5\pi}{6}\right)^{-1} = \left(\cos\left(\pi - \dfrac{\pi}{6}\right)\right)^{-1} = \left(\dfrac{-\sqrt{3}}{2}\right)^{-1} = \dfrac{d}{dy}\left(f_2^{-1}\left(\dfrac{1}{2}\right)\right)$

(iii) $\quad \left(\cos\left(\dfrac{-\pi}{6}\right)\right)^{-1} = \left(\cos\left(-\pi - \dfrac{\pi}{6}\right)\right)^{-1} = \left(\dfrac{-\sqrt{3}}{2}\right)^{-1} = \dfrac{d}{dy}\left(f_3^{-1}\left(\dfrac{1}{2}\right)\right)$

No local inverse exists at $x = \pi/2$ because this is a critical point – i.e. $f'(\pi/2) = \cos(\pi/2) = 0$.

9. The derivative of the function $f: \mathbb{R}^2 \to \mathbb{R}^2$ defined by

$$\left.\begin{array}{l} u = xe^y \\ v = xe^{-y} \end{array}\right\}$$

is the matrix

$$\begin{pmatrix} \dfrac{\partial u}{\partial x} & \dfrac{\partial u}{\partial y} \\[2mm] \dfrac{\partial v}{\partial x} & \dfrac{\partial v}{\partial y} \end{pmatrix} = \begin{pmatrix} e^y & xe^y \\ e^{-y} & -xe^{-y} \end{pmatrix} \tag{1}$$

Local inverses exist except at critical points – i.e. where

$$0 = \begin{vmatrix} \dfrac{\partial u}{\partial x} & \dfrac{\partial u}{\partial y} \\[2mm] \dfrac{\partial v}{\partial x} & \dfrac{\partial v}{\partial y} \end{vmatrix} = \begin{vmatrix} e^y & xe^y \\ e^{-y} & -xe^{-y} \end{vmatrix} = -2x$$

Thus local inverses exists at all points $(x, y)^{\mathrm{T}}$ with $x \neq 0$. The derivative of a local inverse is given by

$$\begin{pmatrix} \dfrac{\partial x}{\partial u} & \dfrac{\partial x}{\partial v} \\[2mm] \dfrac{\partial y}{\partial u} & \dfrac{\partial y}{\partial v} \end{pmatrix} = \begin{pmatrix} \dfrac{\partial u}{\partial x} & \dfrac{\partial u}{\partial y} \\[2mm] \dfrac{\partial v}{\partial x} & \dfrac{\partial v}{\partial y} \end{pmatrix}^{-1} \tag{2}$$

$$= \begin{pmatrix} e^y & xe^y \\ e^{-y} & -xe^{-y} \end{pmatrix}^{-1} = \dfrac{1}{-2x}\begin{pmatrix} -xe^{-y} & -xe^y \\ -e^{-y} & e^y \end{pmatrix}$$

From (1) we observe that

$$\left(\dfrac{\partial u}{\partial x}\right)_y = e^y$$

and from (2) that

$$\left(\dfrac{\partial x}{\partial u}\right)_v = \dfrac{1}{2}e^{-y}$$

10. (i) $xy = \dfrac{1}{4}$

(ii) $\begin{pmatrix} u - 10 \\ v + 2 \end{pmatrix} = \begin{pmatrix} -1 & 6 \\ -2 & -1 \end{pmatrix}\begin{pmatrix} x + 1 \\ y - 3 \end{pmatrix}$

(iii) $y^2 - x = -\frac{1}{4}$

$x^2 - y = -\frac{1}{4}$

At $(\frac{1}{2}, \frac{1}{2})^{\mathrm{T}}$, $\nabla\mathbf{u} = \begin{pmatrix} -1 \\ 1 \end{pmatrix}$, $\nabla\mathbf{v} = \begin{pmatrix} 1 \\ -1 \end{pmatrix}$ which are parallel, so the contours touch at this point.

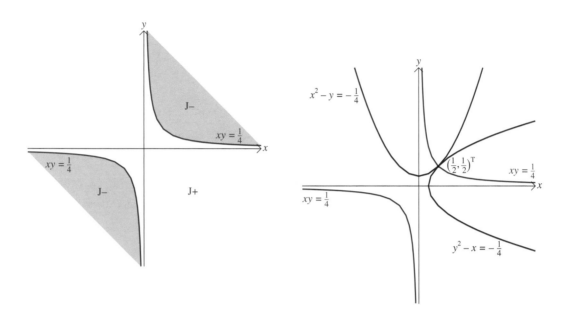

12.

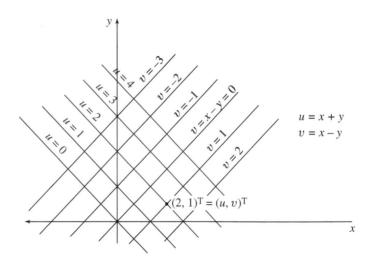

$$\frac{\partial}{\partial x} = \frac{\partial u}{\partial x}\frac{\partial}{\partial u} + \frac{\partial v}{\partial x}\frac{\partial}{\partial v}$$

$$= \frac{\partial}{\partial u} + \frac{\partial}{\partial v}$$

Hence

$$\frac{\partial^2 z}{\partial x^2} = \frac{\partial}{\partial x}\left(\frac{\partial z}{\partial u} + \frac{\partial z}{\partial v}\right) = \frac{\partial}{\partial u}\left(\frac{\partial z}{\partial u} + \frac{\partial z}{\partial v}\right) + \frac{\partial}{\partial v}\left(\frac{\partial z}{\partial u} + \frac{\partial z}{\partial v}\right)$$

$$= \frac{\partial^2 z}{\partial u^2} + 2\frac{\partial^2 z}{\partial u \partial v} + \frac{\partial^2 z}{\partial v^2}$$

14. $\dfrac{\partial}{\partial x} = \dfrac{\partial u}{\partial x}\dfrac{\partial}{\partial u} + \dfrac{\partial v}{\partial x}\dfrac{\partial}{\partial v}$ and $\dfrac{\partial}{\partial y} = \dfrac{\partial u}{\partial y}\dfrac{\partial}{\partial u} + \dfrac{\partial v}{\partial y}\dfrac{\partial}{\partial v}$

$$\begin{pmatrix} \dfrac{\partial u}{\partial x} & \dfrac{\partial u}{\partial y} \\ \dfrac{\partial v}{\partial x} & \dfrac{\partial v}{\partial y} \end{pmatrix} = \begin{pmatrix} \dfrac{\partial x}{\partial u} & \dfrac{\partial x}{\partial v} \\ \dfrac{\partial y}{\partial u} & \dfrac{\partial y}{\partial v} \end{pmatrix}^{-1} = \begin{pmatrix} 1 & 1 \\ 1 & -1 \end{pmatrix}^{-1} = -\frac{1}{2}\begin{pmatrix} -1 & -1 \\ -1 & 1 \end{pmatrix}$$

Hence

$$\frac{\partial}{\partial x} = \frac{1}{2}\frac{\partial}{\partial u} + \frac{1}{2}\frac{\partial}{\partial v}$$

$$\frac{\partial}{\partial y} = \frac{1}{2}\frac{\partial}{\partial u} - \frac{1}{2}\frac{\partial}{\partial v}$$

These equations can also, of course, be obtained by observing that

$$\left.\begin{aligned} u &= \tfrac{1}{2}x + \tfrac{1}{2}y \\ v &= \tfrac{1}{2}x - \tfrac{1}{2}y \end{aligned}\right\}$$

$$\frac{\partial^2 z}{\partial x^2} = \frac{\partial}{\partial x}\left(\frac{\partial z}{\partial x}\right) = \frac{\partial}{\partial x}\left(\frac{1}{2}\frac{\partial z}{\partial u} + \frac{1}{2}\frac{\partial z}{\partial v}\right)$$

$$= \frac{1}{4}\frac{\partial^2 z}{\partial u^2} + \frac{1}{2}\frac{\partial^2 z}{\partial u \partial v} + \frac{1}{4}\frac{\partial^2 z}{\partial v^2}$$

$$\frac{\partial^2 z}{\partial y^2} = \frac{\partial}{\partial y}\left(\frac{\partial z}{\partial y}\right) = \frac{\partial}{\partial y}\left(\frac{1}{2}\frac{\partial z}{\partial u} - \frac{1}{2}\frac{\partial z}{\partial v}\right)$$

$$= \frac{1}{4}\frac{\partial^2 z}{\partial u^2} - \frac{1}{2}\frac{\partial^2 z}{\partial u \partial v} + \frac{1}{4}\frac{\partial^2 z}{\partial v^2}$$

Thus

$$\frac{\partial^2 z}{\partial x^2} - \frac{\partial^2 z}{\partial y^2} = \frac{\partial^2 z}{\partial u \partial v}$$

16.

$$\frac{\partial}{\partial r} = \frac{\partial x}{\partial r}\frac{\partial}{\partial x} + \frac{\partial y}{\partial r}\frac{\partial}{\partial y}$$

$$= \cos\theta\frac{\partial}{\partial x} + \sin\theta\frac{\partial}{\partial y}$$

$$\frac{\partial}{\partial \theta} = \frac{\partial x}{\partial \theta}\frac{\partial}{\partial x} + \frac{\partial y}{\partial \theta}\frac{\partial}{\partial y}$$

$$= -r\sin\theta\frac{\partial}{\partial x} + r\cos\theta\frac{\partial}{\partial y}$$

It follows that

$$\frac{\partial f}{\partial r} = \cos\theta\frac{\partial f}{\partial x} + \sin\theta\frac{\partial f}{\partial y} = \langle \nabla f, \mathbf{e}_1 \rangle$$

$$\frac{1}{r}\frac{\partial f}{\partial \theta} = -\sin\theta\frac{\partial f}{\partial x} + \cos\theta\frac{\partial f}{\partial y} = \langle \nabla f, \mathbf{e}_2 \rangle$$

where $e_1 = (\cos\theta, \sin\theta)^T$ and $e_2 = (-\sin\theta, \cos\theta)^T$ are the unit vectors indicated below.

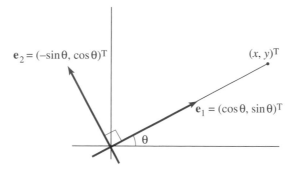

Observe that $\langle \nabla f, e_1 \rangle$ is the directional derivative of f in the e_1 direction.

Thus $\dfrac{\partial f}{\partial r}$ is the rate at which f increases at $(x, y)^T$ in the *radial* direction.

Similarly $\dfrac{1}{r}\dfrac{\partial f}{\partial \theta}$ is the rate at which f increases at $(x, y)^T$ in the *tangential* direction.

19. Introduce polar coordinates into the equation

$$\frac{\partial^2 z}{\partial x^2} + \frac{\partial^2 z}{\partial y^2} = 0$$

We have

$$\frac{\partial}{\partial x} = \frac{\partial r}{\partial x}\frac{\partial}{\partial r} + \frac{\partial \theta}{\partial x}\frac{\partial}{\partial \theta}$$

$$\frac{\partial}{\partial y} = \frac{\partial r}{\partial y}\frac{\partial}{\partial r} + \frac{\partial \theta}{\partial y}\frac{\partial}{\partial \theta}$$

Recall that

$$\left.\begin{array}{l} x = r\cos\theta \\ y = r\sin\theta \end{array}\right\}$$

and hence

$$\begin{pmatrix} \dfrac{\partial r}{\partial x} & \dfrac{\partial r}{\partial y} \\[2mm] \dfrac{\partial \theta}{\partial x} & \dfrac{\partial \theta}{\partial y} \end{pmatrix} = \begin{pmatrix} \dfrac{\partial x}{\partial r} & \dfrac{\partial x}{\partial \theta} \\[2mm] \dfrac{\partial y}{\partial r} & \dfrac{\partial y}{\partial \theta} \end{pmatrix}^{-1} = \begin{pmatrix} \cos\theta & -r\sin\theta \\ \sin\theta & r\cos\theta \end{pmatrix}^{-1}$$

It follows that

$$\begin{pmatrix} \dfrac{\partial r}{\partial x} & \dfrac{\partial r}{\partial y} \\[2mm] \dfrac{\partial \theta}{\partial x} & \dfrac{\partial \theta}{\partial y} \end{pmatrix} = \frac{1}{r(\cos^2\theta + \sin^2\theta)} \begin{pmatrix} r\cos\theta & r\sin\theta \\ -\sin\theta & \cos\theta \end{pmatrix}$$

$$= \begin{pmatrix} \cos\theta & \sin\theta \\[2mm] -\dfrac{\sin\theta}{r} & \dfrac{\cos\theta}{r} \end{pmatrix}$$

Thus

$$\frac{\partial}{\partial x} = \cos\theta \frac{\partial}{\partial r} - \frac{\sin\theta}{r}\frac{\partial}{\partial\theta}$$

$$\frac{\partial}{\partial y} = \sin\theta \frac{\partial}{\partial r} + \frac{\cos\theta}{r}\frac{\partial}{\partial\theta}$$

We conclude that

$$\frac{\partial^2 z}{\partial x^2} = \frac{\partial}{\partial x}\left(\frac{\partial z}{\partial x}\right) = \left(\cos\theta\frac{\partial}{\partial r} - \frac{\sin\theta}{r}\frac{\partial}{\partial\theta}\right)\left(\cos\theta\frac{\partial z}{\partial r} - \frac{\sin\theta}{r}\frac{\partial z}{\partial\theta}\right)$$

$$= \cos^2\theta \frac{\partial^2 z}{\partial r^2} + \frac{\cos\theta\sin\theta}{r^2}\frac{\partial z}{\partial\theta} - \frac{\cos\theta\sin\theta}{r}\frac{\partial^2 z}{\partial r\,\partial\theta} + \frac{\sin^2\theta}{r}\frac{\partial z}{\partial r}$$

$$- \frac{\sin\theta\cos\theta}{r}\frac{\partial^2 z}{\partial\theta\,\partial r} + \frac{\sin\theta\cos\theta}{r^2}\frac{\partial z}{\partial\theta} + \frac{\sin^2\theta}{r^2}\frac{\partial^2 z}{\partial\theta^2}$$

$$\frac{\partial^2 z}{\partial y^2} = \frac{\partial}{\partial y}\left(\frac{\partial z}{\partial y}\right) = \left(\sin\theta\frac{\partial}{\partial r} + \frac{\cos\theta}{r}\frac{\partial}{\partial\theta}\right)\left(\sin\theta\frac{\partial z}{\partial r} + \frac{\cos\theta}{r}\frac{\partial z}{\partial\theta}\right)$$

$$= \sin^2\theta \frac{\partial^2 z}{\partial r^2} - \frac{\sin\theta\cos\theta}{r^2}\frac{\partial z}{\partial\theta} + \frac{\sin\theta\cos\theta}{r}\frac{\partial^2 z}{\partial r\,\partial\theta} + \frac{\cos^2\theta}{r}\frac{\partial z}{\partial r}$$

$$+ \frac{\cos\theta\sin\theta}{r}\frac{\partial^2 z}{\partial\theta\,\partial r} - \frac{\cos\theta\sin\theta}{r^2}\frac{\partial z}{\partial\theta} + \frac{\cos^2\theta}{r^2}\frac{\partial^2 z}{\partial\theta^2}$$

Hence the given partial differential equation reduces to

$$\frac{\partial^2 z}{\partial r^2} + \frac{1}{r}\frac{\partial z}{\partial r} + \frac{1}{r^2}\frac{\partial^2 z}{\partial\theta^2} = 0$$

Chapter 8. Implicit functions

Exercise 8.4

1.   Write $\phi(x, y) = y^2 - x^2(1 - x^2)$. Then $\dfrac{\partial\phi}{\partial y} = 2y$.

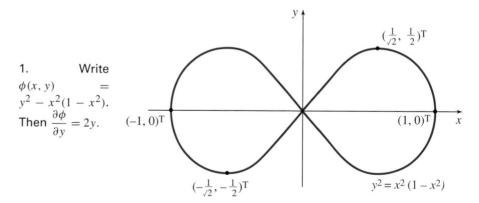

This is nonzero for $(x, y)^{\mathrm{T}} = (\xi, \eta)^{\mathrm{T}}$ if and only if $\eta \neq 0$. Thus a unique local differentiable solution $y = g(x)$ exists at each $(\xi, \eta)^{\mathrm{T}}$ satisfying $\eta^2 = \xi^2(1 - \xi^2)$ when $\eta \neq 0$. Also

$$g'(x) = \frac{dy}{dx} = -\left(\frac{\partial\phi}{\partial y}\right)^{-1}\left(\frac{\partial\phi}{\partial x}\right)$$

$$= -\frac{1}{2y}(-2x(1-x^2) - x^2(-2x))$$

$$= \frac{x}{y}(1 - 2x^2)$$

The local solution $y = g_1(x)$ at $(1/\sqrt{2}, 1/2)^T$ is given by

$$y = g_1(x) = \sqrt{(x^2(1-x^2))} \quad (0 \le x \le 1)$$

and the local solution $y = g_2(x)$ at $(-1/\sqrt{2}, -1/2)^T$ is given by

$$y = g_2(x) = -\sqrt{(x^2(1-x^2))} \quad (-1 \le x \le 0)$$

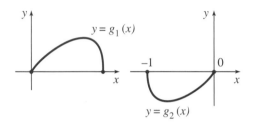

3. The point $(x, y)^T = (-1, 1)^T$ lies on the curve $\sin(\pi x^2 y) = 1 + xy^2$ because

$$\sin(\pi(-1)^2 1) = \sin \pi = 0$$
$$1 + (-1)(1)^2 = 1 - 1 = 0$$

Write $\phi(x, y) = \sin(\pi x^2 y) - 1 - xy^2$. Then

$$\frac{\partial \phi}{\partial y} = \pi x^2 \cos(\pi x^2 y) - 2xy$$
$$= -\pi + 2 \ne 0$$

when $(x, y)^T = (-1, 1)^T$.

Thus a unique local differentiable solution $y = g(x)$ exists at $(x, y)^T = (-1, 1)^T$ and

$$g'(x) = \frac{dy}{dx} = -\left(\frac{\partial \phi}{\partial y}\right)^{-1}\left(\frac{\partial \phi}{\partial x}\right)$$
$$= \frac{-1}{\pi x^2 \cos(\pi x^2 y) - 2xy}(2\pi xy \cos(\pi x^2 y) - y^2)$$

Thus

$$g'(-1) = \frac{-1}{2 - \pi}(2\pi - 1) = \frac{2\pi - 1}{\pi - 2}$$

5.

$$\frac{\partial z}{\partial x} = -\frac{6z^2 x^2 + yz}{5z^4 + 4zx^3 + y^4 + xy}$$

$$\frac{\partial z}{\partial y} = -\frac{4zy^3 + xz}{5z^4 + 4zx^3 + y^4 + xy}$$

7(i) Indifference curve: $x + y^{1/2} = c$

Write $\phi = x + y^{1/2} - c$ and differentiate implicitly to give

$$1 + \frac{1}{2} y^{-1/2} \frac{dy}{dx} = 0$$

so

$$\frac{dy}{dx} = -2y^{1/2} < 0$$

and so the curve slopes downwards. Differentiating again,

$$\frac{d^2 y}{dx^2} = -y^{-1/2} \frac{dy}{dx}$$
$$= -y^{-1/2}(-2y^{1/2}) = 2 > 0$$

Hence the curve is convex.

8. Write

$$\phi(x, y, z) = \sin(yz) + \sin(zx) + \sin(xy)$$

Then

$$\frac{\partial \phi}{\partial z} = y \cos(yz) + x \cos(zx)$$

We need that $\phi(\xi, \eta, \zeta) = 0$ but $\phi_z(\xi, \eta, \zeta) \neq 0$. An appropriate choice is $(\xi, \eta, \zeta)^T = (0, \pi, 1)^T$. At this point a unique local differentiable solution $z = g(x, y)$ exists. We have

$$g'(x, y) = \left(\frac{\partial z}{\partial x}, \frac{\partial z}{\partial y} \right) = -\left(\frac{\partial \phi}{\partial z} \right)^{-1} \left(\frac{\partial \phi}{\partial x}, \frac{\partial \phi}{\partial y} \right)$$

$$= \frac{-1}{y \cos(yz) + x \cos(zx)} (z \cos(zx) + y \cos(xy), z \cos(yz) + x \cos(xy))$$

Thus

$$g'(0, \pi) = \frac{-1}{-\pi}(1 + \pi, -1) = \left(1 + \frac{1}{\pi}, -\frac{1}{\pi} \right)$$

10. Write

$$\phi(x, y, z) = \begin{pmatrix} \phi_1(x, y, z) \\ \phi_2(x, y, z) \end{pmatrix} = \begin{pmatrix} xy^2 + y^2 z^3 + z^4 x^5 - 1 \\ zx^2 + x^2 y^3 + y^4 z^5 + 1 \end{pmatrix}$$

Then

$$\det \begin{pmatrix} \dfrac{\partial \phi_1}{\partial x} & \dfrac{\partial \phi_1}{\partial y} \\ \dfrac{\partial \phi_2}{\partial x} & \dfrac{\partial \phi_2}{\partial y} \end{pmatrix} = \begin{vmatrix} y^2 + 5x^4 z^4 & 2xy + 2yz^3 \\ 2zx + 2xy^3 & 3x^2 y^2 + 4y^3 z^5 \end{vmatrix}$$

$$= \begin{vmatrix} 6 & 0 \\ 0 & -1 \end{vmatrix} = -6$$

when $(x, y, z)^T = (1, 1, -1)^T$. It follows that a unique local differentiable solution

$$\begin{pmatrix} x \\ y \end{pmatrix} = \mathbf{g}(z) = \begin{pmatrix} g_1(z) \\ g_2(z) \end{pmatrix}$$

exists at $(x, y, z)^T = (1, 1, -1)^T$ and

$$\mathbf{g}'(z) = \begin{pmatrix} \dfrac{dx}{dz} \\ \dfrac{dy}{dz} \end{pmatrix} = - \begin{pmatrix} \dfrac{\partial \phi_1}{\partial x} & \dfrac{\partial \phi_1}{\partial y} \\ \dfrac{\partial \phi_2}{\partial x} & \dfrac{\partial \phi_2}{\partial y} \end{pmatrix}^{-1} \begin{pmatrix} \dfrac{\partial \phi_1}{\partial z} \\ \dfrac{\partial \phi_2}{\partial z} \end{pmatrix}$$

$$= - \begin{pmatrix} y^2 + 5x^4 z^4 & 2xy + 2yz^3 \\ 2zx + 2xy^3 & 3x^2 y^2 + 4y^3 z^5 \end{pmatrix}^{-1} \begin{pmatrix} 3y^2 z^2 + 4z^3 x^5 \\ x^2 + 5y^4 z^4 \end{pmatrix}$$

Thus

$$\mathbf{g}'(-1) = - \begin{pmatrix} 6 & 0 \\ 0 & -1 \end{pmatrix}^{-1} \begin{pmatrix} -1 \\ 6 \end{pmatrix}$$

$$= \frac{1}{6} \begin{pmatrix} -1 & 0 \\ 0 & 6 \end{pmatrix} \begin{pmatrix} -1 \\ 6 \end{pmatrix} = \frac{1}{6} \begin{pmatrix} 1 \\ 36 \end{pmatrix} = \begin{pmatrix} \frac{1}{6} \\ 6 \end{pmatrix}$$

12. Write

$$\left. \begin{aligned} \phi_1 &= x^2 + u + e^v \\ \phi_2 &= y^2 + v + e^w \\ \phi_3 &= z^2 + w + e^u \end{aligned} \right\}$$

We need to calculate

$$\det \begin{pmatrix} \dfrac{\partial \phi_1}{\partial u} & \dfrac{\partial \phi_1}{\partial v} & \dfrac{\partial \phi_1}{\partial w} \\ \dfrac{\partial \phi_2}{\partial u} & \dfrac{\partial \phi_2}{\partial v} & \dfrac{\partial \phi_2}{\partial w} \\ \dfrac{\partial \phi_3}{\partial u} & \dfrac{\partial \phi_3}{\partial v} & \dfrac{\partial \phi_3}{\partial w} \end{pmatrix} = \begin{vmatrix} 1 & e^v & 0 \\ 0 & 1 & e^w \\ e^u & 0 & 1 \end{vmatrix} = 1 + e^{u+v+w}$$

Since this determinant is never zero, a unique local differentiable solution

$$\begin{pmatrix} u \\ v \\ w \end{pmatrix} = \mathbf{g}(x, y, z) = \begin{pmatrix} g_1(x, y, z) \\ g_2(x, y, z) \\ g_3(x, y, z) \end{pmatrix}$$

exists at every point $(x, y, z, u, v, w)^T$ which satisfies the equations. Also

$$\mathbf{g}'(x, y, z) = \begin{pmatrix} \dfrac{\partial u}{\partial x} & \dfrac{\partial u}{\partial y} & \dfrac{\partial u}{\partial z} \\ \dfrac{\partial v}{\partial x} & \dfrac{\partial v}{\partial y} & \dfrac{\partial v}{\partial z} \\ \dfrac{\partial w}{\partial x} & \dfrac{\partial w}{\partial y} & \dfrac{\partial w}{\partial z} \end{pmatrix}$$

$$= - \begin{pmatrix} \dfrac{\partial \phi_1}{\partial u} & \dfrac{\partial \phi_1}{\partial v} & \dfrac{\partial \phi_1}{\partial w} \\ \dfrac{\partial \phi_2}{\partial u} & \dfrac{\partial \phi_2}{\partial v} & \dfrac{\partial \phi_2}{\partial w} \\ \dfrac{\partial \phi_3}{\partial u} & \dfrac{\partial \phi_3}{\partial v} & \dfrac{\partial \phi_3}{\partial w} \end{pmatrix}^{-1} \begin{pmatrix} \dfrac{\partial \phi_1}{\partial x} & \dfrac{\partial \phi_1}{\partial y} & \dfrac{\partial \phi_1}{\partial z} \\ \dfrac{\partial \phi_2}{\partial x} & \dfrac{\partial \phi_2}{\partial y} & \dfrac{\partial \phi_2}{\partial z} \\ \dfrac{\partial \phi_3}{\partial x} & \dfrac{\partial \phi_3}{\partial y} & \dfrac{\partial \phi_3}{\partial z} \end{pmatrix}$$

$$= -\begin{pmatrix} 1 & e^v & 0 \\ 0 & 1 & e^w \\ e^u & 0 & 1 \end{pmatrix}^{-1} \begin{pmatrix} 2x & 0 & 0 \\ 0 & 2y & 0 \\ 0 & 0 & 2z \end{pmatrix}$$

$$= -\frac{1}{1 + e^{u+v+w}} \begin{pmatrix} 1 & -e^v & e^{v+w} \\ e^{u+w} & 1 & -e^w \\ -e^u & e^{u+v} & 1 \end{pmatrix} \begin{pmatrix} 2x & 0 & 0 \\ 0 & 2y & 0 \\ 0 & 0 & 2z \end{pmatrix}$$

$$= -\frac{1}{1 + e^{u+v+w}} \begin{pmatrix} 2x & -2ye^v & 2ze^{v+w} \\ 2xe^{u+w} & 2y & -2ze^w \\ -2xe^u & 2ye^{u+v} & 2z \end{pmatrix}$$

Chapter 9. Differentials

Exercise 9.5

2. $504.71

3. We have

$$f_x\, dx + f_y\, dy + f_z\, dz = 0$$
$$g_x\, dx + g_y\, dy + g_z\, dz = 0$$

Also

$$dy = h'(x)\, dx$$

One can eliminate dz from the first two equations and compare the result with the third equation. Alternatively, one can observe that the condition for the three linear equations to have a nontrivial solution for dx, dy and dz is

$$\begin{vmatrix} f_x & f_y & f_z \\ g_x & g_y & g_z \\ h'(x) & -1 & 0 \end{vmatrix} = 0$$

Hence

$$h'(x)\begin{vmatrix} f_y & f_z \\ g_y & g_z \end{vmatrix} - (-1)\begin{vmatrix} f_x & f_z \\ g_x & g_z \end{vmatrix} + 0 \begin{vmatrix} f_x & f_y \\ g_x & g_y \end{vmatrix} = 0$$

and so

$$h'(x) = -\begin{vmatrix} f_x & f_z \\ g_x & g_z \end{vmatrix} \bigg/ \begin{vmatrix} f_y & f_z \\ g_y & g_z \end{vmatrix} = \frac{f_z g_x - f_x g_z}{f_y g_z - f_z g_y}$$

5. We have

$$dx = u\, du - v\, dv$$
$$dy = v\, du + u\, dv$$

and hence

$$du = \left(\frac{u}{u^2 + v^2}\right) dx + \left(\frac{v}{u^2 + v^2}\right) dy$$

But

$$du = \left(\frac{\partial u}{\partial x}\right)_y dx + \left(\frac{\partial u}{\partial y}\right)_x dy$$

Thus

$$\left(\frac{\partial u}{\partial x}\right)_y = \frac{u}{u^2 + v^2} \qquad \left(\frac{\partial u}{\partial y}\right)_x = \frac{v}{u^2 + v^2}$$

7. There are functions $k\colon \mathbb{R}^2 \to \mathbb{R}, l\colon \mathbb{R}^2 \to \mathbb{R}, m\colon \mathbb{R}^2 \to \mathbb{R}$ and $n\colon \mathbb{R}^2 \to \mathbb{R}$ such that

$$T = k(u, p) = l(p, v)$$

and

$$p = m(T, v) = n(T, u)$$

Hence

$$dT = \left(\frac{\partial T}{\partial u}\right)_p du + \left(\frac{\partial T}{\partial p}\right)_u dp$$

$$dT = \left(\frac{\partial T}{\partial p}\right)_v dp + \left(\frac{\partial T}{\partial v}\right)_p dv$$

$$dp = \left(\frac{\partial p}{\partial T}\right)_v dT + \left(\frac{\partial p}{\partial v}\right)_T dv$$

$$dp = \left(\frac{\partial p}{\partial T}\right)_u dT + \left(\frac{\partial p}{\partial u}\right)_T du$$

The condition that this system of linear equations have a nontrivial solution for du, dT, dp and dv is that the determinant given in the question is zero.

9. We have

$$dx = u\,du - v\,dv$$
$$dv = \frac{u}{v}du - \frac{1}{v}dx$$

Hence

$$\delta v \approx \frac{u}{v}\delta u - \frac{1}{v}\delta x$$

Chapter 10. Sums and integrals

Exercise 10.10

1. Pascal's triangle as far as the relevant row is given below.

$$
\begin{array}{ccccccccccccccccc}
&&&&&&&& 1 \\
&&&&&&& 1 && 1 \\
&&&&&& 1 && 2 && 1 \\
&&&&& 1 && 3 && 3 && 1 \\
&&&& 1 && 4 && 6 && 4 && 1 \\
&&& 1 && 5 && 10 && 10 && 5 && 1 \\
&& 1 && 6 && 15 && 20 && 15 && 6 && 1 \\
& 1 && 7 && 21 && 35 && 35 && 21 && 7 && 1 \\
1 && 8 && 28 && 56 && 70 && 56 && 28 && 8 && 1
\end{array}
$$

The required expansion is therefore

$$(1 + x)^8 = 1 + 8x + 28x^2 + 56x^3 + 70x^4 + 56x^5 + 28x^6 + 8x^7 + x^8$$

3.

The sum $\sum_{k=0}^{n-1} f\left(\dfrac{k}{n}\right)\dfrac{1}{n}$ rep-
resents the shaded area in the diagram and hence is an approximation to the area $\int_0^1 f(x)\,dx$

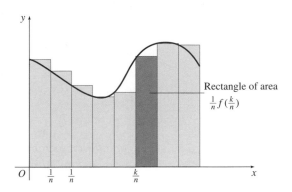

Rectangle of area
$\dfrac{1}{n}f(\tfrac{k}{n})$

This approximation improves as n increases (*provided that f is continuous*). In the limit, we have

$$\int_0^1 f(x)\,dx = \lim_{n\to\infty}\sum_{k=0}^{n-1} f\left(\frac{k}{n}\right)\frac{1}{n}$$

Applying this result in the case when $f(x)=e^x$, we obtain

$$\int_0^1 e^x\,dx = \lim_{n\to\infty}\sum_{k=0}^{n-1} e^{k/n}\frac{1}{n}$$

Using the formula for the sum of a geometric progression,

$$\sum_{k=0}^{n-1} e^{k/n}\frac{1}{n} = \frac{1}{n}\sum_{k=0}^{n-1}(e^{1/n})^k$$

$$= \frac{1}{n}\left\{\frac{(e^{1/n})^n-1}{e^{1/n}-1}\right\} = \frac{1}{n}\left\{\frac{e-1}{e^{1/n}-1}\right\}$$

We therefore have to evaluate

$$\lim_{n\to\infty}\left\{\frac{e^{1/n}-1}{1/n}\right\}$$

But this is the same as

$$\lim_{h\to 0}\left\{\frac{e^h-1}{h}\right\} = \lim_{h\to 0}\left\{\frac{e^h-e^0}{h}\right\}$$

which is the value of the derivative of the exponential function at the point 0. Since this is equal to one, it follows that

$$\int_0^1 e^x\,dx = \lim_{n\to\infty}\sum_{k=0}^{n-1} e^{k/n}\frac{1}{n} = e-1$$

5. (i) By Exercise 7.6.2,

$$\frac{d}{dy}(\sinh^{-1} y) = \frac{1}{\sqrt{1+y^2}}$$

Hence

$$\int_{-1/2}^{1/2} \frac{dy}{\sqrt{1+y^2}} = [\sinh^{-1} y]_{-1/2}^{1/2} = \sinh^{-1}\left(\frac{1}{2}\right) - \sinh^{-1}\left(-\frac{1}{2}\right)$$

(ii) By Exercise 7.6.3

$$\frac{d}{dy}(\tanh^{-1} y) = \frac{1}{1-y^2} \quad (-1 < y < 1)$$

Hence

$$\int_{-1/2}^{1/2} \frac{dy}{1-y^2} = [\tanh^{-1} y]_{-1/2}^{1/2} = \tanh^{-1}\left(\frac{1}{2}\right) - \tanh^{-1}\left(-\frac{1}{2}\right)$$

An alternative method is to use partial fractions which will yield the answer in terms of logarithms.

7. The graph of $y = \sec^2 x$ has an unpleasant *discontinuity* at $x = \pi/2$ which lies in the middle of the range of integration. The fundamental theorem of calculus therefore does *not* apply in this case.

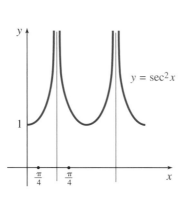
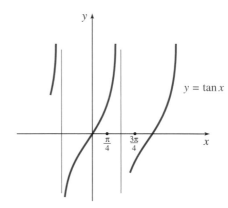

13. We have

$$x^3 = x(x^2 + x - 2) - x^2 + 2x$$
$$= x(x^2 + x - 2) - (x^2 + x - 2) + 3x - 2$$

Hence

$$\frac{x^3}{x^2 + x - 2} = (x - 1) + \frac{3x - 2}{x^2 + x - 2}$$

Alternatively, one can use 'long division' as below:

$$
\begin{array}{r}
x - 1 \\
x^2 + x - 2 \overline{) x^3 } \\
\underline{x^3 + x^2 - 2x} \\
-x^2 + 2x \\
\underline{-x^2 - x + 2} \\
3x - 2
\end{array}
$$

Thus

$$\int_2^3 \frac{x^3}{x^2 + x - 2}\, dx = \int_2^3 (x - 1)\, dx + \int_2^3 \frac{3x - 2}{x^2 + x - 2}\, dx$$

But

$$\frac{3x-2}{x^2+x-2} = \frac{3x-2}{(x+2)(x-1)} = \frac{A}{x+2} + \frac{B}{x-1}$$

where

$$A = \lim_{x \to -2} (x+2) \frac{(3x-2)}{(x+2)(x-1)} = \frac{-8}{-3} = \frac{8}{3}$$

$$B = \lim_{x \to 1} (x-1) \frac{(3x-2)}{(x+2)(x-1)} = \frac{1}{3}$$

Hence

$$\int_2^3 \frac{x^3}{x^2+x-2} \, dx = \left[\frac{1}{2}(x-1)^2 + \frac{8}{3} \ln(x+2) + \frac{1}{3} \ln(x-1) \right]_2^3$$

$$= \left[\frac{1}{2}(x-1)^2 + \frac{1}{3} \ln(x+2)^8 (x-1) \right]_2^3$$

$$= \left(\frac{1}{2}4^2 + \frac{1}{3} \ln 5^8 \cdot 2 \right) - \left(\frac{1}{2}1^2 + \frac{1}{3} \ln 4^8 \cdot 1 \right)$$

$$= 7\tfrac{1}{2} + \frac{1}{3} \ln(5^8 \cdot 2) - \frac{1}{3} \ln(4^8)$$

15. We have

$$7 - 2x - x^2 = 8 - 1 - 2x - x^2 = 8 - (1+x)^2$$

Hence

$$\int_0^1 \frac{dx}{\sqrt{(7-2x-x^2)}} = \int_0^1 \frac{dx}{\sqrt{(8-(1+x)^2)}}$$

$$= \left[\arcsin \frac{1+x}{\sqrt{8}} \right]_0^1$$

$$= \arcsin \left(\frac{2}{\sqrt{8}} \right) - \arcsin \left(\frac{1}{\sqrt{8}} \right)$$

17. We have

$$\int_0^{\pi/3} \cos^3 \theta \, d\theta = \int_0^{\pi/3} \frac{\cos 3\theta + 3\cos\theta}{4} \, d\theta$$

$$= \frac{1}{4} \left[\frac{\sin 3\theta}{3} + 3 \sin\theta \right]_0^{\pi/3}$$

$$= \frac{1}{4} \left(\frac{\sin \pi}{3} + 3 \sin \frac{\pi}{3} \right) = \frac{3\sqrt{3}}{8}$$

19. An obvious change of variable to try is $x = \sqrt{y}$. Then

$$dx = \tfrac{1}{2} y^{-1/2} \, dy$$

and so

$$\int_1^2 \frac{\sqrt{(1+\sqrt{y})}}{\sqrt{y}} \, dy = \int_1^{\sqrt{2}} \sqrt{(1+x)} 2 \, dx$$

$$= \left[\frac{4}{3}(1+x)^{3/2} \right]_1^{\sqrt{2}}$$

$$= \frac{4}{3}(1+\sqrt{2})^{3/2} - \frac{4}{3}(2)^{3/2}$$

21. We have

$$\sin\theta = 2\sin\frac{1}{2}\theta \cos\frac{1}{2}\theta = 2\tan\frac{1}{2}\theta \cos^2\frac{1}{2}\theta$$

$$= 2\tan\frac{1}{2}\theta \frac{1}{1+\tan^2\frac{1}{2}\theta} = \frac{2t}{1+t^2}$$

(because $1 + \tan^2\frac{1}{2}\theta = \sec^2\frac{1}{2}\theta$). Thus

$$\cos\theta = \{1 - \sin^2\theta\}^{1/2} = \left\{ 1 - \frac{4t^2}{(1+t^2)^2} \right\}^{1/2}$$

$$= \left\{ \frac{1 - 2t^2 + t^4}{(1+t^2)^2} \right\}^{1/2} = \left\{ \left(\frac{1-t^2}{1+t^2} \right)^2 \right\}^{1/2} = \frac{1-t^2}{1+t^2}$$

Note also that, since $t = \tan\frac{1}{2}\theta$,

$$dt = \frac{1}{2}\sec^2\frac{1}{2}\theta \, d\theta = \frac{1}{2}(1 + \tan^2\frac{1}{2}\theta) \, d\theta$$

and so

$$d\theta = \frac{2\,dt}{1+t^2}$$

Using the above results, we obtain

(i) $$\int_{\pi/4}^{\pi/2} \frac{d\theta}{\sin\theta} = \int_{\tan(\pi/8)}^{1} \frac{1+t^2}{2t} \frac{2\,dt}{1+t^2}$$

$$= \Big[\ln t \Big]_{\tan(\pi/8)}^{1} = -\ln\left\{ \tan\left(\frac{\pi}{8}\right) \right\}.$$

Note that this result may also be directly obtained from Exercise 10.10.6.

(ii) $$\int_0^{\pi/4} \frac{d\theta}{\cos\theta} = \int_0^{\tan(\pi/8)} \frac{1+t^2}{1-t^2} \frac{2\,dt}{1+t^2}$$

$$= \int_0^{\tan(\pi/8)} \left(\frac{1}{1-t} + \frac{1}{1+t} \right) dt$$

$$= \Big[-\ln(1-t) + \ln(1+t) \Big]_0^{\tan(\pi/8)}$$

$$= \ln\left\{ \frac{1 + \tan\left(\frac{\pi}{8}\right)}{1 - \tan\left(\frac{\pi}{8}\right)} \right\}$$

This result can also be expressed in terms of $\tanh^{-1} y$ as in Exercise 10.10.5.

(iii) $\displaystyle\int_{\pi/6}^{\pi/3} \frac{d\theta}{\sin\theta + \cos\theta} = \int_{\tan(\pi/12)}^{\tan(\pi/6)} \frac{(1+t^2)}{(1+2t-t^2)} \frac{2\,dt}{(1+t^2)}$

$\displaystyle\qquad = \int_{\tan(\pi/12)}^{\tan(\pi/6)} \frac{1}{2\sqrt{2}} \left(\frac{1}{t-\sqrt{2}-1} - \frac{1}{t+\sqrt{2}-1} \right) dt$

$\displaystyle\qquad = \left[\frac{1}{2\sqrt{2}} \ln \frac{t-\sqrt{2}-1}{t+\sqrt{2}-1} \right]_{\tan(\pi/12)}^{\tan(\pi/6)}$

Rather than using partial fractions, it is also possible to complete the square by writing $1 + 2t - t^2 = 2 - (t-1)^2$.

23. Write $\phi(x) = x^3 + x^2 + x + 1$. Then

$$\int_0^1 \frac{3x^2 + 2x + 1}{x^3 + x^2 + x + 1}\,dx = \int_0^1 \frac{\phi'(x)}{\phi(x)}\,dx$$
$$= [\ln\phi(x)]_0^1 = \ln 4$$

25. We have

(i) $\displaystyle\int_0^\pi x^3 \cos x\,dx = [x^3 \sin x]_0^\pi - \int_0^\pi 3x^2 \sin x\,dx$

$\displaystyle\qquad = -[3x^2(-\cos x)]_0^\pi + \int_0^\pi 6x(-\cos x)\,dx$

$\displaystyle\qquad = -3\pi^2 - [6x \sin x]_0^\pi + \int_0^\pi 6\sin x\,dx$

$\displaystyle\qquad = -3\pi^2 + 6[-\cos x]_0^\pi$

$\displaystyle\qquad = -3\pi^2 + 12$

(ii) $\displaystyle\int_0^X x^2 e^{-x}\,dx = [x^2(-e^{-x})]_0^X - \int_0^X 2x(-e^{-x})\,dx$

$\displaystyle\qquad = -X^2 e^{-X} + [2x(-e^{-x})]_0^X - \int_0^X 2(-e^{-x})\,dx$

$\displaystyle\qquad = -X^2 e^{-X} - 2X e^{-X} + [2(-e^{-x})]_0^X$

$\displaystyle\qquad = -X^2 e^{-X} - 2X e^{-X} - 2e^{-X} + 2$

Exercise 10.15

1. From Exercise 2.15.6 we have

$$x^{n-1} e^{-x} < \frac{(n+1)!}{x^2} \quad (x > 0)$$

We may therefore take $g(x) = (n+1)!/x^2$ in the comparison test (§10.12) since the integral

$$\int_1^\infty \frac{dx}{x^2}$$

converges (Example 10.18). We deduce that

$$\int_1^\infty x^{n-1} e^{-x}\,dx$$

also converges. The convergence of the integral over the range $[0, \infty]$ then follows.

If $n \geq 1$,

$$\Gamma(n+1) = \int_0^\infty x^n e^{-x}\, dx = [x^n(-e^{-x})]_0^\infty - \int_0^\infty nx^{n-1}(-e^{-x})\, dx$$

$$= n\int_0^\infty x^{n-1}e^{-x}\, dx = n\Gamma(n)$$

This calculation is based on the use of integration by parts in the case of an integral with finite limits of integration (§10.9). We have

$$\int_0^X x^n e^{-x}\, dx = [x^n(-e^{-x})]_0^X - \int_0^X nx^{n-1}(-e^{-x})\, dx$$

$$= -X^n e^{-X} + \int_0^X nx^{n-1}e^{-x}\, dx$$

Thus

$$\int_0^\infty x^n e^{-x}\, dx = \lim_{X\to\infty}\int_0^X x^n e^{-x}\, dx$$

$$= -\lim_{X\to\infty} X^n e^{-X} + n\lim_{X\to\infty}\int_0^X x^{n-1}e^{-x}\, dx$$

$$= 0 + n\int_0^\infty x^{n-1}e^{-x}\, dx$$

Note that $X^n e^{-X} \to 0$ as $X \to \infty$ (Exercise 2.15.6).

Finally, observe that

$$\Gamma(n+1) = n\Gamma(n) = n(n-1)\Gamma(n-1) = \cdots = n!\Gamma(1)$$

and

$$\Gamma(1) = \int_0^\infty e^{-x}\, dx = \lim_{X\to\infty}\int_0^X e^{-x}\, dx = \lim_{X\to\infty}(1 - e^{-X}) = 1$$

3. We begin by noting that $x^y = e^{y\ln x}$ (§2.6.2). Thus

$$\frac{\partial}{\partial y}(x^y) = (\ln x)x^y$$

$$\frac{\partial^2}{\partial y^2}(x^y) = (\ln x)^2 x^y$$

$$\frac{\partial^n}{\partial y^n}(x^y) = (\ln x)^n x^y$$

It follows that

$$\frac{d^n}{dy^n}\int_0^1 x^y\, dx = \int_0^1 \frac{\partial^n}{\partial y^n}x^y\, dx = \int_0^1 (\ln x)^n x^y\, dx$$

But

$$\int_0^1 x^y\, dx = \left[\frac{x^{y+1}}{y+1}\right]_0^1 = \frac{1}{y+1}$$

and so

$$\frac{d^n}{dy^n}\int_0^1 x^y\, dx = \frac{d^n}{dy^n}\frac{1}{y+1} = \frac{(-1)^n n!}{(y+1)^{n+1}}$$

The result follows on taking $y = 0$.

5. We have

$$\lim_{n\to\infty} \int_0^1 (n+1)x^n \, dx = \lim_{n\to\infty} \left[(n+1)\frac{x^{n+1}}{(n+1)} \right]_0^1 = \lim_{n\to\infty} 1 = 1$$

But, if $0 \leqslant x < 1$,

$$\lim_{n\to\infty} (n+1)x^n = 0$$

and so

$$\int_0^1 \left(\lim_{n\to\infty} (n+1)x^n \right) dx = \int_0^1 0 \, dx = 0$$

7. We have

$$\int_{a(y)}^{b(y)} f(x,y) \, dx = \int_\xi^{b(y)} f(x,y) \, dx + \int_{a(y)}^\xi f(x,y) \, dx$$

$$= \int_\xi^{b(y)} f(x,y) \, dx - \int_\xi^{a(y)} f(x,y) \, dx$$

Hence

$$\frac{d}{dy} \int_{a(y)}^{b(y)} f(x,y) \, dx = \left(f(b(y),y)b'(y) + \int_\xi^{b(y)} \frac{\partial f}{\partial y}(x,y) \, dx \right)$$

$$- \left(f(a(y),y)a'(y) + \int_\xi^{a(y)} \frac{\partial f}{\partial y}(x,y) \, dx \right)$$

and so

$$\frac{d}{dy} \int_{a(y)}^{b(y)} f(x,y) \, dx = \{ f(b(y),y)b'(y) - f(a(y),y)a'(y) \}$$

$$+ \int_{a(y)}^{b(y)} \frac{\partial f}{\partial y}(x,y) \, dx$$

9. We know that

$$\frac{1}{1-X} = 1 + X + X^2 + X^3 + \cdots$$

provided $-1 < X < 1$ (Example 10.16). By taking $X = -x^2$, it follows that

$$\frac{1}{1+x^2} = 1 - x^2 + x^4 - x^6 + \cdots$$

provided $-1 < x < 1$. Thus, if $-1 < y < 1$,

$$\arctan y = \int_0^y \frac{dx}{1+x^2}$$

$$= \int_0^y \{ 1 - x^2 + x^4 - x^6 + \cdots \} \, dx$$

$$= y - \frac{y^3}{3} + \frac{y^5}{5} - \frac{y^7}{7} + \cdots$$

11. We use the formulas for the radius of convergence R given in §10.14.

(i) $\quad \dfrac{1}{R} = \lim_{n\to\infty} \left| \dfrac{a_{n+1}}{a_n} \right| = \lim_{n\to\infty} \dfrac{(n+1)^2}{n^2} = \lim_{n\to\infty} \left(1 + \dfrac{1}{n} \right)^2 = 1$

Thus $R = 1$.

(ii) $\quad \dfrac{1}{R} = \lim_{n \to \infty} \left| \dfrac{a_{n+1}}{a_n} \right| = \lim_{n \to \infty} \dfrac{2^{n+1}}{2n} = 2$

Thus $R = \frac{1}{2}$.

(iii) $\quad \dfrac{1}{R} = \lim_{n \to \infty} |a_n|^{1/n} = \lim_{n \to \infty} (n^n)^{1/n} = \lim_{n \to \infty} n = \infty$

The formula we have obtained here is interpreted to mean that $R = 0$ – i.e. the power series converges only for $x = 0$.

Chapter 11. Multiple integrals

Exercise 11.6

1. When changing the order of integration in a repeated integral it is first necessary to identify the region D of integration. Wherever possible, it is a good idea to sketch this region:

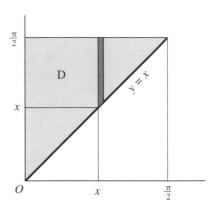

The region D is found by observing that x varies between 0 and $\pi/2$ while, for any given value of x in this range, y varies between x and $\pi/2$. The diagram illustrates the situation when the order of integration is reversed.

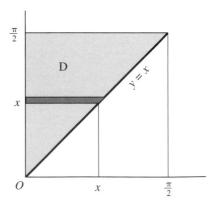

We obtain

$$\int_0^{\pi/2} \int_x^{\pi/2} \frac{\sin y}{y}\, \mathrm{d}y\, \mathrm{d}x = \int_0^{\pi/2} \int_0^{y} \frac{\sin y}{y}\, \mathrm{d}x\, \mathrm{d}y$$

$$= \int_0^{\pi/2} \frac{\sin y}{y} [x]_0^y\, \mathrm{d}y$$

$$= \int_0^{\pi/2} \sin y\, \mathrm{d}y = [-\cos y]_0^{\pi/2} = 1$$

3. The diagrams show that both repeated integrals are equal to

$$\iint_D f(x, y)\, dx\, dy$$

where D is the region illustrated.

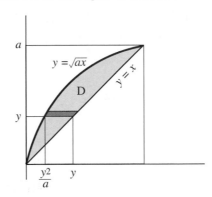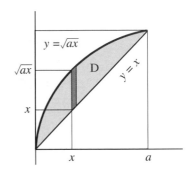

It follows that

$$\int_0^a y \int_{y^2/a}^y \frac{dx}{(a-x)\sqrt{ax-y^2}}\, dy = \int_0^a \int_x^{\sqrt{ax}} \frac{y\, dy}{\sqrt{ax-y^2}} \frac{dx}{a-x}$$

$$= \int_0^a [-(ax-y^2)^{1/2}]_x^{\sqrt{ax}} \frac{dx}{a-x}$$

$$= \int_0^a \{(ax-x^2)^{1/2} - (ax-ax)^{1/2}\} \frac{dx}{a-x}$$

$$= \int_0^a \left\{\frac{x}{a-x}\right\}^{1/2} dx$$

With some ingenuity this can be evaluated neatly (put $x = a\sin^2\theta$) but we shall introduce the uninspired and clumsy change of variable

$$u^2 = \frac{x}{a-x}$$

which transforms the integral to

$$2a \int_0^\infty \frac{u^2}{(1+u^2)^2}\, du.$$

We know that

$$\int_0^\infty \frac{dy}{1+y^2} = [\arctan y]_0^\infty = \frac{\pi}{2}$$

Hence, writing $y = u\sqrt{t}$, we obtain that

$$\int_0^\infty \frac{du}{1+tu^2} = \frac{\pi}{2} t^{-1/2}$$

Thus

$$\int_0^\infty -\frac{u^2}{(1+tu^2)^2}\, du = \int_0^\infty \frac{\partial}{\partial t}\left(\frac{1}{1+tu^2}\right) du$$

$$= \frac{d}{dt}\int_0^\infty \frac{du}{1+tu^2}$$

$$= \frac{d}{dt}\left(\frac{\pi}{2} t^{-1/2}\right) = -\frac{\pi}{4} t^{-3/2}$$

Taking $t = 1$, we deduce that

$$2a \int_0^\infty \frac{u^2}{(1+u^2)^2}\, du = \frac{1}{2}\pi a$$

5. We have

$$\frac{\partial}{\partial b} \int_a^\infty I(\alpha, b) = \frac{\partial}{\partial b} \int_0^\infty d\alpha \int_0^\infty (\sin bx) e^{-\alpha x}\, dx\, d\alpha$$

$$= \frac{\partial}{\partial b} \int_0^\infty (\sin bx) \int_a^\infty e^{-\alpha x}\, d\alpha\, dx$$

$$= \frac{\partial}{\partial b} \int_0^\infty (\sin bx) x \left[-\frac{e^{-\alpha x}}{x} \right]_a^\infty d$$

$$= \frac{\partial}{\partial b} \int_0^\infty \left(\frac{\sin bx}{x} \right) e^{-ax}\, dx$$

$$= \int_0^\infty \frac{\partial}{\partial b} \left(\frac{\sin bx}{x} \right) e^{-ax}\, dx$$

$$= \int_0^\infty \left(\frac{x \cos bx}{x} \right) e^{-ax}\, dx = J(a, b)$$

Thus

$$J(a, b) = \frac{\partial}{\partial b} \int_a^\infty \frac{b}{\alpha^2 + b^2}\, d\alpha$$

$$= \frac{\partial}{\partial b} \left[\arctan \frac{\alpha}{b} \right]_a^\infty$$

$$= \frac{\partial}{\partial b} \left\{ \frac{\pi}{2} - \arctan \frac{a}{b} \right\}$$

$$= -\frac{1}{1 + a^2/b^2} \left(-\frac{a}{b^2} \right) = \frac{a}{a^2 + b^2}$$

7. The curves $x^2 + 2y^2 = 1$, $x^2 + 2y^2 = 4$, $y = 2x$ and $y = 5x$ are mapped onto $u = 1, u = 4, v = 2$ and $v = 5$ respectively. This allows us to identify Δ as the region indicated below:

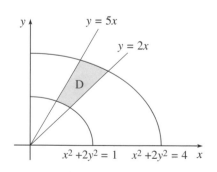
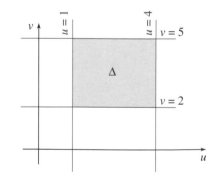

The appropriate Jacobian is

$$\frac{\partial(x, y)}{\partial(u, v)} = \left\{ \frac{\partial(u, v)}{\partial(x, y)} \right\}^{-1} = \begin{vmatrix} \dfrac{\partial u}{\partial x} & \dfrac{\partial u}{\partial y} \\ \dfrac{\partial v}{\partial x} & \dfrac{\partial v}{\partial y} \end{vmatrix}^{-1} = \begin{vmatrix} 2x & 4y \\ -\dfrac{y}{x^2} & \dfrac{1}{x} \end{vmatrix}^{-1}$$

$$= \left(2 + \frac{4y^2}{x^2}\right)^{-1} = \frac{1}{2(1 + 2v^2)}$$

Hence

$$\iint_D \frac{y}{x} \, dx \, dy = \iint_\Delta \frac{v}{2(1 + 2v^2)} \, du \, dv$$

$$= \int_1^4 du \int_2^5 \frac{v}{2(1 + 2v^2)} \, dv$$

$$= \int_1^4 du \frac{1}{2} \left[\frac{1}{4} \ln(1 + 2v^2)\right]_2^5$$

$$= \frac{1}{8} \ln \frac{51}{9} [u]_1^4 = \frac{3}{8} \ln \left(\frac{17}{3}\right)$$

9. The region D in the $(x, y)^{\mathrm{T}}$ plane which is mapped onto the square Δ in the $(u, v)^{\mathrm{T}}$ plane enclosed by the lines $u = 1, u = e, v = 1$ and $v = e$ is sketched below. Note that the curves need to be drawn rather carefully if the correct shape for D is to be obtained. Even more care is necessary in the consideration of the critical points.

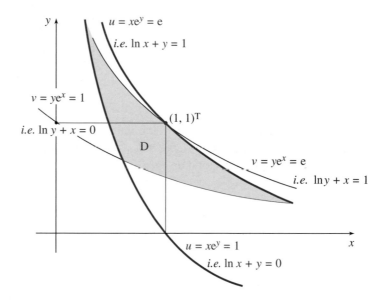

The Jacobian of f is

$$\frac{\partial(u, v)}{\partial(x, y)} = \begin{vmatrix} e^y & xe^y \\ ye^x & e^x \end{vmatrix} = e^{x+y} \begin{vmatrix} 1 & x \\ y & 1 \end{vmatrix} = (1 - xy)e^{x+y}$$

and hence f has critical points on the line $xy = 1$. It is important that none of these critical points lie *inside* the region D. Suppose that a point $(x, y)^{\mathrm{T}}$ on the curve $xy = 1$ lies inside D. Then $x > 0, y > 0, xy = 1, xe^y < e$ and $ye^x < e$. Thus

$$1 = xy < e^{1-y}e^{1-x} = e^{2-1/x-x}$$

and so, taking logarithms,

$$0 < 2 - \frac{1}{x} - x$$

$$x + \frac{1}{x} < 2$$

But it is easy to check that the minimum of the left hand side for $x > 0$ is equal to 2. Thus the inequality cannot hold. It follows that no critical point of f lies *inside D* (although the critical point $(1, 1)^{\mathrm{T}}$ is on the boundary of D).

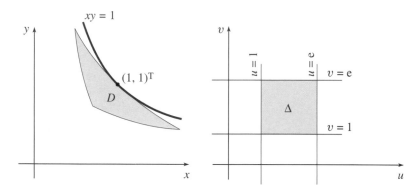

In practice of course the problem of the location of the critical points is usually ignored until it is noticed that the answers being obtained cannot possibly be correct.

We now consider the given double integral

$$\iint_D (x^2 y^3 - x^3 y^4) e^{4x+3y}\, dx\, dy$$

$$= \iint_\Delta x^2 y^3 (1 - xy) e^{4x+3y} \{(1 - xy) e^{x+y}\}^{-1}\, du\, dv$$

$$= \iint_\Delta x^2 y^3 e^{3x} e^{2y}\, du\, dv$$

$$= \iint_\Delta u^2 v^3\, du\, dv$$

$$= \int_1^e u^2\, du \int_1^e v^3\, dv$$

$$= \left[\frac{1}{3} u^3\right]_1^e \left[\frac{1}{4} v^4\right]_1^e$$

$$= \frac{1}{12}(e^3 - 1)(e^4 - 1)$$

11. The region Δ in this case is the brick shaped region determined by the inequalities

$$\left.\begin{array}{c} 0 \le r \le R \\ -\pi < \theta \le \pi \\ 0 \le \phi < \pi \end{array}\right\}$$

To see this, one must observe that, as $(r, \theta, \phi)^{\mathrm{T}}$ varies subject to these inequalities, the corresponding vector $(x, y, z)^{\mathrm{T}}$ visits each point in the sphere of radius R and centre $(0, 0, 0)^{\mathrm{T}}$ exactly once.

The appropriate Jacobian is

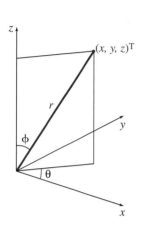

$$\frac{\partial(x, y, z)}{\partial(r, \theta, \phi)} = \begin{vmatrix} \dfrac{\partial x}{\partial r} & \dfrac{\partial x}{\partial \theta} & \dfrac{\partial x}{\partial \phi} \\[2mm] \dfrac{\partial y}{\partial r} & \dfrac{\partial y}{\partial \theta} & \dfrac{\partial y}{\partial \phi} \\[2mm] \dfrac{\partial z}{\partial r} & \dfrac{\partial z}{\partial \theta} & \dfrac{\partial z}{\partial \phi} \end{vmatrix}$$

$$= \begin{vmatrix} \cos\theta\sin\phi & -r\sin\theta\sin\phi & r\cos\theta\cos\phi \\ \sin\theta\sin\phi & r\cos\theta\sin\phi & r\sin\theta\cos\phi \\ \cos\phi & 0 & -r\sin\phi \end{vmatrix}$$

$$= \cos\phi \begin{vmatrix} -r\sin\theta\sin\phi & r\cos\theta\cos\phi \\ r\cos\theta\sin\phi & r\sin\theta\cos\phi \end{vmatrix}$$

$$\quad -r\sin\phi \begin{vmatrix} \cos\theta\sin\phi & -r\sin\theta\sin\phi \\ \sin\theta\sin\phi & r\cos\theta\sin\phi \end{vmatrix}$$

$$= r^2\cos^2\phi\sin\phi \begin{vmatrix} -\sin\theta & \cos\theta \\ \cos\theta & \sin\theta \end{vmatrix}$$

$$\quad -r^2\sin^3\phi \begin{vmatrix} \cos\theta & -\sin\theta \\ \sin\theta & \cos\theta \end{vmatrix}$$

$$= r^2\cos^2\phi\sin\phi\,(-\sin^2\theta - \cos^2\theta)$$

$$\quad -r^2\sin^3\phi\,(\cos^2\theta + \sin^2\theta)$$

$$= -r^2\sin\phi\,(\cos^2\phi + \sin^2\phi) = -r^2\sin\phi$$

Thus

$$\iiint\limits_{D} dx\,dy\,dz = \iiint\limits_{\Delta} r^2\sin\phi\,dr\,d\theta\,d\phi.$$

Note that we take the modulus of the Jacobian.
Hence

$$\iiint\limits_{D} dx\,dy\,dz = \int_0^R r^2\,dr \int_{-\pi}^{\pi} d\theta \int_0^{\pi} \sin\phi\,d\phi$$

$$= \int_0^R r^2\,dr \int_{-\pi}^{\pi} d\theta[-\cos\phi]_0^{\pi}$$

$$= 2\int_0^R r^2\,dr\,[\theta]_{-\pi}^{\pi}$$

$$= 4\pi\left[\frac{1}{3}r^3\right]_0^R = \frac{4\pi R^3}{3}$$

13. See §13.7.

Chapter 12. Differential equations of order one

Exercise 12.13

1. (i) Order 3; degree 1.

2. (i) $x^3\dfrac{dy}{dx} + x^2 y + 1 = 0$

(ii) $\dfrac{d^2y}{dx^2} - \dfrac{dy}{dx} - 2y = 0$

(iii) $y = 2x\dfrac{dy}{dx}$

3. (i) $(4+x)\dfrac{dy}{dx} = y^3$

$$y^{-3}\,dy = \dfrac{dx}{4+x}$$

$$\int y^{-3}\,dy = \int \dfrac{dx}{4+x} + c$$

$$-\dfrac{1}{2}y^{-2} = \ln(4+x) + c$$

(ii) $(xy+y)\dfrac{dy}{dx} = (x-xy)$

$$\dfrac{y\,dy}{1-y} = \dfrac{x\,dx}{x+1}$$

$$\int \dfrac{y\,dy}{1-y} = \int \dfrac{x\,dx}{x+1} + c$$

But

$$\int \dfrac{y\,dy}{1-y} = -\int \left(\dfrac{1-y}{1-y} - \dfrac{1}{1-y}\right)dy$$

$$= -y - \ln(1-y)$$

$$\int \dfrac{x\,dx}{x+1} = \int \left(\dfrac{1+x}{1+x} - \dfrac{1}{1+x}\right)dx$$

$$= x - \ln(1+x)$$

Hence

$$-y - \ln(1-y) = x - \ln(1+x) + c$$

or

$$\left(\dfrac{1+x}{1-y}\right) = ae^{x+y}$$

(iii) $x^2y\dfrac{dy}{dx} = e^y$

$$ye^{-y}\,dy = \dfrac{dx}{x^2}$$

$$\int ye^{-y}\,dy = \int x^{-2}\,dx + c.$$

But

$$\int ye^{-y}\,dy = -ye^{-y} - \int -e^{-y}\,dy$$

$$= -ye^{-y} - e^{-y}$$

and so

$$-ye^{-y} - e^{-y} = -x^{-1} + c$$

or

$$x(y+1)e^{-y} = 1 + ax$$

(iv) $\dfrac{1}{y}\dfrac{dy}{dx} = (\ln x)(\ln y)$

$$\frac{dy}{y \ln y} = (\ln x)\, dx$$

$$\int \frac{dy}{y \ln y} = \int (\ln x)\, dx + c$$

But

$$\int \frac{dy}{y \ln y} = \ln(\ln y),$$

$$\int (\ln x)\, dx = x \ln x - x$$

Check these results by differentiating. Hence

$$\ln \ln y = x \ln x - x + c$$

4. (i) $x^3 y = x^3 e^x + c$

(ii) We first write the equation in standard form as below.

$$\frac{dy}{dx} - \frac{2xy}{1+x^2} = x^2$$

The integrating factor is

$$\mu = e^{\int P\, dx} = e^{-\int (2x/(1+x^2))\, dx} = e^{-\ln(1+x^2)} = \frac{1}{1+x^2}$$

Multiplying through by μ, we obtain

$$\frac{1}{1+x^2}\frac{dy}{dx} - \frac{2x}{(1+x^2)^2} y = \frac{x^2}{1+x^2}$$

$$\frac{d}{dx}\left(\frac{y}{1+x^2}\right) = 1 - \frac{1}{1+x^2}$$

$$\frac{y}{1+x^2} = \int \left(1 - \frac{1}{1+x^2}\right) dx + c$$

$$y = (1+x^2)(x - \arctan x) + c(1+x^2)$$

(iii) $y \cos x = -\dfrac{x \cos(2x)}{4} + \dfrac{\sin(2x)}{8} + c$

6. (i) Homogeneous; solution:

$$e^{\frac{-2y}{x}} = -2\ln(cx^3)$$

(iii) Separable; solution:

$$-y^2 e^{-y} - 2y e^{-y} - 2e^{-y} = x^2 - \frac{1}{x^2} + c$$

(v) Linear; solution:

$$y = e^{3x}(c - x - x^2)$$

7. (i) Not linear. We have

$$\frac{\partial M}{\partial y} = 3x^2 \qquad \frac{\partial N}{\partial x} = 3x^2$$

and hence the equation is exact. Solution: $x^3 y - 3x^2 + y^2 = c$.

(ii) We have

$$\frac{\partial M}{\partial y} = 3 \qquad \frac{\partial N}{\partial x} = -1$$

and so the equation is not exact. It is linear, since we can write it in the form

$$\frac{dy}{dx} - \frac{3y}{x} = x^4$$

The integrating factor is

$$\mu = e^{\int P\,dx} = e^{-\int (3/x)\,dx} = e^{-3\ln x} = \frac{1}{x^3}$$

Multiplying through by μ we obtain

$$\frac{1}{x^3}\frac{dy}{dx} - \frac{3y}{x^4} = x$$

$$\frac{d}{dx}\left(\frac{1}{x^3}y\right) = x$$

$$\frac{y}{x^3} - \int x\,dx + c$$

$$y = \frac{1}{2}x^5 + cx^3$$

9. (i) Exact; solution:

$$x\cos y + y\sin x = c$$

(iii) Linear; solution:

$$\frac{y}{x^4} = c - \frac{1}{4x^4} - \frac{1}{3x^3}$$

(v) Homogeneous and exact; solution:

$$x + 2ye^{\frac{x}{y}} = c$$

(vii) Homogeneous; solution:

$$cx - e^{-\frac{2y}{x}}(x + 2y) = 3x\ln x$$

10. (i) Exact for all $n \in \mathbb{Z}$; solution:

$$\frac{x^{n+1}}{n+1} - \frac{3}{2}xy = c \quad (n \neq -1)$$
$$\ln|x| - \frac{3}{2}xy = c \quad (n = -1)$$

(ii) Separable only when $n = 0$; solution:

$$2x - 3xy = c$$

(iii) Linear for all $n \in \mathbb{Z}$.

(iv) Homogeneous of degree one if and only if $n = 1$; solution:

$$x^2 - 3xy = c.$$

13. We have $du = \sec^2 x \, dx$ and $dv = -\sin y \, dy$ and so the given equation reduces to

$$(3u - 2v) \, du - u \, dv = 0.$$

This is a linear equation which we write in the standard form

$$\frac{dv}{du} + \frac{2}{u} v = 3$$

The integrating factor is

$$\mu = e^{\int P du} = e^{\int (2/u) \, du} = e^{2 \ln u} = u^2$$

Multiplying through by u^2, we obtain

$$u^2 \frac{dv}{du} + 2uv = 3u^2$$

$$\frac{d}{du}(u^2 v) = 3u^2$$

$$u^2 v = u^3 + c$$

i.e.

$$(\tan x)^2 (\cos y) = (\tan x)^3 + c$$

15. (ii) Let $\arctan y = u$. Solution:

$$x = \arctan y - 1 + ce^{-\arctan y}$$

17. The condition for exactness gives

$$g(y) \cos x + (y + 2)g'(y) \cos x = 3g(y) \cos x$$

$$\frac{g'(y)}{g(y)} = \frac{2}{y + 2}$$

$$g(y) = (y + 2)^2$$

Solution:

$$(y + 2)^3 \sin x = c$$

22. From example 7.10 we know that the given change of variable reduces the partial differential equation to the form

$$\frac{\partial f}{\partial u} = \frac{1}{4}$$

The general solution is therefore

$$f = \tfrac{1}{4}u + g(v)$$

i.e.

$$f = \tfrac{1}{4}(x^2 + y^2) + g(x^2 - y^2)$$

where g is an arbitrary function.

23. Transformed equation: $\dfrac{\partial f}{\partial v} - 2vf = 0$ Solving, using the integrating factor e^{-v^2},

$$e^{-v^2} f = g(u)$$
$$f(x, y) = g(y^2 - x^2)e^{x^2}$$

24. Transformed equation:

$$\frac{\partial f}{\partial v} = \frac{4u}{v^2}$$
$$f = \frac{-4u}{v} + g(u)$$
$$f(x, y) = g(xy) - 4y$$

Chapter 13. Complex numbers

Exercise 13.4

1. (i) $z = \{(1 - 2i) + (-2 + 3i)\}(-3 - 4i) = (-1 + i)(-3 - 4i)$

$$= (3 + 4) + i(-3 + 4) = 7 + i$$

Thus $\mathrm{Re}\, z = 7$ and $\mathrm{Im}\, z = 1$.

(ii) $z = (1 - 2i)(-2 + 3i) - (-3 - 4i) \qquad = (-2 + 6) + i(4 + 3)$

$$+ 3 + 4i = (4 + 3) + i(7 + 4) = 7 + 11i$$

Thus $\mathrm{Re}\, z = 7$ and $\mathrm{Im}\, z = 11$.

(iii) $z = \dfrac{(1 - 2i) - (-3 - 4i)}{(-2 + 3i)} = \dfrac{4 + 2i}{-2 + 3i} = \dfrac{(4 + 2i)(-2 - 3i)}{(-2 + 3i)(-2 - 3i)}$

$\qquad = \dfrac{(-8 + 6) + i(-4 - 12)}{(4 + 9) + i(-6 + 6)} = \dfrac{-2 - 16i}{13}$

Thus $\mathrm{Re}\, z = -\frac{2}{13}$ and $\mathrm{Im}\, z = -\frac{16}{13}$.

(iv) $z = \dfrac{(1 - 2i)(-3 - 4i)}{(-2 + 3i)} = \dfrac{(-3 - 8) + i(6 - 4)}{(-2 + 3i)}$

$\qquad = \dfrac{(-11 + 2i)(-2 - 3i)}{(-2 + 3i)(-2 - 3i)} = \dfrac{(22 + 6) + i(-4 + 33)}{(4 + 9) + i(-6 + 6)} = \dfrac{28 + 29i}{13}$

Thus $\mathrm{Re}\, z = \frac{28}{13}$ and $\mathrm{Im}\, z = \frac{29}{13}$.

The moduli of z_1, z_2 and z_3 are given by

$$|z_1| = \{1^2 + (-2)^2\}^{1/2} = \{1 + 4\}^{1/2} = \sqrt{5}$$
$$|z_2| = \{(-2)^2 + 3^2\}^{1/2} = \{4 + 9\}^{1/2} = \sqrt{13}$$
$$|z_3| = \{(-3)^2 + (-4)^2\}^{1/2} = \{9 + 16\}^{1/2} = 5$$

Some diagrams are helpful in calculating the arguments of z_1, z_2 and z_3. We have

$$\arg z_1 = \theta_1 = -\psi_1 = -\arctan 2$$
$$\arg z_2 = \theta_2 = \pi - \psi_2 = \pi - \arctan\left(\tfrac{3}{2}\right)$$
$$\arg z_3 = \theta_3 = -\pi + \psi_3 = -\pi + \arctan\left(\tfrac{4}{3}\right)$$

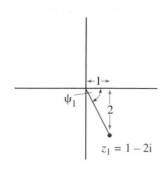
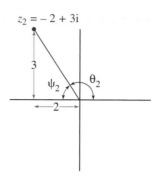
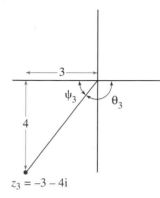

3. $-i, -1/2 + i\sqrt{3}/2, 1/2 - i\sqrt{3}/2, -5\sqrt{3}/2 - i5/2, e^{-3}\cos(5\sqrt{7}) - ie^{-3}\sin(5\sqrt{7})$

4. (i) $\dfrac{1}{i} = \dfrac{i}{i^2} = -i$

   (ii) $i^4 = i^2 \cdot i^2 = (-1)(-1) = 1$

   (iii) $i^3 = i^2 \cdot i = (-1)i = -i$

   (iv) $|1 + i| = \{1^2 + 1^2\}^{1/2} = \sqrt{2}$ $\quad \arg(1 + i) = \arctan\left(\dfrac{1}{1}\right) = \dfrac{\pi}{4}$

5. (i) 192

6. (i) $z_1 = r_1 e^{i\theta_1} z_2 = r_2 e^{i\theta_2} z_1 z_2 = r_1 r_2 e^{i(\theta_1 + \theta_2)}$

8.

$$z = \cos(3\theta) + i\sin(3\theta) = (\cos\theta + i\sin\theta)^3$$
$$= \cos^3\theta + 3i\cos^2\theta\sin\theta + 3i^2\cos\theta\sin^2\theta + i^3\sin\theta$$
$$\operatorname{Re} z = \cos^3\theta - 3\cos\theta(1 - \cos^2\theta) = 4\cos^3\theta - 3\cos\theta$$

10. $z_1 = r_1 e^{i\theta_1} z_2 = r_2 e^{i\theta_2}$

$$|r_1 e^{i\theta_1} + r_2 e^{i\theta_2}| = |r_1\cos\theta_1 + r_2\cos\theta_2 + i(r_1\sin\theta_1 + r_2\sin\theta_2)|$$
$$= \sqrt{(r_1^2\cos^2\theta_1 + r_2^2\cos^2\theta_2 + 2r_1 r_2\cos\theta_1\cos\theta_2 + r_1^2\sin^2\theta_1}$$
$$\overline{+ r_2^2\sin^2\theta_2 + 2r_1 r_2\sin\theta_1\sin\theta_2)}$$
$$= \sqrt{r_1^2 + r_2^2 + 2r_1 r_2\cos(\theta_1 - \theta_2)}$$
$$\leq \sqrt{r_1^2 + r_2^2 + 2r_1 r_2} = r_1 + r_2 \text{ since } \cos(\theta_1 - \theta_2) \leq 1$$

Exercise 13.8

1. (ii) To solve $z^5 = 1$, we write $1 = e^{\pi i}$. The roots may then be expressed as

$$z_0 = e^{\pi i/5}$$
$$z_1 = e^{3\pi i/5}$$
$$z_2 = e^{5\pi i/5} = e^{\pi i} = -1$$
$$z_3 = e^{7\pi i/5}$$
$$z_4 = e^{9\pi i/5}$$

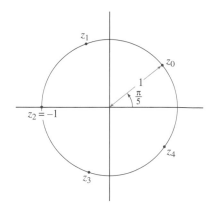

(iii) To solve $z^2 = i$, we write $i = e^{i\pi/2}$. The roots may then be expressed as

$$z_0 = e^{i\pi/4} = \left(\frac{1+i}{\sqrt{2}}\right)$$
$$z_1 = e^{i\pi/4}e^{i\pi} = -e^{i\pi/4} = -\left(\frac{1+i}{\sqrt{2}}\right)$$

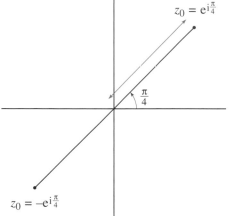

4. (i) $z^3 - 3z^2 - 10z = z(z^2 - 3z - 10)$
$$= z(z-5)(z+2)$$
(ii) $P(z) = z^4 + z^3 - 3z^2 - 5z - 2$

Observe that $P(0) = -2 \neq 0$, $P(1) = -8 \neq 0$, but $P(-1) = 0$. Thus $z = -1$ is a root (and $z = 0$ and $z = 1$ are not). It follows that

$$P(z) = (z+1)Q(z)$$

By the technique of §13.6 we find that $Q(z) = z^3 - 3z - 2$. Again $z = -1$ is a root since $Q(-1) = 0$. Thus

$$Q(z) = (z+1)R(z)$$

By the same technique we obtain $R(z) = z^2 - z - 2 = (z-2)(z+1)$. The required factorisation is therefore

$$P(z) = (z+1)^3(z-2)$$

Chapter 14. Differential and difference equations

Exercise 14.11

1. (ii) $(3D^3 + 5D^2 - 2D)y = 0$

$D(3D^2 + 5D - 2)y = 0$

$D(3D - 1)(D + 2)y = 0$

The general solution is
$$y = A + Be^{x/3} + Ce^{-2x}$$

(iii) $P(D)y = (D^3 - 4D + D + 6)y = 0$
We have $z = -1$ is a root of $P(z) = 0$. Thus
$$P(z) = (z + 1)Q(z) = (z + 1)(z^2 - 5z + 6)$$

It follows that
$$P(D)y = (D + 1)(D - 2)(D - 3)y = 0$$

and so the general solution is
$$y = Ae^{-x} + Be^{2x} + Ce^{3x}$$

(iv) $P(D)y = (D^4 - 2D^3 - 3D^2 + 4D + 4)y = 0$
Again $z = -1$ is a root of $P(z) = 0$. Thus
$$P(z) = (z + 1)Q(z) = (z + 1)(z^3 - 3z^2 + 4)$$

by the table below. Observe that $z = -1$ is also a root of $Q(z) = z^3 - 3z^2 + 4$ and so
$$Q(z) = (z + 1)R(z) = (z + 1)(z^2 - 4z + 4) = (z + 1)(z - 2)^2$$

Hence
$$P(D)y = (D + 1)^2(D - 2)^2 y = 0$$

and so the general solution is
$$y = A_0 e^{-x} + A_1 x e^{-x} + B_0 e^{2x} + B_1 x e^{2x}$$

n	4	3	2	1	0	-1
p_n	1	-2	-3	4	4	0
q_n	0	1	-3	0	4	0
r_n	0	0	1	-4	4	0

(v) $P(D)y = (D^5 + 3D^4 + 7D^3 + 13D^2 + 12D + 4)y = 0$
Again $z = -1$ is a root of $P(z)$ and so
$$P(z) = (z + 1)Q(z) = (z + 1)(z^4 + 2z^3 + 5z^2 + 8z + 4)$$

But $z = -1$ is also a root of $Q(z)$. Hence

$$Q(z) = (z+1)R(z) = (z+1)(z^3 + z^2 + 4z + 4)$$

Yet again $z = -1$ is a root of $R(z)$ and so

$$R(z) = (z+1)S(z) = (z+1)(z^2 + 4) = (z+1)(z - 2i)(z + 2i)$$

Thus

$$P(D)y = (D+1)^3(D - 2i)(D + 2i)y = 0$$

The general solution is therefore

$$y = A_0 e^{-x} + A_1 x e^{-x} + A_2 x^2 e^{-x} + B e^{2ix} + C e^{-2ix}$$

If only real solutions are of interest, we may rewrite this as

$$y = A_0 e^{-x} + A_1 x e^{-x} + A_2 x^2 e^{-x} + b \cos 2x + c \sin 2x$$

2. The general solution is

$$y = e^x (A \cos 3x + B \sin 3x)$$

The required particular solution is

$$y = e^x \sin 3x$$

5. (ii) $P(E)y_x = (E^2 + 2E - 3)y_x = 0$
$(E + 3)(E - 1)y_x = 0$

The general solution is

$$y = A(-3)^x + B$$

(iii) Write $z_x = y_{x+2}$. The equation then reduces to

$$z_{x+2} + 2z_{x+1} + z_x = 0$$

i.e. $P(E)y_x = (E^2 + 2E + 1)z_x = 0$ which has general solution
$(E + 1)^2 z_x = 0$

$$z_x = A_0(-1)^x + A_1 x(-1)^x$$

Thus

$$y_{x+2} = A_0(-1)^x + A_1 x(-1)^x \quad (x = 0, 1, 2, \ldots)$$

and we may choose $y_0 = B$ and $y_1 = C$ where B and C are arbitrary.

(iv) $P(E)y_x = (E^2 + E + 1)y_x = 0$
The roots of $P(z) = z^2 + z + 1 = 0$ may be obtained from the formula

$$z = \frac{-1 \pm \sqrt{(1 - 4)}}{2}$$

We obtain that

$$P(E)y_x = \left(E - \left(-\frac{1}{2} + \frac{i\sqrt{3}}{2}\right)\right)\left(E - \left(-\frac{1}{2} - \frac{i\sqrt{3}}{2}\right)\right)y_x = 0$$

and so the general solution is

$$y_x = A\left(-\frac{1}{2} + \frac{i\sqrt{3}}{2}\right)^x + B\left(-\frac{1}{2} - \frac{i\sqrt{3}}{2}\right)^x.$$

This may be expressed more neatly by observing that

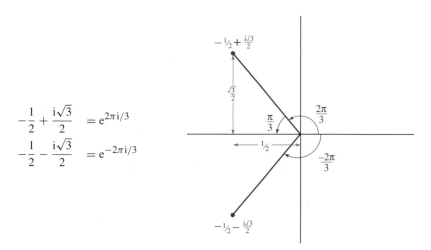

$$-\frac{1}{2} + \frac{i\sqrt{3}}{2} = e^{2\pi i/3}$$

$$-\frac{1}{2} - \frac{i\sqrt{3}}{2} = e^{-2\pi i/3}$$

The general solution is therefore

$$y_x = Ae^{2\pi ix/3} + Be^{-2\pi ix/3}$$

If only real solutions are of interest, this may be rewritten as

$$y_x = a\cos\left(\frac{2\pi x}{3}\right) + b\sin\left(\frac{2\pi x}{3}\right).$$

(v) $P(E)y = (E^3 - 3E^2 + 9E + 13)y_x = 0$
Once more $z = -1$ is a root of $P(z)$. Thus $P(z) = (z+1)Q(z) = (z+1)(z^2 - 4z + 13)$.
The roots of $Q(z)$ may be obtained from the formula

$$z = 2 \pm \sqrt{(4 - 13)}$$

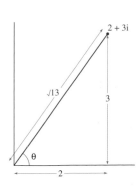

We therefore have

$$P(E)y_x = (E+1)(E - 2 - 3i)(E - 2 + 3i)y_x = 0$$

The general solution is therefore

$$y_x = A(-1)^x + B(2 + 3i)^x + C(2 - 3i)^x$$

which we may rewrite as

$$y_x = A(-1)^x + B13^{x/2}e^{i\theta x} + C13^{x/2}e^{-i\theta x}$$

or, if only real solutions are of interest, as

$$y_x = A(-1)^x + b13^{x/2}\cos(\theta x) + c13^{x/2}\sin(\theta x)$$

where $\theta = \arctan\left(\frac{3}{2}\right)$.

6. (i) $y_x = A(-1+3\mathrm{i})^x + B(-1-3\mathrm{i})^x$
$$= 10^{x/2}(a\cos(\theta x) + b\sin(\theta x)) \text{ where } \cos\theta = -1/\sqrt{10}, \sin\theta = 3/\sqrt{10}$$

7. In each case, the complementary function is the general solution of

$$P(\mathrm{D})y = (3\mathrm{D}^3 + 5\mathrm{D}^2 - 2\mathrm{D})y = 0$$

i.e.
$$\mathrm{D}(3\mathrm{D} - 1)(\mathrm{D} + 2)y = 0$$

The complementary function is therefore

$$y = A + B\mathrm{e}^{x/3} + C\mathrm{e}^{-2x}$$

† The arguments for particular solutions are given in detail. Alternatively, as in (ii), the methods of §14.1 may be used.

†

(i) $\mathrm{D}(3\mathrm{D} - 1)(\mathrm{D} + 2)y = \mathrm{e}^x$
Since $q = \mathrm{e}^x$ is a solution of $(\mathrm{D} - 1)q = 0$, a particular solution $p(x)$ of our equation is to be found among the solutions of

$$(\mathrm{D} - 1)\mathrm{D}(3\mathrm{D} - 1)(\mathrm{D} + 2)y = 0$$

which has general solution

$$y = A + B\mathrm{e}^{x/3} + C\mathrm{e}^{-2x} + K\mathrm{e}^x$$

From the form of the complementary function, we know that A, B and C may be chosen freely and the simplest choice is $A = B = C = 0$. We therefore seek a particular solution of the form

$$p(x) = K\mathrm{e}^x$$

To evaluate K we substitute in $(3\mathrm{D}^3 + 5\mathrm{D}^2 - 2\mathrm{D})y = \mathrm{e}^x$. This gives

$$3K\mathrm{e}^x + 5K\mathrm{e}^x - 2K\mathrm{e}^x = \mathrm{e}^x$$
$$6K = 1$$
$$K = \tfrac{1}{6}$$

The required general solution is therefore

$$y = \tfrac{1}{6}\mathrm{e}^x + A + B\mathrm{e}^{x/3} + C\mathrm{e}^{-2x}$$

(ii) $\mathrm{D}(3\mathrm{D} - 1)(\mathrm{D} + 2)y = \cos x$
We seek a particular solution of the form

$$p(x) = j\cos x + k\sin x.$$

Substituting in $(3\mathrm{D}^3 + 5\mathrm{D}^2 - 2\mathrm{D})y = \cos x$, we obtain

$$j(3\sin x - 5\cos x + 2\sin x) + k(-3\cos x - 5\sin x - 2\cos x) = \cos x$$
$$(5j - 5k)\sin x + (-5j - 5k - 1)\cos x = 0$$

Thus $5j - 5k = 0$ and $-5j - 5k - 1 = 0$ – i.e.

$$\left.\begin{array}{c} j = -\dfrac{1}{10} \\[2mm] k = -\dfrac{1}{10} \end{array}\right\}$$

The required general solution is therefore

$$y = -\frac{1}{10}\cos x - \frac{1}{10}\sin x + A + Be^{x/3} + Ce^{-2x}$$

(iii) $D(3D - 1)(D + 2)y = x$

Since $q = x$ is a solution of $D^2 q = 0$, a particular solution $p(x)$ of our equation is to be found among the solutions of

$$D^3(3D - 1)(D + 2)y = 0$$

which has general solution

$$y = A_0 + A_1 x + A_2 x^2 + Be^{x/3} + Ce^{-2x}$$

As in (i) we take $A_0 = B = C = 0$ and seek a particular solution of the form

$$p(x) = A_1 x + A_2 x^2$$

Substituting in $(3D^3 + 5D^2 - 2D)y = x$, we obtain

$$-2A_1 + A_2(5 \cdot 2 - 2 \cdot 2x) = x$$
$$(-2A_1 + 10A_2) + x(-4A_2 - 1) = 0$$

Thus $A_1 = 5A_2$ and $4A_2 + 1 = 0$ – i.e.

$$\left.\begin{array}{c} A_1 = -\frac{5}{4} \\[2mm] A_2 = -\frac{1}{4} \end{array}\right\}$$

The required general solution is therefore

$$y = -\tfrac{5}{4}x - \tfrac{1}{4}x^2 + A + Be^{x/3} + Ce^{-2x}$$

(iv) $D(3D - 1)(D + 2)y = e^{-2x}$

Since $q = e^{-2x}$ is a solution of $(D + 2)q = 0$, a particular solution $p(x)$ of our equation is to be found among the solutions of

$$D(3D - 1)(D + 2)^2 y = 0$$

which has general solution

$$y = A + Be^{x/3} + C_0 e^{-2x} + C_1 x e^{-2x}$$

As in (i) we take $A = B = C_0 = 0$ and seek a particular solution of the form

$$p(x) = C_1 x e^{-2x}$$

To evaluate C_1 we substitute in $(3D^3 + 5D^2 - 2D)y = e^{-2x}$. Since

$$D(xe^{-2x}) = e^{-2x} - 2xe^{-2x}$$

$$D^2(xe^{-2x}) = -2e^{-2x} - 2e^{-2x} + 4xe^{-2x} = -4e^{-2x} + 4xe^{-2x}$$

$$D^3(xe^{-2x}) = 8e^{-2x} + 4e^{-2x} - 8xe^{-2x} = 12e^{-2x} - 8xe^{-2x}$$

we obtain that

$$3C_1(12e^{-2x} - 8xe^{-2x}) + 5C_1(-4e^{-2x} + 4xe^{-2x}) - 2C_1(e^{-2x} - 2xe^{-2x})$$
$$= e^{-2x}$$

$$e^{-2x}\{C_1(36 - 20 - 2) - 1\} + xe^{-2x}\{-24 + 20 + 4\} = 0$$

and hence

$$C_1 = \frac{1}{14}$$

The required general solution is therefore

$$y = \frac{1}{14}xe^{-2x} + A + Be^{x/3} + Ce^{-2x}$$

(v) $y = A + Be^{x/3} + Ce^{-2x} - \dfrac{1}{10}\cos x - \dfrac{1}{10}\sin x + \dfrac{1}{14}xe^{-2x}$

9.

$$P(E)y_x - (E^3 \quad 5E^2 + 8E - 4)y_x = 2^x + x$$

We begin by finding the complementary sequence. This is the general solution of

$$P(E)y_x = (E^3 - 5E^2 + 8E - 4)y_x = 0$$

Since $z = 1$ is a root of $P(z)$, we obtain that $P(z) = (z - 1)Q(z) = (z - 1)(z^2 - 4z + 4) = (z - 1)(z - 2)^2$.
Thus

$$P(E)y_x = (E - 1)(E - 2)^2 y_x = 0$$

which has general solution

$$y_x = A + B2^x + Cx2^x$$

This is the required complementary sequence.

We now seek a particular solution p_x. Since $q = 2^x + x$ is a solution of $(E - 2)(E - 1)^2 q = 0$, p_x is to be found among the solutions of

$$(E - 1)^3(E - 2)^3 y_x = 0$$

which has general solution

$$y_x = A_0 + A_1 x + A_2 x^2 + B_0 2^x + B_1 x2^x + B_2 x^2 2^x$$

From the form of the complementary sequence, we know that A_0, B_0 and B_1 may be chosen freely and it is simplest to choose $A_0 = B_0 = B_1 = 0$. We then seek a particular solution of the form

$$p_x = A_1 x + A_2 x^2 + B_2 x^2 2^x$$

Substituting in $(E^3 - 5E^2 + 8E - 4)y_x = 2^x + x$, we obtain

$$A_1\{(x+3) - 5(x+2) + 8(x+1) - 4x\}$$
$$+ A_2\{(x+3)^2 - 5(x+2)^2 + 8(x+1)^2 - 4x^2\}$$
$$+ B_2\{(x+3)^2 2^{x+3} - 5(x+2)^2 2^{x+2} + 8(x+1)^2 2^{x+1} - 4x^2 2^x\}$$
$$= 2^x + x$$

i.e.

$$(A_1 - 3A_2) + (2A_2 - 1)x + (8B_2 - 1)2^x = 0$$

Thus $A_1 = 3A_2$, $2A_2 = 1$ and $8B_2 = 1$ - i.e.

$$\left.\begin{array}{l} A_1 = \frac{3}{2} \\[4pt] A_2 = \frac{1}{2} \\[4pt] B_2 = \frac{1}{8} \end{array}\right\}$$

The required general solution is therefore

$$y_x = \tfrac{3}{2}x + \tfrac{1}{2}x^2 + \tfrac{1}{8}x^2 2^x + A + B_0 2^x + B_1 x 2^x$$

11. The general solution of the differential equation of question 1(ii) is

$$y = A + Be^{x/3} + Ce^{-2x}$$

We must find values of A, B and C so as to satisfy the given boundary conditions. If $B > 0$, $y \to +\infty$ as $x \to \infty$. If $B < 0$, $y \to -\infty$ as $x \to +\infty$. Since $y \to 1$ as $x \to +\infty$, we must have $B = 0$. Then $y \to A$ as $x \to +\infty$ and hence $A = 1$. Finally we need that $y(0) = 0$. This means that

$$0 = 1 + Ce^0 = 1 + C$$

and so $C = -1$.

The required solution is therefore

$$y = 1 - e^{-2x}$$

13. $(D-3)y = 0$ $(D+1)^3 y = 0$ $(D-3)(D+1)^3 y = 0$
General solution

$$y = (A_0 + A_1 x + A_2 x^2)e^{-x} + Be^{3x}$$

16. The equilibrium solution $\mathbf{y} = \mathbf{p}$ is found by solving

$$\begin{pmatrix} 1 & -1 \\ 1 & 1 \end{pmatrix}\begin{pmatrix} p_1 \\ p_2 \end{pmatrix} + \begin{pmatrix} -1 \\ -1 \end{pmatrix} = \begin{pmatrix} 0 \\ 0 \end{pmatrix}$$

which gives $p_1 = 1$ and $p_2 = 0$.

To discuss stability, we need the eigenvalues of the matrix. These are found by solving

$$\begin{vmatrix} 1 - \lambda & -1 \\ 1 & 1 - \lambda \end{vmatrix} = (1 - \lambda)^2 + 1 = 0$$

which gives $\lambda_1 = 1 + i$ and $\lambda_2 = 1 - i$. Since $\text{Re}\,\lambda_1 > 0$ (and $\text{Re}\,\lambda_2 > 0$), the system is *unstable*.

The general solution may be found by calculating eigenvectors **u** and **v** corresponding to λ_1 and λ_2. We have

$$\begin{pmatrix} 1 & -1 \\ 1 & 1 \end{pmatrix} \begin{pmatrix} u_1 \\ u_2 \end{pmatrix} = (1+i) \begin{pmatrix} u_1 \\ u_2 \end{pmatrix} \qquad \begin{pmatrix} 1 & -1 \\ 1 & 1 \end{pmatrix} \begin{pmatrix} v_1 \\ v_2 \end{pmatrix} = (1-i) \begin{pmatrix} v_1 \\ v_2 \end{pmatrix}$$

i.e.   i.e.

$$\left. \begin{array}{l} u_1 - u_2 = u_1 + iu_1 \\ u_1 + u_2 = u_2 + iu_2 \end{array} \right\} \qquad \left. \begin{array}{l} v_1 - v_2 = v_1 - iv_1 \\ v_1 + v_2 = v_2 - iv_2 \end{array} \right\}$$

and so   and so

$$u_1 = iu_2 \qquad\qquad\qquad v_1 = -iv_2$$

We make the choice $u_2 = 1$ and $v_2 = 1$ which yields the eigenvectors

$$\mathbf{u} = \begin{pmatrix} i \\ 1 \end{pmatrix} \text{ and } \mathbf{v} = \begin{pmatrix} -i \\ 1 \end{pmatrix}$$

The general solution is then

$$y_1 = 1 + \alpha i e^{(1+i)x} - \beta i e^{(1-i)x}$$
$$y_2 = 0 + \alpha e^{(1+i)x} + \beta e^{(1-i)x}$$

which we may express in the form

$$y_1 = 1 + (A \cos x - B \sin x) e^x$$
$$y_2 = 0 + (B \cos x + A \sin x) e^x.$$

We now take account of the boundary conditions $y_1(0) = 1$ and $y_2(0) = 1$. These give

$$\left. \begin{array}{l} 1 = 1 + A \\ 1 = 0 + B \end{array} \right\}$$

The required solution is therefore

$$y_1 = 1 - e^x \sin x$$
$$y_2 = e^x \cos x$$

Note that, far from converging to the equilibrium solution, these oscillate infinitely.

Exercise 14.13

1. (i) 1 (ii) 1 (iii) 1 (iv) 2 (v) 2 (vi) 3

3. (i) $y_{x+1} - y_x = 2x$

                $\Delta y_x = 2x$

                $y_x = x(x-1) + c$

   (ii) $y_{x+1} - y_x = 2^x$

                $\Delta y_x = 2^x$

                $y_x = 2^x (2-1)^{-1} + c$

                $y_x = 2^x + c$

   (iii) $y_{x+1} + 2y_x = x$

Using the method of Example 14.25, we write

$$(-2)^{-x-1}y_{x+1} - (-2)^{-x}y_x = (-2)^{-x-1}x$$
$$\Delta((-2)^{-x}y_x) = -\tfrac{1}{2}x(-\tfrac{1}{2})^x$$

Thus

$$\left(-\tfrac{1}{2}\right)^x y_x = -\tfrac{1}{2}\left\{-\tfrac{4}{9}\left(-\tfrac{1}{2}\right)^{x+1} - \tfrac{2}{3}x\left(-\tfrac{1}{2}\right)^x\right\} + c$$
$$y_x = -\tfrac{1}{9} + \tfrac{1}{3}x + c(-2)^x$$

Note that the result

$$\Delta\left(\frac{-r^{x+1}}{(r-1)^2} + \frac{xr^x}{(r-1)}\right) = xr^x \qquad (r \neq 1)$$

is quoted in Example 14.3.2.

(iv)
$$y_{x+1} - 2y_x = 3^x$$
$$2^{-x-1}y_{x+1} - 2^{-x}y_x = 2^{-x-1}3^x$$
$$\Delta(2^{-x}y_x) = \tfrac{1}{2}\left(\tfrac{3}{2}\right)^x$$
$$2^{-x}y_x = \tfrac{1}{2}\left(\tfrac{3}{2} - 1\right)^{-1}\left(\tfrac{3}{2}\right)^x + c$$
$$y_x = 3^x + c2^x$$

5. Making the indicated change of variable, we obtain

$$2^{z_{x+1}}2^{z_x} = 2$$

i.e.

$$z_{x+1} + z_x = 1$$
$$(-1)^{-x-1}z_{x+1} - (-1)^{-x}z_x = (-1)^{-x-1}$$
$$\Delta((-1)^{-x}z_x) = (-1)^{-x-1}$$
$$(-1)^{-x}z_x = -(-2)^{-1}(-1)^{-x} + c$$
$$z_x = \tfrac{1}{2} + c(-1)^x$$
$$y_x = \log_2\left(\tfrac{1}{2} + c(-1)^x\right)$$

7. (i) The appropriate difference equation is

$$y_{x+1} = y_x + \frac{10}{100}y_x$$
$$y_{x+1} - (1.1)y_x = 0$$

This is solved as in Example 14.25. We obtain the general solution

$$y_x = c(1.1)^x$$

To evaluate c the boundary condition

$$y_{100} = 1\,000\,000$$

is used. Then

$$c = \frac{1\,000\,000}{(1.1)^{100}} = 72.57$$

But $y_0 = c$ and so the amount originally invested is $72.57 (approximately).

(ii) In this problem the boundary condition is $y_0 = 10$ and so $c = 10$. The average held over the first X years is

$$\frac{1}{X}(y_0 + y_1 + \cdots + y_{X-1}) = \frac{10}{X}((1.1)^0 + \cdots + (1.1)^{X-1})$$

$$= \frac{10}{X}\left(\frac{(1.1)^X - 1}{(1.1) - 1}\right) = \frac{100}{X}((1.1)^X - 1)$$

We therefore need to solve the equation

$$15.93 = \frac{100}{X}((1.1)^X - 1)$$

Trial and error shows the solution to be $X = 10$. The amount held at the beginning of the $(X+1)$th year is therefore

$$y_{11} = 10(1.1)^{11} = 28.53$$

Appendix

Greek letters used in the text

Upper case	Lower case	
	α	alpha
	β	beta
Γ	γ	gamma
Δ	δ	delta
	η	eta
Θ	θ	theta
	λ	lambda
	μ	mu
	ν	nu
	ξ	xi
	π	pi
	ρ	rho
Σ	σ	sigma
	ϕ	phi

The symbol ∇ called 'nabla' used for the gradient vector is not a Greek letter, but nabla is the greek word for a type of harp.

Index